IBM® PS/2
Technical Guide

HOWARD W. SAMS & COMPANY
HAYDEN BOOKS

Related Titles

Hard Disk Management Techniques for the IBM®
Joseph-David Carrabis

IBM® PC AT User's Reference Manual
Gilbert Held

IBM® PC & PC XT User's Reference Manual, Second Edition
Gilbert Held

The Waite Group's Desktop Publishing Bible
James Stockford, Editor, The Waite Group

Personal Publishing with PC PageMaker®
Terry Ulick

Best Book of: Lotus® 1-2-3®, Second Edition
Alan Simpson

Lotus® 1-2-3® Financial Models
Elna Tymes and Tony Dowden with Charles E. Prael

Best Book of: WordPerfect® 4.2
Vincent Alfieri

Best Book of: WordPerfect® 5.0
Vincent Alfieri

Best Book of: WordStar® (Features Release 4)
Vincent Alfieri

Best Book of: dBASE II®/III®
Ken Knecht

Best Book of: Framework™
Alan Simpson

Best Book of: Multiplan™
Alan Simpson

Best Book of: Symphony™
Alan Simpson

Best Book of: Microsoft® Works for the PC
Ruth Witkin

For the retailer nearest you, or to order directly from the publisher, call 800-428-SAMS. In Indiana, Alaska, and Hawaii call 317-298-5699.

IBM® PS/2
Technical Guide

Caroline M. Halliday
James A. Shields

HOWARD W. SAMS & COMPANY
A Division of Macmillan, Inc.
4300 West 62nd Street
Indianapolis, Indiana 46268 USA

© 1988 by Caroline M. Halliday and James A. Shields

FIRST EDITION
FIRST PRINTING—1988

All rights reserved. No part of this book shall be reproduced, stored in a retrieval system, or transmitted by any means, electronic, mechanical, photocopying, recording, or otherwise, without written permission from the publisher. No patent liability is assumed with respect to the use of the information contained herein. While every precaution has been taken in the preparation of this book, the publisher and author assume no responsibility for errors or omissions. Neither is any liability assumed for damages resulting from the use of the information contained herein.

International Standard Book Number: 0-672-22628-6
Library of Congress Catalog Card Number: 88-60991

Acquisitions Editor: *James S. Hill*
Development Editor: *Jennifer Ackley*
Manuscript Editor: *Marie Butler-Knight*
Illustrator: *T.R. Emrick*
Cover Illustrators: *Debi Stewart and Linda Leviton*
Photographers: *Larry Nickel and Hal Rummel*
Indexer: *Ted Laux*
Technical Reviewers: *Steven Armbrust, Mary L. DeWolf, and Ed McNierney*
Compositor: *Shepard Poorman Communications Corp.*

Printed in the United States of America

For Linda and Stephen

Contents

Preface   xi

   Acknowledgments   xii
   Trademarks   xii

1 Introduction   1

   The IBM PC   1
   The IBM XT   7
   The IBM AT   8
   The Compaq Deskpro 386   11
   PC Software   13
   IBM Personal System/2   14
   Graphics System Overview   22
   Monitors   23
   Math Coprocessor   24
   Mouse   24
   Removable Media Device Options   24
   Hard Disks   25
   Data Migration   26
   Add-In Boards for the Model 30   26
   Add-In Boards for the Micro Channel   27
   Operating System Software   28
   Printers   29
   Summary   29

2 The Model 30   31

   Functional Description   33
   The MCGA   41
   Physical Description   45
   Installation and Setup   53
   Advanced Diagnostics   54
   Compatibility and Performance   55
   Model 30 Options   61
   Summary   61

3 The Model 25 — 63

- Functional Description — 66
- Physical Description — 67
- Setup and Configuration — 74
- Performance — 74
- Model 25 Options — 78

4 Personal System/2 Family Architecture — 79

- The System Board — 80
- Micro Channel Architecture — 94
- BIOS (Basic Input/Output System) — 119
- POST (Power-On Self Test) — 124

5 Personal System/2 Video Subsystem Architecture — 127

- Physical Description — 132
- Video Modes — 136
- Memory Arrangement — 146
- Hardware Registers — 148
- Summary — 152

6 The Model 50 — 153

- Features — 153
- Physical Description — 155
- Functional Description — 162
- Security — 178
- Installation and Use — 179
- Installing Options — 191
- Warranty and Service — 194
- Documentation — 194
- Disk Cache Program — 195
- Performance and Compatibility — 201
- Summary — 205

7 The Model 60 — 207

- Features — 207
- Physical Description — 210
- Functional Description — 218
- Security — 223
- Installation and Use — 223
- Installing Options — 224
- Problem Diagnosis — 226
- Warranty and Service — 235
- Documentation — 236
- Performance and Compatibility — 237
- Summary — 241

8 The Model 80 — 243

Features	244
Physical Description	248
Functional Description	257
Installation and Configuration	265
Security	266
Documentation	267
Performance and Compatibility	267
Summary	274

9 Personal System/2 Options — 275

Monitors	276
Model 30 Display Adapter Requirements	283
Display Adapter 8514/A	286
Math Coprocessors	290
LAN Boards	291
Host Connection Boards	294
Communications Boards	294
Mouse	296
Memory Boards	297
External 5.25-Inch Diskette Drive	298
Other IBM Mass Storage Options	300
Laser Printer Controller Card	302
Other Printers	302
Third-Party Hardware	305
Personal System/2 Clones	315

10 Disk Operating System (DOS), Version 3.3 — 317

History	317
Features	319
Packaging and Documentation	322
Updates	323
Installation and Use	324
Enhanced User Interfaces	333

11 Operating System/2 (OS/2) — 335

System Architecture	337
OS/2 Extended Edition	355
Languages	359
Packaging and Documentation	363
Installation	364
Using OS/2	373
Summary	383

12 AIX Personal System/2 Operating System — 385

Features — 385
Terminal Support — 388
Languages — 388
Connectivity — 391
Installation — 394
Market Position — 394

13 Control Software for the Model 80 — 397

Concurrent DOS 386 — 399
PC-MOS/386 — 406
DESQview/386 — 413
Windows/386 — 417
Exploiting the Power of the Model 80 — 423

14 Connectivity — 425

Media and File Interchange — 426
Asynchronous Communications — 428
Local Area Networks — 430
5250 Communications — 437
3270 Communications — 440
What Next? — 460

A Performance Testing — 463

B Benchmark Machines — 471

C 80286/80386 Instruction Sets — 473

Index — 483

Preface

This book is about the IBM Personal System/2 family of computers. It is a technical book that describes the overall architecture of the family, along with the characteristics and capabilities of the individual computers. It contains detailed information about each model as well as performance comparisons (to each other and to competitive products).

The book begins with an overview and market perspective of the Personal System/2 family. The overview, like many sections of the book, will be of interest to both new and advanced computer users.

A technical discussion of the entry level Models 25 and 30 follows, including functional characteristics, physical layout, and installation and use. The capabilities of these PC-like computers and their performance running popular application software products is examined.

The technical features that make up the architecture of the Personal System/2 family—including system board features, the Micro Channel expansion bus, the VGA video subsystem, and the compatibility BIOS and advanced BIOS—are presented in complete detail.

The book supplies extensive technical descriptions of the Models 50, 60, and 80. The functional characteristics, physical layout, and installation and use of each model is described. The performance of each machine is examined and compared with that of competitive machines and other machines in the family. The overall capabilities and future potential of each model is discussed.

The architecture, installation, and use of IBM's Operating System/2—the new multi-tasking operating system for use on Models 50, 60, and 80, the AT, and XT 286—is considered. The enhancements found in DOS 3.3, the version of PC DOS for the Personal System/2 computers, are covered. Personal System/2 AIX, IBM's UNIX-based operating system for the Model 80, and operating systems and application managers (often referred to as control programs) that allow the simultaneous execution of multiple DOS applications on the Model 80 are also examined.

The subject of connecting Personal System/2 computers to other personal computers and to host computers is another key part of the book. Media and file interchange is discussed along with asynchronous communications, local area networks, and communications with IBM System/3x and System/370 host computers. Extensive information is provided on available hardware and software solutions.

Optional equipment available for use with the Personal System/2 family, both from IBM and third-party manufacturers, is also considered.

In summary, this book provides comprehensive operational and technical information about the Personal System/2 family and associated products, information which is essential to today's computer users.

Acknowledgments

We would like to acknowledge the following people and thank them for their assistance with this book: Mary DeWolf, Paula Jaworski, Steven Armbrust, Ed McNierney, and IBM Corporation.

Trademarks

All terms mentioned in this book that are known to be trademarks or service marks are listed below. In addition, terms suspected of being trademarks or service marks have been appropriately capitalized. Howard W. Sams & Company cannot attest to the accuracy of this information. Use of a term in this book should not be regarded as affecting the validity of any trademark or service mark.

Postscript is a registered trademark of Adobe Systems Incorporated.

PageMaker is a registered trademark of Aldus Corporation.

Apple is a registered trademark, and Apple Macintosh and Apple Macintosh Plus are trademarks of Apple Corporation.

dBASE III and dBASE III PLUS are trademarks of Ashton-Tate.

AST is a registered trademark, and Rampage/2, Advantage/2, SuperPak, Advantage/2-386, and AST Premium/286 are trademarks of AST Research, Inc.

Atari is a trademark of Atari, Inc.

1dir + and Bourbaki are trademarks of Bourbaki, Inc.

CHIPSlink is a trademark of Chips and Technologies, Inc.

Commodore is a trademark of Commodore Corporation.

COMPAQ, COMPAQ DESKPRO, COMPAQ DESKPRO 286, COMPAQ PORTABLE 286, COMPAQ PLUS, and COMPAQ PORTABLE II are registered trademarks, and COMPAQ PORTABLE 386, COMPAQ DESKPRO 386/20, COMPAQ DESKPRO 386, and COMPAQ PORTABLE III are trademarks of Compaq Computer Corporation.

CompuServe is a registered trademark of CompuServe Information Services, an H & R Block Company.

DCA is a registered trademark and IRMA is a trademark of Digital Communications Associates, Inc.

VT100 is a registered trademark of Digital Equipment Corporation.

Digital Research is a registered trademark, and Concurrent, CP/M, and GEM are trademarks of Digital Research, Inc.

Preface

Dow Jones News/Retrieval is a registered trademark of Dow Jones & Company, Inc.

Everex, EVGA, and RAM II 2000 are trademarks of Everex Systems, Inc.

Hayes and AT command set are registered trademarks of Hayes Microcomputer Products, Inc.

Hercules is a trademark of Hercules Computer Technology.

IDEAssociates, IDEAcomm 5251, IDEAcomm 5251/Share, IDEAcomm 5251/Gateway, IDEAcomm 5250/RemoteMC, IDEAcomm 5250/Modem, IDEAcomm 5251/MC, IDEAcomm 3278/MC, IDEA Minicomm, IDEA Supercomm, IDEA Supermax, IDEAmax/MC, and IDEAshare are trademarks of IDEAssociates, Inc.

Intel is a registered trademark of Intel Corporation.

INed, INmail, and INnet are trademarks of Interactive Systems Corporation.

Personal System/2, RT, RT PC, RT Personal Computer, Proprinter, Proprinter II, Operating System/2, Operating System/2 Standard Edition, OS/2, Advanced Interactive Executive (AIX), AIX PS/2, IBM Personal Computer XT, IBM Personal Computer XT 286, Micro Channel, NetView, TopView, PCjr, and SolutionPac are trademarks, and IBM, Personal Computer AT, IBM 3270 Personal Computer AT, Quietwriter, and Multi-Color Graphics Array (MCGA) are registered trademarks of International Business Machines Corporation.

MERGE/386 is a trademark of Locus Computing Corporation.

Logitech and Logimouse are trademarks of Logitech, Inc.

Lotus 1-2-3 is a registered trademark of Lotus Development Corporation.

MCI Mail is a trademark of MCI Corporation.

MicroFocus is a trademark of MicroFocus, Ltd.

Matrox is a trademark of Matrox.

WordStar is a registered trademark of MicroPro International Corporation.

Microsoft, Microsoft Windows, MS-DOS, MS, XENIX, and Windows/386 are registered trademarks of the Microsoft Corporation.

NEC and Multisync are registered trademarks of NEC Corporation.

RamQuest 50/60, Orchid Designer VGA, and Orchid Designer VGA-2 are trademarks of Orchid Technology.

The Norton Utilities is a trademark of Peter Norton Computing.

Princeton Graphics is a trademark of Princeton Graphics Systems.

Quadram is a registered trademark, and Quadboard PS/Q, QuadPort PS/Q, ProSync, PS/Q, Ultra VGA, QuadHPG, QuadMEG PS/Q, QuadMaster, QuadVGA, and MainLink II are trademarks of Quadram.

DESQview is a trademark of Quarterdeck Office Systems.

Rodime is a trademark of Rodime.

PC-MOS/386 is a trademark of The Software Link, Inc.

Sony and MultiScan are trademarks of Sony Corporation of America.

RapidRAM 2 is a trademark of STB Systems, Inc.

MicroRAM, MicroRAM 50/60, MicroRAM AD, and MicroRAM M80 are trademarks of Tecmar, Inc.

Texas Instruments is a trademark of Texas Instruments.

Lancard A-II is a trademark of Tiara Systems, Inc.

VisiCalc is a trademark of Visicorp.

SIMM (single in line memory module) is a trademark of Wang Labs.

WordPerfect is a trademark of WordPerfect Corporation.

Ethernet is a registered trademark of Xerox Corporation.

XTREE is a trademark of Executive Systems, Inc.

xiii

1

Introduction: the PC Market and the New Generation

Auspicious events often occur with little fanfare. Such was the case with the August 1981 introduction of the IBM Personal Computer. To most there was little to indicate that this was anything other than the introduction of yet another computer from the world's largest computer manufacturer.

The IBM Personal Computer did not have the trappings of previous historic offerings. It was simple and unassuming; it did not promise to be an "all around" computer as the System/360 had, nor was it advertised as the computer of the coming decade as was the System/370. In a significant departure from past practice, IBM seldom referred to the product by its model number, preferring instead to call it by name. In short order, however, the name IBM Personal Computer was shortened by the world at large to the now famous "IBM PC." Its popularity has exceeded all expectations and has revolutionized the computer industry.

The IBM PC

The IBM PC was developed by the Entry Systems Division of IBM in less than eighteen months. To meet this ambitious schedule the PC was designed and built using commonly available parts and devices—a strategy often used by smaller computer companies, but seldom employed by IBM. In subsequent years, this enabled third-party vendors to produce PC clones, which also helped to popularize the machine.

IBM decided that the machine's architecture would be built around a 16-bit central processing unit (CPU), even though popular personal computers of the time such as Apple, Atari, and Commodore used 8-bit CPUs. The CPU chosen for use in the PC was the Intel 8088, operating at 4.77 megahertz (MHz). This processor was a derivative of Intel's popular 8086 processor.

The Intel 8086 is a single chip arithmetic and logic processor that was introduced in 1978. The 8086 employs 16-bit addresses and operands and, through the use of a segmented addressing scheme, provides access to up to 1,048,576 bytes (1 MB) of memory. The 8088 is functionally identical to the 8086, but has an 8-bit rather than a 16-bit data bus interface, and was announced in 1979. IBM chose the 8088 for use as the PC's CPU in order to provide 16-bit functionality with minimal effect on overall system cost.

The PC, shown in Figure 1-1, was designed as a desktop unit consisting of a system unit with a detachable keyboard and video display. Devices such as printers and modems were attached to optional adapters installed in the system unit via cables. The 8088 microprocessor and components essential to its effective operation were contained on a single system board mounted horizontally on the bottom of the system case. On the original entry level PC these components included 16,386 bytes (16 KB) of dynamic RAM (random access memory), with sockets for an additional 48 KB of RAM, a four-channel *direct memory access* (DMA) controller, and an eight-level interrupt controller. Also included were a system clock, a three-channel timer/counter, a keyboard controller, a cassette tape controller, and a 40-pin socket, which IBM later announced was for the installation of an optional Intel 8087 math coprocessor. The system board also included an input/output (I/O) channel connected to five 8-bit system expansion slots. These expansion slots were provided for the installation of I/O adapters (including the system video adapter) and RAM memory boards.

Perhaps more significant than any of these, however, was the 40 KB of socket-mounted read only memory (ROM). It was the IBM copyrighted contents of this memory that made a collection of open market electronic parts uniquely an IBM PC.

System Architecture

Modern computers employ several layers of software between the processor and the user. The lowest level of such software is the *basic input/output system* (BIOS). This system is a collection of routines that can be called to control I/O devices and provide basic system services. In mini and mainframe computers these routines are generally part of the operating system that is loaded into system memory from an external I/O device, such as a disk or tape drive, by a bootstrap loader resident in system ROM.

In order to alleviate the need to load this software from an external device each time the system is initialized and to conserve the small amount of system RAM available on the base model PC, IBM chose to store the PC's BIOS in one of the five 8 KB ROM chips located on the system board. The remaining four ROM

Introduction

Figure 1-1 The IBM Personal Computer. *The 8088-based IBM PC was the first of the new generation of IBM computers to change the industry.* Photo courtesy of International Business Machines Corporation.

chips contained an implementation of the Microsoft BASIC-80 interpreter. Thus, the PC as a stand-alone device contained all software necessary to develop and execute BASIC programs. External devices were, of course, required to store and retrieve programs. These are discussed later.

System ROM and RAM co-exist in the 8088's 1 MB address space. Originally, IBM limited system RAM to the lower 256 KB of the address space, reserving the address space between 256 KB and 1 MB (40000H and FFFFFH) for future RAM expansion, video buffers, system ROM, and adapter board ROM. This limit was later increased to allow 640 KB of system RAM. System BIOS and BASIC ROM are addressed beginning at F0000H (see Figure 1-2).

Adapters and Devices

The video buffers referred to above did not exist in system RAM. As is now common practice, they existed as special purpose RAM mounted on the video adapter board installed in one of the PC's I/O expansion slots. Two video adapters were offered for installation in the PC—the *Monochrome Display and Printer Adapter* (MDA), which also provided a parallel printer port, and the *Color Graphics Adapter* (CGA). The MDA displayed 25 80-column lines of high resolution text, but no graphics, on the green phosphor monochrome display used by IBM for its Displaywriter word processors. The Color Graphics Adapter provided 640-pixel

| Memory Map | Address | Function |

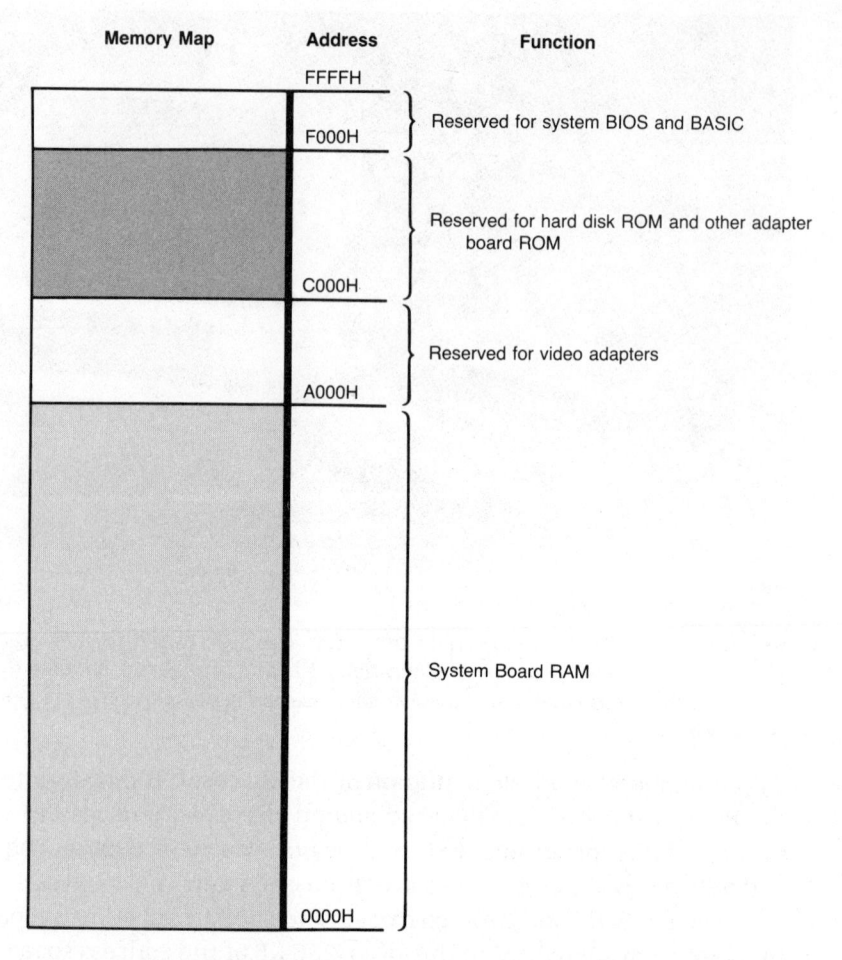

```
              FFFFH
              ┐
              │  Reserved for system BIOS and BASIC
              F000H
              ┐
              │  Reserved for hard disk ROM and other adapter
              │  board ROM
              C000H
              ┐
              │  Reserved for video adapters
              A000H
              ┐
              │
              │
              │  System Board RAM
              │
              │
              0000H
```

Figure 1-2 PC Address Space. *DOS can address up to 1 MB of contiguous memory. The addresses above 640 KB are used by the system for functions such as video buffers and BIOS.*

by 200-pixel, black and white graphics and 320-pixel by 200-pixel, 4-color graphics modes. Text displayed consisted of 25 lines and 40 or 80 columns of medium resolution characters (with a character cell of 8 by 8 pixels versus 9 by 14 pixels with the MDA) in up to 16 foreground and 8 background colors. This adapter supported both composite video and *red-green-blue* (RGB) interface color displays, as well as television sets with a user-supplied radio frequency (RF) modulator. Initially, IBM offered no color monitor for use with this adapter, but later introduced the IBM Personal Computer Color Display for that purpose.

Although not required for the PC's operation, very few systems were sold without diskette drives. The original PC was offered with one or two internal diskette drives as an option. These drives could read or write 160 KB of information on a single surface of a 5.25-inch diskette. Drives with dual read/write heads, which boosted this information storage capacity to 320 KB per diskette, were later offered. Subsequent software changes yielded the current capacity of 360 KB per double-sided, double-density diskette. The diskette drives were controlled by an adapter that mounted in one of the system's five I/O expansion slots. This controller could also control two external 5.25-inch diskette drives via a connector located on its rear mounting bracket.

Since most PCs were sold with two expansion slots occupied—one by the video adapter and another by the diskette drive adapter—at most, three slots were left available. Memory expansion boards originally offered by IBM used 16K by 1 bit memory chips and had a maximum capacity of 64 KB. This limited the amount of memory that could be installed in the expansion slots to a total of 192 KB, and thus the system memory to a total of 256 KB. In due time, memory boards became available from IBM and third-party manufacturers that used 64K by 1 bit memory chips to provide 256 KB or more memory per board, thus easily allowing the installation of a full 640 KB of RAM.

Compatibles and Clones

By the end of 1982 over a quarter of a million PCs had been sold. But these were no longer only IBM versions of the machine. The lucrative PC clone market had begun to arrive, with over twenty PC compatibles being displayed at the Fall Comdex (a national computer dealers show) of 1982. The acceptance of the PC and its compatibles was to change the shape of business and industry. It was no longer necessary to design or use such specific, customized equipment for industrial control, and office automation became available to everyone rather than being exclusive to large corporations. It is only necessary to reflect back a few years to the state of the world without PCs to realize the dramatic effect that they have had on our daily lives.

At the end of 1982, the compatible manufacturers were touting their wares, but it was not certain who would produce a quality PC compatible. Only a few realized the significance of true compatibility with the IBM machines. Some companies assumed that the end user would be more interested in the functionality of the machine than in its ability to behave exactly like an IBM PC.

In the next two years this attitude was to change. As more applications were developed for the IBM PC, it became important to be able to claim that a PC-compatible computer could run the same applications as an IBM PC, without modification. With a new generation of computer users came a new requirement for the hardware, as end users had a choice of hardware vendors as well as software vendors.

One of the most successful PC clones arrived in rather strange clothing from Compaq Computer Corporation in November 1982. Compaq did what many vendors have since done: they added value to the basic PC role model offered by IBM. Seeing the need for a portable computer with full IBM PC functionality, in less than a year Compaq built an IBM PC-compatible computer that, if not portable, was at least transportable. The system, shown in Figure 1-3, consisted of a single unit containing the computer, one or two diskette drives, a dual-mode monitor, and a detachable keyboard. Subsequent improvements in technology have made this machine appear heavy in retrospect, but at the time people really did carry them around. In 1983 the Compaq computer was on the front edge of a revolution. Now the same machine is commonplace, if not outdated.

Figure 1-3 The Compaq Portable. *The full functionality of the PC was maintained in a portable clone of the PC.*
Photo courtesy of Compaq Computer Corporation.

This immensely successful computer set the standard for IBM compatibility. Its dual-mode display adapter and monitor provided compatibility with both the IBM Monochrome Display and Printer Adapter and Color Graphics Adapter systems. This machine also established and maintained the standard for portable computers as well. IBM's portable personal computer, finally introduced in March 1984, was less sturdy, offered fewer full-length expansion slots, and provided only CGA display compatibility.

Many competitors to Compaq also provided top quality compatible machines and made fortunes from the PC market. However, many of the companies that made only PC products, such as Seequa and Columbia Data Products, did not survive in the longer term, as Compaq did. Their machines were good but the business was ultimately too competitive. Other manufacturers, such as Wang and Tandy, were successful during the bad times partially because PC clones were only a part of their product line. The industry now looks first to Compaq as the compatible manufacturer that will keep up with IBM in the PC market; but in 1983 they had not established the reputation and standing they now have and deserve.

The compatible manufacturers added to the PC's capabilities by providing extra disk drives, extra RAM, parallel ports, and serial ports as part of the system board. Extra storage capacity seemed to be the leading requirement. This was accomplished both by the addition of RAM on the system boards and the addition of hard disks.

As these features became more generally available, people started looking to IBM to add them to the PC. Remember that at this time compatibles were available rather than universally accepted. The IBM label was important for corporate applications and clones were only just beginning to be recognized.

While extra memory was useful, the feature that many business users soon began to long for was additional disk storage. Hard disks with capacities much greater than flexible diskettes were available from third-party manufacturers, but due to the PC's somewhat limited 63.5-watt power supply, these had to be mounted and powered externally.

In March 1983, some eighteen months after the introduction of the PC, IBM offered a solution by introducing the IBM Personal Computer XT. This machine quickly became known as the IBM PC XT or IBM XT. At this time the influence of third-party compatible manufacturers was probably not impacting IBM's decisions, but this situation was to change in the next couple of years.

The IBM XT

While the XT with its hard disk may not truly have been extended technology, as IBM claimed, it did retain total software compatibility with the PC while remedying some of the PC's shortcomings. Its outward appearance was the same as that of the PC, but its interior contained several useful enhancements. It featured a new system board that used the same 4.77 MHz 8088 and 8087 processors as the PC; however, it came with 128 KB of RAM, expandable to 256 KB on the system board. The PC keyboard was retained, but the cassette interface was dropped. A new power supply rated at 130 watts was provided to power the standard 360 KB diskette drive, 10 MB hard disk, and expansion slots.

Eight expansion slots were provided instead of the PC's five. Of these eight slots, six were full length, and two could only accommodate short cards. Since the hard disk drive controller occupied a full expansion slot, the standard asynchronous communications adapter occupied one short slot, and the diskette controller and the video controller each occupied one full length slot, an XT had a maximum of three full length slots and one short slot available.

IBM also offered an expansion chassis containing a power supply and the

XT's 10 MB disk for use with existing PCs. This unit, which had the same size and appearance as a PC chassis could contain one or two 10 MB hard disks. It was also usable with the XT for those who desired more than one diskette drive as well as a hard disk, or more than one hard disk. If a system had more than one hard disk, both disks had to be in the expansion unit. The expansion unit also provided up to six additional expansion slots; however, because of timing dependencies, only certain boards could be used in the expansion unit's slots.

The third-party vendors were quick to follow IBM's lead and offer hard disks for their machines, if they hadn't already done so. The announcement of DOS 2.0 at the same time as the XT, however, had a greater influence on the market than the XT itself. DOS 2.0 offered many more features than the earlier versions and provided a better interface for applications. The number of software packages also grew rapidly in the next year, providing end users with programs that could really fill their needs.

Word processors became less primitive and the concept of a computerized spreadsheet was accepted. Today, software continues to improve and stretch the capabilities of machines, but at that time users were happy with the available application software even if it didn't exploit all of the hardware capabilities of the PC. PC users were no longer dependent on the MIS department to produce simple reports. Mass mailings of brochures and personalized letters became a reality. These were times of change, but even more changes were still to come.

Many firms now manufacture and sell IBM-compatible PCs and cost-effective multi-function boards. The firms that have prospered the most are the ones that have maintained the highest degree of quality and compatibility, while at the same time providing competitive pricing and added value. Compaq attempted to do this with the June 1984 introduction of its Deskpro desktop computer. This computer featured a 4.77/7.1 MHz 8086 with an 8-bit PC-style bus.

The future of this machine as the industry leader was, however, shortlived. In August 1984 IBM introduced a machine that represented a significant advance in the technology.

The IBM AT

By the time the AT was introduced, speculating about upcoming PC products from IBM had become a favorite pastime in the computer industry. Pundits who had been disappointed with the announcement of the XT were considerably more enthusiastic about the introduction of the AT, shown in Figure 1-4.

Everything about the AT was bigger and better—from its key-lockable system case to the enlarged Return and Shift keys on the keyboard, to its 6 MHz Intel 80286 processor. The 16-bit 80286 processor was introduced by Intel in 1982. It was designed to be compatible with the 8086 when operating in real-address mode and to provide enhanced functionality when operating in the pro-

Introduction

Figure 1-4 The IBM PC AT. *The growth of the PC market was reflected in the 80286-based AT.*
Photo courtesy of International Business Machines Corporation.

tected virtual-address mode. When operating in real mode the processor could access up to 1 MB of memory. Protected mode supported multi-tasked operation and the execution of tasks whose memory size exceeds the size of physical memory. In protected mode the processor could address up to 16 MB of physical memory and each task could have a virtual address space of up to 1,073,741,824 bytes (1 gigabyte) mapped into physical memory.

The 6 MHz 80286 was supported by a well-matched set of support devices including a 4 MHz 80287 math coprocessor, a seven-channel direct memory access (DMA) controller, and a sixteen-level interrupt controller. System ROM consisted of two 32 KB modules with sockets for two more; system board RAM was either 256 KB on the base model or 512 KB on the enhanced model. Other added features were a real-time clock and *complementary metal oxide silicon* (CMOS) configuration memory in place of most system board configuration switches.

The AT provided a 16-bit I/O channel with eight expansion slots, six of which were for PC-style, 8-bit adapters or the new 16-bit expansion adapters, and two of which were for 8-bit adapters. The one adapter standard with all ATs was the Western Digital hard disk and diskette drive controller. This controller occupied a 16-bit slot and controlled up to two diskette and two hard disk drives. The AT used half-height diskette drives with a capacity of 1.2 MB or 360 KB. A 1.2 MB drive was standard; an optional second drive of either type could be added. As is often the case with new IBM products, it was easy to transfer data to the AT, but not from it. Diskettes written by the 1.2 MB diskette unit could not be read reliably on the PC's or XT's 360 KB diskette drives, even though the 1.2 MB unit could read 360 KB diskettes.

Enhanced model ATs included a 20 MB hard disk and a board that provided one serial and one parallel port. The connector for the serial port was a 9-pin male D-shell connector rather than the more common 25-pin connector seen on PCs. The 20 MB hard disk was mounted in a bay to the left of the diskette drive bay and was sheltered from view by the cover's grille. The hard disk and the three half-height devices mounted in the diskette drive bay were secured using a rail and retainer bracket system that greatly eased installation and disassembly, compared with the side screw attachment method used on the PC and XT.

The AT did not remain static. Its operating speed was increased in April 1986, with the speed of the CPU and I/O expansion bus being increased to 8 MHz and the speed of the 80287 math coprocessor being increased to 5.33 MHz. Also, the optional 20 MB disk drive was replaced by a standard 30 MB unit, and an optional 101-key keyboard with 12 function keys and a separate cursor pad was added.

In much the same way as it had introduced the IBM Color Display with the XT, IBM announced the Enhanced Graphics Adapter (EGA) and Enhanced Color Display at the same time as the AT. The capabilities offered by this pair of products proved very popular because they provided high resolution text (with 9 by 16 character cells) as well as high resolution graphics (640 by 350 pixels with 16 colors). Although the capabilities of the EGA subsystem were popular, the actual IBM products enjoyed less success than compatible products that soon became available at substantially lower cost. This, to a lesser extent, is also what happened with the PC and AT computers. After IBM's successful entry into the microcomputer market, other firms capitalized on the open architecture of the PC and its successors to offer lower priced "clones" that were IBM compatible.

Compaq was among many manufacturers to follow IBM's lead in producing 80286-based machines. They introduced the AT-compatible Deskpro 286 as well as the Compaq Portable 286, in April 1985. By this time, Compaq had a solid reputation as one of the industry leaders. In their first year of operation, Compaq's sales totaled $111 million. This was followed by $329 million in their second year.

In February 1986, the Compaq Portable II was introduced and Compaq's third year sales were reported as over $500 million. In April of 1986, the 500,000th Compaq computer was manufactured; in the same month the company was listed as a Fortune 500 company. It should be remembered that this company has made its money from PC- and AT-compatible machines, and does not have other product lines in other markets.

Compaq is not alone in the compatible market but they are the most successful. The market clamored for PCs, and those with foresight (and more money) chose AT clones over PC compatibles. Steadily the market share for clones increased from an insignificant force that IBM could ignore to a substantial mass. The user buying the machines became more knowledgeable, and the demand on the manufacturers increased. The industry had a shakeout and many companies did not survive; however, the market continued to grow, albeit at a slightly slower rate than the initial explosive rates of 49 percent per year. As a result, a variety of machines is currently available. The user can now make an informed choice, where only a few years ago there was no choice.

Compaq and others were successful at following IBM's lead and improving on the IBM PC standard; however, it was Compaq's September 1986 introduction of the Deskpro 386 that marked the first successful introduction of new PC processor technology by a vendor other than IBM.

The Compaq Deskpro 386

The Compaq Deskpro 386 features a 16 MHz Intel 80386 processor and, at the time of its introduction, supported either an optional 4 or 8 MHz 80287 math coprocessor. (It now supports an 80387 that runs at the same speed as the processor—either 16 or 20 MHz.) The Deskpro 386, shown in Figure 1-5, features expansion slots compatible with those of the 8 MHz IBM AT, thus maintaining compatibility with IBM offerings while at the same time providing enhanced performance.

The 32-bit 80386 processor, introduced by Intel in 1985, was designed to be compatible with the 8086 when operating in real-address mode and to provide enhanced functionality when operating in the protected virtual-address mode. When operating in real mode the processor can access up to 1 MB of memory. Protected mode supports multi-tasked operation and the execution of tasks whose memory size exceeds the size of physical memory. In protected mode the processor can address up to 4 gigabytes of physical memory, and each task can have a virtual address space of up to 64 terabytes (one terabyte equals 1,099,511,627,776 bytes) mapped into physical memory. The 80386 also provides a virtual 8086 mode that allows the protected, multi-tasked execution of programs that are written for execution on an Intel 8086/8088 microprocessor.

Chapter 1

Figure 1-5 The Compaq Deskpro 386. *Compaq did not wait for IBM to introduce an 80386-based machine, but did wait for them to define a 32-bit bus standard for the general expansion slots.*
Photo courtesy of Compaq Computer Corporation.

The Deskpro 386's 80386 microprocessor—coupled with its 32-bit, high speed, static column RAM and AT-compatible I/O expansion bus—makes it a machine that significantly enhances performance while maintaining AT compatibility. Its protected mode allows it to run applications that are available for use with protected mode, multi-tasking operating systems. More importantly, its virtual 8086 mode allows it to run multiple PC applications without applications interfering with each other. This ability may prove crucial to the success of future systems, because the most important factor in the recent success of the PC industry has not been the computers themselves, but the abundance of software that has been developed for use with them.

PC Software

It has been said that the availability of an IBM PC version of VisiCalc from Software Arts assured the success of the IBM PC. While there is some truth to that statement, the authors of the tens of thousands of software packages currently available for the IBM PC and compatibles probably feel the same about their product.

Packages such as VisiCalc, Lotus 1-2-3, WordStar from MicroPro, and WordPerfect from WordPerfect Corporation have been successful because they provide people with the means to do more work or to do the same work more quickly and efficiently. The utility of these packages is universal, and the ability of a computer to run them is a mark of its compatibility with the IBM PC. Manufacturers of larger computers often provide enhancements to languages and software packages offered with and for use on their computers. This strategy, which is often simply a strategy to protect the vendor's market base, has not been successful in the PC arena. PC vendors must make their computers sufficiently IBM compatible to properly run popular applications software, and software vendors must make their software sufficiently independent of particular IBM PC models to run on all IBM PCs and compatibles. With everyone involved working seriously toward this goal (and most are), the result is possibly the greatest improvement in standardization and productivity since the nineteenth century invention of interchangeable machine parts.

While it is true that most equipment vendors take software compatibility seriously, no one has taken it more seriously than IBM. Some may say that innovations in the PC family have been far too few and new models have offered only small incremental enhancements, but software compatibility has been maintained all along the way. This was easy with the introduction of the XT, since it offered no new technology; neither was it much of a problem with the introduction of the AT, as long as the AT was run in real mode. Applications run on all of the machines (AT, PC, and XT), as long as the respective machines are equipped with the appropriate video displays and peripherals. The net effect of running an application on an AT rather than a PC or XT is simply that it runs faster.

Where software compatibility begins to get more complicated, however, is in the area of enhanced capabilities. The true test of a good computer design is its ability to provide new capabilities without detrimental impact on existing applications. Such was the problem that IBM faced in designing a successor to the PC and the AT. This problem—coupled with the possibility that IBM might conclude that the PC's open architecture made it too easy to copy—led to varied speculation on what the "new PC" might be like. Many industry observers postulated that future PCs would feature a more closed architecture, featuring high speed processors, large memory capacity, and video and communications functions integrated on the system board.

Chapter 1

As is often the case in such matters, many of the forecasts were correct, but often not in exactly the way the forecasters had anticipated. Under the circumstances, it could hardly have been any other way. For in April 1987, IBM announced not a new PC, but the IBM Personal System/2 family—a whole new line of computers that would improve and expand on the capabilities provided by the PC and the AT, while at the same time maintaining a high degree of compatibility with them.

IBM Personal System/2

The introduction of the Personal System/2 family of computers ended months of speculation about the future of the PC family. Rumors that IBM was making a "clone killer" were rampant. Compaq, with its September 1986 announcement of the Deskpro 386, had caused the industry to speculate wildly about IBM's plans for the future. Compaq's conservative approach, which did not include an extension to the 16-bit AT bus for the 32-bit microprocessor, was seen as an acknowledgment that IBM should be the one to expand the bus.

But the announcement held surprises for many industry analysts. IBM decided not to extend the PC bus. They announced a new generation of machines, and provided peripheral solutions for the new systems while reinforcing and expanding support for the existing PC family. The announcements fell into several general categories: systems, monitors, software, add-in boards, and peripheral devices.

Models 30, 50, 60, and 80 were announced as the four new members of the Personal System/2 family. The base model, the Model 30, is an 8086-based machine with a PC-like architecture that offers faster performance than an XT, at a lower cost. Models 50, 60, and 80 are the major players in the Personal System/2 family. They introduce a new Micro Channel bus architecture that fully supports the 80286 and the 80386 processors, and offers tremendous potential for the future. The Model 50 is a desktop computer, like the Model 30, but is an 80286-based machine with performance comparable to that of an AT. The Model 60 is a floor-standing version of the Model 50, with additional expansion features. The floor-standing Model 80 is the current top of the line machine. It is based on the 80386 microprocessor and has full 32-bit features. Figure 1-6 shows the desktop Personal System/2 products; Figure 1-7 shows Models 60 and 80.

IBM added the Model 25 to the Personal System/2 family after the initial product announcements. The Model 25 is a repackaged Model 30 with a display incorporated into the system unit. Its styling is similar to the styling of the Apple Macintosh. It is targeted at the educational market, which is price sensitive. With a list price of $1,350 for the base model, the Model 25 should be obtainable with an educational discount for under $1,000. The IBM PC Convertible, introduced in

Introduction

April 1986, is also considered part of the Personal System/2 family, mainly because of its standard 3.5-inch diskette drives.

Figure 1-6 Personal System/2 Models 25, 30, and 50. *The Model 25 (left) and Model 30 (right) are the Personal System/2 machines with PC-style expansion slots, while the Model 50 is the low end of the Micro Channel based machines.* Photo by Bill Schilling.

Figure 1-7 Personal System/2 Models 60 and 80. *The Model 60 (left) and Model 80 (right) are the higher performance machines in the Personal System/2 family.* Photo by Bill Schilling.

Common Design Features

The Personal System/2 machines have been designed as a family. All share a philosophy of streamlined design, minimal installation or upgrade requirements, and extensive integration. As each computer in the family is considered, it is obvious that it is an extension of the previous model. This leads to the conviction that the Model 80 may be the current top of the line machine, but it will not hold that position forever. Each of the current models has a market niche; moreover, the design provides room for the product line to grow as new Intel microprocessors become available, without compromising compatibility with the past.

All of IBM's announcements have reinforced this family concept. The machines have an integral video system, and although the Model 30's is a subset of the video system on the other models, any of the four Personal System/2 displays that were announced can be attached to any of the models.

Two major operating systems have been endorsed by IBM for these machines. PC DOS has been upgraded to run on the new machines as version 3.3. Additionally, a new operating system—OS/2—has been disclosed. This requires a machine that has an 80286 or 80386 microprocessor. Among its many new features, OS/2 supports multi-tasking and can address more than 640 KB of memory—two of the current restrictions on DOS. New versions of languages such as C, Pascal, Fortran, and Cobol are available, along with a new Macro Assembler version. New versions of the PC LAN software have been released, as well as new micro-mainframe programs such as the 3270 Workstation Program.

Add-in boards for the Micro Channel bus architecture have been released. These include memory expansion boards, LAN boards, and other communication boards. An optical disk drive and hard disk upgrade are also available from IBM.

Several peripheral devices are also in the announced products. These include four new printers—three dot matrix and one letter quality—and an external 5.25-inch diskette drive.

In addition to its new product announcements, IBM has shown its commitment to the new line by stating that it will not manufacture any more PCs or ATs and will only fill existing orders. All existing machines will be fully supported by IBM but new orders will be for the Personal System/2 family.

The realm of products that make up the new line is extensive. The implications on the marketplace are being felt now and will create an impact for several years. It is important to try to grasp the full effect of this family and to realize that these products are shaping the industry just as the PC did in 1981.

The new personal computer family is more unified than the PC family. In the PC family the AT was not designed at the same time as the PC; as a result, compromises had to be made in the design of the AT to maintain true compatibility with the past.

The Personal System/2 machines are a true family of machines, with each successive model improving and expanding the scope of the one before. They all have similar sleek styling and are lighter in weight than the comparable PC models. The footprints of the desktop versions, Models 30 and 50, have been reduced by about 25 percent from the PC and AT. The floor-standing Models 60 and 80 slide conveniently under a desk and take up comparatively little room.

All models consist of a system unit and keyboard, and work with any of the four Personal System/2 displays. The keyboard uses the 101-key Enhanced Keyboard layout that IBM has now made its standard.

The system boards in the units provide more devices and functions than those of the PC family machines. Each has a serial port, a parallel port, and a pointing device port. The video system is also an integral part of the system board. On the Model 30, the video system is the Multi-Color Graphics Array (MCGA), a subset of the Video Graphics Array (VGA) on Models 50, 60, and 80.

Each system board has an expansion bus with slots that can be used for system expansion. Again, the Model 30 is slightly different; the expansion slots in the Model 30 are similar to those in a PC. Models 50, 60, and 80 incorporate the Micro Channel architecture, creating a new specification for add-in board manufacturers to follow.

The integrated nature of the system board means that the typical user will probably not require these expansion slots. Initially, they may be needed only for configurations requiring networking or micro-to-mainframe communications. In the long run, however, the expansion slots may be necessary for additional memory and coprocessors, as new multi-tasking operating systems become available.

Each of the machines has an automatic voltage-sensing power supply that adjusts to work with the U.S. 120 V_{ac} supply or the European 240 V_{ac} system. Due to improvements in design—use of newer integrated circuit (IC) technology and more *very large scale integration* (VLSI) on the system board—the power requirements of the machines are less than those of the PC or AT.

The Personal System/2 family has standardized on a new media size of 3.5-inch diskettes for its removable media. These diskettes are more durable than the 5.25-inch diskettes, but for the millions of people who use PCs the media incompatibility can be an organizational nightmare. Each of the models can be purchased with a hard disk option. That the hard disk has become a commodity rather than a luxury item reflects growth in the industry.

The numbering method for the Personal System/2 models has a logical system. The part number for each computer starts with 85, signifying that it is a Personal System/2 machine. The next two digits are 30 for the Model 30, 50 for the Model 50, and so on for each model. The three-digit extension shows the type of disk drives that are installed in the machine. The rightmost digit is the number of 3.5 inch diskette drives; for example, the 8530-002 has two diskette drives and the 8530-021 has one. The other two digits signify the size of the hard disk in ten's of megabytes (MB); in the 8530-021 the hard disk has a 20 MB capacity and in the 8580-111 the hard disk has a 115 MB capacity.

Each of the machines in the Personal System/2 family is available with slightly different options and each fulfills a different market need. A brief description of each machine shows in very general terms how the product line looks.

Model 30

The Model 30 can be considered a bridge machine from the PC family to the Personal System/2 family. It uses the PC bus architecture and a microprocessor that is similar to the 8088 used in the PC and XT machines. It has the Personal System/2 styling and high level of integration, with many features standard on the system board. It has a smaller footprint and a lower price than the PC/XT, along with better performance.

In many ways the Model 30 is a repackaged fast PC/XT. It is based on the 8086 microprocessor, running at 8 MHz with zero wait states. The 8086 is similar in architecture to the 8088 except that it has a 16-bit data and address path. A faster math coprocessor is offered to match the faster speed of the main microprocessor. The expansion bus of the machine has the same architecture as the XT family, but it has only three 8-bit slots.

For a bare bones PC this scarcity of slots would be a serious problem, but the Personal System/2 family of machines has many integral features. To this end the Model 30 has its video adapter as part of the system board. The video system, the MCGA, emulates all of the modes of the Color Graphics Adapter (CGA). In addition, four expanded modes are available, including a 640 by 480 2-color mode. This video system will drive any of the four Personal System/2 analog monitors. The choice of monitor is a decision separate from the choice of the machine.

A typical PC configuration would include a real-time clock to maintain the time and date for the machine and at least one serial and one parallel port. The Model 30 has a serial port, a parallel port, and a pointing device port, along with a battery-backed clock, as standard on the system board. This frees the expansion slots for more specialized applications, such as a LAN connection board or a micro-to-mainframe communication board.

The Model 30 is available in two different configurations. A dual 720 KB 3.5-inch diskette system, part number 8530-002, is $1,695. The system with one 720 KB disk drive and a 20 MB hard disk, part number 8530-021, is $2,295. The choice of 720 KB disk drives is inconsistent with all of the other machines. The PC family uses 5.25-inch diskettes and the rest of the Personal System/2 family uses 1.44 MB, 3.5-inch diskettes. The Model 30's diskettes can be read on the other Personal System/2 machines, but it cannot read 1.44 MB diskettes. This raises some doubt as to which family the machine belongs.

Whether the Model 30 is a true member of the Personal System/2 family is an interesting question, but if the PC Convertible is to be considered a part of the

HOWARD W. SAMS & COMPANY

Bookmark

DEAR VALUED CUSTOMER:

Howard W. Sams & Company is dedicated to bringing you timely and authoritative books for your personal and professional library. Our goal is to provide you with excellent technical books written by the most qualified authors. You can assist us in this endeavor by checking the box next to your particular areas of interest.

We appreciate your comments and will use the information to provide you with a more comprehensive selection of titles.

Thank you,

Vice President, Book Publishing
Howard W. Sams & Company

COMPUTER TITLES:

Hardware
- ☐ Apple 140
- ☐ Macintosh I01
- ☐ Commodore I10
- ☐ IBM & Compatibles I14

Business Applications
- ☐ Word Processing J01
- ☐ Data Base J04
- ☐ Spreadsheets J02

Operating Systems
- ☐ MS-DOS K05
- ☐ OS/2 K10
- ☐ CP/M K01
- ☐ UNIX K03

Programming Languages
- ☐ C L03
- ☐ Pascal L05
- ☐ Prolog L12
- ☐ Assembly L01
- ☐ BASIC L02
- ☐ HyperTalk L14

Troubleshooting & Repair
- ☐ Computers S05
- ☐ Peripherals S10

Other
- ☐ Communications/Networking M03
- ☐ AI/Expert Systems T18

ELECTRONICS TITLES:
- ☐ Amateur Radio T01
- ☐ Audio T03
- ☐ Basic Electronics T20
- ☐ Basic Electricity T21
- ☐ Electronics Design T12
- ☐ Electronics Projects T04
- ☐ Satellites T09

- ☐ Instrumentation T05
- ☐ Digital Electronics T11

Troubleshooting & Repair
- ☐ Audio S11
- ☐ Television S04
- ☐ VCR S01
- ☐ Compact Disc S02
- ☐ Automotive S06
- ☐ Microwave Oven S03

Other interests or comments: _____

_____

Name _____
Title _____
Company _____
Address _____
City _____
State/Zip _____
Daytime Telephone No. _____

A Division of Macmillan, Inc.
4300 West 62nd Street   Indianapolis, Indiana 46268

22628

Bookmark

BUSINESS REPLY CARD

FIRST CLASS PERMIT NO. 1076 INDIANAPOLIS, IND.

POSTAGE WILL BE PAID BY ADDRESSEE

HOWARD W. SAMS & CO.
ATTN: Public Relations Department
P.O. BOX 7092
Indianapolis, IN 46209-9921

NO POSTAGE
NECESSARY
IF MAILED
IN THE
UNITED STATES

HOWARD W. SAMS & COMPANY

HOWARD W. SAMS & COMPANY
HAYDEN BOOKS

Hard Disk Management Techniques for the IBM®
Joseph-David Carrabis
ISBN: 0-672-22580-8, $22.95

IBM® PC AT User's Reference Manual
Gilbert Held
ISBN: 0-8104-6394-6, $29.95

IBM® PC & PC XT User's Reference Manual, Second Edition
Gilbert Held
ISBN: 0-672-46427-6, $26.95

The Waite Group's Desktop Publishing Bible
James Stockford, Editor, The Waite Group
ISBN: 0-672-22524-7, $24.95

Personal Publishing with PC PageMaker®
Terry Ulick
ISBN: 0-672-22593-X, $18.95

Best Book of: Lotus® 1-2-3®, Second Edition
Alan Simpson
ISBN: 0-672-22563-8, $21.95

Lotus® 1-2-3® Financial Models
Elna Tymes and Tony Dowden with Charles E. Prael
ISBN: 0-672-48410-2, $19.95

Best Book of: WordPerfect® 4.2
Vincent Alfieri
ISBN: 0-672-46581-7, $21.95

Best Book of: WordPerfect® 5.0
Vincent Alfieri
ISBN: 0-672-48423-4, $21.95

Best Book of: WordStar® (Features Release 4)
Vincent Alfieri
ISBN: 0-672-48404-8, $19.95

Best Book of: dBASE II®/III®
Ken Knecht
ISBN: 0-672-22349-X, $21.95

Best Book of: Framework™
Alan Simpson
ISBN: 0-672-22421-6, $21.95

Best Book of: Multiplan™
Alan Simpson
ISBN: 0-672-22336-8, $16.95

Best Book of: Symphony™
Alan Simpson
ISBN: 0-672-22420-8, $21.95

Best Book of: Microsoft® Works for the PC
Ruth Witkin
ISBN: 0-672-22626-X, $21.95

To order, return the card below, or call 1-800-428-SAMS. In Indiana call (317) 298-5699.

Please send me the books listed below.

Title	Quantity	ISBN #	Price

☐ Please add my name to your mailing list to receive more information on related titles.

Name (please print) _____
Company _____
City _____
State/Zip _____
Signature _____
(required for credit card purchase)
Telephone # _____

Subtotal _____
Standard Postage and Handling **$2.50**
All States Add Appropriate Sales Tax _____
TOTAL _____

Enclosed is My Check or Money Order for $_____
Charge my Credit Card: ☐ VISA   ☐ MC   ☐ AE
Account No.   Expiration Date _____
☐☐☐☐ ☐☐☐☐ ☐☐☐☐ ☐☐☐☐

22628

Place
Postage
Here

Howard W. Sams & Company

Dept. DM
4300 West 62nd Street
Indianapolis, IN 46268-2589

Personal System/2 family then the Model 30 certainly is a member. If the general design features of the Model 30 are considered, its membership is valid. Although it cannot be a fully fledged member of the family due to its PC style of bus, in most respects it is closer to being a Personal System/2 than a PC. The degree of integration, with parallel and serial ports and the video adapter part of the system board, tends to make it a part of the Personal System/2 family. The one element, however, that should establish true membership in one family or the other is the Model 30's removable media. As this is not the same as on either the PC or other PS/2 machines in its standard configuration, the Model 30 cannot be considered a true member of either family or a true bridge machine.

Model 25

The Model 25 was announced in July 1987 after the first four models had been shown. This machine offers the video display as part of the system unit instead of as a separate option. It is similar to the Model 30 in terms of performance and capabilities. The microprocessor is an 8086 running at 8 MHz with zero wait states, and the video system is the MCGA. There are only two XT-style expansion slots instead of the three available on the Model 30. Only one of the expansion slots is full size; the other is 8 inches long. The rest of the system board is as fully equipped with I/O devices as that of the Model 30. There is a serial port, a parallel port, and a pointing device port standard with the machine. The Model 25 does not have a battery-backed real-time clock, which is available on the Model 30.

This model has a choice of keyboards: a Space Saving version without a separate numeric keypad or the standard 101-key Enhanced Keyboard. There is also a choice of 12-inch analog display in the system: a monochrome or color version is available. The part numbers for the monochrome version are 8525-001 with the Space Saving keyboard and 8525-G01 with the Enhanced Keyboard. In a departure from normal Personal System/2 numbering practice, the color version has part number 8525-004 with the Space Saving keyboard and 8525-G04 with the Enhanced Keyboard.

The memory capabilities of the machine are more limited than those of the Model 30. As standard, the machine has 512 KB of RAM instead of the 640 KB that is standard on the Model 30. An expansion kit can be purchased to upgrade the system board RAM to 640 KB. No hard disk is available with the machine. A single 720 KB, 3.5-inch diskette drive is standard and one additional diskette drive can be added as an option.

The presence of this machine in the family reinforces the position of the Model 30 in the Personal System/2 series. It offers a viable option for users who do not require expansion capabilities for their machines. Once the machine is purchased there is no real growth path for the user, but it does supply adequate functionality for many end user situations.

Model 50

The Model 50 is a desktop computer that is a replacement for the AT. The microprocessor is a 10 MHz 80286, which is a faster version of the CPU in the AT. Although the AT and the Personal System/2 family run the same application software, the architecture beneath the covers is different.

The system consists of the typical three parts: the system unit, keyboard, and monitor. The keyboard is the Enhanced Keyboard that is standard across the product line (except for the Model 25). The system unit has a footprint that is smaller than the AT by approximately 25 percent. There is only one version of the system available; it has a 20 MB hard disk and a 1.44 MB, 3.5-inch diskette drive. The part number is 8550-021. An additional 1.44 MB, 3.5-inch diskette drive can be added as an option.

The system board has a serial port, parallel port, and pointing device port as standard. The video system is integral to the system board as well and is the full-featured VGA. This is a superset of the Enhanced Graphics Adapter (EGA) available for the PC family. It fully emulates the CGA and the EGA but offers several new modes, including 640 by 480 pixel, 16-color graphics; 320 by 200 pixel, 256-color graphics; and 720 by 400 pixel text. There is 1 MB of RAM on the system board and two levels of BIOS in 128 KB of ROM. The two BIOS areas are called CBIOS and ABIOS. CBIOS is a compatibility BIOS and ABIOS supports multi-tasking operating systems and extended memory addressing to 16 MB.

There are four expansion slots on the system board. One of these is occupied by the disk controller. All are Micro Channel expansion slots and require an expansion board that is designed to work with the new architecture. A real-time clock is also supplied as standard on the board.

This machine is an excellent standard and a worthy successor to the AT. Its computational performance exceeds that of the AT, and it is fully functional. It is limited in that the hard disk is only 20 MB in size, but as a node on a network it should be adequate. A power user who needs more storage space for programs is likely to require more expansion slots as well and so would choose a Model 60 or 80.

Model 60

The Model 60 is a floor-standing version of the Model 50. It has the same 10 MHz 80286 microprocessor and similar features. Two versions are available. The 8560-041 has a 40 MB hard disk and the 8560-071 has a 70 MB hard disk. The system again consists of three parts: the system unit, keyboard, and monitor.

The system unit is similar in size to an AT, but is floor standing, with two swivel feet that support it in the upright position. It has a 1.44 MB, 3.5-inch diskette drive and a hard disk as standard. There is room for a second 3.5-inch diskette drive below the first and room inside the machine to add an additional hard disk. A second 40 MB disk can be added to the Model 8560-041; a 70, 115, or 314 MB disk can be added to the Model 8560-071. As an alternative option, an internal optical disk drive can be installed in the second disk drive position.

The system board is fully equipped. It has the VGA as its display subsystem, a battery-backed clock, and serial, parallel, and pointing device ports. There are eight 16-bit Micro Channel expansion slots with one occupied by the disk controller. The machine contains 1 MB of RAM and the same two levels of BIOS as on the Model 50.

The Model 60 is the floor-standing replacement for the AT. It provides additional performance, as its microprocessor runs at 10 MHz, and is suitable for all applications. Its large storage capacity enables it to be a network server as well as an excellent workstation. However, due to its cost ($5,295 for the 8560-041) the additional expansion features may not justify it over the Model 50; the Model 80 may be more suitable.

Model 80

There are two versions of the Model 80, each with different media options. One has a 16 MHz 80386 and the other a 20 MHz 80386. The two models have different memory architectures but are basically similar in other ways. The 16 MHz version has two options: the 8580-041 has a 40 MB hard disk and the 8580-071 has a 70 MB hard disk. The 20 MHz version has a 115 MB (8580-111) or 314 MB (8580-311) hard disk. The machine is a full 32-bit machine with lots of potential. Current operating systems and applications are not designed to make full use of this machine's potential, but as new operating systems and applications evolve, this machine will be more than adequate to handle them.

Again, the system board for these machines has a serial, parallel, and pointing device port. The video subsystem is the VGA and the keyboard is the Enhanced version. A battery-backed real-time clock is also standard on the machine. The 128 KB of ROM BIOS includes both CBIOS and ABIOS. The memory on the 16 MHz machine is located in two small boards that plug into the main board. Each board contains 1 MB of 80-nanosecond RAM. On the 8580-041 model 1 MB is standard, and on the 8580-071 2 MB is standard. On the 20 MHz machine the memory is located on two small boards that plug into the main board, but in this case each board contains 2 MB of RAM. One board is supplied as standard.

The Micro Channel expansion slots on the Model 80 are 16- and 32-bit slots. There are three 32-bit slots and five 16-bit slots with one 16-bit slot used by the disk controller. The 32-bit slots can be used to house 16-bit boards, if desired, in the same way that the AT slots can be used to accommodate 8-bit PC expansion boards.

A second 1.44 MB, 3.5-inch diskette drive can be located beneath the first on the front of the machine. A second hard disk can be installed in the machine, as on the Model 60, or an internal 5.25-inch optical disk drive can be installed.

The power supply on the Model 80 (225 watts) is larger than on the other models and has a self-start feature that enables the machine to reboot after a power failure. This is useful for unattended operation such as for a network server or remote data acquisition applications.

Graphics System Overview

The VGA standard on the Personal System/2 machines provides compatibility with the past and moves the display system slightly into the future. The MCGA that is standard on the Model 30 is a subset of the VGA and offers BIOS level compatibility with the CGA, in addition to some new modes. The VGA offers BIOS level compatibility with the Monochrome Display and Printer Adapter (MDA), the CGA, and the EGA.

The MCGA has four additional operating modes in addition to the CGA compatibility. The first is the 320 by 200 pixel graphics mode with 256 colors from a palette of 256K colors. This is double scanned on the monitor to produce an image that is 320 dots wide by 400 high, with each of the 200 vertical dots appearing twice—one beneath the other. The second is the 640 by 480 pixel, 2-color graphics mode. Both the first and second graphics modes have a unity aspect ratio, so if a circle is drawn, the image that is drawn appears to be circular and not an ellipse. This is not the case with graphic images drawn on a CGA. There are two text modes on the MCGA: 40 columns by 25 rows in 16 colors or 80 columns by 25 rows in 16 colors. The character size for these two modes is 8 pixels horizontally by 16 vertically.

The VGA supports all of the modes that are available on the MCGA and also offers compatibility with the CGA, EGA, and MDA. However, the character size for the text modes is 9 pixels horizontally and 16 vertically. One additional graphics mode is available in 640 by 480 pixels with 16 colors. This produces a resolution similar to that of enhanced EGA cards on the market, but is not produced in the same fashion and can have a different palette from the enhanced EGA cards.

The Model 30, or even a PC or AT, can be given VGA performance by using the Personal System/2 Display Adapter. This is a display adapter board that plugs into a PC expansion slot and allows all of the features of the VGA to be used.

The 8514/A Display Adapter and its associated 8514 memory expansion kit expand the display capabilities of Models 50, 60, and 80. This Micro Channel board allows all of the modes of the VGA with any of the monitors in the family.

When the 8514 monitor is used with the 8514/A Display Adapter, Professional Graphics Controller (PGC) compatibility is obtained and a 1024 by 768 pixel, 16-color mode is available. The number of colors available in this mode increases to 256 with the memory expansion option installed. Two additional text modes are available with the 8514/A Display Adapter. One displays 85 columns and 38 rows with a character size of 12 pixels horizontally by 20 vertically, and the other displays 146 columns by 51 rows with a character size of 7 pixels horizontally by 15 vertically. Also, if the memory expansion is present, 256 colors are possible at a resolution of 640 by 480 pixels on all monitors.

In addition, the 8514/A provides many features for high quality imaging applications, such as high speed bit-block transfers, scissoring, filled areas, and hardware-assisted line drawing. But the cost of the board is proportional to the features provided. The 8514/A is $1,290, the memory expansion kit is $270, and the 8514 monitor is $1,550.

The VGA graphics standard is not as advanced as the industry might have expected, but it does provide excellent compatibility at a BIOS level with previous display adapters. The advances are ideal for the business environment, as they include a greater range of colors, but performance has not significantly increased.

Monitors

The video subsystem on the Personal System/2 machines (the MCGA and the VGA) is a new standard. The output of the video port is analog and not digital like the CGA or EGA. Likewise, the monitors that are supported by the machines are not the same as the old monitors, including the analog monitor that is required by the PGC. There are four monitors in the range: 8503, 8512, 8513, and 8514.

The 8503 is a 12-inch diagonal monochrome monitor that displays white on black or black on white. It can also be used to produce 64 shades of gray for graphics applications. It has a horizontal dot addressability of 320, 640, and 720, and a vertical dot addressability of 350, 400, and 480, which is the limit for the VGA. The 200-line modes of the VGA are double scanned to give 400 lines on the screen. The monitor has a tilt-and-swivel stand as standard and is the lowest price monitor in the family at $250. This monitor is best suited to text applications in which the graphics features are only used occasionally.

The 8512 is a 14-inch diagonal color monitor with a stripe pitch of 0.41 mm. This produces pixels that are made up of three vertical stripes instead of a triad of three dots, as found on the traditional type of monitor. It produces color images at all of the resolutions that are available on the VGA. The horizontal dot addressability is 320, 640, and 720; the vertical dot addressability is 350, 400, and 480. The 200-line modes of the VGA are double scanned on the monitor, giving 400 lines of dots that make up the image. The optional tilt-and-swivel stand can be snapped onto the base of the monitor. The price of the monitor is $595 and the stand is $35. This monitor is also best suited for applications in which the graphics feature is used only occasionally.

Chapter 1

The 8513 is a 12-inch diagonal color display with a dot pitch of 0.28 mm. This produces a crisper image than that of the 8512, but the dots are smaller. The horizontal and vertical dot addressability is the same as that of the 8512. The tilt-and-swivel stand is standard on the monitor, which has a price of $685. This monitor produces crisp graphics output and is desirable if the graphics feature is significant.

The 8514 is a 16-inch diagonal color display. It works with the VGA in all of its modes and has the additional capability of a maximum addressability of 1024 by 768. This can be obtained by using the 8514/A Micro Channel Display Adapter, which increases the performance of the VGA. This monitor also includes a standard tilt-and-swivel base and is priced at $1,550. It is suitable for high resolution graphics applications such as computer aided design.

Math Coprocessor

All of the new machines have a math coprocessor option that can be installed in a socket on the system board. The Model 30 uses an 8087 that is rated for the 8 MHz clock speed of the main CPU. Models 50 and 60 use a 10 MHz 80287, and the Model 80 uses a 80387 rated at 16 MHz for the 8580-041 and 8580-071 and at 20 MHz for the 8580-111 and the 8580-311. It is important to use parts that are rated for the appropriate speed. Although a part that is rated at a lower speed may appear to work in a fast system, spurious data can be obtained at any time. In addition, the 20 MHz 80387 will not operate in the 16 MHz machine.

The AT's math coprocessor runs at only two-thirds the speed of the CPU. On the Personal System/2 machines the coprocessor operates at the same speed as the CPU. Calculations that use the Model 50's 10 MHz coprocessor perform nearly twice as fast as on the AT's 5.33 MHz coprocessor.

Mouse

A mouse is available for the Personal System/2 family. This is a two-button serial device with a connector that attaches to the pointing device port of the system unit. The cable is 9 feet long, allowing ample length to reach the desktop from the floor-standing machines. Alternative sources for a mouse include Logitech and Microsoft.

Removable Media Device Options

The 3.5-inch diskette is the standard size for all removable media in the Personal System/2 family. On Models 30 and 25 this disk drive is a 720 KB version; on all other models a 1.44 MB drive is supplied. The Model 30 can be purchased with one or two 3.5-inch drives as standard. If two drives are purchased, the machine cannot contain a hard disk. The Model 25 is supplied with one 3.5-inch drive as standard and a second can be purchased as an option. Models 50, 60, and 80 are supplied with one drive and a second can be purchased as an internal option.

All of the Personal System/2 machines can use an external 5.25-inch diskette drive. The external drive has its own power supply and requires an internal adapter card to be installed in the system unit. It is, however, only a 360 KB drive and will not read data from diskettes that have been generated on 1.2 MB diskettes, even if the diskette is formatted at 360 KB. This means that to transfer data between machines using 5.25-inch diskettes, ATs must have a 360 KB diskette drive.

On the Model 30 the necessary adapter card is half length and occupies one expansion slot. On Models 50, 60, and 80 the adapter card is a Micro Channel version and takes one expansion slot. The method of installation is fairly simple but seems slightly perverse. The external drive when configured is drive B. The installation procedure uses the spare internal drive connector and links it to the adapter card. The adapter card has a 37-pin D-type connector (the same as on the PC version of the external disk drive) that is connected to the external drive.

An external optical disk drive can be attached to any of the Personal System/2 machines. Up to 200 MB of data can be stored on a 5.25-inch removable disk. This is write-once-read-many-times data, due to the technology involved. The disk drive requires an adapter installed in an expansion slot of the machine. Two drives can be attached to the adapter card in a machine. If the drive is to be installed in a Model 30, the part number is IBM 3363 A01 for the first drive and adapter card. For the Micro Channel version, the part number is IBM 3363 A11 for the adapter card. The second drive has the same part number—IBM 3363 B01—for any machine.

On Models 60 and 80, an internal optical disk drive (feature number 8700) is available. This fits into the position that would be occupied by the second hard disk, and requires an expansion slot for the adapter card. The 5.25-inch media is then removed through the front panel opening below the 3.5-inch diskette drives.

A Micro Channel version of an adapter card to drive the IBM 6157 Streaming Tape Drive is available as feature number 4160. This board occupies a single slot in the system unit.

Hard Disks

Each of the models in the Personal System/2 family except the Model 25 can be purchased with a hard disk. In the Model 30 the hard disk has a 20 MB capacity and an integral hard disk controller. It takes the place of the second 3.5-inch diskette drive. The Model 50 has a 20 MB hard disk with a separate disk controller that occupies one of the Micro Channel expansion slots. The Model 60 has either a 44 MB or a 70 MB hard disk. The 44 MB disk uses the same hard disk interface as the PC family, but the 70 MB disk is an *enhanced small device interface* (ESDI) device.

In the Personal System/2 family, all hard disks that are 70 MB or larger are ESDI types. This improves the performance of the disk, but affects the upgrade paths available for the Model 60 and Model 80. If the 8560-041 version of the Model 60 or the 8580-041 version of the Model 80 is purchased, a second hard disk must be the 40 MB, feature number 3046. The Model 60 part number 8560-071 and the Model 80 part numbers 8580-071 and 8580-111 can add a second 70 MB, 115 MB, or 314 MB hard disk. All additional hard disks can use the disk controller that is standard in the machines.

Data Migration

With 3.5-inch media standard for the Personal System/2 family and the 5.25-inch disk drive available only as an external option, a logical concern is the transfer of data between existing PCs and ATs and the Personal System/2 machines. Apart from the various, not totally desirable, additional drive options on the machines, there is a Data Migration Facility (part number 5003). This is a one-way system that allows the transfer of data from a PC or AT to the Personal System/2 machines. The Data Migration Facility contains the necessary communication software and a connector adapter that plugs into the Centronics connector of an IBM printer cable (not supplied), to allow the cable to be connected to a parallel port on the second machine.

Add-In Boards for the Model 30

Most expansion boards for the PC work in the Model 30 if they can be physically installed and can operate at the higher CPU speed of 8 MHz instead of the PC's 4.77 MHz. A display adapter card is available that provides all of the features of the VGA on the Model 30 or the PC or AT when a Personal System/2 display is attached. The VGA features are not available on PCs and ATs without this expansion board. The Model 30 only contains a subset of these features (called the MCGA) without this expansion board.

A speech adapter board adds voice capability to the Model 30. Speech can be encoded or decoded via this adapter. Two techniques are used: *Continuously Variable Slope Delta* (CVSD) modulation and *Linear Predictive Coding* (LPC). This board cannot be used in the same machine as the 3278/79 Emulation Adapter.

A PC Music Feature is available to enable a Model 30 or a PC or AT to create stereo FM synthesized sound. The board simulates 336 different voices or instruments and plays up to 8 notes simultaneously. One or two cards can be installed in a machine. Three different outputs are available: headphones, left and right stereo channels, and *Musical Instruments Digital Interface* (MIDI).

For users of the 3270 Workstation Program there is a 2 MB Expanded Memory Adapter that can be used in the PC family and in the Model 30. This adapter provides memory for up to six PC DOS sessions, as well as *Lotus Intel Microsoft* (LIM) specification expanded memory and RAM disk features.

The Token-Ring PC Adapter and the Token-Ring PC Adapter II introduced for the PC family also work on the Model 30. To fill out the product line, additional network boards are also available. The PC Network Adapter II is provided for broadband network use. It should be used instead of the original PC Network Adapter, as it takes advantage of the additional processor speed available on the Model 30. The PC Network Baseband Adapter performs a similar role for baseband network applications.

An adapter for the new Pageprinter printer is available for the Model 30 and the PC and AT. Due to power supply limitations, other expansion boards cannot be installed in the machine while the Pageprinter adapter is installed. The Pageprinter is the first desktop laser printer that IBM has announced for the PC market.

Add-In Boards for the Micro Channel

Because the architecture for the Micro Channel expansion bus is different from the PC bus, new expansion boards are necessary for Models 50, 60, and 80. The typical PC multi-function boards are not necessary due to the high degree of integration on the system board. The expansion slots are necessary, however, for additional memory and communication applications. The line of expansion boards available will continue to grow as third-party vendors bring their offerings to market.

The IBM 80286 memory expansion board has 512 KB of RAM as standard. A total of 2 MB of RAM is possible on the board, providing extended memory for the machine. Up to three 512 KB memory kits can be added to the 512 KB of RAM standard on the memory expansion board. Only one slot in the expansion bus is occupied when the board is fully populated. This board is separate from the memory that is on the system board. The 80386 memory expansion board for the Model 80 comes standard with 2 MB of RAM. Up to two additional 2 MB memory modules can be added, for a total of 6 MB per board.

Various serial data ports can be added to the Micro Channel. The Multi-Protocol Adapter/A provides a serial data channel that can be used for asynchronous, bisynchronous, HDLC, or SDLC full-duplex or half-duplex communications. The Dual Async Adapter/A has two serial ports that can be programmed for communication speeds between 50 and 19,200 bps, across up to 50 feet of cable. Up to seven serial ports can be configured in Models 50, 60, or 80 using this expansion board—one port on the system board and three sets of two on Dual Async boards.

A 300/1200 Internal Modem/A is available. It is Hayes compatible and supports the AT command set. The detachable cable provided is 7 feet long; the board occupies one slot in the system unit.

A Micro Channel version of the Expanded Memory Adapter for the PC is the Expanded Memory Adapter/A. In addition to providing all of the features of the PC Expanded Memory Adapter, this board provides extended memory for use with the OS/2 operating system.

The Token-Ring Network Adapter/A is a 4-megabits-per-second Micro Channel version of the Token-Ring Network Adapter II board for the PC. The broadband network is supported by the PC Network Adapter II/A and the baseband network by the PC Network Baseband Adapter/A.

IBM System 36/38 connectivity is obtained via the System 36/38 Workstation Emulation Adapter/A. 3270 connection is also provided on a Micro Channel expansion board.

Two scanner attachments with a Micro Channel interface are available. The 3117 Adapter/A allows a 3117 scanner to be attached to the Personal System/2 machine. The High Speed Adapter/A allows the connection of a 3118 scanner or a 3117 with an extension unit.

Operating System Software

The latest upgrade to PC DOS is version 3.3. This is the minimum operating system that will run on the Personal System/2 machines. It has several features that were not present in previous versions, in addition to support of the new systems. These features include changes and additions to DOS commands, new DOS interrupts, and changes to the default values in the configuration settings. The upgrade is compatible with existing software and offers desirable enhancements, such as a means of permanently setting the clock without having to find the original setup disk, and improved BACKUP and RESTORE commands.

The new operating system that IBM will be supporting for the future is Operating System/2 (OS/2). This runs only on machines with an 80286 or 80386 microprocessor. DOS has severe limitations as application programs get larger and users want to run multiple applications at one time. The PRINT command in DOS, for example, runs in the background, but it is obvious when the computer is gathering information to print, because all foreground work stops abruptly. The concept of a multi-tasking operating system is that it allows several applications to run at once without affecting each other. In addition, DOS cannot address more than 1 MB of memory directly. As application programs get bigger, the interim solution of expanded memory that maps areas of memory will not be satisfactory. The new operating system extends this 1 MB limit.

There are several versions of OS/2: OS/2 Standard Edition, Versions 1.0 and 1.1, and OS/2 Extended Editions 1.0 and 1.1. The Standard Edition Version 1.0 extends the 1 MB memory limitation to 16 MB, the maximum that the 80286 can address. Several applications can be run at once on the machine and the user can switch between them at will. Most existing DOS applications are able to run under OS/2, although only as they did before—one at a time. Version 1.1 enhances the look of the product by supplying a Presentation Manager and graphics. This is an icon-based interface that allows menu selections to be chosen with the mouse, in a manner similar to Microsoft Windows or the Apple Macintosh.

The extended edition of OS/2 includes a relational database and a communications manager as part of the operating system. The Extended Edition Versions 1.0 and 1.1 are to be available in the second half of 1988.

IBM has also announced that the AIX operating system will be provided for use on the Model 80. This is the UNIX operating system that is on the RT PC. IBM will produce a version of AIX for the Model 80 that will be application source code compatible with the version on the RT.

Printers

A Pageprinter laser printer is available to mate with the Pageprinter adapter for the Model 30. IBM plans to offer a Micro Channel version of this adapter for the Personal System/2 machines. The intelligence for the printing process is on the adapter, and the printer is basically an engine.

The Proprinter II Model 002 dot matrix printer offers a superset of the features provided by the Proprinter XL. In addition to all of the Proprinter XL functions, it offers a fast 12-pitch draft font and a different near letter quality font. The Quietwriter III is a letter quality printer that can print all-points-addressable graphics. It is quieter and faster than its Quietwriter Model 2 predecessor and offers additional modes and fonts. The Proprinter X24 is a new dot matrix printer that has a 24-wire head. This prints clearer characters than previous models could; each letter can be printed in a matrix that is 24 dots high instead of 9 dots high (as on the Proprinter II).

Summary

The Personal System/2 family and all of its associated products make up a versatile platform for the present. The technology and architecture allow growth for the future. The implications of the new standard are far reaching. Although the PC and AT market will not disappear nor will DOS-based applications, IBM will no longer be a major player in that aspect of the market. With more add-in boards for the Micro Channel becoming available and the possibility of clones not far behind, the industry will follow the new standard, while continuing (at least for the near term) to use PC class machines.

2

The Model 30

The Model 30 is a strange member of the Personal System/2 family. It does not have the Micro Channel expansion bus and its 3.5-inch diskette drives do not have the same maximum capacity as those of Models 50, 60, and 80. On the other hand, it is not a member of the PC family either. Its styling is similar to that of the Personal System/2 family rather than that of the PC. (See Figure 2-1.) The system board with its integrated serial and parallel ports is not reminiscent of the IBM PC, although many PC clones have supplied such standard features. It does have remnants of its PC heritage with the PC expansion bus, but does not have 5.25-inch diskette drives. The machine could be called the IBM PC/2, because it has many of the features missing from the original PC that the PC clone manufacturers have provided. This designation indicates how the machine appears in relation to the PC family, but does not justify its existence as part of the Personal System/2 family.

However, the machine can also be considered a bridge machine, with one foot in each camp. Its 3.5-inch diskettes can be read directly by the other members of the Personal System/2 family, even though Models 50, 60, and 80 can use higher density diskettes. Most of the expansion boards that can be used in the PC (but not AT) family of machines can run in the Model 30. The limitations are physical size and the faster speed of the processor in the Model 30.

The Model 30 is low in cost compared to the Model 50 and has all of the standard features that would be required for a low-end user. The machine is composed of three parts: the system unit, the keyboard, and a monitor. Any of the monitors available for the Personal System/2 machines can be used. The keyboard is similar in layout to the AT Enhanced Keyboard, and is the same as the keyboards on Models 50, 60, and 80. The system unit itself contains the system board and disk drives.

The system board has an integral video system called the Multi-Color Graphics Array (MCGA), which is a subset of the Video Graphics Array (VGA) on Models 50, 60, and 80. This provides the basic video modes necessary for text

Chapter 2

Figure 2-1 The Model 30. *The 8086-based Model 30 is the successor to the PC XT.* Photo courtesy of International Business Machines Corporation.

processing and low-end graphics. The MCGA is compatible with the Color Graphics Adapter (CGA) but not the Enhanced Graphics Adapter (EGA). The text mode of the MCGA has a character box 8 pixels wide by 16 pixels high, producing clear characters with serifs that are easy to read.

Also on the system board is a serial port, a parallel port, and a pointing device port. This is similar to the other members of the Personal System/2 family. There are only three expansion slots in the Model 30, but the presence of a full 640 KB of RAM on the system board and the ports reduces the need for expansion for many end users. The expansion slots are needed for expansion requirements, such as network adapters or a micro-to-mainframe link, and any expanded memory that may be desirable for a particular application.

The system board has an 8 MHz 8086 microprocessor as its CPU. The 8086 is the same as the 8088 microprocessor that is in a PC except that it has a 16-bit address and data path to the system board memory. The expansion bus is only 8 bits wide, as in the PC and XT. The performance of the machine is approximately twice that of an IBM PC XT. IBM has stated that it will not continue to manufacture the PC family although, as is standard with IBM products, support for existing machines will continue to be available.

The Model 30 has proved to be a popular machine from early sales figures, but this could well be due to the lack of expansion boards for the other members of the Personal System/2 family. The long term forecast for the Model 30 machines really depends on the future of PC operating systems. The machine itself is fully functional, but does not run the OS/2 operating system that IBM is selling as the successor to DOS. OS/2 makes use of the protected mode of the 80286 microprocessor; the 8086 does not have this mode of operation.

If DOS remains the prevalent operating system for personal computers (and its use on several million PCs guarantees its life for a few years), there is still room for the Model 30. However, if the machine is considered part of an environment that will standardize on the Micro Channel and run OS/2 or an equivalent operating system, then the Model 30 is a machine already past its prime.

Functional Description

The Model 30 system has many of the features of the PC and several that should have been on the PC. Figure 2-2 shows the system block diagram. The system board contains 640 KB of RAM, which is the full complement for the DOS operating environment. A video subsystem, a serial port, and a parallel port, as well as the keyboard and pointing device interfaces, are integrated into the system board. There are three XT-style expansion slots and a power supply that provides all of the necessary voltages for the system.

The system can be described in terms of functional blocks. The 8086 operates in parallel with its 8087 math coprocessor and is controlled by a VLSI CPU support gate array. There is 640 KB of local RAM and 64 KB of ROM. This subsystem is attached to the I/O channel and there are six additional integrated I/O functions on the system board. These are the MCGA video subsystem, the 3.5-inch diskette drive interface, the hard disk connector, the serial port, the parallel port, and the real-time clock. The hard disk controller is attached to the hard disk assembly itself and is not on the system board. The three expansion slots, which take standard PC expansion boards, are not located on the system board itself but in an adapter that is installed in the system board.

The 8086 Processor

The 8086 on the system board runs at 8 MHz, with zero wait states required to access system board RAM. This results in a 500 nsec cycle time (four clock

Chapter 2

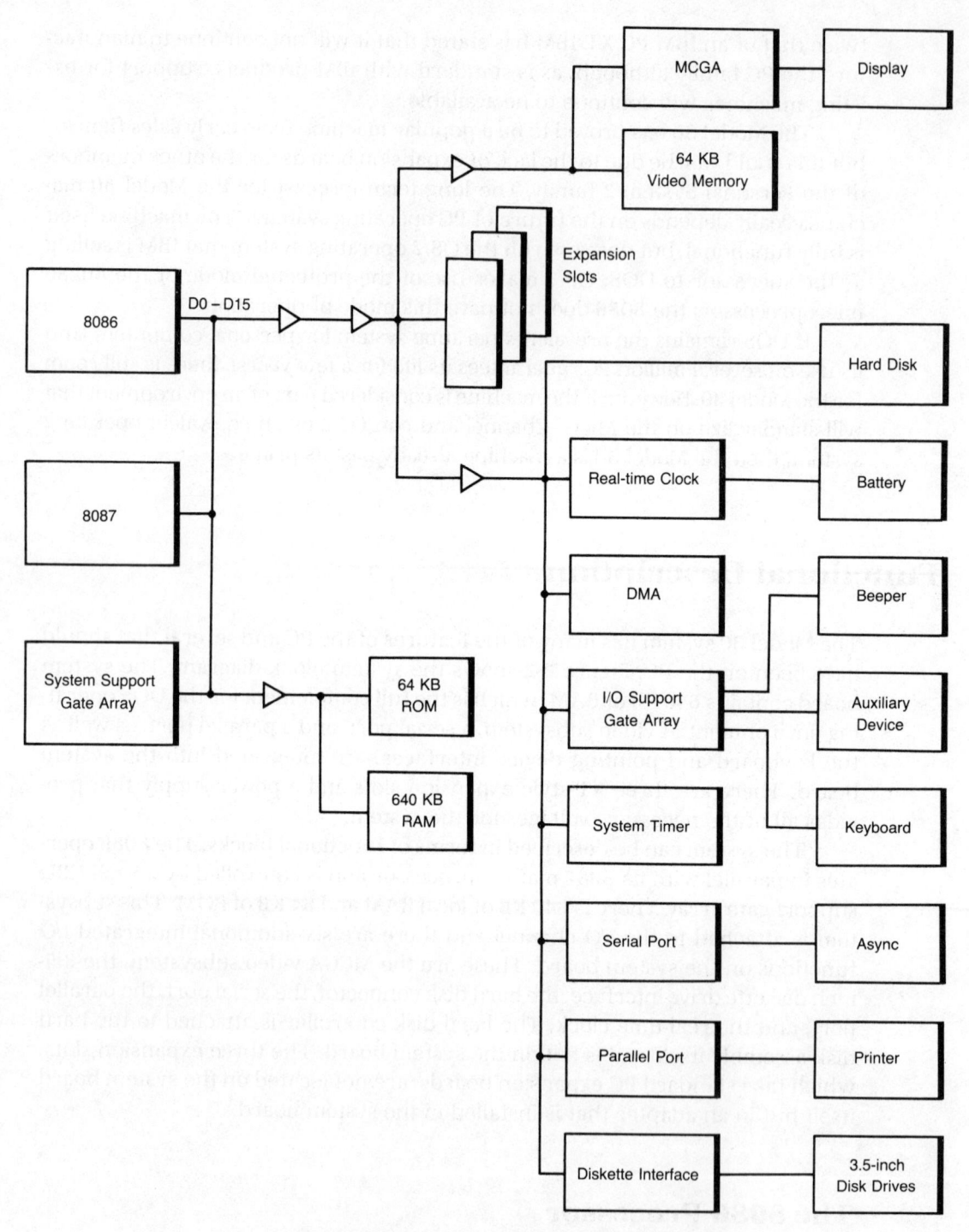

Figure 2-2 Model 30 System Block Diagram. *The architecture of the Model 30 is comparable to that of the PC, even though the physical implementation has been changed.*

cycles) for local memory read or write. Accesses to the bus take longer, however. Four wait states, giving a cycle time of 1 microsecond, are required to perform an I/O transfer or memory access that is 8 bits wide. The bus is only 8 bits wide, as on the XT, so transfers that are programmed as 16 bits wide on the microprocessor and that require devices on the bus to be addressed are translated by the system board logic as two 8-bit transfers. Devices that require longer cycle times can insert extra wait states through the use of the *I/O Channel Ready* signal. The logic on the system board automatically inserts any necessary wait states, to ensure that the address setup times for the system are at least as long as those obtained on an 8088-based system.

The 8087 Math Coprocessor

The system board has a socket for the installation of an 8087 math coprocessor. Applications that take advantage of the coprocessor's capabilities perform arithmetic functions several times faster than when the 8086 itself is used. The math coprocessor runs at 8 MHz (68 percent faster than the 4.77 MHz operation of the XT's 8087).

System Memory

Up to 1 MB of memory can be addressed by the system. The address map for the system consists of the following:

- ▶ 640 KB of RAM from the address 00000H to 9FFFFH
- ▶ 128 KB of video buffer from A0000H to BFFFFH
- ▶ 192 KB reserved for basic input/output system (BIOS) on the I/O channel from C0000H to EFFFFH
- ▶ 64 KB for system ROM from F0000H to FFFFFH

In the Model 30, 64 KB of ROM contains the system POST (power on self test), BIOS and a BASIC interpreter, and 128 characters of video dot patterns.

Gate Arrays

There are five gate arrays on the system board: two for the video subsystem, two for the CPU support, and one for the 3.5-inch diskette controller. The CPU support gate arrays are the system support gate array and the I/O support gate array. The video subsystem has a video memory controller gate array and a video formatter gate array.

System Support Gate Array

The bus controller, memory controller and parity checker, bus conversion logic and wait state generator, and the system clock generator are in the system support gate array. This integrated circuit (IC) is one of the three bus masters contained in the Model 30, the other two being the 8086 and the 8087. The 8086 requests and grants the bus via the CPU RQ/GT line, and the 8087 requests and grants the bus via the NPU RQ/GT line. The system support gate array generates the memory refresh bus cycles and the DMA bus cycles and so has the highest priority on the bus. The method used by the IC to take control of the bus varies, depending on whether an 8087 is present in the system.

If a math coprocessor is not present, the system support gate array requests the bus from the microprocessor via the CPU RQ/GT line. The microprocessor grants the bus by pulsing the same line. With a math coprocessor present in the system, the 8087 is first asked to give up the bus. If it is in control of the bus at that time, the system gate array is granted control via the NPU RQ/GT line. If the 8086 is in control, the 8087 requests the bus from the 8086, obtains it, and then gives it up to the system support gate array.

The Model 30 Technical Reference manual warns that if an in-circuit emulator is being used instead of the 8086, it is possible to damage the system support gate array if the request/grant lines are out of synchronization.

The memory on the system board is refreshed once every 4 milliseconds (ms). This takes nine clock cycles, each 125 nanoseconds in length, and is performed through a dedicated refresh channel in the system support gate array. The parity checker in this array checks all of the 640 KB of RAM and signals a parity error, when found, via the *– parity check* signal.

The system RAM and ROM on the system board itself can support 16-bit wide transfers. However, the I/O devices and expansion connectors are only 8 bits wide to maintain compatibility with the PC family of machines. The system support gate array contains the bus conversion logic and wait state generator that enables the conversion of any transfers to I/O devices or the expansion bus to 8 bits, inserting the required wait states at the same time. The IO CH RDY line is monitored by the wait state generator to determine the number of wait states needed.

The other section of the system support gate array is the system clock generator. This takes an input from a 48 MHz crystal and divides it down to give an 8 MHz clock cycle with a 33 percent duty cycle for the system. The 1.84 MHz clock signal for the serial port is also derived in this gate array.

I/O Support Gate Array

The I/O support gate array is the second VLSI component that is part of the microprocessor control circuitry. It contains chip select logic, the keyboard and pointing device controller, and the I/O ports. The chip select logic can control six

The Model 30

different areas of the system board: the serial port, the parallel port, the video controller, the 3.5-inch diskette controller, the hard disk controller, and the real-time clock.

The chip select logic works in combination with the read/write planar control register at address 65H. With no bit set, the read and write operations are made to the I/O channel; however, if a bit is set in the register, the relevant section is selected. Table 2-1 shows the bit assignments for this register. The real-time clock cannot be disabled even though it is controlled by the chip select logic.

Table 2-1 Planar Control Register Bit Assignments. *This register is used in association with the chip select logic to access a particular I/O device.*

Bit	Function
0	Hard disk chip select
1	Parallel port chip select
2	Video chip select
3	3.5-inch disk controller chip select
4	Serial port chip select
5	Reserved—set to 0
6	Reserved—set to 0
7	Parallel port output enable

The keyboard and pointing device ports are interchangeable on the Model 30. The controller for these ports is located in the I/O support gate array. This logic takes the serial data, checks its parity, and then puts the data in the I/O port buffer address 60H for the system to use.

There are three system timers on the Model 30—channels 0 through 2. Channels 0 and 2 have the same function as the timers on the PC: channel 0 is used for the time of day function and channel 2 can be programmed to sound the speaker through bit 0 of I/O port 61H. The additional channel is reserved for use with diagnostics.

The I/O support gate array contains circuitry that emulates an 8237 DMA controller and its support logic. The DMA controller can support four 20-bit channels of DMA, three of which are available on the I/O bus. The DMA channel assignments are

DRQ0 not available

DRQ1 not used

DRQ2 diskette

DRQ3 hard disk

Chapter 2

The DMA controller runs at 4 MHz and performs an 8-bit transfer in six clock cycles (1.5 microseconds).

The interrupt controller performs in the same way as on the PC, despite its different packaging. There are eight levels of edge-triggered interrupts, with interrupt level 0 having the highest priority and level 7 the lowest. Level 0 is used by the timer for the "system clock tick." The keyboard, pointing device, and real-time clock are on interrupt level 1 and have a BIOS interrupt handler at interrupt 71H. The nonmaskable interrupt (NMI) is generated from three signals: the 8087 – *interrupt*, the memory controller's – *parity*, and the – *I/O channel check*. The video interrupt level is IRQ2, even though the video BIOS routines do not use this interrupt. Interrupt level 3 is not used by the system, IRQ4 is used for the serial port, IRQ5 for the hard disk, and IRQ6 for the 3.5-inch diskette drive. The parallel port interrupt level is IRQ7, even though the parallel port BIOS routines do not use this interrupt.

Although there are many advantages to the level-sensitive interrupts that are part of the Micro Channel architecture, the PC bus architecture with its edge-triggered interrupts is a usable system. It is possible to share interrupts with other devices and IBM describes a standard method that has been available for several years. Many third-party add-on manufacturers are not using a system that allows interrupts to be shared, so consequently this practice is not prevalent in the PC industry, even though the concept is sound.

Diskette Gate Array

Part of the control circuitry for the 3.5-inch diskette drive is contained in the diskette gate array on the system board itself. This controller consists of the logic necessary for reading data on the diskette, writing data to the diskette, and decoding for internal registers. The rest of the control circuitry consists of a phase detector and amplifier along with a voltage controlled oscillator, making up a phase-locked loop. The drive itself is connected via a ribbon cable that is 40 wires wide. Up to two drives can be driven from the daisy-chained cable; all power and logic signals are included in the ribbon.

Hard Disk Controller

The controller for the hard disk is located on the hard disk; the system board only contains the interface circuitry. A 44-pin connector provides the power connections and the I/O channel signals for the hard disk. The I/O read and write signals, reset, data signals, IO CH RDY, IRQ5, and DMA request and acknowledge are the same as their corresponding signals on the I/O channel. There are five additional signals that are not on the regular I/O channel. The planar control register is used to activate the disk card select (– *disk cs*) signal. The – *disk installed* signal is active when a hard disk is present in the system. The three registers that are on the hard disk controller are accessed via address lines A0 through A2.

Serial Port

The asynchronous serial port and its controller are an integral part of the system board with a standard RS232C interface. The baud rate can be programmed to any value between 50 bits per second (bps) and 9600 bps via the programmable baud rate generator. The characters that are sent to the port can be 5, 6, 7, or 8 bits in length. The number of stop bits can be programmed to be 1, 1.5, or 2. The start, stop, and parity bits are automatically added to data that is sent from the port and removed from data that is sent to the port. The most obvious difficulty with an asynchronous port attached to a synchronous computer is in ensuring that the data is collected at the port without any loss of characters. The serial port controller is double buffered, with the system being able to access one character of information while the next character is being collected by the port. The controller supplies the necessary modem control signals: clear to send, request to send, data set ready, data terminal ready, ring indicate, and carrier detect.

The serial port is configured as COM1, I/O port addresses 3F8H through 3FFH, and uses IRQ4. The data string that is sent consists of the start bit, the least significant bit of the data through to the most significant bit of the data, the parity bit, and then the stop bit.

The serial port is compatible with the serial port on the PC. The eight registers in the serial port controller transmit and receive the data at the port. The most significant bit of the line control register (the divisor latch access bit) is used to select the divisor latches in the programmable baud rate generator. If the divisor latch is not selected, address 3F8H is the transmitter holding register when it is *written* to, and is used to contain the character that is to be sent from the port. If the divisor latch is not selected and address 3F8H is *read* from, address 3F8H is the receiver buffer register instead of the transmitter holding register. The receiver buffer register is used to contain the character that has been received at the port. If the divisor latch is selected, address 3F8H is used to load the required value in the low byte of the baud rate generator and address 3F9H is used to load the required value in the high byte of the baud rate generator. If the divisor latch is not selected, address 3F9H is used as the interrupt enable register.

There are four types of controller interrupts: modem status, receiver line status, transmitter holding register empty, and the received data available status. Each of these interrupts, if enabled in this register, is capable of setting the chip interrupt output signal. If an interrupt is disabled in this register, the values in the modem status register (address 3FEH) and the line status register (address 3FDH) are still set and the other functions are performed; only the chip interrupt output signal is disabled.

Chapter 2

Address 3FAH is used as the interrupt identification register. This register prioritizes the four types of interrupts so that, when the serial port is accessed by the CPU, the interrupt with the highest priority is serviced. The receiver line status has the highest priority, then the received data ready, the transmitter holding register empty, and finally the modem status.

Address 3FBH is the line control register and contains the format of the asynchronous communication channel. The word length, parity, and stop bits are defined in this register. It is always available for the main system to read, to determine the current settings for the port. Address 3FCH is the modem control register and address 3FFH is a scratch register.

Parallel Port

The parallel port is also an integral part of the system board. It is addressed by the system as LPT1 and can accept data bytes that are 8 bits wide. There are three registers for this port:

- ▶ The data latch at address 378H
- ▶ The printer control register at address 37AH
- ▶ The printer status register at address 379H

The data latch register performs different functions depending on whether it is written to or read. When written to, it contains the data that is to be sent from the port; when read, it contains the data that is received by the port.

The printer control register has five control signals: the interrupt request enable, device select signal, the initialization of the device signal, the automatic line feed signal, and the strobe signal for clocking data to the device. The printer status register contains five status signals: device busy, the acknowledge signal, paper out signal, device selected, and error signals.

Video System

The video system for the unit is a new standard, unlike any provided for the PC family. The MCGA is located on the system board itself and the monitor is plugged into a connector on the rear panel of the machine. The MCGA provides software compatibility with the CGA, so any applications that are written for the CGA work on the Model 30. None of the monitors that operate on the PC can be used though. The MCGA output is analog and so an analog monitor is necessary. (See Chapter 9 for additional information.) The PC monitors—such as the Monochrome Display, Color Display, and Enhanced Color Display—will not work, nor will the analog display for the Professional Graphics Adapter.

The MCGA

The MCGA consists of several parts. A memory controller gate array and a video formatter gate array control the video circuitry. The video data is stored in 64 KB of multiport dynamic memory. On the CGA, the character information is stored in ROM that is a part of the character generator. On the Model 30, there are ROM-resident fonts in the BIOS for compatibility reasons, but the MCGA has a loadable character generator that consists of 8 KB of static RAM. This is used as an intermediate location for character fonts that are to be displayed. The color palette is 256 by 18 bits and there are three 6-bit, digital-to-analog converters to translate the colors for the display.

Video Circuitry

The video memory controller uses its index register and twenty-two data registers to control memory. An index value is loaded in the memory controller index register at address 3D4H and points to a particular data register. The data that is to be used by this register is loaded into address 3D5H. The full list of registers is in Table 2-2; their functions include cursor start position, sync pulse width, and number of characters to load.

The video formatter register duplicates and adds to the functions found in the 6845 on the CGA. Addresses 3DDH, 3DEH, and 3DFH contain three new registers: two are reserved and one (3DDH) is for extended mode control. The color palette is controlled via the four video formatter registers at addresses 3C6H to 3C9H.

The character generator can be loaded from RAM. The desired character fonts are loaded starting at address A0000H. Up to four 8-pixel by 16-pixel, 256-character sets can be loaded into RAM, and two of these can be loaded into the character generator at a time.

The color palette has an 8-bit read address register, an 8-bit write register, and 256 18-bit read/write data registers. The data registers consist of three 6-bit sections. Each color element—red, green, and blue—is loaded in turn into the data register. This can be performed one register at a time or a block at a time. Although it is possible to access the MCGA registers directly, the palette should be changed through the BIOS interface for compatibility with other application programs. Four BIOS calls through INT 10H are provided for this purpose:

- ▶ **Set a single color register (AL=10H).** BX contains register to set, DH contains red value, CH contains green value, and CL contains blue value.
- ▶ **Set a block of color registers (AL=12H).** ES:DX is the pointer to table of color values, with table format of red, green, blue, red, green, blue.

Chapter 2

BX contains first color register to set and CX contains number of color registers to set.

▶ **Read a single color register (AL=15H).** BX contains color register to read. On return, DH contains red value, CH contains green value, and CL contains blue value.

▶ **Read a block of color registers (AL=17H).** ES:DX is the pointer to table for color values. BX contains first color register to read and CX contains number of color registers to read.

Table 2-2 Data Registers for Video Memory Controller. *The memory controller uses an index register to point to one of the data registers.*

Index Number	Description
00	Total no. of characters in horizontal scan line
01	No. of horizontal characters displayed
02	Character position for horizontal sync to become active
03	Sync pulse width, both horizontal and vertical
04	Total no. of vertical scan lines (8 lsb)
05	No. of horizontal scan lines in the vertical scanning interval
06	No. of vertical characters displayed
07	Line count for vertical sync to become active (8 lsb)
08	Reserved
09	No. of scan lines per character
0A	Horizontal character count before cursor displayed
0B	Horizontal character count before cursor becomes inactive
0C	Starting address of video display memory (8 msb)
0D	Starting address of video display memory (8 lsb)
0E	Cursor position address (8 msb)
0F	Cursor position address (8 lsb)
10	Mode control register: displays type, some graphics mode selection
11	Interrupt control register: controls IRQ2 and shows interrupt status
12	Character generator interface and sync polarity, or display sense
13	Character font pointer
14	No. of characters to load into character generator in one vertical retrace
20	Reserved

The MCGA on the Model 30 provides compatibility with the IBM Color Graphics Adapter (CGA) and two new video graphics modes: a 320-pixel by 200-pixel mode with 256 colors and a 640-pixel by 480-pixel mode with 2 colors, both from a palette of 262,114 (256K). The 320 by 200 mode is double scanned on the monitor to produce 400 lines displayed with 200 lines of information. Each line is painted twice on the screen to produce a dramatic color screen effect. This double scanning is done for all CGA graphics modes as well as the two new modes.

Video Modes

There are nine video modes available on the MCGA. The first seven are similar to those available on the CGA. Table 2-3 shows a list of the modes and their types. Note that mode pairs 0 and 1, 2 and 3, and 4 and 5 perform similar functions. For example mode 0 is the black and white version of mode 1. On the CGA, the color burst signal is turned off.

Table 2-3   MCGA Video Modes. *The MCGA is a subset of the functions that are available on the VGA. It can emulate the CGA but not the MDA or the EGA, as the VGA can.*

Mode No. (Hex)	Display Size	Character Size	Maximum Pages	Description
0, 1	320 × 400	8 × 16	8	Text
2, 3	640 × 400	8 × 16	8	Text
4, 5	320 × 200	8 × 8	1	APA graphics
6	640 × 200	8 × 8	1	APA graphics
11	640 × 480	8 × 16	1	APA graphics
13	320 × 200	8 × 8	1	APA graphics

Modes 11 and 13 are new for the Model 30; the others are all CGA compatible.

In the text modes, BIOS modes 0 and 1, the character box is 8 pixels wide by 16 high. The fonts that are displayed are taken from the character generator, which can be loaded from RAM. Up to four sets of 8 by 16 character boxes, each with 256 characters, can be stored in the video buffer starting at address A0000H. The character generator is loaded using BIOS call AH = 11H. The character generator can be loaded with up to two of the fonts that are in the video buffer and programmed to display either or both fonts.

For graphics mode 11H, the character box is also 8 pixels by 16 pixels and the image loaded as a character can have all 16 lines set. This mode is not double scanned and produces a unity aspect ratio for pixels. This means that if a square is drawn on the screen by a program selecting a certain number of pixels high by the same number of pixels wide, the image on the display is also square.

For all of the other graphics modes (4, 5, 6, and 13H), the image is double scanned and the character box is only 8 by 8.

The video buffer organization varies from mode to mode. For the CGA text modes (0, 1, 2, and 3), the arrangement is the same as for the CGA. The character code and attribute byte form a 2-byte character definition. The character code is in the even byte and the attribute for that character is stored in the following odd byte. The character code references one of the 256 characters in the character generator. The attribute byte is split into two 4-bit addresses: one for the foreground color and one for the background color. This means that 16 colors are usually available for the foreground color and 16 for the background color. However, there are two exceptions. If character blinking is enabled, only 8 colors are available for the background color, because the most significant bit of the attribute byte is used for the blink setting. If 512 characters are required from the character generator instead of 256, the most significant bit of the low-order 4 bits is used to select which of the two character sets is to be used for that character, leaving only 8 colors for the foreground.

Bit 5 of the CGA mode control register sets the blink mode. To gain access to 512 characters in the character generator, index 12H of the sync polarity register and bit 4 of the character generator interface should be set.

Graphics modes 4, 5, and 6 use the same memory mapping convention as the CGA. The odd scan lines are stored in successive bytes starting at address B8000H, and the even scan lines are in successive bytes starting at address BA000H. The image is presented on the screen as a double scanned image by displaying the first scan line information twice on rows 1 and 2, then the second scan line information twice on rows 3 and 4, and so forth.

Modes 4 and 5 are 4-color modes that use pairs of bits to select color. Bits 4 and 5 of the CGA border control register are used to look up the selected colors. Bit 5 selects the color palette set (0 or 1) and bit 4 the intensity of the set chosen (1 for intensified and 0 for normal). If the pair of bits for the pixel is set to 00, the background color is displayed for this pixel. The background color is the value of bits 3 through 0 of the border control register. For any other value in the pixel bit pair, the palette address generated is composed of bit 4 of the border control register, the pixel bit pair, and bit 5 of the border control register. The default value for bits 4 and 5 of the border control register is 1. In this situation the palette selected is the CGA-compatible palette.

Graphics mode 6 is 640 pixels by 200 pixels with 2 colors from a palette of 256K. The address of the first scan line is B8000H and the second is BA000H. Each bit represents one pixel on the scan line. The default color selection is the same as on the CGA. The background color is stored in palette address 00H and is black in its default state. The foreground color can be obtained in one of two ways. If bit 2 of the CGA model control register is set to a 1, then the foreground color is obtained from palette address 7H, which in its default state is white. If this bit 2 is set to a 0, then the foreground color is taken from the address that is pointed to by the CGA border control register's least significant 3 bits.

The two new graphics modes of the MCGA use linear address mapping. The video buffer starts at address A0000H and continues through AFFFFH. For mode 11H (640 by 480 pixels with 2 colors from a palette of 256K), each bit defines a pixel on the screen. The color selection is also the same as mode 6. In effect, mode 11H is the same as mode 6 but the character cell is 8 by 16 and the scan lines are addressed linearly in memory. Mode 13H (320 by 200 pixels with 256 colors from a palette of 256K) uses the same memory address mapping as mode 11H. One byte of data represents one pixel on the screen.

The digital-to-analog converter takes the value in the digital palette and converts it to the analog component of the color displayed. If the attached display is a monochrome monitor, BIOS automatically converts the values set in the digital palette to a red, green, and blue value that provides 1 out of a choice of 64 gray shades. The weighting used is 30 percent red, 59 percent green, and 11 percent blue. The new values are loaded into the color register instead of the original data and the original data is not preserved.

Physical Description

The Model 30 consists of three main parts and four cables. The system unit and the monitor both have power cables. The monitor has a signal cable that plugs into the system unit and the keyboard has a detachable cable that also plugs into the system unit.

The Model 30 uses any of the Personal System/2 monitors; monitors are described in detail in Chapter 9, because they are not a standard part of any particular machine. The Model 30 keyboard has the same layout as the 101-key Enhanced Keyboard available for late model XTs and ATs. The keyboard cable has the same connector at the keyboard end, but has a 6-pin miniature DIN connector on the system unit end that is different from that used on the AT. The keyboard itself is interchangeable, enabling an AT-compatible Enhanced Keyboard to replace the IBM version. This may be desirable for some users since the IBM keyboards have a positive tactile feel, whereas the Compaq keyboards and many of the Keytronics keyboards have a far softer touch.

The system unit weighs only 15.7 pounds as opposed to the 32 pounds of the XT. This weight reduction is due largely to its plastic case (the XT has a nickel-plated copper case). This apparently simple change in construction to rigid plastic illustrates the quality design of these new units. At first glance the change is elementary, providing a cost reduction in manufacturing and a desirable weight reduction. However, FCC electromagnetic interference (EMI) regulations have changed since the design of the XT. The allowable amount of interference that can emanate from any electrical unit is now far less than before. This type of interference is typically seen as snow on a television screen or heard as a buzzing noise on a radio. Metal cases are the best shield for this type of interference, so the change to plastic would not appear to be desirable.

This is not to say that the new machines do not meet the more rigid specifications. The Model 30 is not simply a PC motherboard in a new case. Many design considerations have reduced the EMI and enabled the Model 30 to fully meet FCC regulations. The Model 30 system board is designed to generate far less EMI than the PC in the first place, and the plastic case has a thin metallic coating sprayed onto the inside to provide shielding.

Two versions of the Model 30 are available: the 8530-002 with two 3.5-inch,

Chapter 2

720 KB disk drives and the 8530-021 with one 3.5-inch, 720 KB disk drive and a 20 MB hard disk. The 8530-002 costs $1,695 and the 8530-021 costs $2,295. The disk drives are located at the front of the system unit. The second 3.5-inch drive position in the lower priced version is occupied by the hard disk in the alternative configuration.

System Unit Exterior

The Model 30 system unit is a sleekly styled box that is 16 inches long, 15.6 inches deep, and 4 inches high. It is smaller than the XT's system unit, which is 19.6 inches long, 16.1 inches deep, and 5.5 inches high. This footprint gives the unit a small dimension across the front but the depth is still substantial. If the machine had been less deep and perhaps a little higher it would fit into modular furniture a little more comfortably, but would not be as cleanly laid out inside.

The rigid plastic case consists of two parts. The top part composes the top surface and the sides of the unit. It is cream colored with a gray strip at the base of each side where the screws that hold the case together are located. The base of the unit is a metal chassis with a gray plastic rear panel and a cream colored plastic front panel attached. The unit is held together with four straight slotted screws, two on either side of the unit.

The power on and off switch is located on the front of the unit at the right-hand side. It is a red toggle switch that is recessed into the front molding to prevent switching the machine on or off accidentally. The off position for the switch is indicated by an open circle molded into the case. The on position is indicated by a straight line logo that is also molded into the case. To the left of the on/off switch is the disk drive position for either the second 3.5-inch diskette drive or the hard disk. Alongside the second disk drive position is the standard 3.5-inch drive. The IBM Model 30 logo is located near the top left corner of the front panel. The part number and serial number are conveniently located on the front of the machine beneath the on/off switch.

The keylock switch is located on the side of the system unit near the power on/off switch, rather than on the front as it is on the AT. The switch is used to lock the cover onto the unit and to disable the keyboard, in the same fashion as the keylock on the AT.

The left side of the unit contains a line of ventilation holes. This is the side of the unit that holds the expansion boards; to ensure good air flow and adequate cooling of the unit, it should not be placed flush against a wall.

Rear Panel Connectors

Several recessed connectors, shown in Figure 2-3, are mounted in the rear panel of the Model 30. The standard 3-pin power connector for the power cable is located at the left end of the rear panel, as you look at the rear of the unit. The cooling fan is part of the power supply and is located beside the power connector. The system board connectors occupy the bottom right two-thirds of the rear panel. These connectors are labelled from left to right with the numbers 1 through 5 molded into the rear panel.

Figure 2-3   Rear Panel of the Model 30. *The keyboard and auxiliary device ports are interchangeable on the Model 30.* Photo by Bill Schilling.

The first two connectors are 6-pin miniature DIN connectors for the keyboard and pointing device. The documentation states that the first connector is for the keyboard, but in fact either connector can be used for either device and the machine automatically compensates. These two connectors are the only two on the rear panel that can be confused with each other, so the interchangeability of the connectors is a nice design feature.

The third connector is a female 25-pin D-type connector for the parallel port. The other 25-pin D-type connector is male and is for the serial port. The choice of a 25-pin connector for the serial port may seem strange, since the AT uses a 9-pin connector. According to IBM, its policy is to use a 25-pin connector if there is room; otherwise, the 9-pin version is used.

The final connector at the right end of the rear panel is a female 15-pin subminiature D-type video display connector. The pins in this connector are arranged in three rows of five, with one of the pins on the center row blanked out.

All of the D-type connectors have the standard hexagonal threaded standoffs on each end, to allow the connectors on the external cables for the unit to be screwed to the system unit.

The rear connectors for the expansion slots are oriented in a horizontal direction above the monitor and serial port connectors. As the unit is shipped, there are three plastic blanks in the expansion board positions. If the end of the expansion board has a connector that needs to protrude through the rear panel, the blank is removed when the board is installed.

Inside the System Unit

The cover of the system unit is removed by unscrewing four captive screws, two at the base of each side of the unit. These mounting screws are held in position with a spring. As the screw is undone it pops out slightly and is held in the cover. Once the screws are loosened, the cover can be slid slightly towards the rear of the unit and lifted straight up. The cover has a single sheet of painted aluminum across the top as a shield against EMI.

Installation of the top cover is the reverse of the removal process. The cover is set on top of the unit with the bevelled lip towards the front and the ventilation grill on the left-hand side. The chassis has a gap behind a metal tab for each mounting screw to pass through. The cover is slid forward about a half-inch, until the beveled lip fits inside the front panel and the mounting screws go under the metal tabs. The four screws are then tightened to complete assembly.

When the cover is removed, all of the elements that compose the system unit can be seen. (See Figure 2-4.) The keylock switch is located at the far right side of the unit at the front, and has a twisted-pair cable that snakes through a stress relief bracket and onto the system board. The power supply is on the right side of the unit at the rear. It is linked to the on/off switch on the front panel by a lever mechanism. There is a red toggle switch on top of the power supply in the front right-hand corner. This switch has a metal rod attached that runs from the power supply to the front panel. When the on/off switch is operated from the front panel, both this rod and the switch on the power supply itself are moved.

The front half of the unit has four major sections. At the far right is the on/off switch and its lever mechanism. Next is the bay that holds either the hard disk or the second 3.5-inch diskette drive. Beside the hard disk drive bay is the mounting bay for the first 3.5-inch diskette. At the left side of the unit is the section that holds the expansion cards. There is a metal bar that runs from the front of the unit to the back at the top of this section. This bar provides additional rigidity and structural support for the chassis.

Diskette Drive Removal

The system board is located at the bottom of the chassis and extends across the left two-thirds of the case. This position means that the first 3.5-inch diskette drive is mounted over the system board and the hard disk mounting bay is to the right of the system board. The 3.5-inch diskette drive can be taken out of the unit by removing the ribbon cable connector on the rear of the drive and the nylon stud on the side of the drive. The front bezel for the drive is levered off from the front of the unit. This is done by placing a screwdriver between the top

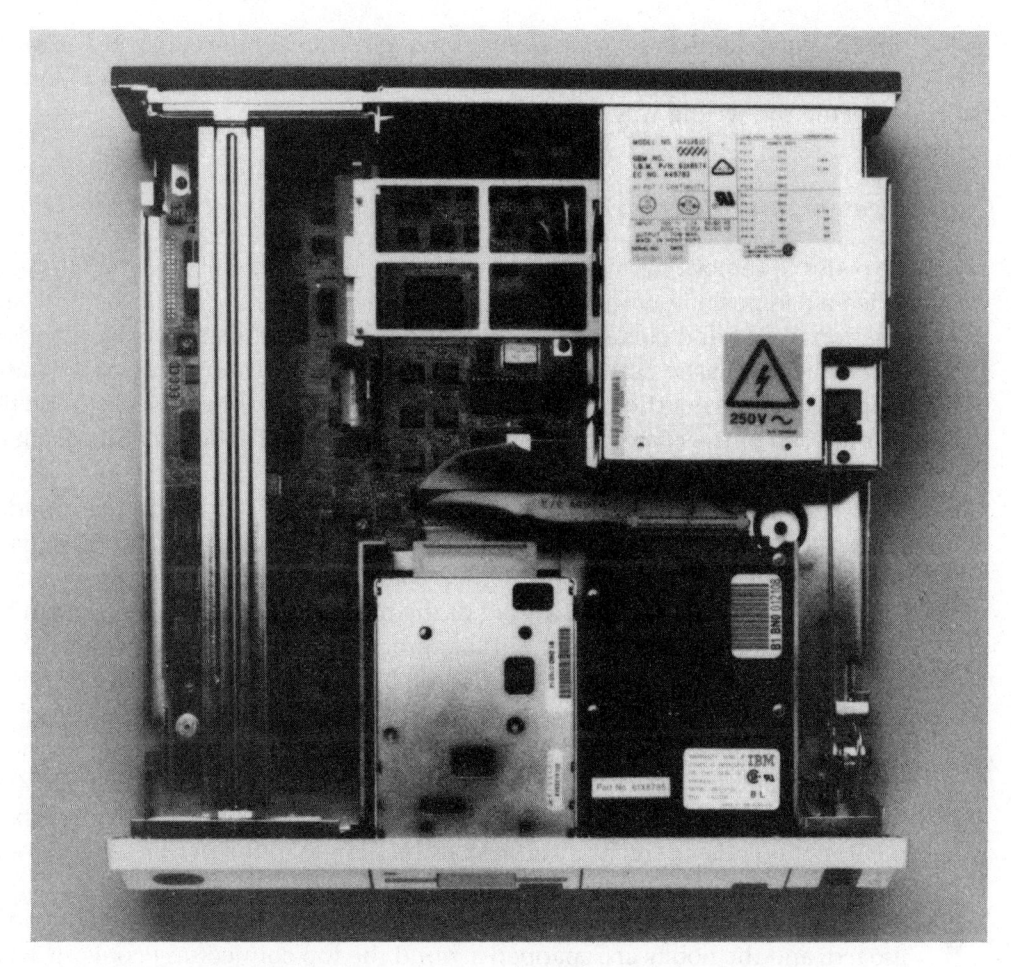

Figure 2-4 Inside the System Unit. *The PC-style expansion boards are mounted horizontally into the chassis. The adapter board to connect them to the system board is located centrally towards the rear of the unit.* Photo by Bill Schilling.

of the bezel and the front panel and twisting slightly. The bezel is held by two tabs on the top edge of the bezel. When these are released, the bezel tips forward and can be lifted out from the unit. The drive itself is removed through the front panel by lifting a tab that is located at the front of the drive, at its base, and sliding the drive forward.

The second 3.5-inch diskette drive is removed and installed in a similar fashion, except that the nylon retainer is on the right side of the drive instead of the left. The hard disk installation and removal is the same as the second 3.5-inch diskette drive, except that the attached ribbon cable is different. The 3.5-inch diskette drives use a 40-pin connector and the hard disk uses a 44-pin connector. Both cables are installed in the unit as standard and are preshaped and positioned so that the correct connector for the drive is obvious. The orientation

Chapter 2

for the connectors is also apparent from the routing of the cable, but as is typical with IBM connectors, there is a key insert preventing the connector from being inserted the wrong way.

System Board Removal

In order to remove the system board, the plastic rear panel must be snapped off. It is held in position with five retaining plastic studs and a single locating stud at the top right-hand corner of the panel. The blade of a screwdriver is inserted between the plastic rear panel and the inner metal rear panel, and rotated to pop the studs from their mounting holes. Once this has been done at the power supply end of the rear panel and centrally above the keyboard and parallel port connectors, the plastic panel can be removed.

The system board has a single 80-pin connector that accepts an adapter board for the expansion bus. This adapter board contains three PC expansion slots and the battery backup for the real-time clock. Because of this arrangement, the PC adapters are inserted in the machine parallel to the system board instead of perpendicular to it. This adapter is supported at the base by the connector in the system board and at the top by a plastic bracket that extends from the top of the power supply to the top of the adapter. One end of this bracket has two hooks that wrap around the top PC expansion slot connector. The other end has two small flanges that hold it in two holes in the side of the power supply. The bracket is removed by squeezing the top of the hooks that are around the connector, then raising the bracket so that the flanges in the power supply are at an angle to come out of their slots. To reconnect, the flanges in the brackets are placed into the power supply, the bracket is lowered onto the top of the adapter board, and the hooks are snapped around the top connector. Problems with reseating the cover may occur if these hooks are not firmly attached to the top connector.

Once the bracket and the plastic rear panel are removed, the system board can be taken out of the unit. First, all cables need to be removed from the system board. There is one cable for the 3.5-inch diskette drives, one for the hard disk, one for the keylock switch, and two for the power supply. Eight screws hold in the system board: one is on the rear panel above the keyboard connector and seven are on the board itself. When these are removed, the board can be slid about a half-inch towards the front of the unit, then pulled out of the left side of the case.

System Board Components

The system board, shown in Figure 2-5, consists mostly of surface-mount devices (SMD), and is densely populated. There are fewer components on this board than the PC system board, because the functions of many support ICs

The Model 30

have been incorporated into custom VLSI components. This has both improved the reliability of the board and reduced its size.

A few components on the board have pins that extend through the board. These are nearly all either socketed ICs or discrete components that have a

Figure 2-5 Model 30 System Board. *The video system, the MCGA, is an integral part of the system board and is a subset of the VGA found on Models 50, 60, and 80.* Photo by Bill Schilling.

higher power rating than the SMDs. The ROM ICs are socketed, as is the 8086 microprocessor. An additional socket alongside the 8086 is for the 8087 math coprocessor. The math coprocessor can be installed after any expansion boards have been removed. The only other socketed component is the INMOS digital-to-analog converter (DAC) for the video subsystem. The printed circuit board is multi-layer with an internal ground plane, giving the board its brown color.

Two connectors for the DC power are located on the edge of the board that is closest to the power supply. Both are 6-pin connectors, keyed to prevent accidental interchange. The connectors for the hard disk cable and the 3.5-inch disk drive cable are located on the right side of the board, so that when the unit is assembled they are just visible behind the first 3.5-inch disk drive. The 44-pin hard disk connector is farther to the rear of the unit than the 40-pin 3.5-inch diskette drive connector.

The connector that accepts the expansion slot adapter is located centrally on the board towards the rear. The expansion slot adapter fits vertically into this connector, allowing PC-style expansion boards to be plugged into the unit in a horizontal orientation. The PC expansion boards are held in position by a card guide fixed to the metal chassis inside the front panel of the unit. The typical rear panel bracket slides into the metal chassis at the rear of the unit. If the expansion board has a connector on the rear bracket (for example, for a serial port), the plastic rear panel needs to be modified. There are three plastic blanks in the rear panel that, according to IBM's Guide to Operations, can be broken out of the molding by levering with a screwdriver. This is not easy to do with the rear panel still installed as illustrated. The blanks can be twisted out more easily by hand, after removing the rear panel.

The board that accepts the PC expansion board also contains the battery for the real-time clock. This is a 3-volt lithium battery that is soldered into the board. It has a life of several years, but replacement requires a new adapter board.

The speaker, called a beeper by IBM, is located in the front left-hand corner of the board. It is not the same as the speaker in the PC and AT, being in a metal can and mounted on the system board itself instead of behind the front panel. The two ROM ICs are beside the beeper. Each is a 32K by 8 bit ROM in a 28-pin DIP socket.

The 640 KB of RAM on the system board is located immediately to the rear of the beeper and ROMs. The first 128 KB of RAM is in SMD soldered onto the system board. The upper 512 KB is on two single-in-line packaged (SIL) boards. The first 128 KB of RAM is arranged in four 64K by 4 bit ICs and two 64K by 1 bit ICs. The SIL boards that contain the rest of the RAM are the same form factor as those used in other models of the Personal System/2 family, but the different model boards are not interchangeable. Each board consists of three RAM ICs mounted together with decoupling capacitors on a small board, which has a single gold-plated edge connector. This board fits into a socket that holds the board at about a 30-degree angle to the main system board. These SIL boards are similar to those used in the XT 286.

The Model 30

Installation and Setup

The Personal System/2 machines have been designed to simplify installation and setup. The switches in the system itself have been eliminated; only PC expansion boards have switches that require setting. The system is contained in three packing boxes. Two boxes contain the monitor and the system unit. The last box, called the IBM Personal System/2 Model 30 Starter Kit, contains the keyboard, Starter Diskette, and documentation. There are two pieces of documentation: the Setup Instructions and the Guide to Operations.

The Setup Instructions are supplied on a single sheet of paper that provides a mostly pictorial, step-by-step guide to installation. There are thirteen simple steps. For example, step 1 illustrates removing the units from the packing cases and step 2 checks that the power switch on the system unit is turned off. The final steps show how to insert a Model 30 Starter Diskette and turn on the machine. Only step 13 is more advanced; it states that a copy of the Starter Diskette should be made and the operating system installed, with no further information. This clear, concise installation sheet is ideal. Even experienced installers will probably take the few seconds necessary to skim this sheet and ensure that all parts are correct and available, and it is not intimidating to new users.

The other documentation package is the Personal System/2 Guide to Operations and Starter Diskette. This Guide to Operations is very different from the tomes supplied with the PC. It is a slim book of approximately 70 pages that explains the basic installation of the machine and expansion boards, as well as of the Starter Diskette. Again, as in the single sheet setup guide, the pages are clear and simple, so an inexperienced user should feel comfortable. However, the installation of a math coprocessor is not described, which is disappointing since it is included in the equivalent manuals for the other machines in the family, including the Model 25. The basic installation instructions are similar to those on the single sheet, with ample descriptions to explain anything that isn't clear from the pictures.

Next in the Guide to Operations is a picture of the front of the unit with the system unit, display, keyboard, and all visible switches labelled. The rear view picture that follows shows each of the rear connectors and their functions.

The Starter Diskette operation is described in detail for beginners in the Guide to Operations. This disk is a junior version of the disk that is used for system setup in Models 50, 60, and 80, where adapters are initialized and address conflicts highlighted. There are two parts to the Starter Diskette routines. The first part, called Learning About Your System, is a pictorial tutorial of the system hardware, the different types of software, and basic system testing. The second part of the Starter Diskette is the services section, which consists of diagnostic tests and system utilities.

The tutorial is organized into several sections, which IBM calls chapters. The display is in 320-pixel by 400-pixel mode, with 256 colors from a palette of

Chapter 2

256K. It illustrates through a simple application the dramatic increase in clarity of pictures created in this new video mode. Although this could be considered simply an excuse to show off the new video system, the demonstration is a clear and concise method of introducing a new user to the machine.

The system is described at two levels: as a simplistic system unit, monitor, and keyboard; and as a microprocessor, RAM, and ROM. The concept of formatting a diskette is explained as well as the correct orientation for inserting a disk into the drive. This information is useful because these simple concepts are so often forgotten by new users (if they were ever introduced in the first place), and this tutorial is friendly and clear. The software introduction section of the program introduces application programs such as a spreadsheet or database by displaying simple sample screens. The sample screens are interactive; for example, in the spreadsheet screen the value in a cell can be changed, and other cells that are affected by the change are adjusted.

Chapter 2 explains the hardware basics, highlighting on screen the various items that have further descriptions available. For example, each element on the front panel of the system unit is described, such as the power on/off switch and the disk drive lights. The keyboard section even includes a simple game to explain the use of cursor keys.

The services section of the Starter Diskette allows diagnostic tests to be run on the system. A disk can be copied or formatted, the hard disk can be parked in preparation for moving the system, and the date and time can be set. The diagnostic tests are similar to those performed by the advanced diagnostics, except that the system is checked in an automated fashion. The display shows the number of the test that is being performed and prompts for user input for the display and keyboard tests. A blank preformatted diskette is necessary to test the disk drive. The user is not warned of this prior to the start of the system checkout. If the user attempts to proceed by removing the diskette from the drive, the message `ERROR SYSTEM UNIT 601` is displayed. This is a pity, because the rest of the diagnostic test is a step-by-step guide.

Any error messages that the system reports during this test are the same as those found in the advanced diagnostics provided with the Hardware Maintenance Service manual. This enables the user to perform basic system checking and provide the repair service with constructive help.

Advanced Diagnostics

The advanced diagnostics program, which is available at extra cost with the Model 30 Hardware Maintenance Service manual, is similar in look and feel to the advanced diagnostics on the AT. It comes on a single diskette and provides a method for testing different parts of the system unit individually and creating a disk file that logs the errors that are found, as well as copying or formatting a disk. The hard disk can be formatted through this program or through the corresponding section on the Starter Diskette.

Compatibility and Performance

The Model 30 has three expansion slots for the installation of PC-compatible expansion boards. There appear to be few restrictions on using third-party boards. First, the expansion board that is to be added must be able to cope with the 8 MHz bus speed. PC-compatible manufacturers increased the speed of the clock in their machines several years ago and nearly all expansion boards now work at the faster speed. However, there are bound to be PC expansion boards around that were designed without considering the 8 MHz bus.

The other caution concerns boards that contain serial or parallel ports. The Model 30 has a serial port that is configured as COM1 and a parallel port that is LPT1. Any expansion board containing a port, and this includes internal modems, must be able to address a different port than COM1 or LPT1, or be capable of disabling the conflicting port.

Software compatibility with application software that runs on the PC is excellent, once the data has been transferred. The biggest stumbling block is copy protected diskettes. Any program that requires a key diskette in drive A for starting operation or during operation is a problem. Although an external 5.25-inch diskette drive can be added to the system, it cannot be addressed as drive A. For major application programs that still use a copy protection scheme, an update is necessary. This update will probably be available on 3.5-inch diskettes so the data transfer will be less painful. For game programs that are copy protected, however, it seems unlikely that an update will be viable. The cost to the manufacturer to update to the 3.5-inch media is probably not worth the effort. Many game programs do not run under DOS and use an alternative method of formatting the disk, such as the P-system, further reducing the cost effectiveness of an upgrade.

Generally speaking, if a program can be installed onto a hard disk and has no copy protection scheme that requires a key disk, it can be transferred and run on the Model 30. Data and program migration is a headache (see Chapter 14); however, once transferred, most programs will run. The video compatibility is good, so applications that are written expressly to support the MCGA are not necessary if CGA graphic resolution is adequate. The MCGA text modes that are equivalent to the CGA provide a more distinct character, with a character cell 8 pixels wide by 16 pixels high—similar to the PC monochrome adapter with its character cell of 9 by 14 pixels. This in itself may provide sufficient clarity for many applications and negate the need for an upgrade in the video system.

The performance of the Model 30 was evaluated by measuring its performance on a typical application—for example, a spreadsheet recalculation or a sort exercise on a database—against the performance of other machines running the same program. When taken individually, the results of these tests do not show how a Model 30 will perform when compared with other PCs, but when viewed in a larger perspective they give a balanced view of expected performance. Each tested machine had a default configuration; for example, the number of files or buffers was the DOS default. Full details of the benchmarking tests are given in Appendix A. Appendix B gives a description of the other test machines that were used for comparative purposes.

The relative speed of the machine while performing standard application program functions was compared with that of the IBM PC XT, IBM AT Model 339 (8 MHz), and its two closest Personal System/2 siblings—the Model 25 and Model 50. Third-party machines tested include the Compaq Deskpro 286 (8 MHz), which is equivalent to the IBM AT.

There are five main categories that application software falls into: word processing, database management, spreadsheet, graphics, and communications. The communications function does not vary significantly with the speed of the machine; so although the compatibility of the machine is important for communications packages, they do not provide any feel for the machine's performance.

The applications that were used for comparative purposes include:

Word processor	WordPerfect from WordPerfect Corporation
Database	dBASE III PLUS from Ashton-Tate
Spreadsheet	Lotus 1-2-3 from Lotus Development Corporation
Language compiler	Macro Assembler from Microsoft Corporation
Graphics	PC-KEY-DRAW from OEDWARE

The tests performed were analyzed in several different ways. For the Model 30, the relevant comparisons were the results that depended on hard disk performance or diskette drive performance when a low density diskette was used, as well as the non-disk-dependent results. The non-disk-dependent results are shown in Table 2-4. These tests were performed by the application program only in RAM. The 640 KB of DOS memory that is available on all of the tested machines can hold the WordPerfect document without the program having to swap data between RAM and disk to move around the document.

The WordPerfect document results are for the time taken to move from the top of a long document to the bottom, the time taken to move from the bottom of the same document to the top, and the time to perform a search and replace operation where each incidence of the letter j is replaced with the letter k. As can be seen from the table, the presence or absence of the math coprocessor in the machine did not affect the results of the tests. Moving around a document does not involve arithmetic functions, so this result is not surprising.

The results show that the XT performed the word processing tests at 0.45 times the speed of the Model 30. The Model 25 was the same speed as the Model 30, and the AT and the Compaq Deskpro 286 were about 1.8 times faster. Models

Table 2-4 Non-Disk-Dependent Benchmark Results. *The Model 30 was markedly faster than an XT and equivalent to a Model 25 in performance.*

Test Machine	Equipment	WordPerfect Move Top to Bottom	Move Bottom to Top	Search and Replace	Lotus 1-2-3 Recalc—Add	Recalc—Multiply	Recalc—Mix
XT	No math	109	82	129	18	30	200
	Math	109	82	129	18	18	19
Model 25	No math	48	37	58	8	14	91
	Math	48	37	58	8	8	9
Model 30	No math	49	38	58	8	14	91
	Math	49	38	58	8	8	9
AT 339	No math	28	21	32	5	8	45
	Math	28	21	32	5	5	7
Compaq 286	No math	28	21	34	5	7	45
	Math	28	21	34	5	5	7
Model 50	No Math	22	16	27	4	6	36
	Math	22	16	27	4	4	5

All results are in seconds.

25 and 30 have similar architectures so similar results would be expected. The XT is driven by an 8-bit wide 8088 running at 4.77 MHz, whereas the Model 30 has an 8086 microprocessor running at 8 MHz. Remember that the 8086 is the same as an 8088 except that it has a 16-bit wide path to system memory. The AT and Deskpro 286 run at 8 MHz also, but with an 80286 microprocessor. A minimum cycle time on the 8086 consists of four clock cycles, and on the 80286 the minimum cycle time is two clock cycles, resulting in slower times for the Model 30. The Model 50 is an 80286-based machine, also running at 10 MHz, and the results show the anticipated performance of over twice the speed of the Model 30.

The Lotus 1-2-3 non-disk-dependent performance tests were recalculation exercises on three different types of spreadsheets. The presence or absence of a math coprocessor in the machine affected the resulting times; the degree to which the coprocessor improved speed reflects the complexity of the spreadsheet. The simple addition spreadsheet, where each cell (1000 in all) was the sum of the previous cell plus 1, showed ratios similar to the WordPerfect test. The Model 25 and 30 performed at the same speed. The XT was 0.44 times as fast as the Model 30; the AT and the Deskpro 286 were 1.6 times faster than the Model 30. The Model 50 was twice as fast as the Model 30 for the same test. The improvement in performance due to the presence of a math coprocessor was negligible.

When a spreadsheet recalculation was performed that involved 1000 multiply/divide functions and the use of Pi, the math coprocessor had an effect.

The performance of the Model 30 was 1.75 times faster with the 8087 present than without its presence. On the Model 30 the math coprocessor runs at the same speed as the microprocessor (8 MHz). The presence of the math coprocessor allows Lotus 1-2-3 to make use of its faster arithmetic functions. When compared without math coprocessors, the XT was 0.46 times as fast as the Model 30 and Model 25, the AT and Deskpro 286 were 1.75 times faster, and the Model 50 was 2.3 times faster. When compared with the math coprocessor installed, the XT was 0.44 times as fast as the Model 30 and 25, the AT and Deskpro 286 were 1.6 times faster, and the Model 50 was twice as fast as the Model 30.

The final spreadsheet recalculation involved a formula that included, among other functions, square roots and natural logarithms on 500 cells. The math coprocessor had an even more significant effect than in the second sample spreadsheet. The Model 30 was 10 times faster with the math coprocessor present than without. When the Model 30 without a math coprocessor was compared with the other machines, the XT was 0.46 times as fast, the Model 25 had a similar performance, and the AT and Deskpro 286 were twice as fast. The Model 50 was 2.5 times faster than the Model 30 for the same test. The math coprocessor changed the ratios slightly. The XT was 0.47 times as fast as the Model 30, the Model 25 had a similar performance, the AT and Deskpro 286 were 1.29 times faster, and the Model 50 was 1.8 times faster.

The other tests used to evaluate the test machines were disk dependent. The results obtained from using the hard disk in each machine are given in Table 2-5. The Model 25 is not included, as it does not offer a hard disk as an option. The time taken to load the word processing file (the same file used for the other word processing tests) was measured, as was the time taken to load the addition spreadsheet into Lotus 1-2-3.

Note that the only machine tested using a disk caching program was the Model 50, as this is supplied with the machine. Many commercial disk caching programs could be used to similarly improve the performance of the other reviewed machines. The ratios obtained for loading the files into the word processor and the spreadsheet were not the same, showing the application-specific tendency of this test. When viewed as a whole, however, a range of performance can be assessed. If the time taken to load a file for the Model 30 is considered a ratio of one, the XT performed about half as fast, the AT and the Deskpro 286 performed 68 percent faster, and the Model 50 performed 2.3 times faster.

The time taken by the computers to perform a database sorting function is typical of application functions that involve many features of the machine. The file needs to be read from disk, sorted in memory, and written out again to disk. For sorts of a reasonable size, such as the tested condition, this process requires interim steps where the database needs to access the disk often, either to obtain the rest of the records or to write out a partially sorted database. The results, seen in Table 2-5, show a sample database that was sorted in two different ways. The XT performed this function 0.37 times as fast as the Model 30, the AT was

1.29 times faster, the Deskpro 286 was 1.53 times faster, and the Model 50 was 1.13 times faster without cache and 1.61 times faster with cache. The Model 50 hard disk is a lower performance unit than would be expected for a machine of its class, and this is reflected in the results. The XT has a hard disk that was reasonable when it was released, but the technological advances since its release can be seen in the test results.

Table 2-5 Hard Disk Benchmark Results. *The Model 30 is not supplied with the standard IBM cache program that is available on Models 50, 60, and 80.*

Test Machine	Equipment	WordPerfect Load File	Lotus 1-2-3 Load File	dBASE III PLUS Sort on Title	dBASE III PLUS Sort on Disk ID	PC-KEY-DRAW Run Sample Macro	MASM Assemble File
XT	No cache	39	90	88	92	1441	40
Model 30	No cache	25	44	32	34	655	16
AT 339	No cache	12	29	25	26	526	11
Compaq 286	No cache	11	28	21	22	477	9
Model 50	No cache	10	27	29	29	409	7
	Cache	10	20	20	21	405	7

All results are in seconds.

The Microsoft Macro Assembler (MASM) was used to demonstrate the relative performances of the machines for an assembly process. The Model 30 was 2.5 times faster than the XT, 0.69 times as fast as the AT, 0.56 times as fast as the Deskpro 286 (which has a slightly faster disk than the AT), and 0.43 times as fast as the Model 50. This assembly process is dependent on the system architecture for the most part, but since the results have to be written out to the hard disk, the relative performances of the hard disks are also reflected in the results.

The final benchmarking test was for a graphics program, PC-KEY-DRAW. The sample macro that is supplied with the program was run. This requires a small amount of disk access, but the results obtained mostly reflect the relative performances of the display subsystems in the machines. PC-KEY-DRAW operates in CGA or CGA emulation mode on each of the machines. The Model 30 with its MCGA performed 2.2 times faster than the XT with its CGA, 0.73 times as fast as the Compaq Deskpro 286 with an EGA adapter, 0.8 times as fast as the AT 339 with an EGA, and 0.62 times as fast as the Model 50 with a VGA.

The same tests that were run on the hard disk in each machine were also run on the diskette drive for the machines. These results are obviously slower than those obtained for the hard disk, because a diskette drive is inherently slower than a hard disk. Although relative performance is significant, it should be noted

Chapter 2

that the XT, AT, and Deskpro 286 have 5.25-inch diskette drives and the Personal System/2 machines have 3.5-inch drives. The results are shown in Table 2-6.

Table 2-6 Diskette Benchmark Results. *The Personal System/2 machines have 3.5-inch diskette drives and the PC-compatible machines have 5.25-inch diskette drives.*

Test Machine	WordPerfect Load File	Lotus 1-2-3 Load File	dBASE III PLUS Sort on Title	dBASE III PLUS Sort on Disk ID	PC-KEY-DRAW Run Sample Macro	MASM Assemble File
XT	73	159	232	239	1480	48
Model 25	50	140	219	224	697	36
Model 30	54	140	219	224	704	36
AT 339	43	115	193	184	572	30
Compaq 286	43	115	181	182	546	27
Model 50	45	134	209	211	480	22

All results are in seconds.

The time taken to load a file into an application varied with the machine. The Model 30 was 1.2 times faster than the XT, and 0.83 times as fast as the AT and the Deskpro 286. The Model 50 was only slightly faster (1.11) than the Model 30. The Model 25 was slightly faster at loading a file into WordPerfect than the Model 30. As there is no difference in their respective architectures, this difference can be attributed to the different tolerances of the two machines tested.

The results of the database sort also show less of a difference between the machines when their performances using the diskette drives are considered. The Model 30 was 1.06 times faster than an XT, but was slower than the AT, Deskpro 286, and the Model 50 by factors of 0.85, 0.81, and 0.95 respectively. The 5.25-inch diskette drives performed slightly better than their 3.5-inch counterparts.

The Macro Assembler results show that the Model 30 was 1.33 times faster than the XT, 0.83 times as fast as the AT, 0.75 times as fast as the Deskpro 286, and 0.61 times as fast as the Model 50. The disk-dependent factor of this test was less significant than in some of the other benchmarks.

The results from the graphics test show that the Model 30 was 2.1 times faster than the XT, 0.81 times as fast as the AT, 0.78 times as fast as the Deskpro 286, and 0.68 times as fast as the Model 50. Again, the disk dependency was less of a factor in this test than in the database sort.

The overall results from the Model 30 show that it was a much better performer than the XT, in general performing twice as fast, although in certain diskette-dependent tests only slightly faster. The Model 30 was slower than the AT and the Deskpro 286 for all of the tests run, but only by a factor of about 20 percent. The Model 50 performed better than the Model 30, but the weakness of the hard disk in the Model 50 can be seen in the results. The Model 30 is comparable to the Model 25 except that it can have a hard disk—a great advantage to many users.

Model 30 Options

Two additional options, which are not suitable for the PC family, can be added to the Model 30: a math coprocessor and a speech adapter. The 8087-2 math coprocessor is a higher rated part than the one used in the PC or XT, running at 8 MHz. It is inserted into the socket on the system board for use. The speech adapter encodes and decodes speech, using *continuously variable slope delta* (CVSD) modulation techniques and *linear predictive coding* (LPC) for synthesis. This is an expansion board that takes up a single slot, but it is not designed to work on the PC.

Summary

The Model 30 is a well-designed, fast replacement for the PC. Functionally it is a desirable machine, because it has all the features required for most configurations integrated into the system board. It is higher in cost than many PC clones but it does have the IBM label. If the user's requirements do not include the next generation of operating systems (that require an 80286-based system) and the 3.5-inch diskettes are not a problem, this machine is adequate for many applications. The sales figures reflect its popularity in the marketplace, but only time will show whether this popularity continues when all other machines in the family and Micro Channel add-in boards are available in sufficient quantity to meet the demand for these products.

There are arguments against widespread use of the Model 30, however. The price of AT clones is competitive with this machine and many of them offer better performance, just as many features, and 5.25-inch diskette drives. However, even though these machines contain 80286s, many of them are not sufficiently IBM compatible to run OS/2. The Model 50 is an 80286-based machine with Micro Channel architecture and more integrated features than the Model 30, and it runs OS/2. Overall, the Model 30 is a nice machine that is already past its prime. Had it been released when the rumors about the PC2 were prevalent, it would have been an excellent stepping stone to the Personal System/2 family of machines.

3

The Model 25

The Personal System/2 Model 25 is the other machine in the Personal System/2 family that does not quite fit the design concept. It is, however, a valid machine, providing a reasonably compact unit with performance similar to the Model 30. Like the Model 30 it does not have Micro Channel expansion capabilities, but it is a lower cost alternative. Unlike the Model 30, it does not have a hard disk or a real-time clock.

The Model 25 combines a system unit and monitor into a machine outline that has a passing resemblance to an Apple Macintosh. It has a smaller footprint than either the PC or the Model 30. There are two versions of the basic machine: one with a color monitor and the other with a monochrome monitor. There are two keyboards available for either machine, making a total of four different model numbers. The 8525-001 is the least expensive version available and the 8525-G04 is the most expensive. With the educational discounts that are being offered by IBM, it is possible to obtain a machine for under $1,000 for use in the educational environment. (See Figure 3-1.)

The educational market is an area in which IBM, until the Model 25 announcement, has been unable to gain extensive market acceptance. Competitors such as Apple, with the Macintosh and the Macintosh Plus, have been the dominant players in the market. Now IBM has provided a machine that can be considered serious competition for this lower end market.

The Model 25 is an 8086-based machine that runs at 8 MHz. It comes with 512 KB of zero-wait-state memory as standard and has an expansion option to bring the system board memory up to 640 KB. One 720 KB, 3.5-inch disk drive is standard, with room for a second alongside as an option. The system board has an integral serial port, a parallel port, a keyboard port, and a pointing device port, along with an audio earphone connector. There are two PC-style expansion slots—one full length and the other 8 inches long—for additional expansion.

The two keyboard options are the standard 101-key Enhanced Keyboard or

Chapter 3

Figure 3-1 The Model 25. *The 8086-based Model 25 is targeted to the educational market; however, it is not limited to this arena.*
Photo courtesy of International Business Machines Corporation.

a slightly less expensive Space Saving keyboard. As shown in Figure 3-2, the layout of the Space Saving keyboard is similar to that of the Enhanced Keyboard, except that the numeric keypad (usually located on the far right) is missing. It is incorporated into the main part of the keyboard and is accessed via the Num Lock key.

The monitor is an integral part of the system unit and is similar to the analog monitors that are available for the rest of the Personal System/2 family of machines. The Monochrome Display has a 12-inch diagonal tube, like the 8503, and the color version is the same as the 12-inch 8513 Color Display. The video subsystem is contained on the system board and is the same as the Multi-Color Graphics Array (MCGA) used on the Model 30. This provides compatibility with the Color Graphics Adapter (CGA) but not the Enhanced Graphics Adapter (EGA). It also provides extra video modes.

The Model 25 is a compact unit that gives the illusion of being very small,

Figure 3-2 Enhanced Keyboard and Space Saving Keyboard. *The numeric keypad that is present on the right of the Enhanced Keyboard is removed from the Space Saving version. The same keys have been remapped to the main QWERTY area in an arrangement similar to their previous locations.* Photo by Bill Schilling.

because it is narrow across the front. However, it is surprisingly deep, being approximately the same depth as the XT. Like the Model 30, its acceptance in the market has two perspectives. If DOS and DOS-based applications remain prevalent, then the machine is serious competition for the PC clones. However, if OS/2, which uses the protected mode of the 80286, is the way of the future, the life of the Model 25 is short lived, despite its beautiful engineering. The home buyer simply does not need the IBM label on a machine to the extent that the corporate buyer does. Neither this machine nor the Model 30 are the clone killers that rumor suggested IBM was developing. They are solid, reliable, fast PCs that offer a reasonably priced alternative to the no-name clones that are flooding the market.

Chapter 3

Functional Description

The Model 25 is similar in function to the Model 30; however, the system looks different and is not as full featured as its siblings. IBM does not offer a hard disk for the Model 25; a second 3.5-inch diskette drive is the only mass storage expansion option offered. The standard unit comes with only 512 KB of memory. The expansion option is necessary to get the full 640 KB on the system board. However, with its integral monitor, two PC-style expansion slots, and 8 MHz 8086 microprocessor, it is a tidy, self-contained system that can compete with the Model 30.

The system itself is similar to the Model 30. It has an 8 MHz 8086 microprocessor that operates with zero wait states when accessing system board memory. This results in a 500 nsec cycle time for accessing local memory. Accesses to the I/O channel and expansion bus memory are 8 bits wide and are performed with a cycle time of 1 microsecond. The 8086 is capable of performing 16-bit memory accesses. If the memory accessed is located on the bus and not in local memory, the system board has the necessary logic to translate the memory access into two 8-bit transfers, resulting in a cycle time of 2 microseconds.

Alongside the 8086 microprocessor is a socket for the 8087 math coprocessor. The coprocessor is an option that can be added to the machine to improve the time required to perform various mathematical functions. These advantages are seen only with applications that make use of the coprocessor; they are not automatic.

The address map for the system is the same as the map for the Model 30. The 8086 can address 1 MB of memory, including:

▶ 640 KB of RAM from address 00000H to 9FFFFH

▶ 128 KB of video buffer from A0000H to BFFFFH

▶ 192 KB reserved for BIOS on the I/O channel from C0000H to EFFFFH

▶ 64 KB for system ROM from F0000H to FFFFFH

The 64 KB of ROM in the Model 25 contains the system POST (power on self test), BIOS and a BASIC interpreter, and 128 characters of video dot patterns.

The video system on the machine is the same as for the Model 30. The MCGA is attached to the internal monitor and provides all of the video capabilities that are found on the Model 30. The CGA is fully emulated and all applications that support the CGA should operate on the MCGA. Two additional graphics modes are available with the MCGA that were not a feature of the CGA. For units with the color monitor, there is a 320-pixel by 200-pixel mode with 256 colors from a palette of 262,144 (256K), which is internally double scanned to display 320 pixels by 400 pixels. There is also a 640-pixel by 480-pixel mode in 2 colors from the same palette selection of 256K colors. The monochrome units display the same resolution, but with a gray scale for the multi-color modes. The video modes of the MCGA are described in detail in Chapter 2.

The Model 25 consists of a system unit with integral monitor and a keyboard. The cable that connects the keyboard to the system unit is detachable both at the system unit end and at the keyboard end. The Space Saving keyboard that is offered is similar in layout to the IBM Enhanced Keyboard, except that it does not have the numeric keypad on the right-hand side or the LEDs that display the status of the Caps Lock, Num Lock, and Scroll Lock keys. The function of the numeric keypad is provided as a part of the main alphanumeric keypad. The Num Lock key is moved to the Scroll Lock key position and accessed by pressing Shift and Scroll Lock together. This allows the numeric keypad to be accessed at the center of the keyboard. The numbers 0 through 9 are on the keys M, J, K, L, U, I, O, 7, 8, and 9, respectively. The decimal point is in the same position as the period; the plus, minus, and divide keys are in their normal keyboard positions; and the multiply (asterisk) key is in the semicolon position.

This arrangement provides a reasonable solution for many applications but may not be totally acceptable. For applications that do not use the numeric keypad, this rearrangement of keys is not a problem. If a typical numeric keypad application is considered, such as a spreadsheet, the situation is less than ideal. In order to add labels to the spreadsheet, the Num Lock needs to be off; then, for numeric entry it needs to be turned on. The other alternative is to use the conventional typewriter positions for the numbers and not use the numeric keypad at all, but that seems a waste. IBM must have expected the numeric keypad to be used or they would not have provided it on the original PC keyboard.

Physical Description

The system unit of the Model 25 incorporates the monitor as well as the typical system unit components. The size of the unit is approximately 15.1 inches high by 9.5 inches wide by 14.7 inches deep. The color version weighs in at a hefty 37 pounds and the monochrome version is a more respectable 28 pounds. This is not a portable machine; it occupies most of the depth of a standard desk. The case is cream-colored rigid plastic with a gray faceplate around the display screen.

The monitor is located above the system board and disk drives. When the adjustable foot on the base of the machine is fully retracted into the case, the monitor is tilted up about 20 degrees from the horizontal. Adjusting the foot allows the angle of the monitor to be adjusted, either to a fully horizontal position or half way between the extremes. The foot is a spring-loaded, ratchet-type mechanism that is easy to adjust using the convenient hand hold in the rear of the case to lift the rear of the unit. From the lowest position an alternate angle can be selected by gently lifting to the required angle. Once the rear is raised beyond the horizontal position, the foot is released from its latching mechanism and the unit can be fully lowered.

The front of the unit contains the screen, the 3.5-inch diskette drives, the

Chapter 3

power switch, and the display controls. The facias of the disk drive are set back from the front of the monitor, and the controls for the display and the power are located under the display front. The red toggle power switch is on the far right and is a little awkward to find. The LED that indicates when the unit is powered on is in the bottom right-hand corner of the gray facia around the screen.

Beside the power switch is the contrast control. Its location is indicated on the screen facia by a circle that is half filled in. The control itself is a knurled thumbwheel just above the second disk drive position. The brightness control is the leftmost control, but is still on the right-hand side of the unit. Its position is indicated on the screen facia by a sun-like logo in the molding. It too is a knurled thumbwheel that is located under the front of the display and has a detente indicating its central position.

The first 3.5-inch diskette drive—drive A—is located in the left-hand bay beneath the screen. The right-hand bay is empty in the standard configuration and has a blank facia plate. This is replaced by the second 3.5-inch drive when the option is installed.

There are no controls on the top or sides of the unit, but the top contains vents that wrap slightly onto the side for air circulation.

Rear Panel Connectors

The rear panel of the Model 25, shown in Figure 3-3, has six connectors. Viewing the unit from the rear, the standard 3-pin power connector is located half way up on the left side, and the voltage selection switch for 115-volt or 230-volt operation is just above. The other connectors are in a recessed portion of the case along the bottom of the rear panel. The left-hand location is the audio earphone connector that allows the beeper output to be redirected to an earphone. Any ¼-inch diameter audio plug can be used. An earphone with an impedance of between 15 and 35 ohms is specified; however, an impedance level of up to 100 ohms is also satisfactory.

The keyboard connector is labelled with the number 1 and is a 6-pin miniature DIN connector. To the right of the keyboard connector is the pointing device connector, labelled 2, which is also a 6-pin miniature DIN connector. As on the Model 30, these two connectors are interchangeable. The male 25-pin D-type connector to the right of the pointing device connector is the serial port connector. The female 25-pin D-type connector on the far right is the parallel port connector. The serial and parallel port connectors each have hexagonal threaded standoffs on either side, allowing attached cables to be clamped securely by a screw.

Also on the rear panel are the end brackets for the expansion slots. These lie in a horizontal orientation above the serial and parallel port connectors. A blanked out hole that would accommodate a 9-pin D-type connector is not mentioned in the documentation.

The Model 25

Figure 3-3 Rear Panel of the Model 25. *The monitor is an integral part of the Model 25, so there is no external video connector. There is an audio earphone connector, however, that is not available on other models.* Photo by Bill Schilling.

Inside the System Unit

The cover for the Model 25 is held in place by two screws on the rear panel at the top of the recessed section. The monitor occupies approximately the top two-thirds of the unit and is well sealed to prevent the high voltages being exposed when the system board is revealed. The system board lies in a horizontal posi-

Chapter 3

tion beneath the monitor. When the case is opened the bottom third of the unit is lowered. (See Figure 3-4.)

Figure 3-4 Inside the System Unit. *The unit is opened by lowering the bottom third away from the top of the unit. The slots on either side guide the opening and closing of the unit.* Photo by Bill Schilling.

The Model 25

The case is opened by turning the unit onto its front so that the screen of the monitor is flat on the desk. The two screws that hold the cover on are then removed. The base of the unit is lifted very slightly and tipped outwards. A pin on either side of the top cover slides along a slot in either side of the base. The base is hinged at the front of the unit and can be lowered to expose the system board with the disk drives beneath.

The system board lies flat along the top of the exposed base. It is 11 inches deep by 8.3 inches wide. The power supply, cooling fan, and monitor are in the top of the unit. The power wires and signal wires run from the top of the unit to the system board. Their connectors are located on the left side of the board, when looking from the rear; the power supply connector is closer to the rear panel than the display connector. The wires for the fan do not go to the fan connector, as described in the IBM documentation, but are linked into the $+5$ volt$_{DC}$ and ground pins on the display connector.

There are three other connectors on the system board. The 40-pin connector on the right side of the board holds a ribbon cable that wraps underneath the system board to the disk drives. The 44-pin connector on the left edge of the board, although illustrated, is not described in the documentation. This connector is actually wired to the I/O support gate array and its configuration suggests a hard disk. When a Model 30 hard disk drive was inserted it did operate as drive C; however, this is not an IBM supported option. The marketing reasons for not offering a hard disk as an option on the machine may stem from concern over cost, or a hard disk may be planned as a future announcement. A technical reason is more likely. A hard disk is one of the few pieces of equipment that require a high level of confidence in functionality. There may be an electrical tolerance problem or electrical noise problems that cause the disk to work unreliably. Another possibility is that the installation of a hard disk may cause the EMI to increase over the level permitted by FCC regulations.

The system board is held down by eight screws on the system board and one on the edge of the rear panel (see Figure 3-5). The board needs to be removed only for replacement. The disk drive cables can be accessed from underneath and the drives are removed by sliding them forward.

The socketed microprocessor is located centrally on the system board, with the socket for the 8087 math coprocessor alongside. The two ROM chips are also socketed on the front end of the board, with the sockets for the RAM expansion option around them. To the left of these ICs is the rest of the RAM. The memory is on two boards. Each board consists of nine RAM ICs that are mounted on a single-in-line packaged (SIL) board that has a single gold-plated edge connector. This board fits into a socket that holds the board at an angle of about 30 degrees to the main system board.

There are five gate arrays on the system board, which are used for microprocessor support, the video subsystem, and the disk controller. The system support gate array is located between the ROMs and the microprocessor. This IC contains the memory controller, system clock generator, wait state generator

Chapter 3

Figure 3-5 Model 25 System Board. *The 512 KB of system board memory can be upgraded without using one of the expansion slots. The RAM is installed (as illustrated) in the sockets around the ROM.*
Photo by Bill Schilling.

and bus conversion logic, and the DMA support. The two video subsystem gate arrays are located between the RAM SILs and the video connector at the left-hand corner of the board, away from the rear connectors. The other microprocessor support IC, called the I/O support gate array, is located close to the keyboard and pointing device connectors. It contains the keyboard and pointing device controller, the chip select logic, interrupt controller, and the I/O ports.

The card edge connector that holds the adapter card for the PC-style ex-

The Model 25

pansion slots is centrally located towards the rear of the system board, as shown in Figure 3-6. The adapter card that fits into this connector has two card edge connectors that accept PC expansion boards. It is held in the system board by a plastic frame that wraps around the board. It is secured to the system by two screws: one on the system board and one through the rear panel.

The ribbon cable for the disk drive runs from and underneath the system board. The area beneath the system board is accessed by removing the adjustable foot. This foot is held in place by two screws. When the screws are removed, the foot can be tipped away from the base of the machine and lifted out. The connector for drive A is in the center of the cable, which daisy chains onto the connector for drive B.

The installation or removal of a 3.5-inch diskette drive itself is performed with the cover open. Each drive is held in place by two screws on the side of the unit. These screws are exposed only when the cover is open. The drives are

Figure 3-6 Expansion Board Adapter. *PC-style expansion boards can be installed in the Model 25 in the same way as in the Model 30. The adapter board plugs into the system board vertically and expansion boards are inserted into the adapter board.*
Photo by Bill Schilling.

replaced by removing the ribbon connector through the base of the machine, then sliding the drive forward with the cover fully open. Installation involves sliding the drive into the drive bay with the cover open, securing it with screws, and then connecting the ribbon cable through the base of the machine. When a second drive is installed for the first time, the blank bezel is removed by levering it out with the cover open.

Setup and Configuration

The single sheet of instructions for assembling the Model 25 consists of nine drawings that illustrate the necessary steps. The only words on the sheet are the copyright notice, but the steps are complete and clear. The voltage selection for the power inlet is indicated in step 4; it shows a map of the USA for the 115 V_{ac} supply and a map of Europe for the 230 V_{ac} supply. The marketing of this unit into non-English-speaking countries is simplified by these pictorial instructions.

In addition to simplified installation instructions, there are also simplified error messages. The machine comes with a Starter Diskette and a Personal System/2 Guide to Operations manual, which provide a guide to the system and are similar to the equivalent disk and manual on the Model 30. The Guide to Operations shows how to install the system and add the options for the machine. It is a well-written manual with clear and simple illustrations.

The Starter Diskette contains a tutorial that provides a pictorial guide to the basic operations of the machine, including the various switches and connectors, and also introduces the word processor and spreadsheet applications. More detailed information on the tutorial is available in Chapter 2.

If the machine is started without a diskette in drive A, a pictorial image of the computer appears with a diskette in front of drive A and the F1 key depressed. This image is similar to the one used for the same purpose on the PC Convertible.

Once installed, the Model 25 operates in a manner similar to the Model 30 and the PC, but with no hard disk there is a lot of 3.5-inch diskette swapping. The lack of a real-time clock is also a slight hindrance to a clear record of generated files. With these shortcomings taken into consideration, the Model 25 is a rugged and economical option for a personal computer. It does not have the growth potential of its older siblings, as it will not run OS/2, but it provides full functionality under DOS.

Performance

The architectures of the Model 25 and the Model 30 are very similar and this is reflected in their performance. A Model 30 with two diskette drives ran performance tests at the same speed as the Model 25. The improvement in the performance of the Model 30 via the hard disk is not, however, possible with the Model 25.

For performance considerations, the Model 25 was compared with the Model 30, the Model 50, and the XT, using their diskette drives. In addition, the Model 25 was compared with the AT 339 (8 MHz) and the Compaq Deskpro 286. These are its closest siblings and its performance was respectable when the hard disk was not a factor.

The performance of the machines was assessed using typical applications: Lotus 1-2-3 as a spreadsheet program, WordPerfect as a word processor, dBASE III PLUS as a database, Microsoft Macro Assembler (MASM) for a programming environment, and PC-KEY-DRAW as a graphics-intensive program. The results fell into two distinct classes: those that were independent of the diskette and those that made use of the diskette drives.

Table 3-1 shows the non-disk-dependent results. Two of the application tests did not require disk activity. Once a file had been loaded into WordPerfect (obviously this required disk activity), tests could be performed that only made use of RAM. The time to move from the top of the document to the bottom of the document and the time for the reverse procedure were measured. The Model 25 performed at the same speed as the Model 30, 2.2 times faster than the XT, but only 0.57 times the speed of the AT and Deskpro 286, and 0.45 times the speed of the Model 50.

Table 3-1 Non-Disk-Dependent Benchmark Results. *The Model 25 was similar in performance to the Model 30 and was a faster machine than the XT.*

Test Machine	Equipment	WordPerfect			Lotus 1-2-3		
		Move Top to Bottom	Move Bottom to Top	Search and Replace	Recalc—Add	Recalc—Multiply	Recalc—Mix
XT	No math	109	82	129	18	30	200
	Math	109	82	129	18	18	19
Model 25	No math	48	37	58	8	14	91
	Math	48	37	58	8	8	9
Model 30	No math	49	38	58	8	14	91
	Math	49	38	58	8	8	9
AT 339	No math	28	21	32	5	8	45
	Math	28	21	32	5	5	7
Compaq 286	No math	28	21	34	5	7	45
	Math	28	21	34	5	5	7
Model 50	No Math	22	16	27	4	6	36
	Math	22	16	27	4	4	5

All results are in seconds.

The time to perform a search and replace operation on the same document produced similar relative results. The Model 25 performed at the same speed as the Model 30, 2.2 times faster than the XT, 0.55 times the speed of the AT, 0.59 times the speed of the Deskpro 286, and 0.47 times the speed of the Model 50. None of the word processing results were affected by the presence or absence of a math coprocessor.

The increase in performance that can be obtained by the use of a math coprocessor is seen in the spreadsheet tests. Three different spreadsheets were created: one contained a repetitive addition formula in each cell, one contained a multiply or divide formula on a constant Pi, and the third contained a more complex formula in each cell, including a natural logarithm and a square root function. The recalculation times for the spreadsheets were affected by the math coprocessor. Although barely noticeable in the addition spreadsheet, the results were dramatic for the complex, mixed-formula spreadsheet. The addition of the math coprocessor in the Model 25 caused the recalculation of the 500 cells in the mixed spreadsheet to be performed over 10 times faster than without.

The relative performances of the machines when a math coprocessor was not present showed that the Model 25 performed as fast as the Model 30. The XT was 0.45 times as fast on average as the Model 25, the AT was 1.95 times faster, the Deskpro 286 was 1.98 times faster, and the Model 50 was 2.45 times faster on average for the three test spreadsheets.

When the respective math coprocessors were used, the relative performances changed. This was due to the different relative performances of the 8087 and 80287. The Model 25 was the same speed as the Model 30. The XT was 0.45 times as fast, the AT and Deskpro 286 were 1.47 times faster, and the Model 50 was 1.92 times faster than the Model 25, when the results from the three spreadsheets were averaged.

As users with hard disks know, a machine that does not have a hard disk is painfully slow, and the constant diskette swapping that is necessary is a burden when applications of any size are being used. The Model 25 does not have a hard disk, nor is one offered as an option. It can accommodate two 3.5-inch diskette drives, however, and each diskette can hold 720 KB of data. The performance of the diskette drives was assessed by using applications that make use of the disk during operation. Table 3-2 shows the comparative results for the Model 25.

The time to load a file into the word processor or the spreadsheet program varied with application as well as with the size of the file. The Model 25 again performed at approximately the same speed as the Model 30. It was 1.22 times faster than the XT, but the AT and Deskpro 286 were 1.2 times faster and the Model 50 was 1.06 times faster than the Model 25. The XT, AT, and Deskpro 286 all have 5.25-inch diskette drives, while the Personal System/2 machines have 3.5-inch diskette drives, which do not seem to perform as well from a relative time viewpoint.

The time to perform a database sort is a typical application program use. The Model 25 did this sort as fast as the Model 30, 1.06 times faster than an XT,

Table 3-2 Low-Density Diskette Benchmark Results. *The Model 25 does not have a hard disk as an option. Up to two 3.5-inch diskette drives can be installed.*

Test Machine	WordPerfect	Lotus 1-2-3	dBASE III PLUS		PC-KEY-DRAW	MASM
	Load File	Load File	Sort on Title	Sort on Disk ID	Run Sample Macro	Assemble File
XT	73	159	232	239	1480	48
Model 25	50	140	219	224	697	36
Model 30	54	140	219	224	704	36
AT 339	43	115	193	184	572	30
Compaq 286	43	115	181	182	546	27
Model 50	45	134	209	211	480	22

All results are in seconds.

0.87 times as fast as the AT, 0.82 times as fast as the Deskpro 286, but a reasonable 0.94 times as fast as the Model 50. The Macro Assembler test made use of the disk to store the assembled information, but also had a high element of processor use, for tasks such as sorting algorithms for the database. On this test the Model 25 was as fast as the Model 30. The XT was 0.75 as fast as the Model 25, the AT was 1.2 times faster, the Deskpro 286 was 1.33 times faster, and the Model 50 was 1.64 times faster.

The graphics program test involved the use of a sample macro that fully exercised the CGA graphics capabilities of the machine. It required some disk access to run but, as can be seen from the results, this was a relatively small factor. The Model 25 used for evaluation had a color monitor rather than monochrome. The Model 25 was marginally faster than the Model 30 for this test. The XT was 2.12 times slower, the AT was 1.22 times faster, the Deskpro 286 with an EGA adapter was 1.28 times faster, and the Model 50 was 1.45 times faster. These results show that the performance of the machines is affected by the display subsystem that they contain. In addition, the mode that is used—in this case CGA emulation—affects performance. Comparative tests use each machine in the same video mode, but these may not be the usual modes of operation for the display systems, so drawing conclusions requires care.

The Model 25 is an excellent machine if the handicap of no hard disk can be tolerated. One of the expansion slots could be used to install a hard disk that is mounted on an adapter board; an external hard disk could also be used with its associated disk controller occupying one of the two precious slots. If a diskette system only is acceptable, the machine is a good performer. It is a more compact unit than the Model 30 and suffers from the same disadvantages as that machine: it cannot run any of the protected mode operating systems and it does not have Micro Channel bus architecture. If the IBM label is important on a machine, the Model 25 is a high performance PC compatible. If the IBM label is not significant, there are many third-party machines that provide similar performance at a comparable price.

Model 25 Options

An 8087 math coprocessor can be purchased for the Model 25 as an option. This improves the performance of the machine for applications that make use of its calculating power.

An optional carrying case is available for the Model 25 at extra cost. It is 14.4 inches wide, 16 inches high, and 16.4 inches deep, making it large enough to carry the machine, disks, manuals, and cables. The case itself is 3.8 pounds and the machine is 28 pounds for the lighter version, making it too heavy to carry around regularly. This option may be impractical for students—a major target market for the machine.

Third-party vendors are starting to become involved with the add-on potential for the Model 25, while making up for inadequacies in the Model 50. Rodime, Inc. in Pepper Pike, Ohio is offering a package that replaces the 80 ms, 20 MB hard disk provided with the Model 50 with a larger and faster 28 ms, 45.5 MB disk. The original Model 50 hard disk can then be installed in the second disk drive position of the Model 25 and linked to a hard disk controller that is installed in one of the two expansion slots.

4

Personal System/2 Family Architecture

The integral nature of the Personal System/2 family lends itself to a discussion of the architecture's common system features, separate from the description of the individual machines. The features common to Models 50 and up only are considered in this chapter, because Models 30 and 25 do not have the Micro Channel architecture.

Each machine contains a system board that provides most of the functionality for the machine. The only necessary feature that is on a separate expansion board is the hard disk controller. As the type of controller varies with the size of the hard disk and with the different models in the family, this design is logical.

All of the systems consist of a system unit and keyboard that will work with any of the four Personal System/2 analog monitors. The system board in each model has many more standard items than the PC-type machines. Each has a keyboard port, a serial port, a parallel port, and a pointing device port. The video system is also an integral part of the system board. On the Model 30 this is the Multi-Color Graphics Array (MCGA), which is described in detail in Chapter 2. On Models 50, 60, and 80 the Video Graphics Array (VGA) is the standard display interface.

Each system board has an expansion bus with slots that can be used for system expansion. Models 50, 60, and 80 incorporate the Micro Channel architecture, creating a new specification for IBM and third-party add-in boards.

The discussion about common system features is split into several sections. System board features, such as the ports and interrupts, are discussed first, followed by the Micro Channel architecture and the BIOS. Chapter 5 discusses the video subsystem. It includes issues such as compatibility and application guidelines.

Chapter 4

The System Board

Figure 4-1 shows a general block diagram for the system board. The microprocessor (80286 for Models 50 and 60, 80386 for the Model 80) is driven by a crystal and its associated clock. The BIOS ROM, an eight-channel DMA controller, and a math coprocessor (80287 for Models 50 and 60, 80387 for the Model 80) are directly attached to the microprocessor. The address, data, and other control lines for the rest of the system are buffered onto the Micro Channel bus. The memory control circuitry can link the local memory either directly to the microprocessor or to the Micro Channel. The disk controller, video graphics control, serial port and parallel port, sixteen-level interrupt system, three-channel timer, keyboard and pointing device controller, and the clock/calendar and CMOS RAM are buffered onto the Micro Channel, rounding out the system.

Many devices used on the system board have a high level of VLSI (very large scale integration). Besides the VGA, several other elements have been made into custom devices. In addition, surface mount devices have been used extensively in the units.

The major components of the system board are discussed in more detail in the following pages. These main components are the microprocessor, the DMA controller, the math coprocessor, the interrupt system, and the system timers. The I/O controllers—audio subsystem, keyboard and auxiliary device, diskette drive, and serial port and parallel port—are also described. The Micro Channel itself, the major new development for these machines, is then considered, along with the BIOS for the machines.

The System Microprocessor

Models 50 and 60 use the Intel 80286 microprocessor as their CPU. There are two modes of operation for this IC—real and protected mode. The real-address mode is the standard mode of operation for all DOS-based applications. In this mode the memory capabilities are similar to those of the 8086 and 8088. DOS-based applications that are designed to run on the 8088- or 8086-based machines can run without modification in this mode. When running DOS as the operating system, the machine has the same memory restrictions as PCs and Models 30 and 25. Up to 1 MB of linear address space is available to the application. The 80286 has a 16-bit data path, as on the 8086, and in Models 50 and 60 the microprocessor is run at 10 MHz.

The 80286 has a different mode of operation, called the protected mode, that is exploited by the OS/2 operating system. This mode of operation allows up to 16 MB of physical address space and greatly increases the size of applications that can be run on the 80286-based AT and on Models 50 and 60. As applications become available for OS/2 and the industry begins to accept it as a suitable alternative to the DOS environment, the architecture of machines will become even

PS/2 Family Architecture

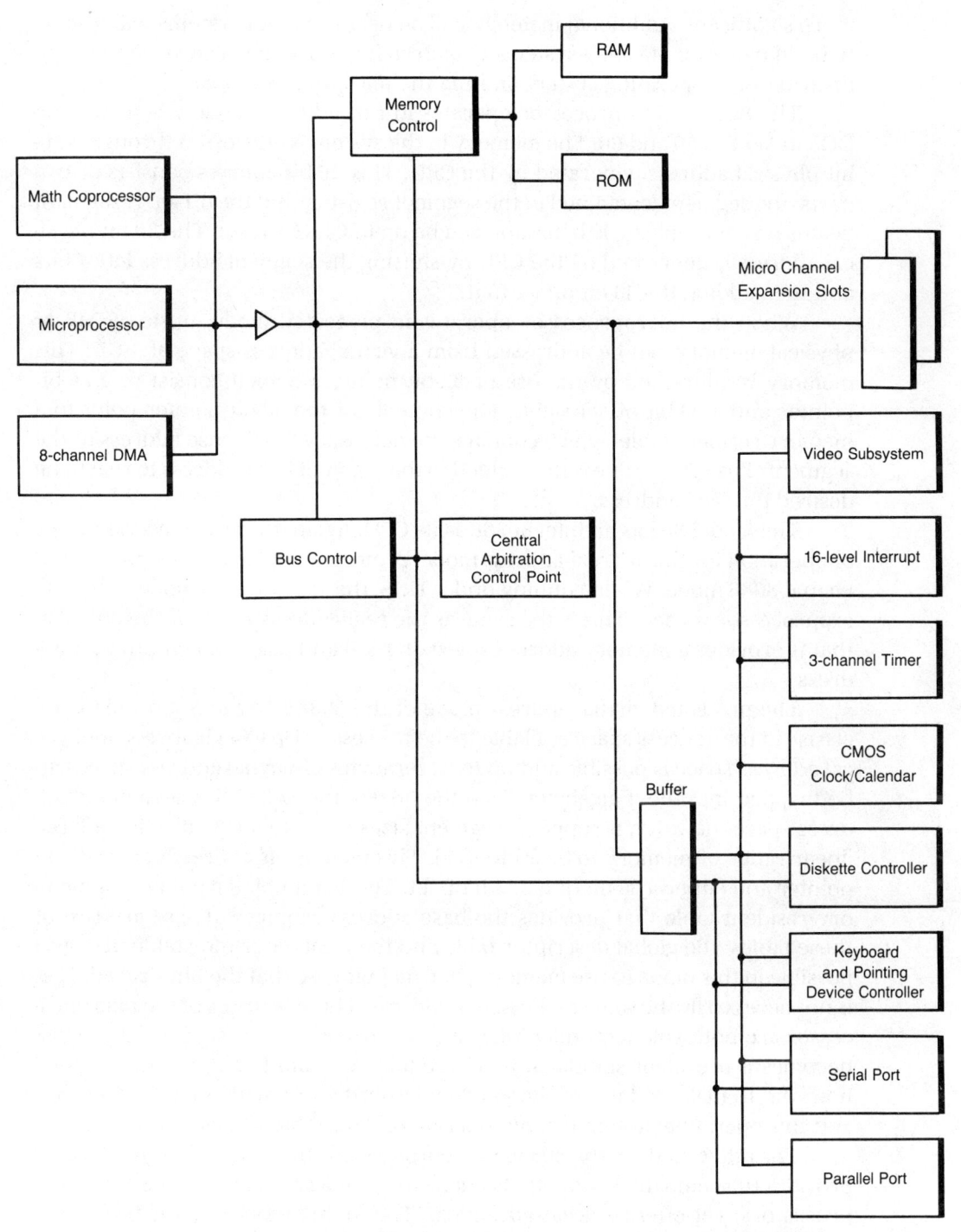

Figure 4-1 General System Block Diagram. *The basic architecture of the Micro Channel based Personal System/2 machines is similar to the PC family, but the physical implementation is different.*

more significant. Additional memory will be necessary, but systems will be able to build true multi-tasking systems that stretch the machine's hardware design, instead of the operating system limiting the machine's potential.

The 80286 microprocessor operates in real-address mode when running DOS in Models 50 and 60. The memory in the system is addressed through a 20-bit physical address generated by the CPU. This 20-bit address consists of two parts: the segment (contained in the segment register) and the offset. A segment begins on a multiple of 16 bytes and can be up to 64 KB in size. The 20-bit physical address is generated by the CPU by shifting the segment address left 4 bits and then adding the 16-bit offset to it.

When the microprocessor operates in protected mode, up to 16 MB of physical memory can be addressed from a virtual address space of 1 GB. This memory is addressed by the use of 32-bit pointers, which consist of a 16-bit pointer and a 16-bit offset value. The contents of the 16-bit pointer point to a memory resident table, which contains the necessary 24-bit base address of the segment. The offset value is then added to the segment base address to reach the desired physical address.

The Model 80 has an Intel 80386 as its CPU. Again, there are several modes of operation for this IC: real-address mode, protected virtual-address mode, and virtual 8086 mode. When running under DOS, the real-address mode of the microprocessor is used. This is the same as the real-address mode of the 80286 in that it provides a memory address space of 1 MB and uses a 20-bit physical address.

The protected virtual-address mode of the 80386 provides a dramatic increase in the address space available to the processor. Up to 4 gigabytes of physical address space is possible and up to 64 terabytes of virtual address space can be mapped into the 4 gigabytes. Like the 80286, the 80386 has segmented address space; however, it supports segment sizes of up to 4 GB, allowing a large linear range of memory to be addressed. This memory is addressed by a 16-bit pointer and either a 16-bit or a 32-bit offset. The 16-bit pointer points to a memory resident table that provides the base address required. There are two of these tables: the global descriptor table and the local descriptor table. It is also possible in this mode to use memory page mapping, so that the physical address is not necessarily the same as the linear address. These features of the microprocessor are not exploited under DOS and will become more significant as OS/2 becomes a prevalent standard. It will require yet another operating system, however, to make full use of the 80386 microprocessor with all of its features, and this operating system will not operate on 80286-based machines.

The other mode of the 80386 microprocessor is the virtual 8086 mode. This provides full compatibility for applications that are written for the 8086 family of processors, yet offers a great advantage. The multi-tasking capabilities of the 80386 can be exploited, because several 8086-based applications can be run at the same time. Products such as Desqview/386 and Windows/386 are available to provide current DOS users with these capabilities. Further information on the use of these and other 386 control programs can be found in Chapter 13.

Cycle Times

The clock cycle for the 10 MHz Models 50 and 60 is 100 ns (nanoseconds); for the 16 MHz versions of the Model 80, it is 62.5 ns. On the 20 MHz versions of the Model 80, the clock cycle time is 50 ns. On each machine except the 16 MHz versions of the Model 80, the refresh controller runs at 10 MHz, giving a refresh clock cycle time of 100 ns. On the 16 MHz version of the Model 80, the refresh controller runs at half the main CPU speed, at 8 MHz, resulting in a refresh cycle time of 125 ns. The refresh overhead on Models 50 and 60 is approximately 7 percent and on the other Personal System/2 machines is less than 5 percent.

On Models 50 and 60, three clock cycles are required to read and write to system board memory; however, suitably designed expansion memory boards can be accessed in as little as two cycles. This minimum is increased to three cycles when data is being transferred between I/O devices and memory. There is a separate read and write cycle for each transfer operation. DMA transfers can be done with a cycle time of 200 ns, because the DMA controller runs at 10 MHz.

On the 16 MHz versions of the Model 80, three clock cycles are required to read and write to system board memory, giving a cycle time of 187.5 ns. For operations where the DMA is in control, the cycle time is 375 ns; for other operations a minimum of 300 ns is required. For I/O operations a total of six wait states is inserted, making a cycle time of 500 ns. The 20 MHz versions of the Model 80 require either zero, one, or two wait states, depending on the type of function to be executed. The default cycle of the Micro Channel has a minimum cycle time of 200 ns. If the 80386 controls the cycle, the cycle time is 100 ns minimum and 300 ns maximum, depending on whether the memory being accessed is on the system board or the Micro Channel. The DMA cycle is 300 ns and other cycle times are a minimum of 300 ns.

Memory Subsystem

Models 50, 60, and 80 have three areas of memory: RAM, ROM, and the RAM associated with the real-time clock. The ROM contains the system power-on self test (POST) code, the compatibility BIOS (CBIOS), advanced BIOS (ABIOS), and BASIC. The CBIOS is compatible with the BIOS used on the PC family of computers; the ABIOS is designed for use with multi-tasking operating systems such as OS/2. The RAM contains the operating system and application software, along with their associated data. The real-time clock has RAM associated with it that is used to store not only date and time for the system, but the setup and configuration information as well.

Chapter 4

Models 50, 60, and 80 do not have the same BIOS configuration as the PC family of machines. Rather than the video BIOS being present on the video adapter, it is present in the system board ROM. The ROM is mapped to addresses 0E0000H and FE0000H on Models 50 and 60 and to addresses 000E0000H and FFFE0000H on the Model 80. These addresses are the top of the first 1 MB of address space and the top of the uppermost 1 MB of address space that the respective machines can address. The ROM in Models 50 and 60 is configured so that it is 16 bits wide, using four 32K by 8 bit modules that occupy 128 KB of memory. In the Model 80, ROM modules of a similar size are used, but they are arranged to be 32 bits wide.

On the 20 MHz versions of the Model 80 (8580-111 and 8580-311), the ROM is only active during POST. After that time, the ROM is copied into RAM and the ROM is then disabled. The addresses are the same whether the ROM or RAM is enabled. On these machines, the ROM runs with four 50 ns wait states when enabled. Once copied to system board RAM, the ROM information is referenced at zero wait states for consecutive references within the same page, or two wait states otherwise. The ROM is enabled and disabled via bit 1, the ROMEN bit, of the memory encoding register at address 0E1H.

Models 50 and 60 have 1 MB of RAM on the system board. On the Model 50 this is arranged as two 512K by 9 bit memory modules, and on the Model 60 there are four 256K by 9 bit memory modules. Although these are referred to as system board memory, because they are on the local bus for the CPU, the memory ICs are not installed directly on the system board as they were in the PC family of machines. The memory ICs are mounted on small circuit boards that plug into the system board. On the Model 80 the memory boards are also on the local system bus, but are larger boards that plug into the system board, measuring 4 inches by 5 inches. On the 16 MHz versions of the machine, up to 2 MB of system board memory can be accommodated on two boards, each containing 1 MB. The memory is 36 bits wide, each 8-bit byte having a parity bit associated with it. On the 20 MHz versions of the Model 80, the system board can contain up to 4 MB of memory on two boards, each containing 2 MB.

The 20 MHz machine can configure this memory in several different ways during Programmable Option Select (POS). If there is one memory board in the unit and a fault is discovered in the first 512 KB of memory, the system deactivates the first 1 MB of memory on the card and only uses the second 1 MB. This provides an additional level of fault tolerance that was not previously available. On the other machines, if a fault is detected, the machine does not pass POST.

The real-time clock contains 64 bytes of RAM that is battery backed. The first 14 bytes of this RAM is used to store the date and time for the system. Models 60 and 80 have an additional 2 KB of RAM that is supported by the same battery. This and the remainder of the RAM in the real-time clock is used to store the system configuration information that is generated in the POS procedure. The POS procedure is detailed later in this chapter.

DMA Controller

The direct memory access (DMA) controller that is present in Models 50, 60, and 80 is an eight-channel device. It is used to perform several functions that free the microprocessor itself for other tasks. The DMA also provides the refresh cycle for the memory in the system. This 625 ns cycle takes an overhead on the system of less than 7 percent.

The DMA controller that is in Models 50, 60, and 80 provides register and program compatibility with the 8237 DMA controller that is found in the AT class of machines. An individual 8237 is not used on the system board; its functions are incorporated in a VLSI component. This new IBM proprietary component has additional capabilities that were not available on the 8237.

The functions of the DMA controller include eight independent DMA channels for 24-bit memory addressing or 16-bit I/O addressing. The DMA controller is used to transfer data from memory to an I/O device or from an I/O device to memory. A DMA transfer involves reading information from the required location, latching it into the DMA controller, and then writing it out to the destination device.

There are forty-one DMA registers in the DMA controller. These registers fall into ten different categories. There is a 24-bit *memory address* register, a 16-bit *I/O address* register, a 16-bit *transfer count* register, and an 8-bit *mode* register for each of the eight channels. The memory address register, which is loaded by the CPU, is incremented or decremented by the mode register. The I/O address register is also loaded by the CPU; its contents do not change during a DMA transfer. The transfer count register is used by the CPU to specify the number of DMA transfers to be executed. The mode register determines the mode of operation of the DMA channel.

One 16-bit *temporary holding* register and one 8-bit *function* register serve all of the channels. The temporary holding register cannot be accessed by the CPU. It is necessary for serial DMA transfers that need to hold data during the transfer. The function register provides extended programming functions as well as 8237-compatible functions.

There are two 4-bit *mask* registers: one serves channels 0 through 3 and the other serves channels 4 through 7. This register allows any channel to be disabled or enabled by the CPU. Two 8-bit *status* registers are split in the same way: one serves channels 0 through 3 and the other serves channels 4 through 7. These registers give information on the current status of the channel. Two 4-bit *ARBUS* registers are used for virtual DMA operations: one for channel 0 and the other for channel 4. The arbitration level of channel 0 and channel 4 can be changed by using extended commands of the function register. The final register is the *refresh* register, which is independent of the DMA.

All DMA commands are 8-bit I/O instructions similar to those on the AT, but there are also additional functions that extend the command set. The eleven extended DMA commands expand the capabilities of the DMA, giving additional control over many of the registers.

Math Coprocessor

The math coprocessor for Models 50 and 60 is the Intel 80287 chip. This component was used as the coprocessor for the AT. It is used by applications that perform a significant number of mathematical operations. This device is commonly used in high resolution graphics when performing rapid recalculation for displaying shapes—for example, the regeneration of drawing views in a CAD application.

The presence of a math coprocessor does not necessarily mean that an application will use it. But when applications use the math coprocessor instructions, performance is several times greater than when the general purpose 80286 is used as the arithmetic calculator.

The speed of the 80287 is 10 MHz, the same as the 80286 in Models 50 and 60. It operates in two modes—real-address and protected mode—in the same way as the 80286 itself. Most applications that can make use of the 8087 can use the 80287 in real-address mode without modification, but there are a few minor exceptions. These arise from the difference in the way the two components handle numeric exceptions.

The 80287 can only execute one numeric instruction at a time, just as an 8087 can. On an 8087, however, it is customary to place a WAIT instruction (indicating that the 80286 processor should wait for the busy line to become inactive) before an ESC (where the 80286 processor escapes to the 80287 extension processor), to ensure that the CPU and the math coprocessor are synchronized and the 8087 is not busy. While this instruction sequence still operates correctly on the 80287, the WAIT instruction is not necessary, because the 80286 monitors the BUSY line until the 80287 indicates it is not busy, then starts the 80287 command.

A WAIT or ESC instruction is necessary after a command to the 80287 that stores a number to memory or loads a number from memory, to ensure that the 80286 does not look at or change the number before it has been read by the coprocessor. This does not apply to four instructions—FSTSW, FSTCW, FLDENV, and FRSTOR—which do not need this WAIT instruction.

Another difference between the 8087 and 80287 is in instruction and data pointers. The 80287 points at prefixes to the saved floating point instruction address instead of just the ESC instruction opcode, as on the 8087.

On PCs and Personal System/2 models that use the 8087, the INT signal goes through an interrupt controller, but with the 80287 the NPX error signal does not. If the starting address of a numeric operand is outside the size of a segment, interrupt 13 occurs. If a second or following word of a floating point operand is outside a segment's size, interrupt 9 results. Interrupt handlers originally designed for use with the PC need two additional error handling routines for these conditions. Interrupt vector 16 should point to the numeric error handler.

Operating systems that make use of the protected mode of the 80286 use the protected mode of the 80287. This mode is entered by executing the SETPM ESC instruction. The protected mode is exited and the real-address mode is reentered by a reset. When the processor is powered up, it initializes in the real-address mode. When the processor is in protected mode, it serves as a protected extension of the 80286 microprocessor. This means that the 80286 memory management and protection rules are followed when status or numeric data is required from memory. When in protected mode, the instruction opcode is not saved and if required should be read from memory. This is not the case with the 8087.

Some of the 80386-based machines that include extensions of the AT bus can use an 80287 math coprocessor, rather than an 80387; however, this does not provide the same performance increase as the 80387. Most manufacturers of these machines have since upgraded their architecture to include an 80387. The 80387 was not readily available at the time the first machines were released, but as its availability has increased, its use has become more common.

IBM announced its 80386-based Model 80 machine with an optional 80387 coprocessor, with the coprocessor available in the third quarter of 1987. The Model 80 uses an 80387 math coprocessor instead of the 80287 used in Models 50 and 60.

The 80387 does not have different modes of operation. The real, protected, and virtual 8086 modes of the 80386 can all be used with the 80387, but the math coprocessor itself does not change modes. Memory accessing is handled by the 80386, not by the coprocessor. If the 80386 is operating in real-address mode, the coprocessor is completely compatible with code that has been written for 8087-based or 80287-based systems. If the 80386 is operating in protected mode, the coprocessor is completely compatible with code that has been written for the 80287 protected mode, whether it was written when the 80287 was operating with an 80286 or an 80386 CPU.

For programs that have been written for the 8087 and are ported to the protected mode 80387 environment, there are four instructions that differ in implementation: FLDENV, FSTENV, FRSTOR, and FSAVE. These instructions have a different operand format. This is not generally a problem for application software, because these instructions are typically used by exception handlers and operating systems.

Three parts of the 80387 operate in parallel, so the next instruction can be loaded while the previous numeric calculation is being performed, resulting in dramatic increases in performance. The additional instructions that are available on the 80387 but not on the 8087 or 80287 include unordered compare instructions, a partial remainder command, and three new transcendental function instructions: cosine, sine, and sine and cosine. It is important to establish the presence of a coprocessor before coprocessor instructions are used at random. This is done with the FINIT, FNINIT, FSTCW, or FSTCW instruction. 8087 and 80287 coprocessor opcodes can be used if EM is set to 1 in the 80386 CR0 control register. If this is not done and another coprocessor instruction is executed with no math coprocessor present, the machine will wait forever for the completion of the instruction and will require rebooting.

System Interrupts

The system interrupts for Models 50, 60, and 80 are supplied by two Intel 8259A interrupt controllers, which each supply eight levels of interrupts. The two chips are cascaded so that IRQ2 of the first 8259A provides access to the second. IBM has reserved IRQ5 on the first IC, and IRQ10, IRQ11, and IRQ15 on the second. In addition to the interrupt levels IRQ0 through IRQ15, there is a nonmaskable interrupt (NMI) which creates the interrupts with the highest priority for the system. These include the parity check interrupt, the channel check, the watchdog timer, and the arbitration timeout or system channel timeout. Despite its name, parts of the NMI functions can be masked, but the watchdog timer and the system channel timeout cannot. The priorities for the interrupt levels are shown in Table 4-1; NMI has the highest priority, then IRQ0 and so forth, on down to IRQ7, which has the lowest priority.

Table 4-1 Interrupt Level Priorities. *The Micro Channel architecture uses level-triggered interrupts instead of edge-triggered interrupts.*

Priority Level (Highest Level First)	Description
NMI	Nonmaskable interrupts: parity, watchdog timer, arbitration timeout, and channel check
IRQ0	System timer interrupt (occurs 18.2 times per second)
IRQ1	Keyboard interrupt
IRQ2	Cascade interrupt (provides access to interrupts IRQ8 to IRQ14)
IRQ8	Real-time clock interrupt
IRQ9	Redirect cascade (allows system to access IRQ2 interrupt handler for compatibility reasons)
IRQ12	Mouse interrupt
IRQ13	Math coprocessor extension interrupt
IRQ14	Hard disk interrupt
IRQ3	Secondary serial port interrupt
IRQ4	Primary serial port interrupt
IRQ6	Diskette drive interrupt
IRQ7	Parallel port interrupt

IRQ5, IRQ10, IRQ11, and IRQ15 are reserved by IBM.

The 8259A interrupt controllers can work in either edge-triggered or level-sensitive mode. On Models 50, 60, and 80, the support circuitry for these controllers on the system board prevents the edge-triggered mode from being entered, and the ICs are always in level-sensitive mode. The interrupts perform as expected on the Personal System/2 machines, with the exception of IRQ9—the redirect cascade interrupt. In another example of IBM's attention to detail, compatibility with the PC family is maintained via this interrupt. For devices that expect to access IRQ2, the request is remapped by the system to IRQ9. When the interrupt occurs, it is handled by IRQ9, terminated, and then passed back to IRQ2, resulting in a transparent redirection and return to the originating device.

System Timers

There are three timers on the system board. Five I/O port addresses are associated with the timers. Two of these timers—channel 0 and channel 2—perform a function similar to that of their counterparts on the AT and PC. Channel 0 is the system timer and channel 2 is the tone generator for the speaker. Two of the ports are the control registers for the timers and the other three are the count registers.

The system timer, often referred to as the system tick, occurs 18.2 times per second. For DOS, the BIOS interrupt 08H services the timer interrupt that is generated by channel 0. The channel 2 timer allows the speaker to be activated in the same way as on the PC and AT, but other facilities are also available for creating an analog sound on the speaker, instead of the digital version previously supplied. The output from channel 2 to the speaker is summed at the audio sum node point with any analog signals that are generated. The audio subsystem is described below. Channel 3 is an additional timer that was not available on the PC and AT. It is a binary 8-bit down counter that has a preset facility.

The control register for channels 0 and 2 is at address 43H and for channel 3 is at address 47H. The initial count is loaded into the count registers and read from the same register. In order to write to the counter, the control word is first written, followed by the initial count with the specified format. To read the counters, the counter latch command is used; the count can then be read from the count registers.

Audio subsystem

As on the PC and AT, channel 2 of the system timer provides a means of sounding the speaker. This entails sending a digital pulse stream at a particular rate to the speaker. Models 50, 60, and 80 not only support this method of producing sound, but they also offer an analog audio system. The audio signal and audio ground

Chapter 4

signals are available on the Micro Channel, enabling expansion boards to make use of this capability. The audio drivers should have a source impedance of 1.2 KOhm. Each audio receiver across the lines should have an input impedance of 7.5 KOhm. There is, however, no volume control on the circuitry; each driver is expected to supply this. The speaker in the system is driven by the total of all of the drivers on the Micro Channel and the output from timer channel 2.

Keyboard

Each of the models in the Personal System/2 family uses the IBM Enhanced Keyboard shown in Figure 4-2 as its standard, except the Model 25, which offers it as an option. This is similar to the 101-key Enhanced Keyboard that was offered on later model XTs and ATs, but the connector that attaches to the system board is different in shape. It is a 6-pin miniature DIN connector. Pins 2 and 6 are reserved on the connector. Pin 4 provides +5 V_{DC} from the system board and pin 3 is ground. Pins 5 and 1 are both input and output pins and are used for the clock and data signals, respectively. The keyboard has an 8042 interface, which is the same as on the AT Enhanced Keyboard. In addition, the scan codes are unchanged.

Figure 4-2 IBM 101-Key Enhanced Keyboard. *The Enhanced Keyboard has a separate numeric keypad and cursor control keys to the right of the main key area. The twelve function keys are located along the top of the keyboard.* Photo by Bill Schilling.

In the system unit itself, the keyboard I/O controller is the 8042. This not only provides the I/O for the keyboard, but also supports an auxiliary device interface on the other 6-pin miniature DIN connector on the rear panel. These connectors are mechanically interchangeable but, unlike the Model 30, they are not functionally interchangeable. The keyboard and mouse will not operate unless they are plugged into the correct connector.

The second half of the 8042 on the system board can be used to interface the system to several different serial devices—for example, a mouse, a digitizer, or even an additional keyboard. The serial data is received and translated. It is then placed as a byte in I/O address 60H. There are two modes of operation for this controller: interrupt or polling. In interrupt mode, the 8042 signals the microprocessor when data is available; in polling mode, the 8042 waits until the microprocessor polls for the information.

The status byte of the 8042 is at address 64H and can be read at any time. The status that can be read from this byte includes parity error, input, output, auxiliary buffers full, or a general timeout. When written to, this same byte can be used to enable and disable the keyboard or auxiliary device as required, translate the keyboard scan codes to the PC scan code set or pass them on without modification, set or reset the system flag bit for the status register, or set the 8042 into interrupt mode. Any interrupt generated by the 8042 is latched externally to the chip until it is serviced by the CPU.

The I/O byte at address 60H is a read only buffer that contains the data received from either the keyboard or the auxiliary device. When the status byte indicates that the output buffer is full, the I/O byte is valid. If the auxiliary device buffer is also indicated full in the status byte, the data is from the auxiliary device and not the keyboard interface. Conversely, the I/O byte and the status register should be written to only when the buffer full indicators are not set.

The keyboard has a password security feature via the 8042 controller. A password can be selected; once enabled, it prevents the 8042 from transmitting the key data to the system unit until the appropriate key sequence is detected. There are three commands associated with this feature: *test installed password* (A4H), *load new password* (A5H), and *enable password* (A6H). *Test installed password* checks if a password is already installed, but does not provide the value for that password. The current password can, however, be overwritten by using the *load new password* command. The format of the password must be sent to the keyboard in the form of scan codes with a null value indicating the end of the password. The commands for the 8042 are written through I/O address 64H.

Diskette Drive Controller

The diskette controller for the 3.5-inch diskette drives is an integral part of the system board on Models 50, 60, and 80. The specifications for the diskette controller are the same for each of the machines, but the methods of connecting the drive to the controller vary with the machine. On the Model 50, there is a card edge connector on the system board that has an adapter board installed in a vertical orientation. This board has the two diskette drive connectors on it. The diskette drives are slid into the top half of the chassis and mate with these connectors without any cable assembly. On Models 60 and 80, a ribbon cable connects the output of the diskette controller to the 3.5-inch drives. On Models 50 and 60, the diskette controller is in a DIL-packaged IC, but on the Model 80 it is a surface-mounted device.

Any new software that is written for the Personal System/2 family of machines should address the diskette drives via the BIOS routines, to enable compatibility with future machines possibly running under different operating systems. This may not be the situation for existing software. The diskette drive controller can support up to two drives and each drive can support two different densities for the magnetic media. The high density mode allows 1.44 MB of data to be stored on a 3.5-inch diskette. The lower density mode is compatible with the Model 30 and the IBM AT 3.5-inch diskette drives. These allow a diskette to be formatted to contain 720 KB of data. The disk drive on Models 50, 60, and 80 is able to read and write to both of these media. When reading or writing in high density mode, the clock rate is 8 MHz; a 4 MHz clock signal is used for the lower density mode. The diskette drive controller has a functional interface between the controller and the diskette drive that is similar to the 5.25-inch diskette drive interface on the PC family of machines.

Parallel Port Controller

The parallel port on Models 50, 60, and 80 is fully compatible with the parallel port on the PC family of machines. However, it does have an additional mode that is not available on the previous machines. This is a bidirectional mode that allows the machine to accept data input through the parallel port. The initial application of this mode for most new users is in association with the Data Migration Facility that allows data to be transferred from existing PCs to the Personal System/2 machines. The Data Migration Facility and its strengths and weaknesses are discussed in Chapter 14.

In order to provide compatibility with the Personal System/2 architecture, the parallel port controller hardware cannot be identical to the PC parallel port, despite its software compatibility. The new controller supports level-sensitive interrupts instead of edge-triggered interrupts, in order to interface with the CPU. The parallel port can be configured as LPT1, LPT2, or LPT3 by selection during the POS setup procedure. This method does not require the adjustment of switches, and conforms with the switchless implementation suggested for all Micro Channel adapters. Selection of LPT1 results in a data address of 03BCH, a status address of 03BDH, and a parallel control address of 03BEH. These addresses are changed to 0378H, 0379H, and 037AH for LPT2 and to 0278H, 0279H, and 027AH for LPT3.

The extended mode of the parallel port is selected during POS. When POS is run, the POS mode bit for the port should be set to a 1 for compatibility mode and a 0 for bidirectional facilities. If the default compatibility mode is chosen, the port behaves in the same manner as the PC. In extended mode, the *write only parallel control direction* bit, when set, signifies the read mode; the port is in write mode if this bit is reset. When the system is reset, by Ctrl-Alt-Del or by power on, POST sets the port into write mode.

The parallel port is an 8-bit-wide device. When operating in compatible mode, data is sent to the output of the port when a write command is executed. However, when in extended mode, a write operation only sends this data to the port if the direction control bit is reset, signifying the write mode. If an application tries to read data from the port and it is set in compatible mode or extended mode with the direction control bit being reset, the data that was last written to the port is presented. The data from an external device attached to the port is available only if the read mode of the extended mode is selected.

The parallel port's read mode allows another computer to send output to a Personal System/2 via the parallel port; the Personal System/2 cannot, however, simultaneously acknowledge receipt of data because it cannot write to the parallel port's read only status port. The status port's signals are the same as on the PC parallel port and include such signals as −BUSY and −ERROR. The parallel control port can be read from and written to except for bit 5, the direction control bit, which is write only. When in extended mode, the 6 least significant bits are taken from the bus during a write command and the 5 least significant bits are presented to the bus during a read command. The situation is the same for the compatibility mode except that the most significant of these bits has no effect.

Serial Port Controller

The serial port controller used on Models 50 and 60 and the 16 MHz Model 80 performs in the same way as the serial port controller on the AT. However, instead of being implemented in hardware with an NS 16450 IC, as on the AT, the more powerful NS 16550 component is used. The FIFO mode that is available on the NS 16550 chip is not implemented, and for all programming purposes, the controller should be considered an NS 16450.

On the Model 80, the NS 16550A is used on the 20 MHz versions. This IC does allow the use of the FIFO mode. The test to determine if a machine has an NS 16550A and supports this mode involves enabling the FIFO mode on the chip and reading bits 6 and 7 of the interrupt identification register at address 0nFAH (where n is 3 for COM1 or 2 for COM2). If bit 6 is a 0 and bit 7 is a 1, the FIFO mode is not supported, but if both bits are set, the IC is an NS 16550A or equivalent chip and the FIFO mode is available.

The bit definitions for three of the registers in the NS16450 register set have been augmented, because of the addition of the FIFO mode. Bit 0 of the interrupt enable register performs the additional function of indicating timeout interrupts when in FIFO mode. Priority 2, a timeout interrupt, is indicated by the setting of bit 3 and bit 2. The reserved bit 7 of the line status register indicates an error in the receiver FIFO register. In addition, the other bits of this register have meanings in FIFO mode that are comparable to their functions, as previously defined for this register. These changes in definitions should not affect previously written software and are only significant if the FIFO mode is to be implemented in future programs.

This controller supports asynchronous communications at transmission rates of between 50 bps and 19,200 bps. Characters can be 5, 6, 7, or 8 bits in length and have 1, 1.5, or 2 stop bits. The start, stop, and parity bits are added to the outgoing data automatically as well as removed from incoming data.

Six signals that can be interpreted by a modem are supported. These are the clear to send and request to send signals, the data set ready and data terminal ready signals, and the ring indicator and data carrier detect signals. The port can also generate and detect line break signals, and detect false start bits. When sending and receiving characters, the port controller uses two buffers. One collects one character and the other the following character. This enables one character to be read while the next is being received, so the precise synchronization needed when a character can only be received after the previous character has been read is not required.

The serial port can be configured as COM1 or COM2, and uses IRQ4 or IRQ3 with each configuration respectively. Bit 3 of the modem control register enables and disables the interrupt feature of this port. The speed of the port is programmed via the programmable baud rate generator. Any value between 1 and 65,535 can be programmed. This value is used to divide the clock input, which runs at 1.8432 MHz, to give the required output rate. This rate is sixteen times the required rate in bits per second. For example, to transmit and receive signals at 50 bps, the baud rate generator should be programmed with a divisor value of 2304. This sets the clock rate to 800 Hz and gives a transmission rate of one-sixteenth of this value. It is theoretically possible to generate a rate greater than 19,200 bps by using a divisor less than 6, but this is not supported by IBM.

Micro Channel Architecture

The decision to change the standard for the expansion boards has been the most talked about element of the Personal System/2 family. It is important to understand that in order for the personal computer to grow and accommodate the types of technological improvements that are and will be available in the next few years, the PC bus is not adequate.

That does not mean that the PC bus is dead, because it is not. The PC and AT are very powerful machines. The 80386-based machines that use an AT bus as the expansion bus will likewise be around for several years. The Micro Channel architecture, however, provides the means to grow for a longer term than the PC bus can offer. This brave decision from IBM is less a change in the bus standard than a discontinuance of the old PC standard.

The Micro Channel architecture can be considered from several different viewpoints: that of the hardware designer, the programmer, and the systems integrator. The physical characteristics of the expansion bus, the pin assignments and timings for the connector, and physical design constraints are of most significance to the hardware designer. The programmer is most interested in the various arbitration levels that are available, interrupt sharing, and the implication for the operating system of the programmable features. The systems integrator is concerned about the overall systems aspects and the interplay between various expansion boards and multiple master systems, as well as the impact of new operating systems and applications that can either provide compatibility with the PC family of machines or exploit the features of the newer machines.

The PC family had an expansion bus that was described in detail by IBM's documentation. The technical reference manuals for each of the machines included BIOS listings and schematics for their respective machines. As a new option or even a new machine was made available, the accompanying documentation was also obtainable. In many ways this had the effect of not only defining the standard, but encouraging the clone market. If, for instance, the PC is compared with the Apple computers, their performances are on a fairly equivalent level . . . but there is not a proliferation of Apple clones. The Apple system has a closed architecture: the information necessary to produce a clone machine is not available and the standard is not available to developers in the same way as the IBM standard. Consequently, although the machines are of a similar caliber, the IBM standard dominates the market.

The open nature of the standard for the PC and AT does have its drawbacks because, although it caused a tremendous general acceptance of the standard, it also caused a stretching of the standard. Any area not explicitly defined was exploited. Resources such as the interrupt levels were in great demand. Despite IBM recommendations for sharing interrupts on the AT, third-party vendors did not comply, because it was likely that other boards in a machine also did not conform; thus, the perception of the end user would be that the board that hogged the interrupt was the one that "worked" correctly, because it was the one that functioned.

In addition, interrupts that were marked "Reserved" by IBM were used by third-party vendors. Interrupt level 3 was reserved in the PC design and used in the AT. This left some third-party vendors with boards that worked in a PC but not in an AT.

The AT design expanded the bus architecture from an 8-bit implementation to an acceptable 16-bit version. However, with the consequent advent of the

80386, users were looking for a 32-bit implementation that would be faster, yet compatible with existing systems and adapters.

The PC bus does not appear to have been designed to accommodate 32-bit or 64-bit growth. Even if everyone had played by the rules precisely, the design is not adequate for expansion. The popularity of the PC was not expected and the resulting industrial revolution is still incredible to consider.

IBM chose to make many decisions when they designed a standard that would accommodate a 32-bit bus. The existing standards could not be ignored, because the sheer inertia of the system would obliterate any chance of success for a new system; yet the existing standard was not adequate. The physical constraints alone were a problem. Expanding the number of pins on a PC/AT expansion board to address 32 bits of data would result in a board that would require a hammer to insert it into its connector.

IBM chose to compromise. They needed to remain as compatible with the past as possible, yet provide expansion for the future. The IBM Personal System/2 family provides almost total compatibility with the past from an application software viewpoint, but requires new hardware expansion boards. The decision to set a new expansion board standard relieved the bus of processor speed limitations. Many PC and AT-style expansion boards do not work reliably with bus speeds over 8 or 10 MHz. With many boards already at their performance limit, it was reasonable to implement a new standard, since these boards needed to be redesigned anyway.

The Micro Channel can be described as the design IBM should have used the first time around. Even though there are no schematics in the documentation, there is adequate, well-defined information that enables boards to be built that fully meet IBM's specifications. There is also adequate information to clone the IBM Personal System/2 machines, but many aspects are patented and copyrighted. The question of the availability of clones is more a legal than a technical issue. IBM is welcoming third-party add-in vendors, but will defend its patents vigorously.

Micro Channel Bus Signals

The Micro Channel architecture is an expansion bus system in the same way as the PC bus. There are connectors on the system board that allow expansion boards to be plugged in and provide additional functionality for the system. (See Figure 4-3.) Each of the expansion boards is a standard size and shape, and all of the interface specifications for these boards have been defined by IBM. Figure 4-4 shows the expansion connector for an 80386 memory expansion board with the full 32-bit implementation of the expansion channel.

The signals that are available on the expansion bus can be divided into five categories. These are the address bus, the data bus, the transfer control bus, an

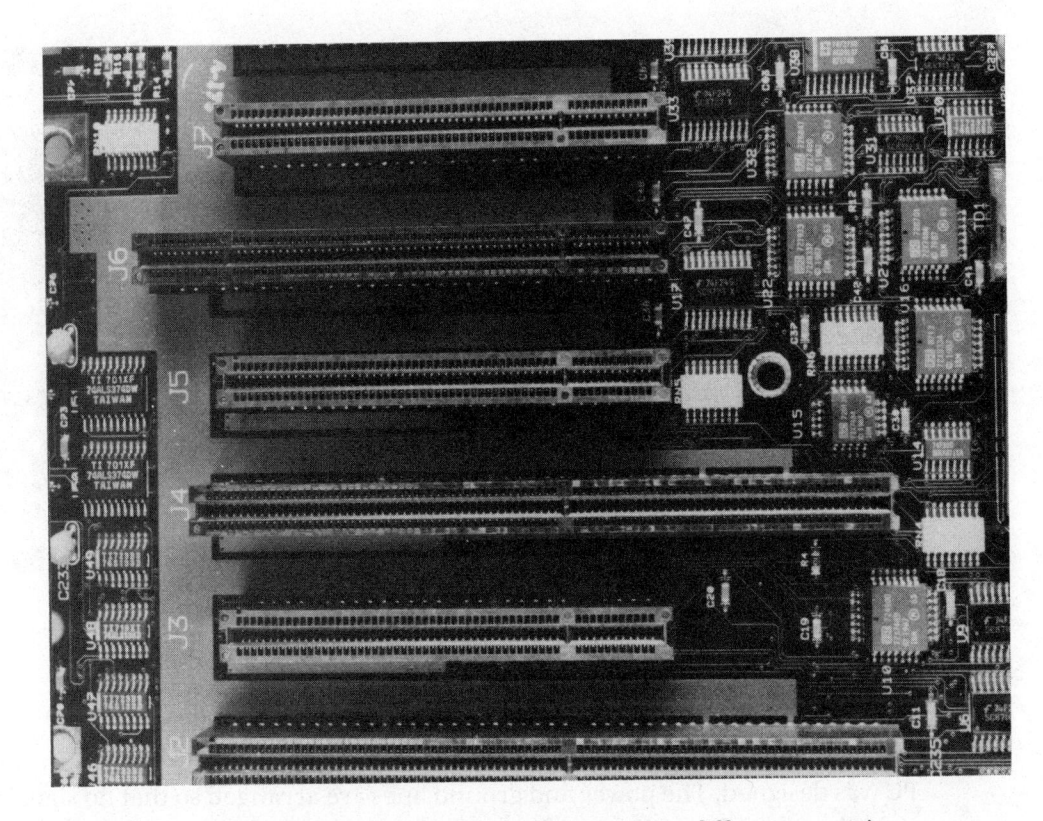

Figure 4-3 Micro Channel Connector. *There are three different expansion connectors for the Micro Channel: 16-bit, 32-bit with Matched Memory extension, or 16-bit with the video extension.*
Photo by Bill Schilling.

arbitration bus, and the control signals. Each of the connectors in the IBM Personal System/2 machines consists of several sections. The shortest is the basic 16-bit connector that has an 8-bit section and a 16-bit extension. In each machine, one of the 16-bit slots has an additional section—the video extension that supplies signals that can extend the basic video capabilities of the machine. On the Model 80, the 32-bit slots have an additional section for the 32-bit extension. Figure 4-5 shows the Micro Channel pin assignments.

The 16-bit connector is a 106-pin card edge connector with 0.05-inch spacing between each pin. There are two reasons for the choice of a 0.05-inch spacing: the connector requires a low insertion force and the spacing between the pin is the same pin spacing as for the surface mount ICs that are used throughout the system board, thus easing board layout.

At first glance, the number of pins in the connector may seem high, but there are many power and ground lines—29 total—to reduce electromagnetic interference (EMI) due to current loops. This has become significant, because

Chapter 4

Figure 4-4 32-bit Memory Adapter. *This 32-bit board includes pins for the 32-bit extension, as well as the Matched Memory extension.*
Photo by Bill Schilling.

FCC regulations regarding EMI are more strict than they were when the original PC was designed. The power and ground lines are arranged so that no signal pin is farther than 0.1 inch from a ground or power pin. Remember that to radio frequency (RF) interference, the DC power pin is effectively a ground. Four of the connector positions are taken up with a small key pin that splits the various sections and aids with location when expansion boards are inserted. The audio ground line is separate from the logic ground line, five positions are reserved, and there are 77 signal lines.

IBM is explicit about the reserved lines, stating "Warning: Any signals shown or described as "Reserved" are not to be driven or received. These signals are reserved to allow compatibility with future implementations of the channel interface. Serious compatibility problems, loss of data, or permanent damage can result to features or the system if these signals are misused."

Table 4-2 shows the sequence for signals to be present on the bus for a basic bus transfer cycle.

Channel Connector and 16-Bit Extension Signals

The signal descriptions below use the following conventions: a signal preceded by a minus sign (–) is active low; a signal without such a designation is active high.

A0 through A23   There are 24 address bits, with A0 being the least significant and A23 being the most significant. If an I/O address width of 16 bits is

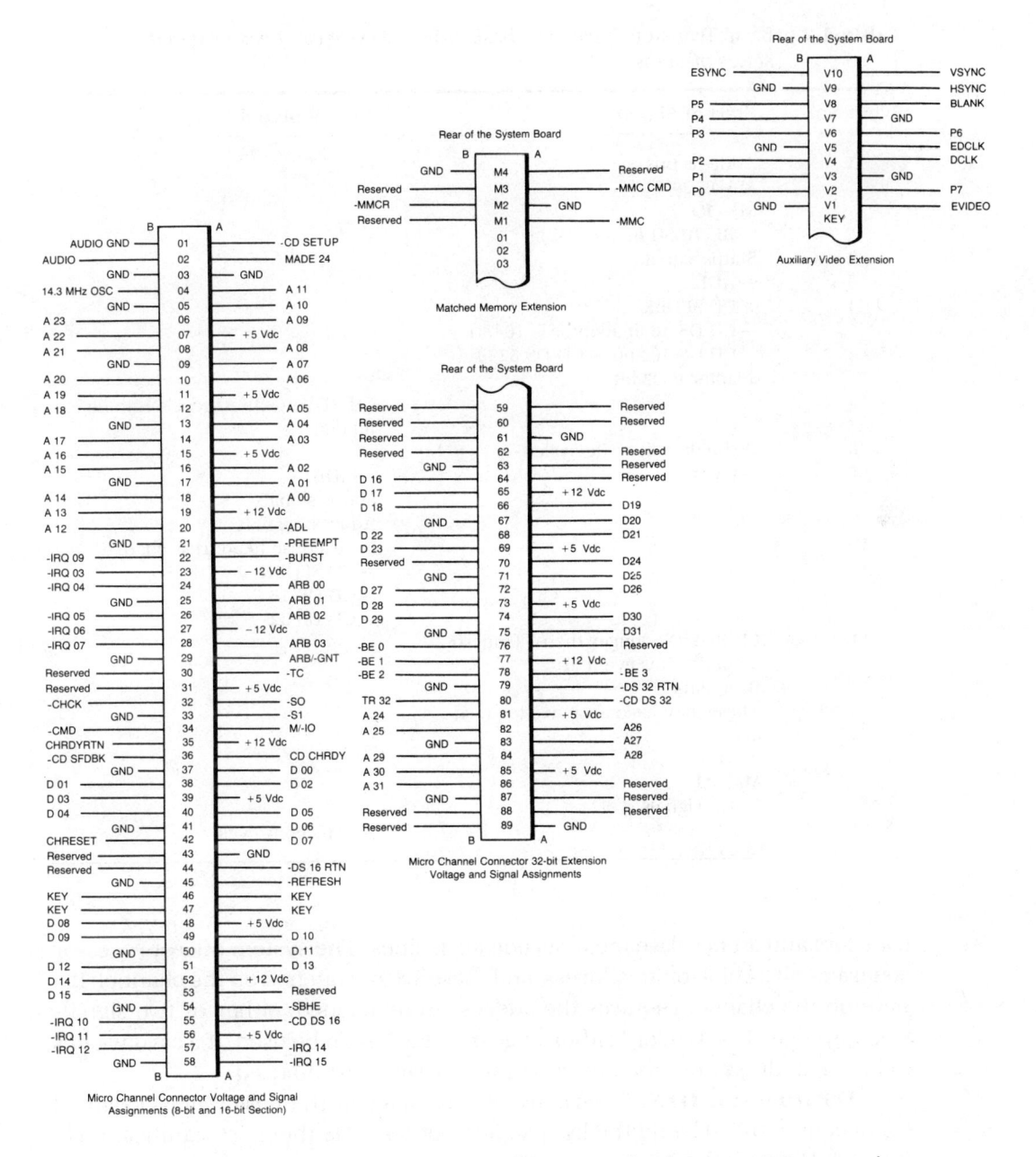

Figure 4-5 Micro Channel Pin Assignments. *The numerous DC power pins aid in the reduction of EMI.*
Figure courtesy of International Business Machines Corporation.

used, it is possible to perform an 8-bit or 16-bit data transfer within a 64 KB range. If the full 24 bits are used as the memory address, an 8-bit or 16-bit memory transfer can be done in a 16 MB range. The lower 16 bits are used for I/O

Table 4-2 Basic Transfer Cycle. *The basic transfer cycle can be considered a series of steps.*

Event No.	Start of Signal	End of Signal
1	Address bus MADE 24 M/ – IO – REFRESH (if applicable)	
2	Status signals	
3	– ADL	
4	– CD SFDBK – CD DS 16 (if adapter is 16-bit) – CD DS 16 and – CD DS 32 (if adapter is 32-bit)	
5		CH RDY inactive if cycle is to be extended
6	Write data (if write cycle)	
7	– CMD	
		– ADL
8		Status signals
9		Address signals
10		In response to address change – CD SFDBK – CD DS 16 – CD DS 32
11	CH RDY (No longer than 3.0 microseconds after event 10)	
12	Read data	
13	These may become active for next cycle: Address M/ – IO Status signals	
14		– CMD ends cycle

transfers and an I/O slave must decode all 16 lines. The system microprocessor generates all 24 bits for an address, and these are not latched on the channel. If a slave on the channel requires the address to be latched, either of two signals (trailing edge of – ADL or leading edge of – CMD) can be used. If a card wishes to put an address onto the bus, a tri-state driver must be used.

D0 through D15 There are 16 data lines in the basic Micro Channel connector, with D0 being the least significant and D15 the most significant. D0 through D7 form the low byte for the system microprocessor and D8 through D15 make up the high byte. If an 8-bit slave is installed on the Micro Channel, communication with the system microprocessor must be through the low order data bits. The high byte pins are located in the 16-bit extension section of the connector. The data on the data pins is valid at different times, depending on whether the cycle is a write cycle or a read cycle. During a write cycle, data is valid whenever – CMD is active; during a read cycle, data must become valid after the leading edge of −CMD but before the trailing edge of −CMD, and must remain valid until after this trailing edge. As with the address lines, D0 through D15 should be driven with tri-state drivers.

−ADL The Address Decode Latch can be used by slaves to latch valid address and status bits. Transparent latches should be used if any latching is necessary and the trailing edge of −ADL or the leading edge of −CMD should be used as the latch point. The driver that is used by the system board to put −ADL on the bus is a tri-state driver.

−CD DS 16 (n) The Card Data Size of 16 indicates that a 16-bit data port is present at the addressed location. Each different connector on the system board has a unique signal line, and a different number (n) is used to indicate the specific connector. The signal that is present on this pin is generated by a 16-bit memory board, I/O, or a DMA slave using a totem-pole driver. The presence of a signal is interpreted by the system board as a valid address decode. This signal is used to communicate with 16-bit slaves. Eight-bit memory, I/O, or DMA slaves do not drive this signal, as it is part of the 16-bit extension.

−DS 16 RTN This signal is called the Data Size 16 Return and is a negative OR logic function of all −CD DS 16 signals on the channel. This results in an active output if any of the boards has an active −CD DS 16 signal. This signal is used by channel resident bus masters to monitor the data size information and must be driven by a bus driver.

−SBHE The System Byte High Enable signal is used to enable and signal the transfer of the high byte of data (D8 through D15). It is driven by a tri-state driver, is located on the 16-bit extension area of the bus, and is used with A0 to distinguish between the low byte and the high byte.

MADE 24 The Memory Address Enable 24 signal is used to indicate that the memory cycle currently in progress has an address of less than 16 MB. All address decodes must include this signal, which should be driven high for the active state by a tri-state driver.

M/−IO This signal line is driven by a tri-state driver and indicates that the current cycle is an I/O cycle when low and a memory cycle when high. This enables the system to differentiate between a memory and an I/O cycle. The signal is used in association with the −S0 and −S1 signals.

−S0, −S1 Status bits 0 and 1, driven by a tri-state driver, are used to indicate the start of a channel cycle and its type. The three signals −S0, −S1, and M/−IO are used together to differentiate between a memory and an I/O cycle. No status cycle is indicated when all three signals are high. The four valid signal combinations for these signals indicate:

I/O write	−S1 high, −S0 and M/−IO low
I/O read	−S0 high, −S1 and M/−IO low
Memory write	−S1 low, −S0 and M/−IO high
Memory read	−S0 low, −S1 and M/−IO high

Chapter 4

The other signal combinations are reserved. Any slave that is on the Micro Channel must fully decode −S0 and −S1. The lines can be latched using the trailing edge of −ADL or the leading edge of −CMD.

An I/O write command requires valid data on the bus from the time that the −CMD signal goes high until after −CMD returns to its low state. This acts as the signal to the I/O slave that the data on the bus should be stored. Any addresses on the bus must be valid before −S0 goes high.

An I/O read command requires the I/O slave to put data on the bus after the −CMD signal goes high, and to be valid when the −CMD signal returns to its low state. Any addresses on the bus must be valid before −S1 goes high.

A memory write command requires valid data on the bus from the time that the −CMD signal goes high until after −CMD returns to its low state. This acts as the signal to the memory that the data on the bus should be read. Any addresses on the bus must be valid before −S0 goes high.

A memory read command requires the memory to put data on the bus after the −CMD signal goes high, and to be valid when the −CMD signal returns to its low state. Any addresses on the bus must be valid before −S1 goes high.

−CMD The Command signal is used to indicate when data is valid on the bus, and should be driven by a tri-state driver. The end of the bus cycle is determined by the trailing edge of the Command signal. For write operations, data is valid for the whole time that the −CMD signal is high. For read operations, data should be put onto the bus after the −CMD signal goes high and be valid when the −CMD signal returns to its low state. Slaves can latch an address on the bus by using the −CMD signal.

−CD SFDBK (n) The Card Selected Feedback signal is made active by the addressed slave card. Each different connector on the system board has a unique signal line, and a different number (n) is used to indicate the specific connector. An active signal by the slave indicates its presence at the specified address. The signal is driven by a totem-pole driver. It can be used during diagnostics and installation to show address conflicts or missing cards due to malfunction. With the exception of slaves that are selected by the −CD SETUP signal, any slave that is selected must drive the −CD SFDBK signal in response to a card selection.

CD CHRDY (n) The Channel Ready signal is made inactive (lowered) to indicate that more time is necessary to complete the channel operation. It is driven by a totem-pole driver. Each different connector on the system board has a unique signal line, and a different number (n) is used to indicate the specific connector. For a read operation, the signal signifies that valid data will be available a specified time after CD CHRDY returns to a high state, and that the data will remain valid for the CPU to sample. This signal can be held low for a maximum of 3.0 microseconds.

CHRDYRTN The Channel Ready Return is driven by a bus driver and allows masters that are resident on the Micro Channel to detect when all of the

PS/2 Family Architecture

other cards on the bus are ready. The signal is a positive AND function of all CD CHRDY signals on the bus. It is active only when all of the cards indicate that they are ready.

ARB0 through ARB3 There are four arbitration priority signals present on the bus. Each of the lines is bidirectional, active high, and must be driven by open collector drivers. This provides sixteen priority levels for arbitration, with ARB0 being the least significant bit and ARB3 the most significant. A card requesting arbitration on the bus can change the state of the arbitration bus after the leading edge of ARB/ − GNT. The cards needing arbitration all monitor the bus status. If a particular card's priority level is less than the bus value, it does not assert its level. The channel is given to the card with the highest priority level; this card does not remove the priority level from the arbitration bus.

ARB/-GNT The arbitration cycle is in progress when this signal is high, and the bus has been granted to the highest priority level that requested the bus when this signal was low. The grant signal is driven with a bus driver from the Central Arbitration Control Point (CACP). The grant signal is sent within a specified time after four signals become inactive. These signals are − S0, − S1, − BURST, and − CMD. Cards that require arbitration must use the arbitration cycle signal to gate their data, address, and control signal off the bus for the duration of the arbitration cycle.

−PREEMPT The Preempt signal is a bidirectional line that requires an open collector driver. It is used by a card to request an arbitration cycle. The card that is granted the bus removes this signal at that time.

−BURST The Burst signal is driven by an open collector drive and is used by a card to indicate that it requires extended use of the bus to transfer a block of data. A card makes this signal active if the extension is required after it has been given the bus, and lowers the signal either during or at the end of the last transfer cycle. This signal is used by the CACP.

−TC The Terminal Count signal is driven by the DMA controller using a tri-state driver. It is only on the bus during DMA operations. It is used by a DMA slave on the bus as the signal that shows that the last transfer in a block to or from the DMA (depending on whether it occurs in a read or write cycle) is occurring.

−IRQ3 through −IRQ7, −IRQ9 through −IRQ12, −IRQ14, and −IRQ15 The missing interrupts are used for other purposes. − IRQ0 is used by the system timer, − IRQ1 is used by the keyboard, and − IRQ2 is used as the cascade interrupt control to access the second half of the interrupts. − IRQ8 is used for the real-time clock and − IRQ13 is used for the math coprocessor exception interrupt, so they are not available on the expansion connector. The interrupt request lines that are on the Micro Channel, however, are driven by an open collector driver. − IRQ10, − IRQ11, − IRQ12, − IRQ14, and − IRQ15 are located in the 16-bit extension section of the Micro Channel connector. Interrupt sharing is possible and is required by the Micro Channel. This means that sev-

Chapter 4

eral cards on the bus can use the same interrupt level. An interrupt is requested by lowering one of the interrupt request lines. This signals to the CPU that an I/O slave on the bus requires servicing. The priorities of the interrupt from the highest priority to the lowest are: −IRQ9, −IRQ10, −IRQ11, −IRQ12, −IRQ14, −IRQ15, −IRQ3, −IRQ4, −IRQ5, −IRQ6, −IRQ7.

−CD SETUP (n) The Card Setup signal is driven with a totem-pole driver by the system board, to individually select a card on the bus. It is used during error recovery procedures and system setup. When −CD SETUP is active, the card ID and configuration data can be read via an I/O read or stored using an I/O write. Each different connector on the system board has a unique signal line, and a different number (n) is used to indicate the specific connector.

−CHCK The Channel Check signal is driven by an open collector driver to indicate a serious error on the bus. A typical error may be a parity error. The signal is driven low to indicate the error and remains low until it is reset by the −CHCK interrupt handler.

AUDIO The audio sum node is the sum of all of the voltages present on this line on the bus. It can be used to transfer audio signals between cards on the bus or to drive the system board audio output from a board on the bus. Its frequency response is 50 Hz to 10 KHz, ± 3 dB. The maximum signal amplitude is 2.5 volts peak to peak, with a DC offset of 0 ± 0.05 volt. The maximum noise level is 0.05 volt peak to peak.

AUDIO GND The Audio Ground is used in association with the AUDIO signal, and is separate from the logical grounds elsewhere in the system.

OSC The Oscillator signal provides a clock signal with a high voltage of greater than 2.3 volts and a low voltage of less than 0.8 volt at a frequency of 14.31818 MHz ± 0.01 percent.

CHRESET The Channel Reset signal is supplied by the system board via a bus driver, and should be used to reset or initialize all boards on the bus. It is active for a minimum specified time during power on or when there is a low line voltage condition. Alternatively, this line can be activated by a program if desired.

−REFRESH The Refresh signal is supplied by the system board and indicates that a memory refresh is occurring. The address lines (A0 through A8) indicate the memory location that is currently being refreshed. IBM notes that this signal should not be used as a timing signal, because its absolute value may vary.

On the Model 80 there is a 32-bit extension to the Micro Channel connector on three of the expansion slots. These extensions consist of two parts: the 32-bit extension and the Matched Memory cycle bus. The 32-bit extension is used to allow transfers that are 32 bits wide instead of 8 or 16 bits. The Matched Memory cycle bus can be used to speed up a bus cycle, when the system is dealing with a memory card that can transfer data to the CPU at a reduced cycle time. The DMA controller on the system is not 32 bits wide; consequently, 32-bit DMA slaves are not supported.

Channel Connector and 32-Bit Extension Signals

A24 through A31 The address lines A24 through A31 are an adjunct to the address lines A0 through A23, which are on the other sections of the Micro Channel connector. A0 is the least significant bit, as before, and A31 is the most significant bit. This allows the system to address a total of 4 GB of memory. These high order address lines cannot be used for I/O operations; I/O operations only use A0 through A15. The address lines are each driven with a tri-state driver and are generated by the system microprocessor. Any slave that requires the address should latch it by using the leading edge of −CMD or the trailing edge of −ADL.

−BE0 through −BE3 The four Byte Enable signals are used to show the width of the data transfer byte that is occurring with a 32-bit slave located on the bus. When a 32-bit slave is being used, data can be transferred at 8-bit, 16-bit, 24-bit, or 32-bit contiguous widths. If TR32 is inactive, these enable signals are driven by the system microprocessor. If the system is controlled by a 16-bit master on the bus and is addressing a 32-bit slave, the Central Translator Logic drives these enable signals when TR32 is active. A 32-bit slave must latch these signals (which are driven by tri-state drivers), if required, as they are unlatched on the bus.

D16 through D31 Data bits D16 through D31 provide the upper 16 bits for the 32-bit data word. These signals are used in association with data signals D0 through D16, which are on the other sections of the Micro Channel connector. D0 is the least significant bit and D31 is the most significant bit. If data is transferred to an 8-bit slave, it is transferred as four 8-bit transfers, using data lines D0 through D7. As with the address lines, all of the data lines are driven with tri-state drivers. The data on the data pins is valid at different times, depending on whether the cycle is a write or read cycle. During a write cycle, data is valid whenever −CMD is active. During a read cycle, data must become valid after the leading edge of −CMD but before the trailing edge of −CMD; it must remain valid until after this trailing edge.

−CD DS 32 (n) The Card Data Size of 32 indicates that a 32-bit data port is present at the addressed location. Each different connector on the system board has a unique signal line, and a different number (n) is used to indicate the specific connector. The signal that is present on this pin is generated by all 32-bit slaves, using a totem-pole driver. The presence of a signal is interpreted by the system board as a valid address decode. This signal should not be driven by 8-bit or 16-bit data transfers by 32-bit slaves.

−DS 32 RTN This signal is called the Data Size 32 Return and is a negative OR logic function of all −CD DS 32 signals on the channel. This results in an active output if any of the boards has an active −CD DS 32 signal. This signal is used by channel resident bus masters to monitor the data size information, and must be driven by a bus driver.

TR 32 The Translate 32 signal is driven by a tri-state driver on a 32-bit

Chapter 4

master or by the Central Translator Logic. When the signal is high, the Central Translator Logic drives the enable signals B0 through B3. If the 32-bit master drives the signal low, the master is responsible for controlling the enable signals B0 through B3. The signal is read by 32-bit slaves in addition to the Central Translator Logic.

Channel Connector Matched Memory Extension Signals

Note that the Matched Memory extension is designed for 32-bit memory boards, but is not supported on the 20 MHz Model 80. On the 20 MHz machine, only a subset of these signals is supported. The − MMC CMD signal is the only one that is driven by the system board and it is the same as the − CMD signal located in the 16-bit section of the expansion channel. The other Matched Memory control signals are not driven onto the channel. The 16 MHz machine fully supports this protocol and provides an increase in performance for memory boards that can handle a memory cycle that is faster than the default.

−MMC   The Matched Memory Cycle signal is driven by a tri-state driver from the system board. It indicates that a board can request a Matched Memory cycle as the system microprocessor is controlling the bus, and can accommodate a faster cycle.

−MMCR   The Matched Memory Cycle Request signal is driven by a 16-bit or 32-bit channel slave to request a faster bus cycle. − ADL and − CMD are inactive throughout the Matched Memory bus cycle. This signal is ignored if it is asserted by the following boards, which are not capable of supporting a faster bus cycle time: 8-bit slave or a 16-bit or 32-bit slave during a DMA cycle or any other bus cycle that does not involve the microprocessor. Matched Memory cycles can only be run by the system board 80386. Each of the 32-bit connectors wires this signal separately to an OR gate on the system board. If the bus cycle proves to be too fast it can be extended by using the CD CHRDY signal.

−MMC CMD   The Matched Memory Cycle Command indicates when data is valid on the bus during a Matched Memory bus cycle. It is not driven during other cycles. The end of the Matched Memory cycle is indicated by the rising edge of this signal. It is similar to CMD, but it is used for the faster cycles.

Auxiliary Video Extension Signals

One of the 16-bit connectors on the Micro Channel includes a video extension, which provides a method by which video adapter boards can interface with the video system on the system board.

VSYNC   The Vertical Sync signal that is used as an output to the display is present on this pin. If the VGA is made inactive, this signal is generated by the expansion board. If the VGA on the system board is active, it drives this signal and presents it to the expansion board.

HSYNC   The Horizontal Sync signal that is used as an output to the display is present on this pin. If the VGA is made inactive, this signal is generated by the expansion board. If the VGA on the system board is active, it drives this signal and presents it to the expansion board.

BLANK The video digital-to-analog converter (DAC) on the system board has its BLANK pin linked to this pin on the expansion connector. The expansion board can drive this pin to an active low, and the DAC will reset the outputs from its analog color output pins to 0 volts.

P7 through P0 The digital video information for each pixel that is to be sent to the system board DAC is output on these pins.

DCLK The DAC uses this signal as the pixel clock for latching video signals P7 through P0. The rising edge of DCLK is used as the latch indication by the DAC. In addition, this is connected to the EXTCLK pin of the VGA, to allow a video expansion board to drive the VGA at a different clock speed.

ESYNC This is the Sync Output Enable signal for the BLANK, VSYNC, and HSYNC signals. An open circuit on this pin is a digital logic high, due to a pullup resistor on the system board.

When ESYNC is high, the VGA on the system board provides the BLANK, VSYNC, and HSYNC signals. When ESYNC is low, the video expansion board provides the BLANK, VSYNC, and HSYNC signals.

EVIDEO This is the Video Output Enable signal for the buffer controlling P7 through P0. An open circuit on this pin is a digital logic high, due to a pullup resistor on the system board.

When EVIDEO is high, the VGA on the system board provides the P7 through P0 signals. When EVIDEO is low, the video expansion board provides the P7 through P0 signals.

EDCLK This is the Output Enable signal for the buffer controlling DCLK. An open circuit on this pin is a digital logic high, due to a pull-up resistor on the system board.

When EDCLK is high, the VGA on the system board provides the DCLK signal. In this configuration, the VGA miscellaneous output register must not select clock source 2. When EDCLK is low, the video expansion board provides the DCLK signal. In this configuration, the VGA miscellaneous output register must select clock source 2. The miscellaneous output register is described more fully in Chapter 5.

Micro Channel Board Physical Specifications

The dimensions for the Micro Channel adapter boards are fully described in the technical reference documentation for the IBM Personal System/2 machines. The shape of the board and the position of the connectors are laid out clearly, in an attempt to standardize the boards *more* than on the PC.

There are several justifications for these restrictive specifications. The FCC regulations for EMI are much stricter than when the PC was designed, so any measures that help reduce overall emissions from the system are advisable. The cooling method used by the system cases involves moving air over the expansion boards. This airflow would be greatly disturbed if, for example, there was a half-length board in the system. In addition, the boards are only mounted rigidly to the case at the rear end of the board; the other end is only supported when the case is closed completely. If the physical dimensions are not strictly adhered to, the case may not close or the board may not be adequately supported. The specification also describes the maximum amount of current that an expansion board should use. This ensures that the current limits of the unit are not exceeded because several boards take more than their fair share of the current available on the expansion bus.

A typical adapter card consists of four major parts: the board itself, the rear mounting bracket, and two handles that are used to insert and remove the board from the expansion slot. The smaller handle at the same end of the board as the rear mounting bracket is called a holder, and the larger knob at the internal end is called a retainer. Both handles are made of polycarbonate UL 94 V-0 and the rear mounting bracket is AISI type 302, ¼ hard stainless steel. The position in which the bottom edge connectors fit into the Micro Channel connectors is defined, and the permissible position for an edge connector on the top edge of the board is also specified. The physical characteristics of the board are shown in Figure 4-6.

The plated edge connectors on the base of the board contain all of the signal lines for the Micro Channel. These contacts should all be the same length on boards that have CMOS ICs present, to ensure that the correct polarity is always applied to the pins during insertion; however, the drawings in the Technical Reference manual describe pins that are longer and have tapered gold contacts for all other boards.

The power specifications for the expansion board are shown in Table 4-3. The maximum current that a board should draw from the 5-volt line is 1.6 amperes, from the +12-volt line is 0.175 amperes, and from the −12-volt line is 40 milliamperes.

The other general design considerations suggested by IBM for the board include safety improvements, and thermal and electromagnetic interference reduction. The maximum height for components on the adapter is 0.6 inch on the component side and 0.078 inch on the other side. The slot with the video extension connector is designed to be used with video adapter boards.

The thermal considerations relate to the cooling method used in the unit. Cables that are attached to the boards for connection to the rear panel or to other boards installed in the unit should be minimized if they cannot be eliminated. Cables can cause restrictions in the air flow and reduce the cooling capabilities of the unit. Components on the board should be spread out so the thermal emissions of the board are as even as possible.

The EMI considerations are among the most important for allowing these units to be fully expandable and compatible with other products on the market.

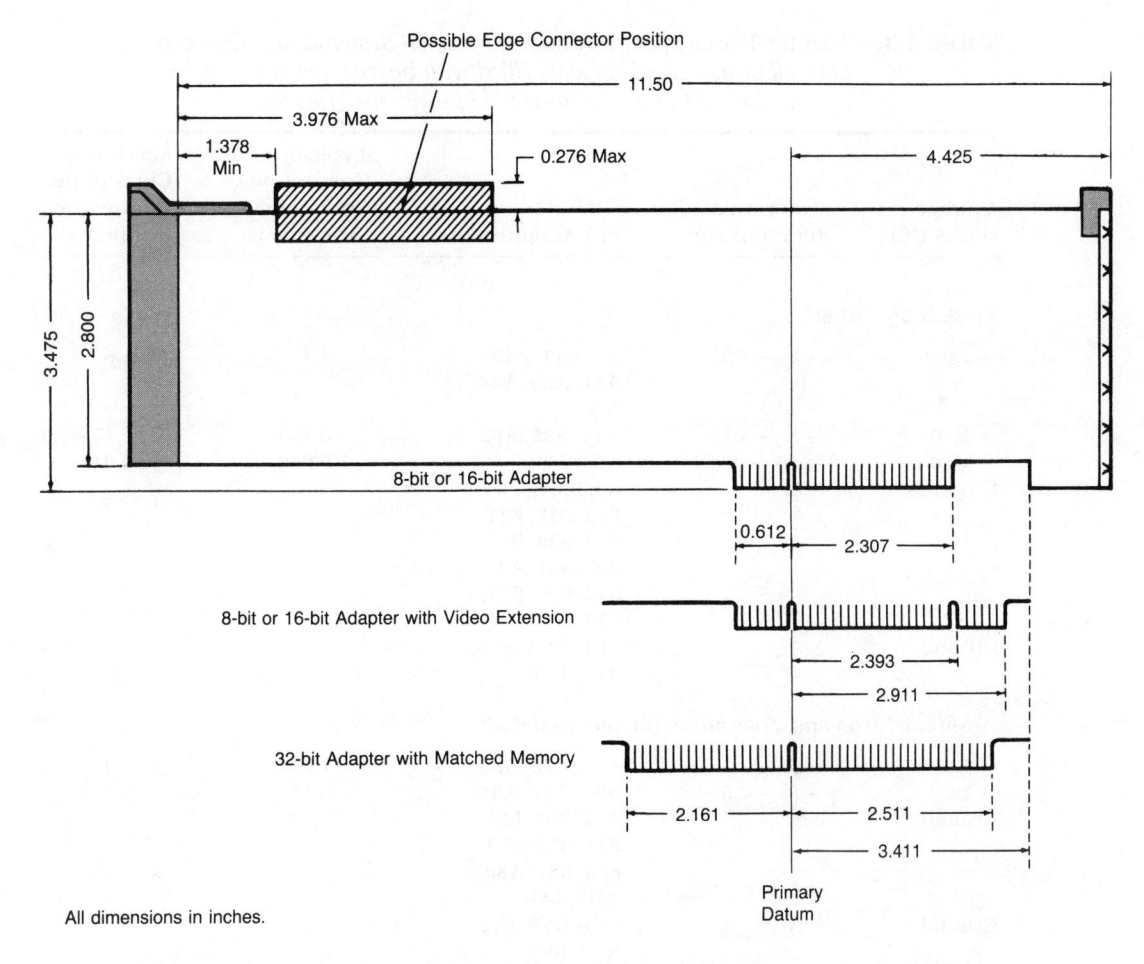

Figure 4-6 Adapter Board Physical Characteristics. *The number of cables that are attached to the board should be minimized, so that EMI and danger of overheating are not increased.*

The rear bracket should not be used as a DC voltage return path, a logic ground connection, or an audio ground connection. The rear bracket is not connected to the Micro Channel directly; it is part of the outer grounding of the system. The adapter board should be isolated from the rear bracket; the rear bracket should be attached to the system unit case; and the board itself should be grounded to the Micro Channel.

All ground pins should be connected to all of the ground pins on the Micro Channel connector, and all +5-volt pins should be connected to all of the +5-volt pins on the Micro Channel connector. If followed, this design rule dramatically reduces EMI, because no signal pin on the connector is farther away from a ground pin than 0.1 inch. The adapter should include an internal power and ground plane, so the minimum number of layers for a board is four.

Decoupling capacitors must be used alongside each surface-mounted

Chapter 4

Table 4-3 Adapter Power Specifications. *A Personal System/2 machine can have all of its expansion slots filled with boards, provided each board adheres to the recommended power specifications.*

Voltage (Volts DC)	Tolerance (Percentage)	Pin Assignments	Typical Current per Connector (Amperes)	Maximum Current per Connector (Amperes)
Models 50 and 60				
+5.0	+5, −4.5	A7, A11, A15, A31, A39, A48, A56	1.4	1.6
+12.0	+5, −4.5	A19, A35, A52	0.100	0.175
−12.0	+10, −9.5	A23, A27	0.040	0.040
Ground	N/A	A3, B3, B5, B9, B13, B17, B21, B25, B29, B33, B37, B41, A43, B45, B50, B54, B58		
Ground	N/A	BV1, AV3, BV5, AV7, BV9		
Additional Pins and Tolerances for the Model 80				
+5.0	+5, −4.5	A69, A73, A85	1.6	2.0
+12.0	+5, −4.5	A65, A77, A81	0.175	0.175
Ground	N/A	A61, B63, B67, B71, B75, B79, B83, B87, A89, AM2, BM4		
Ground	N/A	BV1, AV3, BV5, AV7, BV9		

device (SMD). All dual-in-line packages (DIP), or similar devices that use holes that go through the board, require decoupling capacitors if they are driving or contain edge-triggered logic. The decoupling capacitors should be the low inductance ceramic or layered type, and should have a value of between 0.01 and 0.1 microFarad.

The restrictive use of internal cables will inherently reduce the amount of EMI. The rise and fall times of clocks and strobes should be considered carefully as these can add a surprising amount of radiation to the circuit. Any high current paths within the system should use a return path that is adjacent, in order to gain as much mutual coupling and, consequently, cancellation in the radiated magnetic fields.

External cables that are added to the board to connect to the rear bracket require good shielded cables. The connector on the rear bracket should be a 360-degree, shielded D-type connector or equivalent. Shield connections that are formed by creating a pigtail should be avoided. The shield termination should attach to the outside shield of the connector, and not be brought through the connector and attached to the logic ground or any other ground internal to the system.

Besides the general physical requirements for interference and thermal purposes, the interaction of the board with other boards that may be in the system is important. The I/O address for the Micro Channel is 16 bits wide, and all adapter cards must decode all of the bits, not just enough for their own purposes (which may be only 8 bits). The memory adapters must decode the full 24 bits of the memory address, as well as the MADE 24 signal.

Adapter cards should not use jumpers or switches, but rather registers. Even though IBM's 8514/A Display Adapter disobeys this rule with its memory expansion module, the simplified installation of add-in boards is highly desirable. Remember the memory add-in boards for the PCs with their numerous DIP switches and the installation problems associated with them. The Programmable Option Select (POS) for the IBM Personal System/2 machines is a clean and simple implementation that can easily be supported in an adapter design.

Each card type from each manufacturer is assigned a unique ID that must be driven onto the bus when the card is interrogated. Table 4-4 shows the recommended values for the ID numbers. Pull-up resistors are attached to the bus, so that when no card is inserted in the Micro Channel connector, the CPU sees the ID number FFFFH. As a result, only the ID number bits that are logical zero need to be driven, thus reducing the number of drivers that are necessary on the adapter.

Table 4-4 ID Values for Expansion Boards. *The ID value that is set in the expansion board is used by POS for system configuration purposes.*

Number (Hex)	Description
0000	Reserved (device not ready)
0001 to 0FFF	Bus master
5000 to 5FFF	Direct memory access devices
6000 to 6FFF	Direct program control (including memory-mapped I/O)
7000 to 7FFF	Storage (including multi-function boards with storage)
8000 to 80FF	Video
FFFF	Device not attached

Micro Channel Concepts

The three new concepts that are incorporated into the Micro Channel architecture are Programmable Option Select, level-sensitive interrupts, and multi-device arbitration. In addition, the Micro Channel has a 32-bit extension that

Chapter 4

allows Matched Memory cycles to occur with boards that are capable of supporting a faster bus cycle. Note that the 20 MHz Model 80 does not support the Matched Memory cycle feature.

Programmable Option Select (POS)

The Programmable Option Select (POS) is an identification system for expansion boards. Each board is assigned a unique ID, which is used by the system board for several purposes. The concept of POS is that switches on the expansion boards, which on PC expansion boards set the interrupt level, or the starting memory address can be eliminated. It also allows several cards that perform the same function to be installed. Where possible, primary and secondary port addresses and interrupt levels can be set using POS. If alternates are not available, POS prevents both being active at the same time by disabling one of them. If the primary board later becomes defective, the secondary board can be activated through POS. This allows a system to be built with duplicate expansion boards, which is in some respects fault tolerant, allowing an end user to reset the machine to a usable state until service is obtained. The information from the system board and the expansion boards that make up the POS is obtained from software-writable latches and is stored in RAM that is battery backed.

A major advantage of POS is that it is now possible to identify precisely what cards are installed and in which slots they are located. In addition, any conflict situations can be identified. As integral backup configurations are a suggested board option, many conflicting situations can be resolved. The design concept of POS is to improve the serviceability of the machines and simplify customer setup.

The system unit itself is shipped with a 3.5-inch Reference Diskette that is used to initialize the system at installation. Information required for configuring a particular adapter card is provided with the card and the information is added to the information on the main Reference Diskette. This information is in the form of an adapter description file, diagnostic tests for the adapter, and POST error messages.

There are several configuration utilities on the Reference Diskette. The Automatic Configuration Utility is used for most general installation procedures. The Change Configuration Utility is used for more specialized applications, where the installer wishes to resolve an unusual conflict or change a default installation setting to suit different needs. The Backup and Restore Configuration Utility lets the user make a backup of the system configuration. The Copy an Option Disk Utility allows the user to add new adapter description files for expansion boards that are to be installed in the system.

The Automatic Configuration Utility program is run using the Reference Diskette and is described in Chapter 6 from a functional viewpoint. This program compares the contents of the adapter description files with the data that is in the CMOS RAM. Any adapter that does not have an adapter description file is

PS/2 Family Architecture

disabled. The Set Configuration program that is used to change the contents of the CMOS RAM has two limitations. It can only deal with two different memory blocks for each adapter, but up to sixteen different I/O blocks, or arbitration or interrupt levels can be set.

The adapter description file has a unique filename that consists of the 16-bit card ID that is generated by the adapter. The Automatic Configuration program determines the correct position in CMOS RAM for the POS information and card ID by determining the system type. Knowing the system type enables the program to ascertain the number of slots and the correct RAM location to use for storing the POS data.

The Model 50 has the ID and POS information stored in two different locations within the real-time clock CMOS RAM. Each slot on the Micro Channel uses 2 bytes for the card ID and 4 bytes for the POS data. The Model 60 and the Model 80 store the ID and POS information contiguously in the 2 KB CMOS RAM extension. The automatic configuration only occurs for slots where the adapter board has just been inserted or moved; all previously installed boards are not reinstalled.

The validity of the data that is stored in the CMOS RAM is checked by POST. There is a cyclic redundancy character (CRC) for the 64-byte CMOS RAM and an additional CRC for the 2 KB extension; both are recalculated by the configuration utilities when the installation requirements change. If POST detects a CRC error in this data, it displays a POST error code (161) and then runs Automatic Configuration, if the Reference Diskette is in drive A. The system is then configured so that each adapter is set to the first nonconflicting value that can be found by the Automatic Configuration program. The adapter that is in the first Micro Channel slot is configured first, followed by slot position 2, through to the final slot position in the machine.

The POS should only be set with the Reference Diskette utilities; applications should not use the ID information directly. They should obtain system configuration information through the operating system. Although getting the information directly will only cause incompatibility problems for the system if implemented badly, it will ultimately damage the growth of the industry. It is possible to break the rules, as applications did on the PC, because the CMOS RAM can be set directly; however, by obeying the rules the industry can grow. Users are wanting more and more power from their personal computers and, although the IBM Personal System/2 machines are faster than PCs, their true potential lies in enabling multi-tasking and multi-processing. Applications that break the rules will be ignored as the industry reaches a new level of maturity.

The I/O addresses 0100H through 0107H are used by the adapters and system board setup functions. The I/O address 0096H is used by the adapter as an adapter enable and setup register for POS, and the I/O address 0094H is used by the system board for the purpose of controlling the device unique setup signals. The I/O addresses that are used by the POS registers are shown in Table 4-5. Note that addresses 0095H and 0097H are reserved by IBM.

Chapter 4

Table 4-5 Programmable Option Select Address Space. *One of the purposes of POS is to remove the necessity of internal switches in the system. All configuration is done via software.*

Address Bit			−CD SETUP	POS Register No.	Function
A2	A1	A0			
0	0	0	0	0	Adapter identification byte (least significant)
0	0	1	0	1	Adapter identification byte (most significant)
0	1	0	0	2	Byte 1 of option select data
0	1	1	0	3	Byte 2 of option select data
1	0	0	0	4	Byte 3 of option select data
1	0	1	0	5	Byte 4 of option select data
1	1	0	0	6	Subaddress extension byte (least significant)
1	1	1	0	7	Subaddress extension byte (most significant)

The system is divided into three sections: video subsystem, main system board, and adapters. The setup functions of each section respond to I/O addresses only when their respective setup signals are active. Only one setup signal bit should be active at one time; the others should be set inactive. For the video subsystem, bit 5 of port address 0094H set to a logical 0 is the setup signal bit. The system board setup signal bit is bit 7 of port address 0094H (setup enable register) set to a logical 0. Finally, the adapter setup bit is bit 3 of port address 0096H set to a logical 1. Note that the state of the adapter setup bit has the inverse logic to the equivalent signal for the video subsystem and system board. IBM warns that, apart from contention, actual physical damage may occur if more than one section is activated for setup at the same time.

If an adapter is selected it will respond with the − CD SFDBK signal. This signal is latched by the system board and, because it is then made available on subsequent cycles, can be used by diagnostics programs or the setup facilities to check the operation of a particular adapter at a particular address or DMA port.

The presence of an operating video subsystem, main system board I/O functions, or adapter card is read via the read only register at address 0091H. Bits 1 through 7 are reserved, but bit 0 contains a 1 if the − CD SFDBK signal is obtained.

The system board I/O functions can be considered in three parts: the diskette drive controller, the parallel port, and the serial port. All three are treated as a single device by the system; the VGA is considered a separate device.

When the system board I/O functions are accessed for setup via bit 7 of address 0094H (the setup enable register), the current state of these functions

PS/2 Family Architecture

can be read and modified via the System Board I/O byte, address 0102H. The bit definitions for this byte are as follows.

Bit 7 This bit is used to enable and disable the 8-bit bidirectional mode of the parallel port. POST sets this bit to 1 from the power-on reset state of 0. This bit has negative logic: when the bit is set to 0 the bidirectional feature is enabled and when set to 1 the feature is disabled.

Bits 6 and 5 This bit pair is used to set the address for the parallel port. Bits 6 and 5 set to 1 are a reserved combination. Bits 6 and 5 set to 0 assign the parallel port to parallel port 1, with an address of 03BCH through 03BEH. Bit 6 set to 0 and bit 5 set to 1 assign the parallel port to parallel port 2, with an address of 0378H through 037AH; and bit 6 set to a 1 and bit 5 set to a 0 assign the parallel port to parallel port 3, with an address of 0278H through 027AH.

Bit 3 This bit is used to allocate the serial port. With the bit set to 1, the serial port on the system board is set to serial port 1; with the bit set to 0, the serial port on the system board is set to serial port 2.

Bit 2 This bit is used to enable and disable the serial port. With the bit set to 1, the port is enabled; with the bit set to 0, the port is disabled.

Bit 1 This bit is used to enable and disable the diskette drive interface. With the bit set to 1, the interface is enabled; with the bit set to 0, the interface is disabled. POST always enables this interface.

Bit 0 This bit is used as a control bit for bits 4, 2, and 1 of this byte. With the bit set to 0, the parallel port, the serial port, and the diskette drive interface are disabled. Only when this bit is set to 1 are the values in bits 4, 2, and 1 used to generate the status for the system board I/O features.

The adapter boards are set up using the – CD SETUP (n) signal. This signal is unique for each position on the channel, so each board can be initialized separately. When – CD SETUP is active, the setup read and write operations can be performed.

The address space that is used by the configuration program is shown in Table 4-5. The three least significant bits of the I/O address are decoded to identify the POS register that is to be used. The 8-bit POS registers are 0100H through 0107H, with bytes 0100H and 0101H being read only. If implemented, the other bytes—0102H through 0107H—are read/write registers.

Register 0102H, bit 0 has a defined purpose as an enable and disable bit for the adapter. If this bit is set to 0, the adapter is disabled and only responds to setup read and write functions. Register 0105H has 2 bits that are defined and are not adapter dependent. Bit 7 is the channel check active indicator. This signal is set by the system memory or an adapter to indicate an error on the channel. The nonmaskable interrupt (NMI) handler interrogates this bit on each adapter to determine the source of the error. Bit 6 can be used to provide additional information through bytes 0106H and 0107H. If bit 6—the channel check status indicator—is set to 1, no information is available, but if it is set to 0, there is status information in bytes 0106H and 0107H. This information can be the status itself or a pointer to the address that contains the status information, or a command instructing the system to present the address to another position. If an adapter supports the CHCK bit, this bit must also be supported; but if the CHCK bit is not supported, adapter-unique information can be placed in this bit.

Besides these few bits, the information that is put into the POS bytes can take any form, but there are four fields that are required if the board supports particular features. These are the fairness enable field, the arbitration level field, the device ROM segment address field, and the I/O device address field.

The fairness enable field is used for arbitration purposes. It is set to 1 for the adapter to support the fairness feature and to 0 for this feature to be disabled. When an adapter supports this feature, this bit indicates that the adapter will become inactive if it is preempted from the bus and will not become active again until the − PREEMPT signal is removed from the bus. Use of the fairness feature results in a round-robin participation of adapters on the bus.

The arbitration level field is a 4-bit field that indicates the arbitration level that is to be assigned to the adapter. This is a programmable field that may need to be reassigned at setup or by a diagnostic routine to remove conflicts. Only one adapter can reside at a particular arbitration level.

The device ROM segment address field is provided by all adapters that have memory mapped I/O ROM. This field can be up to 4 bits in length. It specifies the starting address for the ROM at any one of sixteen 8K segments, beginning at C0000H.

The I/O device address field is used by any device that wants to exist on the bus with another device of the same type. This programmable field gives a list of alternative addresses that can be used to address the adapter. This field can then be used to prevent address conflicts by assigning a different address for each conflicting adapter.

The POS system that is used in the IBM Personal System/2 machines with the Micro Channel seems to provide clear and logical design guidelines for the implementation of adapters. The actual success of the standard will depend on adherence to the standard by third-party vendors.

Level-Sensitive Interrupts

The original PC family supported edge-triggered interrupts, but the design requirements for an adapter in the IBM Personal System/2 that supports the Micro Channel is for level-sensitive interrupts. This is one of the many areas where IBM has gone to great lengths to maintain compatibility with the past. New system software is expected to use the level-sensitive interrupts, but existing software that uses positive edge-triggered interrupts continues to be supported.

There are several advantages to using level-sensitive interrupts, including the ability to share interrupts more easily than was possible on the PC. In addition, the logic needed to support level-sensitive interrupts is less than the logic required for edge triggering, and the system can be made less sensitive to transients and noise. With edge-triggered interrupts, the interrupt handler needs to be insensitive enough to ignore noise or transients, yet be sensitive enough to detect an interrupt edge.

The adapters that are installed on the Micro Channel hold the interrupt to an active low until it is reset by the interrupt servicing routine. This means that the service routine should reset the interrupt at the end of its routine. In addition to setting the interrupt level, the adapter supplies an interrupt pending latch that can be read and reset at an I/O address position for the normal servicing routines for the device.

The servicing of interrupts involves the coordination of software and hardware functions. The interrupt is signalled on the bus by the adapter, using an open collector driver, and the interrupt pending latch is set. The interrupt controller then signals the CPU that there is an interrupt pending. The interrupt handler software responds by starting the interrupt handler code. The interrupt handlers check the interrupt pending latch of each adapter in turn until the interrupt (or the first of the interrupts) is discovered. The interrupt is then serviced by the software, the hardware resets the interrupt pending latch, and the hardware level-sensitive interrupt line is raised, resetting the hardware. The service routine completes and returns. If another interrupt is pending, the interrupt handler code is restarted until all interrupts are serviced. The interrupt handler always follows the same sequence and services an interrupt that has the highest priority. This is not necessarily the interrupt that has been pending the longest.

Multi-Device Arbitration

The Micro Channel architecture is designed to allow a variety of adapter boards to coexist on the bus. This includes any board with a microprocessor that fully controls the system when the board has control of the bus. This has been done to a limited extent on the PC by boards such as accelerator boards, but most of them replace the 8088 or 80286 rather than allow both microprocessors to be present (and used) in the system. The Micro Channel bus extends this type of application and allows several microprocessors to be resident and active on the bus, producing a multi-processing system.

Even though several masters can be present in the machine at the same time, only one device can have control of the bus at once. The Micro Channel provides sixteen levels of arbitration, which are administered through a Central Arbitration Control Point (CACP) to allow these subsystems to share the bus.

The arbitration levels that are available to the bus have values between 0 and 15 (FH). The system has two other arbitration levels that are reserved and have a higher priority than the adapters on the bus. The first is referred to as having a value of − 1 and is assigned to the NMI. The second is referred to as having a value of − 2 and is assigned to memory refresh.

The arbitration control is done through seven signals: ARB0 through ARB3, − PREEMPT, ARB/ − GNT, and BURST. The ARB0 through ARB3 signals are set for each adapter when it is installed in the machine; they signify one of sixteen levels of priority for an adapter. The −PREEMPT signal is used to indicate that the adapter requires the bus. The ARB/−GNT signal is used to signify that an arbitration cycle is in progress or that the bus has been granted to a device. The BURST signal is used to indicate that the adapter that has been granted the bus wants to perform multiple transfers across the bus. This is permitted until another adapter activates the −PREEMPT signal.

The CACP is located on the system board, and each adapter that is to be involved with arbitration has logic circuitry on it called a local arbiter. All of the adapters use open collector drivers so that several arbiters can drive the bus at the same time.

The arbitration cycle is requested by an adapter, or several adapters, driving the −PREEMPT signal active low. This causes the device that currently controls the bus to release the bus. The arbitration cycle is started by the CACP raising the ARB/−GNT line. The adapters requesting the bus place their arbitration level values assigned to ARB0 through ARB3 onto the bus. The device that has the lowest assigned value for the arbitration level is awarded the bus.

The arbitration lines are all driven by open collector outputs, so if no boards are driving the line, the output is a logical 1. If several boards are driving the line to a logical 0, the line has a value of a logical 0. Finally, if some of the adapters are driving the signal to a logical 1 and others to a logical 0, the value seen on the bus is a logical 0.

The local arbiter for each adapter is able to determine if there is another adapter on the bus with a lower value than itself. This is done by examining each ARB line in turn, starting with the most significant bit, ARB3. All of the boards examine the value appearing on the bus and compare it with their own arbitration value for that bit. If the value seen on the bus is not the same as the value assigned for that bit, the device removes all of its less significant bits from the bus. The next most significant bit is examined. If there is a match between the value on the bus and the assigned value, the device either continues to drive the bus with its value or resumes driving the bus as appropriate. This ultimately results in the bus containing the lowest arbitration value for the devices competing on the bus.

The arbitration mechanism is clearer with an example. Say two devices request the bus. Board A has an arbitration value of AH (1010 binary) and board B has an arbitration value of 3H (0011 binary). When the arbitration cycle is started by the CACP, each adapter drives the arbitration signals with their respective values: board A drives ARB3 and ARB1 to a logical 1 and board B drives ARB1 and ARB0 to a logical 1. The resulting value that is seen on the bus (due to the open collector outputs) is ARB3 at a logical 0, ARB2 at a logical 0, ARB1 at a logical 1, and ARB0 at a logical 0.

Boards A and B compare the value that is seen on ARB3 with their respective assigned values. Board A discovers that it does not have the same value as the bus, so it stops driving the lower order signals onto the bus. Board B, however, has the same value as the bus for the ARB3 signal, so it does not change its setting. The bus now has ARB3 at a logical 0, ARB2 at a logical 0, ARB1 at a logical 1, and ARB0 at a logical 1. Boards A and B now compare the value on the bus for ARB2 with their values. Both boards match the expected value, so board A resumes driving ARB0. The bus now has ARB3 at a logical 0, ARB2 at a logical 0, ARB1 at a logical 1, and ARB0 at a logical 0. Boards A and B compare the value on the bus for ARB1 with their assigned values and both boards match the expected value, so both continue to drive the bus. When boards A and B compare the value for ARB0, board A discovers a mismatch and stops driving the bus. This results in the value on the bus being 0011, which is the arbitration level assigned to board B, and board B is given control of the bus by the CACP.

Once a device has been awarded control of the bus, it can either operate in its normal mode using the − CMD signal for the cycle or it can use the burst mode. The burst mode is used by devices that need to transfer a large quantity of data at once around the system. A disk controller is a good example of a device that is likely to use the bus in bursts. The BURST signal is a part of the local arbiter circuitry on an adapter card. The adapter (say it is called adapter A) lowers − BURST to activate the burst cycle. If another adapter does not request the bus, it raises the signal after the leading edge of − CMD in the last cycle that it needs to transfer the data across.

However, if another adapter (suppose it is called adapter B) requires the bus, the situation is different. Adapter B signals to the CACP that it requires an arbitration cycle. Adapter A, with − BURST active, completes a partial transfer of the data and makes − BURST inactive so that arbitration can occur. If the fairness feature of the adapter is active on adapter A, it will not participate in the arbitration cycle. The CACP raises the ARB/ − GNT signal when the end of transfer is recognized, arbitration occurs, adapter B gets the bus, and then adapter B removes the − PREEMPT signal.

If an adapter does not give up the bus when a − PREEMPT signal has been active for more than 7.8 microseconds, the ARB/ − GNT signal is driven high immediately and this forcibly removes the adapter from control of the bus. An NMI is then activated and bits 5 and 6 of port 0090H are set. The system microprocessor has control of the bus and the bus remains in an arbitration cycle state until the interrupt is serviced and bit 6 of port 0090H is reset.

BIOS (Basic Input/Output System)

Most computers employ several layers of system software between the processor and the user. The lowest level of such software is the basic input/output system (BIOS). The BIOS is a collection of routines that control I/O devices and provide basic system services. In mini and mainframe computers these routines are usually an integral part of the operating system, which is loaded into system memory from an external I/O device such as a disk or tape drive. On personal computers the BIOS is permanently stored in read only memory (ROM), in order to alleviate the need to load it from an external device each time the system is initialized.

The BIOS provided with Personal System/2 computers is functionally compatible with that used in the IBM PC family of computers. The Personal System/2 Models 50, 60, and 80 actually feature two basic input/output systems—the compatibility BIOS (CBIOS) and the advanced BIOS (ABIOS).

Compatibility BIOS

The CBIOS supports 1 MB of memory address space and is compatible with the PC BIOS. CBIOS services are accessed using software interrupts, generally using assembly language. The ABIOS provides support for OS/2 and other multi-tasking operating systems and supports a 16 MB address space. ABIOS services are accessed by calling routines, from high level language programs if desired, in the same way that a program calls its own routines or operating system routines.

BIOS is contained in system board ROM along with the system's power-on self test (POST) and the BASIC language interpreter. Models 25 and 30, with their PC-compatible BIOS, have 64 KB of ROM; Models 50, 60, and 80, with their two levels of BIOS, have 128 KB of ROM. System ROM is assigned to addresses F0000H through FFFFFH on Models 25 and 30, and E0000H through FFFFFH on Models 50, 60, and 80.

IBM does not provide source listings for the Personal System/2 BIOS code as it has done with the BIOS code for members of the PC family. Instead, IBM has provided a manual (available at extra cost) that gives interface definitions for the CBIOS and the ABIOS. The IBM Personal System/2 and Personal Computer BIOS Interface Technical Reference manual provides information about the system BIOS; information about BIOS code contained in adapter board ROM is provided in separate technical reference manuals.

The BIOS Technical Reference manual describes BIOS software interrupts, parameter passing, data areas, and ROM tables. It also describes how to determine the version date of the system BIOS. Instances where BIOS functions behave differently on different models, and on the same model with different BIOS versions (indicated by date), are noted. Supplements to the manual (to be provided at extra cost) will provide additional BIOS information. The first edition of the BIOS Technical Reference manual (April 1987) contains no information on the ABIOS.

The CBIOS is compatible with the IBM PC BIOS; however, several functions are significantly enhanced in order to support the new hardware functions provided by Personal System/2 computers. Numerous extensions have been made to the INT 10H video function. The CBIOS supports EGA functions that were previously supported by ROM on the EGA adapter, plus additional features and functions provided by the Personal System/2 integrated MCGA and VGA video subsystems. In addition to support for the various new video modes, it also provides subfunctions for determining the video controller and display in use, and determining and saving/restoring the state of the video subsystem.

The INT 13H (diskette and disk) interface is compatible with that provided with the AT. Use of the 3.5-inch diskette drives provided on Personal System/2 computers is transparent to most programs that adhere to previously documented BIOS and DOS interfaces. New applications can make use of three new enhancements. A new subfunction is provided to set the diskette media type for format. The hard disk BIOS interface is maintained as defined on the AT; error return codes have been added to report bad arbitration level, control data address mark, and format failure. A hard disk function to park the hard disk drive heads over the drive's landing zone has also been added.

The asynchronous communications interface (INT 14H) has been enhanced to include new additional features for initialization and port control. *Extended initialize* can be used to set data character length to 5, 6, 7, or 8 bits, and 19,200 bps communications is now supported. *Extended communications port control* allows the modem control register for each line to be read and written; previously only modem and line status could be read.

INT 15H, formerly called *cassette I/O functions* even though it provided a variety of services, is now called *system services*. Services provided on the AT have been maintained, and new subfunctions to support Personal System/2 features have been added. New and enhanced services allow system configuration parameters to be examined and the segment address of the extended BIOS data area to be obtained. They also provide access to the POST error log. The pointing device BIOS interface, watchdog timer control, Programmable Option Select, and ABIOS pre-initialization services are also provided through INT 15H.

In addition to being available on all Personal System/2 products, the *Return System Configuration Parameters* subfunction (C0H) is available for all PC products except PC and PCjr, XTs with a 11/08/82 BIOS date, and ATs with a 1/10/84 BIOS date. Information returned by this subfunction includes model and submodel bytes, BIOS revision level, and flags indicating presence of a real-time clock and second interrupt controller, as well as whether a Micro Channel or PC expansion bus architecture is in use. A flag also indicates whether or not an extended BIOS data area is allocated. See Table 4-6 for a list of system identification information.

Personal System/2 computers support an extended BIOS data area in addition to the 256-byte BIOS data area at 400H—4FFH. The system POST allocates memory for this area as needed, starting at 640 KB and working downward in 1 kilobyte increments. The first byte in the extended BIOS data area is initialized to the length of the area in kilobytes. The current size of the extended BIOS data

Table 4-6 System Identification. *System identification information includes model byte, submodel byte, and BIOS revision level. The BIOS revision level is increased by one for each subsequent release of the code.*

Product	BIOS Date	Model Byte	Submodel Byte	Revision Level
XT	1/10/86	FB	00	01
XT	5/09/86	FB	00	02
AT	6/10/85	FC	00	01
AT	11/15/85	FC	01	00
XT 286	4/21/86	FC	02	00
PC Convertible	9/13/85	F9	00	00
PS/2 Model 25	6/27/87	FA	01	00
PS/2 Model 30	9/02/86	FA	00	00
PS/2 Model 50	2/13/87	FC	04	00
PS/2 Model 60	2/13/87	FC	05	00
PS/2 Model 80	3/30/87	F8	00	00
PS/2 Model 80	10/07/87	F8	01	00

area is 1 KB, but may be expanded in the future. The BIOS memory size determination function (INT 12H) returns the system memory size less the size of the extended BIOS data area. Data for the mouse interface is stored in the extended BIOS data area.

The *return POST error log* subfunction returns a pointer to the POST error log, and the number of entries in the log. This subfunction can also be used to write an error code to the error log.

The pointing device BIOS interface supports the following functions: interface initialization, enable/disable or reset of the pointing device, sample rate setting, resolution and scaling, and reading of status and device type. The Personal System/2 pointing device is disabled at power-on; in order to use a pointing device, the device driver must be installed, the interface initialized, and the pointing device enabled.

System timer, channel 3 (the watchdog timer), can be used to detect that the channel 0 system timer interrupt is not being serviced. The watchdog timer is enabled and its counter is set using an INT 15H subfunction. If IRQ0 is active for more than one period of channel 0's clock output signal, the watchdog timer's counter is decremented. When the count reaches 0, a nonmaskable interrupt is generated.

The *Programmable Option Select* (POS) subfunction returns the base POS adapter register address and puts an adapter that is located in a particular expansion slot in setup mode. It can also be used to take an adapter in a particular slot out of setup mode and enable it.

The keyboard BIOS interface (INT 16H) has been enhanced over that provided on the PC and early ATs to support the new features of the 101-key keyboard. New functions include extended keyboard read, extended keystroke status, and extended shift status.

Advanced BIOS

The ABIOS allows an operating system to submit function requests to it rather than directly controlling system hardware. The CBIOS is designed to provide such services in a single-tasking real-mode environment. The ABIOS, on the other hand, can operate in real mode, protected mode, or in an environment where both are used (such as OS/2 with the DOS compatibility environment).

An operating system makes requests to ABIOS through transfer conventions defined by the ABIOS. These conventions work with common data areas, function transfer tables, and device blocks that link the operating system to the device function routines. These elements reside in system memory and are initialized during ABIOS initialization.

ABIOS must be initialized by the operating system before it can be used. Initialization is performed with the processor operating in real mode, using functions provided by the Personal System/2 CBIOS system services function (INT 15H). INT 15H is first called to build the system parameters table. The information returned in this 32-byte table includes pointers to the ABIOS common start, common interrupt, and common timeout routines, system stack requirements, and the number of system devices. INT 15H is called again with a different subfunction to build a table containing initialization information for each system device. A 24-byte entry is included for each system device; each entry contains the device's ID and the required lengths for the device block, request block, and function transfer table. The operating system then allocates memory, based on the initialization information, and calls ABIOS to build the device blocks and function transfer tables for each device.

Once ABIOS is initialized, requests for service are passed through a parameter block called the request block. Fields in this block identify the device to be accessed, requested operation, and other information including memory locations involved in the data transfer.

An operating system can access ABIOS services using either the *ABIOS Transfer Convention* or the *Operating System Transfer Convention*. Using the ABIOS Transfer Convention and the pointers provided at initialization time, the requestor transfers control to ABIOS via a `CALL FAR INDIRECT` to either the common start, common interrupt, or common timeout routine.

The common start routine is called to start a request. The common interrupt routine is called to resume a multi-staged request; an I/O interrupt usually indicates the completion of a stage. The common timeout routine is called to terminate a request that fails to receive a hardware interrupt within a specified amount of time.

The called common entry routine obtains the device block pointer and function transfer table pointer from the command data area. It then calls the ABIOS device routine, using the routine starting address obtained from the function transfer table.

When using the Operating System Transfer Convention, the operating system itself, rather than the ABIOS common entry routine, determines the starting address of the ABIOS device routine. This allows the operating system to store frequently used ABIOS device function addresses, and re-use them without indexing into the function transfer table. This can provide better performance when handling interrupts from character devices that call a single routine repeatedly.

An operating system that uses ABIOS provides interrupt handlers that receive control when a hardware interrupt occurs. The hardware interrupt handler must retain the logical IDs of all devices that operate on the interrupt level. The interrupt handler processes the interrupt by calling ABIOS at its interrupt entry point once for each active request block of each logical ID that shares the interrupt level, until processing is completed for the logical ID with the interrupt. For each request ABIOS returns a code that indicates if the interrupt was associated with the request block presented.

Once ABIOS has processed all outstanding request blocks for a particular logical ID, the interrupt handler is responsible for managing the end of interrupt processing; ABIOS does not reset the interrupt controller. The operating system can call ABIOS for interrupt processing with interrupts disabled or enabled. If ABIOS disables interrupts, it restores the interrupt flag to its previous condition afterward.

ABIOS provides useful support features for multi-tasking operating systems that would otherwise have to be provided in precious real-mode RAM. Issuing I/O function requests to ABIOS, rather than directly accessing hardware devices, simplifies operating system and device driver code. IBM's Operating System/2 uses ABIOS services; other multi-tasking operating systems offered for use with the Personal System/2 will doubtless do likewise.

POST (Power-On Self Test)

All Personal System/2 computers have a power-on self test similar to that used in all members of the PC family. POST is a series of tests and procedures, stored in system ROM, that receive control when the system is powered on. Once initiated, POST tests all system board devices and initiates devices installed in the system's expansion slots, as well as devices integrated on the system board.

POST supports the use of adapter boards that are initialized and controlled by adapter-specific code contained in ROM modules on the adapter board. POST searches for such ROM by examining the address range C8000H through

E0000H in 2 KB increments. A valid ROM module begins with the values 55H and AAH, followed by a byte specifying the length of the module in 512-byte blocks. An entry point for access via a `CALL FAR` instruction follows immediately afterward.

When POST finds a ROM module, it performs a checksum calculation by adding all the bytes in the module and then checking to see if the sum is zero, modulo 100H. If the sum is zero, the POST performs a CALL to the ROM code so that it can test and initialize the device. If the sum is not zero and thus the checksum is invalid, POST indicates that a ROM error has been detected.

POST, as implemented on the Micro Channel machines (Models 50, 60, and 80), incorporates several features not found in the POST provided with PC family machines and Models 25 and 30. This includes the Programmable Option Select feature, memory relocation, new error message procedures, and a power-on password facility.

Programmable Option Select (POS), described earlier in this chapter, is an identification and configuration feature for adapter boards. It eliminates the need for configuration switches and jumpers on boards, and makes initial system setup and subsequent upgrade quicker, easier, and less prone to error.

POST can disable and relocate memory blocks to form a contiguous block of memory, if it finds bad blocks during memory testing. On Models 50 and 60, this relocation can take place in 16 KB blocks, but can be done only for memory on the 80286 memory expansion board. On the Model 80, 512 KB blocks are used and this procedure can only be performed for system board memory. 1 MB blocks can be enabled and disabled on the expansion board.

Errors detected during POST are displayed as numeric codes. POST also logs them in the extended BIOS data area using an INT 15H function call. A maximum of five codes can be stored in this area. Once all error codes have been displayed, POST sounds two beeps on the speaker and waits for the Reference Diskette to be inserted in drive A.

Inserting the Reference Diskette and pressing the F1 key allows software contained on the Reference Diskette to automatically retrieve the error codes, using a different INT 15H function call, and to automatically display an error message for each error code. See Chapter 6 for additional information.

If the software on the Reference Diskette does not have a built-in error message for a particular error code, it searches the diskette for an error message file. The file must be named **@nnn.PEP**, where **nnn** is the 3-character ASCII representation of the decimal device number (including leading zeroes). The decimal device number appears as the prefix of the 2-digit error number displayed by POST. If the file is not found, a generic error message is displayed.

POST requires the entry of a power-on password before initializing the system, if a power-on password has been stored in CMOS RAM, using the Reference Diskette. An option can be set with the Reference Diskette that allows a system to operate in network server mode. Using this option, the system is initialized with the keyboard locked, and the entry of the power-on password is required

Chapter 4

only if a diskette is in drive A at initialization time. See Chapter 6 for additional information.

The use of a POST in the system architecture ensures that a machine is in a known state before any other software is run. It removes the burden of system initialization from external software and provides a well-defined hardware interface. The additional features provided in the Models 50, 60, and 80 POST make initial system setup and daily operation easier and more reliable.

5

Personal System/2 Video Subsystem Architecture

The Personal System/2 family introduces a new standard video subsystem, different from those used with the PC family. On Models 50, 60, and 80 this is the Video Graphics Array (VGA). On Models 30 and 25 a subset of the VGA functions is supplied in the Multi-Color Graphics Array (MCGA). The modes that are available for the MCGA are discussed in Chapter 2. The VGA is the standard for Personal System/2 machines that incorporate the Micro Channel. The video modes available on these machines are an extension of the existing standards of MDA, CGA, and EGA. They offer new modes with slightly increased resolution and more colors, yet provide compatibility with previous standards.

The video system is an integral part of the system board on all of the machines, instead of being incorporated in add-in boards, as on previous machines. The video connector is on the rear panel of the system units. The expansion of the video capabilities that are offered by the VGA are not as extensive as might be expected. The technology required to produce even higher resolutions is readily available. However, if the needs of the typical user are considered, the additional increase is not cost justified. Many end users are primarily interested in text modes that can give clear text for word processing, database, and spreadsheet work. These needs are filled by the VGA. As display resolution increases, the cost of the monitor required rises too steeply for most users to justify. The computer aided design (CAD) market, which needs high resolution graphics in the 1024-pixel square range, is not a large one. Machines with that type of resolution are often dedicated workstations that are not used for word processing. The CAD operator's needs, for instance, are more closely matched by features provided by add-in boards. One of these boards, the 8514/A, is offered by IBM and is discussed in Chapter 9. These extensions, or even substitutions, to the VGA are possible because of the video extension connector

Chapter 5

that is available on a Micro Channel expansion slot. Figure 5-1 shows a photograph of a display adapter card (8514/A) that has the video extension incorporated into its Micro Channel connector.

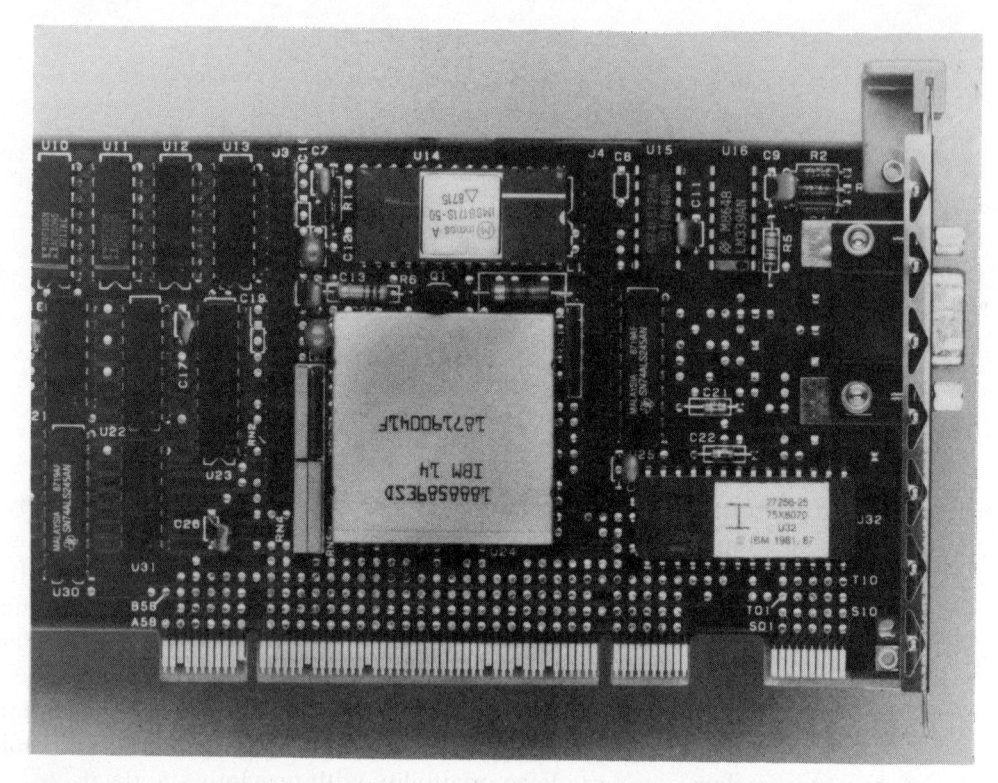

Figure 5-1 8514/A Display Adapter. *One of the expansion slots in each machine incorporates a video extension to the Micro Channel, which can be used for video boards.* Photo by Bill Schilling.

For most standard applications the VGA is likely to be the primary display system, because its resolution for graphics modes is adequate and its text modes give clear characters. The introduction of a new video standard provides a path to the future for applications that will be written to make use of the new features of the VGA, but maintains compatibility with the past by being compatible at a BIOS level with the Monochrome Display Adapter, the Color Graphics Adapter, the Enhanced Graphics Adapter, and the MCGA.

Four new modes are offered by the VGA. The 640-pixel by 480-pixel graphics mode offers 16 colors from a palette of 262,144 (256K), instead of the 2-color mode on the MCGA. The 720-pixel by 400-pixel mode in 16 colors or monochrome; the 360-pixel by 400-pixel, 16-color alphanumeric mode; and the 320-pixel by 200-pixel 256-color graphics mode round out the new modes. As on the MCGA, all 200-line modes are double scanned to produce 400 displayed lines.

Table 5-1 shows the display modes that are available on the VGA. Figures 5-2 through 5-5 show samples of the video capabilities of the IBM monitors.

Table 5-1 VGA Video Display Modes. *The VGA emulates the EGA, CGA, and MDA in addition to providing new enhanced video modes.*

Mode No. (Hex)	Display Size	Character Size	Maximum Pages	Maximum No. Colors	Description
0, 1	320 × 200	8 × 8	8	16	Text
0*, 0*	320 × 350	8 × 14	8	16	Text
0+, 1+	360 × 400	9 × 16	8	16	Text
2, 3	640 × 200	8 × 8	8	16	Text
2*, 3*	640 × 200	8 × 8	8	16	Text
2+, 3+	720 × 400	9 × 16	8	16	Text
4, 5	320 × 200	8 × 8	1	4	APA Graphics
6	640 × 200	8 × 8	1	2	APA Graphics
7	720 × 350	9 × 14	8	Mono	Text Monochrome
7+	720 × 400	9 × 16	8	Mono	Text Monochrome
D	320 × 200	8 × 8	8	16	APA Graphics
E	640 × 200	8 × 8	4	16	APA Graphics
F	640 × 350	8 × 14	2	Mono	APA Monochrome
10	640 × 350	8 × 14	2	16	APA Graphics
11	640 × 480	8 × 16	1	2	APA Graphics
12	640 × 480	8 × 16	1	16	APA Graphics
13	320 × 200	8 × 8	1	256	APA Graphics

Note 1: Modes 0, 1, 2, and 3 are CGA compatible.

Note 2: Modes 0*, 1*, 2*, and 3* are EGA compatible.

Note 3: Modes 0+, 1+, 2+, and 3+ are VGA.

These new modes allow up to 256 colors to be displayed at what, on paper, seems to be a low resolution; however, the visual effect is quite dramatic. The EGA with its 16 colors did not allow shades, but just used 16 strong colors, so an image would be composed of contrasting colors. The VGA allows a far greater selection of colors, up to 256 from a palette of 256K. This allows shaded images to be displayed that are far more satisfying to the human eye, because the images appear to be continuous-tone. For imaging applications where continuous-tone images are needed, the color selection can be in shades of a similar color, resulting in strikingly realistic images.

Expansions have also been made to the text mode capabilities. There is an enhanced RAM-loadable character generator that will accept user defined fonts. The main system ROM for the machine contains the dot patterns for three character sets that can be used. Two of these are the same fonts supplied on the monochrome board, the CGA, and the EGA for compatibility purposes. The third has a 9 by 16 or 8 by 16 (depending on the mode) character cell that is stored in ROM as an 8 by 16 dot image.

Chapter 5

Figure 5-2 8503 Screen Shot. *The least expensive monitor is the monochrome 8503, which can display graphics and gray shades.*
Photo courtesy of International Business Machines Corporation.

In order to display the number of colors, new monitors have been necessary. The original monochrome monitor for the PC, the Color Display, and Enhanced Color Display are digital devices. The red, green, and blue signals (and their intensified counterparts in the case of the Enhanced Color Display) are sent to the monitor as digital signals and converted into an analog form for display within the monitor itself. These signals result in a maximum of 16 colors that can be displayed. With the Enhanced Color Display, there are six signals instead of three, resulting in a total of 64 possible colors. The VGA system requires analog monitors; these accept red, green, and blue signals that have already been translated into an analog value. This allows continuous variation in the intensity of the

Video Subsystem Architecture

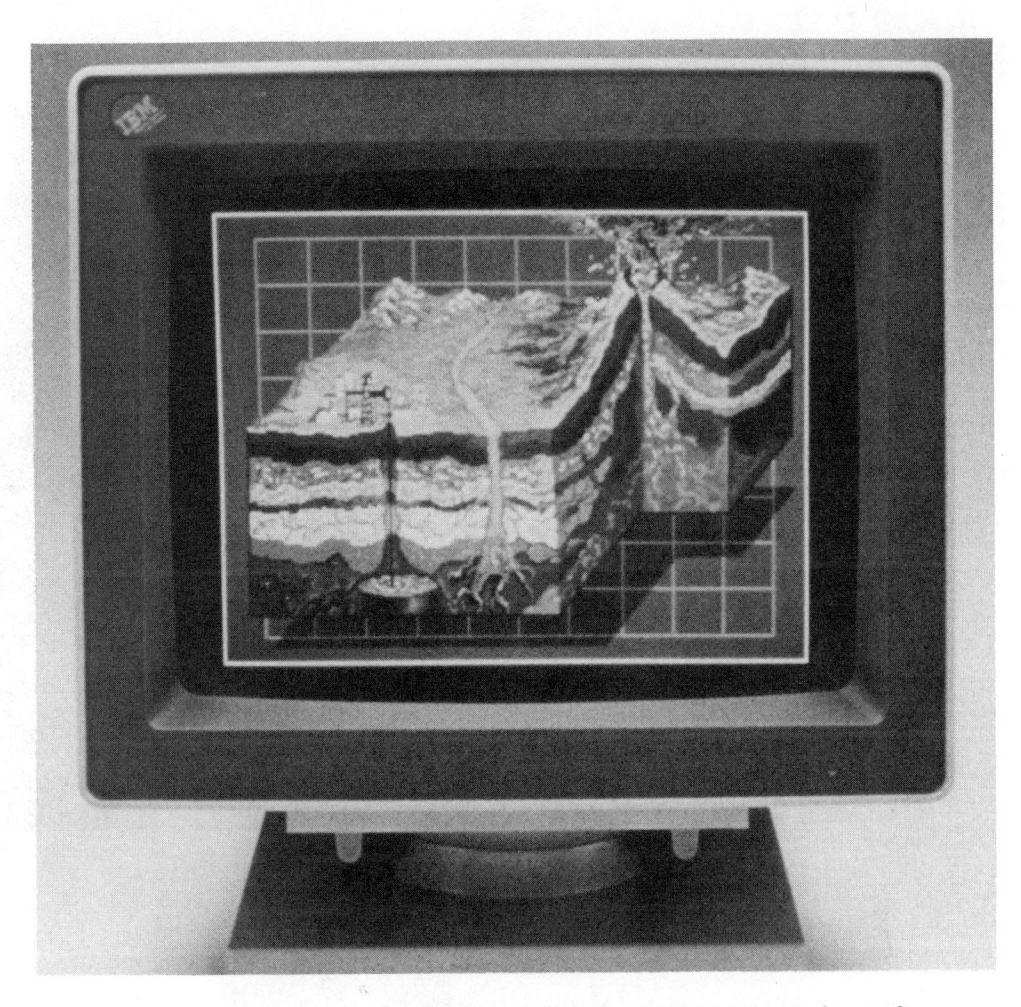

Figure 5-3   8512 Screen Shot. *The 8512 Color Display has a 14-inch diagonal screen with a stripe pitch of 0.41 mm.*
Photo courtesy of International Business Machines Corporation.

color signals and thus more colors or shades. The display that is used with the PC Professional Graphics Controller is an analog monitor, but it does not have the same electrical specifications as the new monitors and therefore cannot be used.

Four analog monitors are offered by IBM for the Personal System/2 machines: the 8503 Monochrome Display, the 14-inch 8512 Color Display, the 12-inch 8513 Color Display, and the 16-inch 8514 Color Display. Each of these displays can be used with each of the video modes offered by the VGA. Even the monochrome monitor supports the 256-color mode, for example, but the image is seen as 64 shades of gray. The monitors that are available for the Personal System/2 machines are described in more detail in Chapter 9.

131

Chapter 5

Figure 5-4 8513 Screen Shot. *The pixels on the 8513 are composed of three dots rather than the three stripes of the 8512. This produces a clearer but smaller pixel.*
Photo courtesy of International Business Machines Corporation.

Physical Description

The VGA subsystem on the Personal System/2 machines includes a large amount of VLSI. The VGA IC itself is the major component, but there are several other components that make up the display subsystem as a whole. The main elements of the video system are the same as for most display systems: video RAM, clock sources for the dot clock, CRT controller, sequencer, graphics controller, attribute controller, and digital-to-analog converter (DAC). The VGA chip incorporates four of these building blocks: the CRT controller, the sequencer, the

Video Subsystem Architecture

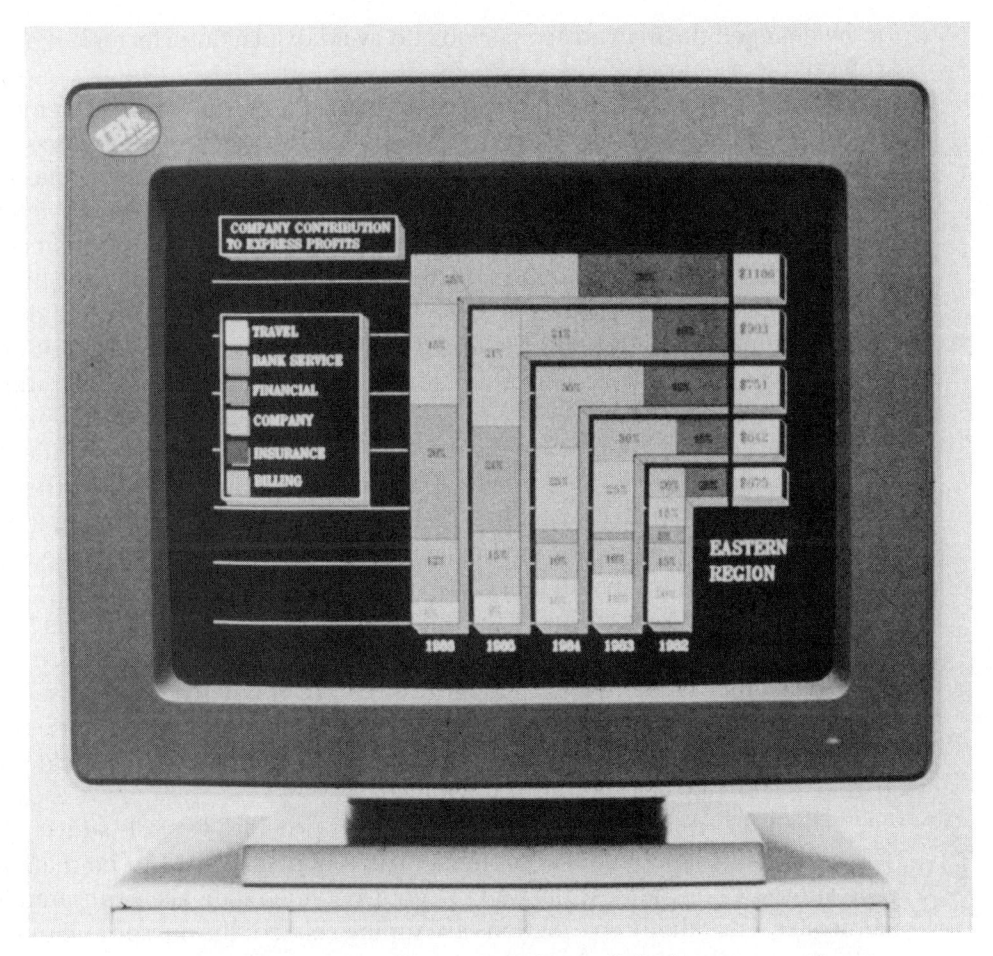

Figure 5-5 8514 Screen Shot. *This 16-inch diagonal display accommodates all of the VGA modes, but is primarily intended for use with the higher resolution 8514/A Display Adapter.*
Photo courtesy of International Business Machines Corporation.

graphics controller, and the attribute controller. For the text modes, the font used is taken from the RAM-loadable font generator. Three fonts are available in ROM; selection of a video mode in the default settings causes one of these character sets to be loaded.

The video subsystem contains 256 KB of video memory, arranged in four bit-planes of 64K by 8 bits. The starting address for this memory is programmable to three different values, to maintain compatibility with previous video adapters. The VGA IC controls the interfacing that is necessary between the system microprocessor and the video RAM. It is not possible to write the RAM directly; all data passes through the VGA. This is an improvement over CGA text mode, where the timing of writing to video RAM was important. Video RAM had to be changed during retrace periods, to avoid display interference, commonly referred to as snow.

The sequencer within the VGA eliminates this problem. This element produces the timing and arbitration for the necessary cycles. The video RAM is used in conjunction with the character clock for active display cycle timings. The active display time is used to control the regenerative memory fetches of data for display. The regenerative buffer is that portion of memory that contains the character and attribute information for text modes; it has a length of two times the number of displayed text rows multiplied by the number of displayed text columns. The video RAM is updated by the microprocessor writing to it. The sequencer performs arbitration between the microprocessor and the active display cycles, by inserting cycles that are dedicated to the microprocessor memory access between active display times. Obviously, writing to video memory is more efficient if done during inactive display times, because there is no contention for the time, but snow is no longer a side effect from using the active period. If memory maps need to be protected from being changed, the memory can be masked using the memory map registers.

The control logic circuitry for the VGA chip includes two crystals that supply the dot clock. These run at 25.175 MHz and 28.322 MHz. The choice of clock for a particular video mode is made through BIOS, when a mode is set for the video system. BIOS sets or resets a bit in the sequencer's clocking mode register. The lower frequency clock produces characters that are 8 bits wide and the higher frequency is used for 9-bit-wide characters.

The CRT controller has four functions. It provides refresh addressing for the video RAM, enabling the contents of this dynamic RAM to be maintained. The addresses of the next portion of required video data are generated in this controller. The other two functions are timing related; the cursor and underline timings, as well as the horizontal and vertical sync timings, are developed in this section.

Although the timings for the microprocessor video memory cycles are generated in the sequencer, the data itself passes through the graphics controller portion of the VGA. During the cycles when the microprocessor is writing to or reading from the video memory, the graphics controller links the dynamic RAM to the bus. During the active display times, this link is changed to connect the dynamic RAM and the attribute controller.

Consider first the CPU updating the video RAM cycle. The timings for this cycle are generated by the sequencer and controlled by the graphics controller. The graphics controller either transfers the data in each direction without modification or can be set to change the data that is being transferred between the memory and the CPU. Functions such as comparing the contents of the memory with a particular color can be done simply by setting a register. Details of the functions of the various VGA registers are discussed later in this chapter.

During the active display times the graphics controller transfers the data to be displayed between the video memory and the attribute controller. The method used depends on the video mode that is selected. For text modes, the data for each pixel and its corresponding attribute data is read into the graphics controller and is sent directly to the attribute controller. For the graphics modes, the information for display is stored in bit-maps; these are converted to serial bit-plane data and sent to the attribute controller for conversion to a color value for the DAC.

Once the data is passed to the attribute controller, it is formatted for input to the DAC. Many attributes can be assigned to a pixel and these are controlled within this section of the VGA. Some of these attributes are the blink bit, which causes pixels to flash on and off when displayed, the underline logic for displaying text characters underlined, and the colors to be assigned to the pixel. The color palette in the VGA always takes an input that is 8 bits wide. For CGA emulation modes, the 4 bits of pixel data are padded to 8 bits. For EGA emulation modes, the 6 bits of pixel data are padded to 8. In the 256-color mode, the intermediate redirection of the data input to the next stage (the DAC) is not required, because the color assignment data is already 8 bits wide.

The 8-bit value that is sent to the DAC from the attribute controller is used as a pointer to the color desired. The DAC contains a look-up table that has 256 entries at any one time. The entry in the look-up table results in an 18-bit value, which is sent to the digital-to-analog converter. This value is then converted to an analog value for each of the three color signals (6 bits of the 18-bit value are used for each color)—red, green, and blue—that the monitor requires. In addition to the 8-bit value, there is a blanking signal input to the DAC. This is used to blank the screen and, if set, results in a zero output from the DAC to the monitor, regardless of the color selected for the pixels.

The contents of the look-up table can be altered by the system microprocessor. Each entry is a register, whose contents are changed by the microprocessor with independent timing to the rest of the video system. The write enable and read enable signals are used by the CPU to access these registers. BIOS commands that can be used to load and read these registers are provided and, wherever possible, should be used. The BIOS can access these registers and automatically provides the necessary gray scales for the monochrome monitor. The DAC itself does not change mode; the color look-up table is modified and only the green signal is used to drive the monitor. This results in one of 64 shades of gray being displayed for all display modes, when a monochrome monitor is acknowledged by the system.

The text fonts are displayed by the use of the character generator in the video subsystem. This is RAM loadable and can contain up to eight 256-character fonts or four 512-character sets. If an alphanumeric mode is selected, one of the ROM fonts is automatically loaded into the character generator. Three fonts are supplied in ROM. These can be loaded into the generator or user-defined fonts can be loaded. The user-defined font can be up to 32 scan lines in height. If 512 characters need to be used, instead of just 256, the character map select register can be programmed so that setting or resetting bit 3 of the attribute

Chapter 5

byte associated with a character cell switches between the first and second half of the character set.

In addition to the system board hardware that provides the video subsystem, there is an expansion to the Micro Channel for video boards. On one slot, the connector is extended to offer signals that can be used by an add-in board to change the video capabilities of the system. The video signals that are available on the Micro Channel are discussed in Chapter 4. The VGA can be enabled or disabled as desired via this connector. This enables other video boards in the system to bypass the VGA and use the DAC for output. When disabled, the VGA ignores video memory and I/O reads and writes; however, the current contents of registers and video memory are preserved.

Video Modes

The VGA offers compatibility with the previous display adapters. Monochrome, CGA, and EGA modes are all supported. In addition, the VGA is a superset of the MCGA video system on Models 25 and 30. The video modes that are supported are shown in Table 5-1. Monochrome compatibility is offered through mode 7 and EGA compatibility through modes 0*, 1*, 2*, 3*, 4, 5, 6, 7, DH, EH, FH, and 10H. CGA compatibility is provided through modes 0, 1, 2, 3, 4, 5, and 6. Modes 11H and 13H are new modes that are also offered on Models 30 and 25. A new mode that is available only on the VGA-based machines is mode 12H, which has the same resolution as mode 11H except that it can simultaneously display up to 16 colors instead of 2 colors.

The BIOS uses INT 10H as the interrupt for video functions. Within INT 10H there are twenty-three different video functions that can be performed. These include reading and writing the cursor position and setting the video mode, as well as determining the display that is attached to the machine. The procedure for accessing one of these routines is the same as on the PC family of machines. The required function number is loaded into the AH register and then INT 10H is executed. The AH register contents are used to vector the interrupt to the desired subroutine in the INT 10H interrupt handler.

In the Personal System/2 machines, the function values are the same as on the PC machines, but the functionality has been extended where appropriate. Each of the functions and a description is listed below.

AH = 00H The set mode function is used to set the video mode of the machine. As with previous machines, the value that is selected in the AL register is the hexadecimal value for the mode required. If the monochrome display is attached to the system, the default video mode is 7H, the monochrome text mode. The default video mode with a color display attached is 3H, the 80-column by 25-row text mode. The Personal System/2 machines do not use the color burst signal that was available on the CGA display adapter. As a result, modes 0, 2, and

5 are the same as modes 1, 3, and 4, respectively. The EGA feature that preserves the video buffer during a mode set, by setting bit 7 of the AL register, is also present on the VGA. As on previous adapters that could support graphics, the cursor is not displayed when the video subsystem is in a graphics mode. When a mode is set by the BIOS, the first 64 color registers are set and the remaining 192 are unspecified, except for the 256-color mode 13H, where all of the registers are defined.

AH = 01H This function sets the cursor type and chooses the top and bottom lines for the cursor that is to be displayed when a text mode is selected. The least significant 4 bits of the CH and CL register pair are used, with CH containing the top line in the character cell for the cursor and CL containing the bottom line. Only one cursor type can be set at one time on the machine. Changing the cursor through this mode overrides any previous setting, even if a different mode was active when the cursor was last changed.

AH = 02H The set cursor position function uses three different registers. The DH and DL register pair is used for the row and column positions, respectively. The BH register indicates the page number for the cursor position. The maximum value that can be placed in this register depends on the video mode. Table 5-1 shows the maximum number of pages for each video mode.

AH = 03H This mode is the inverse of the set cursor position mode. It is used to read the cursor position. Again, the BH register contains the page number, and DH and DL contain the row and column positions, respectively. The cursor type is also reported in CH and CL, where CH contains the top position for the cursor and CL contains the bottom position within the character cell.

AH = 04H The read light pen position is not supported on the VGA or on the Model 30's MCGA. However, if the function is invoked, AH is returned with the value 0 and BX, CX, and DX are all changed.

AH = 05H The AL register sets a new active page for display. As with the set cursor position function, the value that is placed in the register must be valid for the video mode selected. This function is the same as on previous PC video adapters.

AH = 06H This function scrolls the active page up. The register pair CH and CL contain the row and column of the top left-hand corner of the area to be scrolled, and DH and DL contain the row and column of the bottom right-hand corner of the area to be scrolled. The AL register contains the value for the number of lines that are to blanked at the base of the window; the value 0 blanks the entire window. Finally, the BH register indicates the attribute that is to be used on the blank lines.

AH = 07H The scroll active page down function is the inverse of scroll active page up. CH, CL, DH, and DL contain the row and column of the top left-hand and bottom right-hand corners of the window, respectively. The contents of the BH register indicate the attribute to be assigned to any blank lines. The AL register contains the number of lines that are to be blanked at the top of the window.

AH = 08H This function reads the character and its attributes at the current cursor position. The page to be read is indicated by the BH register, which must contain a valid page number for the selected video mode. The AH and AL register pair contain the character and its attributes. The contents of AH, which contains the attributes, are only valid for text modes where attributes are applicable.

AH = 09H This function is the inverse of the read character and attributes function, but is more powerful. It should be used for graphics modes and function 0AH should be used for text modes, if the attribute of the character does not need to be changed. To write a character and its associated attributes to the current cursor position, the BH, BL, CX, and AL registers are used. BH contains the page number to be written to and should contain a valid page number for the video mode that is selected, except for mode 13H. In this 256-color mode, the value in BH is the background color to be displayed.

AL contains the character that is to be displayed, and BL contains its attributes for text modes and its color for graphics modes. In addition, when in graphics modes, setting the most significant bit of BL can be used to perform a logical XOR on the current color value. Again, this feature is not available in the 256-color mode 13H.

The CX register allows the same character to be written to the current and subsequent characters. The value in CX is the number of characters that are to be the same. Note, however, that if the video system is in a graphics mode, the character count should not extend beyond the end of a row or unpredictable results may occur. There is one additional difference in the operation of this function for different graphics modes. The CGA graphics modes 4, 5, and 6 need an additional pointer to reach the second 128 characters of the character set. For other modes, the full 256 characters are in system ROM; but for these three graphics modes, INT 1FH (location 0007CH) is used to obtain the pointer value for the second half of the character set.

AH = 0AH Writing a character to the current cursor position is simplified for text modes if the attribute does not need changing. This function uses the BH register for the current page number and the AL register for the character value. CX can be used to write the same character in successive locations, as with function 09H.

AH = 0BH The set color palette function works on the Personal System/2 machines in the same way as on the EGA. The BH and BL registers are used with this function. If BH is set to 0, BL contains the number of the color to be displayed. If the video mode is 320 by 200 graphics, the background color is set. If the video mode is a text mode, the text itself is displayed in the chosen color. If the 640 by 200 graphics mode is the current display mode, the foreground color is set to the color chosen by the contents of BL. In this case, if the color value is 0, the background color is set. The value of BH can be set to 1 for the 320 by 200 graphics modes. In this video mode, with BH equal to 1, the palette can be selected. Palette 0 contains the green, red, and brown colors. Palette 1 contains cyan, magenta, and white.

AH = 0CH   This function writes a dot on the screen in a graphics mode. AL contains the color value of the dot. In modes where more than one page is available (modes DH, EH, FH, and 10H), BH contains the page number. The DX register contains the row number and the CX register contains the column number, with both values having units of dots. The AL register contains the color value for the dot, unless the most significant bit of the register is set to a 1 when the color value is XORed with the current value of the dot. This XOR function is not performed for the 256-color mode 13H.

AH = 0DH   The value of a dot can be read via this function. DX and CX contain the row and column values, respectively, of the required dot, and BH indicates the current page number. The value for the dot is returned in the AL register.

AH = 0EH   This function causes the system to write in teletype mode to the active page. As is usual in a teletype mode, the bell, carriage return, line feed, and backspace are not printable characters and are executed. The AL register contains the character that is to be written to the screen and the BL register selects the color for the foreground when a graphics mode is selected. The width of the screen is dependent on the current video mode that is set.

AH = 0FH   This function is the inverse of function AH = 00H, allowing the current video mode to be read. AH contains the number of columns on the screen in units of characters rather than dots. AL contains the current video mode number and BH contains the current active page in that mode.

AH = 10H   This function sets the palette registers. These registers can be set for EGA-compatible modes and for VGA-specific modes. There are ten different ways in which this function can be used. The first four are as on the EGA. With AL equal to 00H, the individual palette register pointed to with BL is set to the value in the BH register. With AL equal to 01H, the overscan register is set to the value that is in BH. With AL equal to 02H, all of the palette registers and the overscan register can be set at once. The address that is composed of the contents of ES and DX points to a 17-byte-long table, where the first 16 bytes are the values for the palette registers and the seventeenth byte is the value for the overscan register. With AL equal to 03H and BL equal to 0, the intensify bit is set; setting AL equal to 03H and the least significant bit of BL to a 1 turns off the intensify bit and sets the blinking bit. Characters with this attribute bit set will blink rather than appear in high intensity.

The other values for the AL register in this function are new for the VGA. IBM has reserved the functions with AL equal to 04H, 05H, 06H, 11H, 14H, 16H, 18H, and 19H, so they are not supported. The function with AL equal to 08H allows the overscan register to be read, with the value returned in BH. The palette registers have also been changed so that they can be read. This function with AL equal to 09H works in a similar fashion to its write counterpart. The 17 bytes that are returned after the function are placed in memory, starting at the address that is pointed to by the ES and DX registers combined. The first 16 bytes contain the palette register settings and the last byte contains the overscan value. Individual color registers can be set by using AL equal to 10H. BX contains the number of the color register that is to be set. The three color values for the red, green, and blue components should be in registers DH, CH, and CL, respectively. A block of color registers can be set at once, with AL equal to 12H. Again ES and DX point to the starting address of the color table, which contains the colors for each of the registers in the format of red, green, and blue component in turn for each register. The number of color registers to be set should be in CX and the number of the first color register should be in BX.

Further palette register settings are available by making AL equal to 13H. This mode is called the select color page mode. There are 256 color registers that are available on the VGA. With the exception of the 256-color mode (mode 13H), 64 registers maximum are used at once. As the default, a video mode set through INT 10H and AL equal to 00H causes the first sixty-four color registers to be initialized and made active. The other three blocks of color registers can, however, be set or the registers can be split into sixteen sets of sixteen registers. It is important not only to select the type of paging mode to be used, but to initialize the contents of the registers as well. Setting BL equal to 00H allows the paging mode to be selected. BH equal to 00H selects the 64-color register mode and BH equal to 01H selects the 16-color register mode. If BL is set to 01H, the page number can be chosen. In 64-color register mode, the first block of registers is activated with BH equal to 00H, the second with BH equal to 01H, the third with BH equal to 02H, and the fourth with BH equal to 03H. In 16-color mode, the page numbering is similar: BH equal to 00H selects the first block and so forth, to BH equal to 0FH selecting the sixteenth block.

An individual color register can be read with AL equal to 15H and the BX register set to the number of the register. The red, green, and blue components of the palette register are returned in the DH, CH, and CL registers respectively. When AL is equal to 17H, a block of color registers can be read at once. In this case, the first register to be read is put into register BX and the number of registers to be read is in CX. The ES and DX registers combined provide the address pointer for the values to be placed. Each register's contents consist of 3 successive bytes in this table, with the red, green, and blue components being entered successively. When AL is equal to 1AH, the complementary function of AL equal to 13H is obtained and the current paging mode can be read. The two available modes are the 64-color register mode and the 16-color register mode. The mode setting is returned in BL. BH contains the current page within that mode.

The final palette register setting produces gray shades. When AL is equal to 1BH, BIOS can be made to translate some or all of the color register settings to gray shades for display on the monochrome monitor. The weighted sum that is used takes the value of the red, green, and blue components and multiplies the red portion by 30 percent, the green portion by 59 percent, and the blue portion by 11 percent. The results are added together and the original data in the color register is replaced, with each section having the same calculated value. The monochrome monitor itself only uses the green value, but the blue and red components are also set. BX contains the number of the first color register to be summed and CX contains the number of registers to be changed.

AH = 11H This character generator mode has been available since the arrival of the EGA; thus, it is available for EGA- and VGA-compatible modes. This mode resets the video environment but maintains the regenerative buffer. In general, the function calls with AL equal to 00H, 01H, 02H, and 04H are similar to the calls with AL equal to 10H, 11H, 12H, and 14H; however, the length of the regenerative buffer is recalculated, and the cathode ray tube controller (CRTC) registers are reprogrammed with new values for the number of rows on the screen and the number of characters on a row.

When AL is equal to 00H, a user-generated character set can be loaded into the character generator. The ES and BP registers together point to the character table in RAM. CX contains the number of characters in the set, usually 256 or 512. DX contains the character offset in the table; BL specifies which of the eight registers (character set 0 through 7) are to be loaded; and BH contains the number of bytes in each character.

When AL is set to 01H, the 8 by 14 monochrome-compatible dot pattern can be loaded into the character generator from ROM. BL defines the character generator's block number (character set 0 through 7). With AL set to 02H, the CGA-compatible 8 by 8 character set is loaded into the character generator and again BL specifies the block number.

When AL is set to 03H, the character set to be used is selected. This function is only valid in text modes and has a confusing format. Remember that although character sets with 256 characters are the default setting, by programming the character map select register it is possible to switch to another character set. Then, by setting bit 3 of the character attribute byte to a 1, a character set with 512 characters can be accessed. If the character map select register is not set to switch character sets, the character is displayed as an intense character when bit 3 of the character attribute byte is set. With AL equal to 03H the blocks that are to be used with the attribute bit 3 setting are chosen.

The character set block to be used when bit 3 of the character attribute byte is a 0 should be placed in bits 4, 1, and 0 of the BL register, with bit 4 being the most significant bit and bit 0 being the least significant bit. Bits 5, 3, and 2 of the BL register contain the block number for the character set to be used when bit 3 of the character attribute byte is set. Bit 5 is the most significant bit and bit 2 the least significant. If the second block number is the same as the first, setting bit 3 of the character attribute byte causes the intensified character to be displayed, as if the character map register was not set to switch character sets.

The new additional ROM font, which is stored in ROM as an 8 by 16 dot image and is displayed as a 9 by 16 or 8 by 16 (depending on the mode) character, can be loaded into the character generator by setting AL equal to 04H. BL should be set to contain the block number that is required. This is the default font that is loaded when the machine is started.

The remaining function calls for the character generator should only be used immediately after setting a mode in order to ensure predictable results.

The function calls where AL is equal to 10H, 11H, 12H, and 14H cause the CRTC registers to be reprogrammed and the regenerative buffer length to be changed. The new values are based on checking the number of bytes in a character (which is supplied in the BH register during the function call) and recalculating the number of rows on a screen. The number of rows on the screen is equal to the integer portion of the number of scan lines, divided by the number of bytes per character minus one. The regenerative buffer is then set to be twice the number of rows on the screen, multiplied by the number of columns on the screen.

Five values in the CRTC registers are changed. The maximum scan line value is set to one less than the number of bytes per character. The cursor start point is two less than the number of bytes per character, and the cursor end point is one less than the number of bytes per character. If the current video mode is 7H, the MDA-compatible mode, the underline position is also set to one less than the number of bytes per character. Finally, so that the correct number of rows of characters are displayed, the vertical displacement end value is set. For 200 scan line modes where the image is double scanned, this value is equal to twice the number of rows multiplied by the number of bytes per character minus one. For the 350 and 400 scan line modes, this value is simply the number of rows multiplied by the number of bytes per character minus one. The active page for these modes is always page 0. The register values required and the functions performed are identical to the comparable modes with AL equal to 00H, 01H, 02H, and 04H apart from these recalculations.

When AL equals 20H, the pointer for user-generated graphics characters is set up so that the video subsystem can access them, using INT 1FH. The ES and BP registers together make the address that points at the table. When AL equals 21H, the pointer for user-generated graphics characters is set up so that the video subsystem uses INT 43H. Again, the ES and BP registers make up the address that points to the table. CX contains the number of bytes per character and BL specifies the number of rows. If BL is set to 0, the user can customize the number of displayed rows by setting the DL register. If BL is set to 01H, 14 rows are displayed. If BL is set to 02H, 25 rows are displayed, and if BL is set to 03H, 43 rows can be displayed.

The number of rows to be displayed using the ROM fonts is defined by setting AL equal to 22H, 23H, and 24H. In each case, BL contains the row specifier and has the same syntax as when AL is equal to 21H. The 8 by 14 ROM font, 8 by 8 ROM font, and 8 by 16 ROM font are affected by setting AL equal to 22H, 23H, and 24H respectively.

The last character generator function call available is when AL is equal to 30H. This call returns information on the fonts. The number of bytes per character is returned in CX. DL contains one less than the number of rows of characters on the screen. ES and BP hold the address that points to the table containing the characters. BH should be loaded with a value that indicates the font of interest. BH values equal to 00H through 07H are valid and refer to the following font pointers respectively: current INT 1FH, current INT 43H, 8 by 14 ROM font, 8 by 8 ROM font, top of 8 by 8 ROM font, 9 by 14 alternate ROM font, 8 by 16 ROM font, and 9 by 16 alternate ROM font.

AH = 12H The alternate select function call is used, in part, to determine how much of the memory installed on the video subsystem is available, and to determine the current switch settings. These functions vary on the VGA, because they are dependent on the video mode rather than on the hardware that is installed in the system. If BL is equal to 10H, information is returned on the EGA-compatible modes. On return, if BH is equal to 0, the current video mode is color and the address range for the video registers is 3DxH. If, however, BH is equal to 01H, the video registers are addressed at 3BxH and the current video mode is monochrome. Register BL contains the amount of memory available for EGA emulation. Although there are 256 KB on the VGA, the amount of this memory that is available for use depends on the current video mode. The register pair CH and CL show the software and hardware switch settings respectively. The hardware switch settings on the VGA do not vary, but they can be set by using the software equivalent functions.

There are several other alternate selections possible through the function call AH = 12H. If BL is set to 20H, the alternate print screen routine can be used. If BL is equal to 30H, the number of scan lines that are to be displayed after the next mode set is chosen. This is done by setting AL equal to 0, 1, or 2 for 200, 350, or 400 scan lines, respectively. On returning from the function call, AL equal to 12H indicates that this mode is supported by the system.

If BL is set to 31H, the contents of various color palette registers can be preserved with their present settings and the default palette loading is disabled. With AH equal to 00H and AL equal to 01H, the default palette loading is disabled and the EGA palette registers, the 256 color registers, and the overscan registers cannot be changed during a mode set. To enable setting of these registers during a mode set, make AH and AL equal to 00H for the function call.

The video I/O port and the regenerative buffer for the currently active display can be enabled and disabled by setting BL equal to 32H and AL equal to 0 and 1, respectively. A return value of 12H for AL indicates that this feature is supported. In the same way, the gray shade summing can be enabled and disabled by setting BL equal to 33H and AL equal to 0 and 1, respectively. AL returns a value of 12H if the feature is supported. Cursor emulation is enabled and disabled by setting BL equal to 34H and AL equal to 0 and 1, respectively. AL returns 12H if the function is supported.

It is possible to run two video systems in the Personal System/2 machines. One is the VGA incorporated into the system board and the other is an adapter card that can make use of the video extension in the Micro Channel. If there is no conflict between the two systems and if any given function is only available on one of the boards, then they can coexist. If, however, there is an overlap situation—for example, the BIOS data area is used by both video subsystems—the two displays each need to be selected as required. The VGA can be disabled via a function call and any installed adapter board also needs to be accessible for display switching.

The VGA video subsystem is the secondary video system in a conflict situation and remains disabled until the systems are switched via the alternate select function call. With BL equal to 35H, the display systems can be switched. The AL register is used to select the desired function. The first time display switching is to be activated, AL equal to 0 and 1 is used. Subsequent video switching is done with AL equal to 2 and 3. If AL is equal to 00H, the initial video adapter is turned off, and ES and DX combined point to 128 bytes of RAM that contain the switch states for the video. If AL is equal to 01H, the initial video adapter is turned on. AL equal to 02H disables the active video; again ES and DX point to the switch state buffer. AL equal to 03H enables the inactive video and the contents of ES and DX point to the buffer that contains the saved status information. Again, a return value 12H for AH shows the function is supported. The video screen itself can be switched on and off through the same function call, with BL equal to 36H and AL equal to 0 for screen on and 1 for screen off. AL equal to 12H on return shows that the function is supported.

AH = 13H There are four different ways to write a string of characters to the video subsystem through this function call. If the string contains characters such as carriage return or line feed, they are executed in the same way as they would be in teletype mode and are not displayed. In all cases, the ES and BP registers point to the string that is to be written; CX contains the number of characters to be written (note that this value does not include a count of the attribute bytes); and BH contains the page number that is to be written to. When AL is equal to 0, BL contains the attribute that applies to each of the characters in the string. The characters themselves are stored in the buffer pointed to by ES and BP. For this function call, the cursor is not moved. AL equal to 01H is the same function call except that the cursor is moved. When AL is equal to 02H, the cursor is not moved and the string consists of each character followed by its associated attribute. AL equal to 03H performs the same function except that the cursor is moved. The string arrangement that stores the character along with its attribute is only valid for text modes.

AH = 1AH This function call can be used to read the display type of the monitor attached to the system or to switch between two attached monitors. To read the attached displays, AL should be set to 0. On return, the AL register will contain 1AH to show that the function is supported. BL will give the display code of the active display, and BH the display code for the alternate display. If AL is set to 1, the display code for the active and alternate displays can be written. The AL register returns 1AH to show that the function is supported, BL contains the active display code, and BH contains the alternate display code. The display codes for the four Personal System/2 analog displays attached to the VGA are

07H and 08H for the monochrome and color displays, respectively. A returned value of 0 means that no display is attached. A value of −1 is an unknown display type.

AH = 1BH This new function call for the Personal System/2 machines is a comprehensive list of the functionality and status of the equipment that is being checked. The BX register should be set to 0 to check the information for Models 50, 60, and 80. The ES and DI registers point to the address where the information is to be stored on return from this function. For verification that this function is supported, the AL register contains 1BH on return. A buffer size of 40H bytes is required and the information is stored in known positions relative to the DI value. Table 5-2 lists the information that is returned by the system to this buffer. In addition to the dynamic information that is stored in this buffer, there is a static functionality information table that is pointed to by the table providing the dynamic information. This gives information on the video modes that are supported by the VGA, the different scan line modes that are available, and palette information that does not change with the video mode setting. The IBM Personal System/2 BIOS Interface Technical Reference manual provides further information on these video functions.

Table 5-2 Functionality and Status Call. *This function call returns extensive information about the video system, its current status, and the equipment available.*

Offset (Hex)	Length	Contents
00	Word	Static functionality information offset
02	Word	Static functionality information segment
04	Byte	Current video mode
05	Word	Number of columns on screen
07	Word	Length of regenerative buffer
09	Word	Starting address of regenerative buffer
0B	Word	Cursor position for each page
1B	Word	Cursor type
1D	Byte	Active display page
1E	Word	CRTC address
20	Byte	Contents of 3x8H register
21	Byte	Contents of 3x9H register
22	Byte	Number of rows on screen
23	Word	Character height
25	Word	Active display code
26	Byte	Alternate display code
27	Word	Number of colors
29	Word	Number of display pages
2A	Byte	Scan lines
2B	Byte	Primary character block number
2C	Byte	Secondary character block number
2D	Byte	Miscellaneous status
31	Byte	Amount of video memory
32	Byte	Save pointer state information

AH = 1CH The video state can be saved and restored on the Personal System/2 machines. This will have increasing significance as multi-tasking operating systems become more available, because the need to switch between tasks and to preserve the video state for each task will become important. On the Personal System/2 machines, three areas can be preserved through this function call: the video hardware state, the video BIOS data area, and the contents of the DAC and color registers. The CX register is used to indicate which of these areas is to be preserved, with bits 0, 1, and 2 being set to indicate the preservation of the hardware state, the video BIOS data, and the DAC respectively. If AL is equal to 00H when the function is called, its contents are changed to 1CH on return, to indicate that the function is available. The BX register contains the number of 64-byte blocks that are necessary to preserve the states that have been requested. Then when AL is made equal to 01H, the state can be preserved, CX should contain the requested states, and ES and BX should point to the save area. On return, the function available indication is AL equal to 1CH. This function call alters the current video state, so a restore state operation should be requested to ensure a known video state before continuing. To restore the video state at any time after it has been saved, set AL equal to 02H. CX should contain the states to be reinstated and ES and BX should point to the area where the state was preserved.

All 200-line modes are double scanned on the Personal System/2 machines. This means that the data for the first scan line is displayed on line one of the display; the line underneath it is line two of the display, rather than the second-scan-line-data. The stored image is only 200 lines high, but the displayed image is 400 lines high. Each pixel on the screen can be considered twice as high as the stored pixel image.

Memory Arrangement

To accommodate previous display adapters and provide a more sensible arrangement for the future, the video memory can be arranged in several ways. There are three starting addresses for the video memory. The BIOS automatically programs the VGA with the appropriate starting address during a video mode set. Modes 0H through 6H have a starting address of B8000H, mode 7 has a starting address of B0000H, and modes DH through 13H have a starting address of A0000H. The 256 KB of video RAM is arranged in four 64 KB maps.

The data format for the text modes, modes 0 through 3 and mode 7, is the same as on previous adapters. Each character consists of 2 bytes of data: 1 for the character and 1 for its attribute. The attribute byte can be used to access the second half of a 512-character set if desired. This process is described earlier. The character byte appears at the even numbered address and the attribute byte appears at the odd numbered address, starting at B8000H for modes 0 through 3 and B0000H for mode 7. Modes 0 and 1, which are 40-column modes, require 2,000 bytes of RAM per page, with room for eight pages in the video RAM at once. The other text modes require twice as many bytes to store the character information. Modes 2, 3, and 7 support a border, and up to eight pages can be saved at one time.

The graphics modes have a different memory map arrangement, depending on the video mode. For the modes that are compatible with the CGA—modes 4 through 6—the starting address of the memory map is B8000H. For modes 4 and 5, the pixels are arranged so that each pixel on the screen consists of 2 bits of information. These bit pairs select 1 of 4 colors in the current palette. The bits themselves are arranged in memory so that the data for the first scan line and each subsequent odd-numbered scan line is mapped, starting at B8000H, and the data for the even numbered scan lines is located in memory, starting at address BA000H. For mode 6, each pixel is represented by a single bit in memory. The pixel is displayed as either the foreground or background color, depending on the bit's setting. The memory arrangement is similar to modes 4 and 5 with storage of the odd scan lines starting at address B8000H and the even scan lines starting at BA000H. Video mode 6 supports a setting for the border color. Modes 4, 5, and 6 require 16,000 bytes of RAM for each screenful of pixels, and only one screenful can be stored in video RAM at one time.

With the 16-color graphics modes—DH, EH, 10H, and 12H—the data for the screen is arranged in four bit-planes. The four bit-planes enable 1 of 16 colors to be assigned to each pixel. They are numbered C0 through C3 for the blue, green, red, and intensified components respectively. The first pixel is composed of the most significant bit (MSB) of each of the bit-planes' bytes. So the byte at address A0000H contains the components of the first 8 pixels to be displayed, with the data in the most significant bit (MSB) being the first pixel, and the data in the least significant bit (LSB) being the eighth pixel. The amount of memory occupied by a page of information is different for each of the 16 color graphics modes. Modes DH, EH, 10H, and 12H take buffers that are each 8,000, 16,000, 28,000, and 38,400 bytes long respectively, to store the information for a complete screen of pixels. Up to four pages can be stored at one time for modes DH and EH, but only two pages for mode 10H and one page for mode 12H can be accommodated.

The all-points-addressable monochrome graphics mode is mode FH. The data for this screen mode is stored in two bit-planes, each 28,000 bytes long; one bit represents the presence or absence of a pixel and the other its intensity. The bit-plane C0 stores the bit information and C2 stores the intensity information. Eight pixels are stored in a byte of a bit-plane, with the MSB of the byte being the first pixel to be displayed and the LSB being the eighth pixel to be displayed. Up to two pages can be stored in the video RAM simultaneously.

Mode 11H can provide an image that is 640 pixels wide by 480 pixels high in APA graphics, with 2 colors chosen from a palette of 256K. A border is supported. The data for the pixels is stored in a single bit-plane (C0) with a starting address of A0000H. The format of the pixel storage is the same as for mode 6 in that each bit represents a pixel displayed either in the foreground or background color. However, the map for the page is stored linearly and is not split into odd and even scan lines, as it is for mode 6. This mode requires 38,400 bytes of memory to store a page of pixels. Only one page of data can be stored in the video RAM at one time.

Mode 12H is the 16-color version of mode 11H. For this mode, the data for the pixels is stored in four bit-planes, C0 through C3. Each pixel occupies 1 bit in each bit plane and a border is supported. The four buffers for this data are 38,400 bytes in length and start at address A0000H. Only one page of data can be stored in the video RAM at one time for this mode.

Mode 13H is the 256-color mode, where each byte contains the information for 1 pixel. The data is stored linearly in memory, starting at address A0000H, and when displayed is double scanned. A border is supported. Only one page of data can be stored in the video RAM at once.

Hardware Registers

In addition to the multitude of video function calls that are accessible via BIOS routines, it is also possible to access the registers in the system directly. Many of these registers are similar to the registers available on the EGA, but the form of them may be changed on the VGA. For example, most of those that were write only, can now be read as well. The registers fall into five main areas that correspond to the main functions of the video subsystem: attribute registers, CRTC registers, sequencer registers, graphics registers, and general purpose registers.

Attribute Controller Registers

There are seven attribute controller registers. The *attribute address* register is a read/write register that is used to provide the addresses to load the palette registers or to point to the *attribute data* register. There are sixteen palette registers that can be loaded or read. Each of these is a 6-bit register that acts as an address into the DAC itself. These palette registers should only be updated during the retrace period to prevent snow. The *read/write color select* register contains the data necessary to change between colors in the DAC rapidly. The values loaded are used to provide the high order bits for the output of the attribute controller that is sent to the DAC.

The *attribute mode control* register is used to perform several functions and can be written to or read. One bit selects monochrome emulation or color emulation; another selects text or graphics mode. The enable line graphics character codes allow the eighth bit in a text mode to be replicated into the ninth bit, to allow graphics line drawing characters to appear linked on the screen. This is generally used in the monochrome emulation mode. BIOS automatically sets this bit for the appropriate mode. When this bit is a 0, the ninth bit is displayed as the background color. The color selection for the 256-color mode is different from the selection for other modes. This is partially controlled by the attribute mode control register. In addition, one of the bits in this register is used when a portion of the screen is to be panned.

The attribute control registers also include the *read/write overscan* register, which is used to determine the color for the border that is displayed in some of the video modes. This border is displayed in video modes 2H, 3H, 6H, 7H, and EH through 13H. The border is as wide as a single 80-column character. The *color plane enable* register is used to activate some or all of the color bit-planes stored in the video memory and can be read from or written to. Additionally, this register is used to accept the data that is addressed by the attribute address register.

The *horizontal PEL panning* register is also a read/write register in the attribute controller section of the VGA. The video image can be moved a certain number of pixels to the left across the screen; this register is used to determine the number of pixels. The value that is actually assigned to the register varies with the video mode. For most video modes, the value can be between 0 and 7. For the 0 + , 1 + , 2 + , 3 + , 7, and 7 + video modes, the image can be panned up to 8 pixels at a time, but in the 256-color mode this is restricted to 3 pixels.

CRTC Controller Registers

Within the CRTC are twenty-six registers that are all related to the timing of the image displayed on the screen. The *CRTC address* register points to the required address in the controller, and each of the other registers is indexed from that address. The time for the horizontal and vertical retrace period is calculated from the value that is loaded into the *horizontal total* register and the *start and stop horizontal retrace* registers. This register contains five less than the number of characters in a horizontal scan line plus the retrace period. The *horizontal display enable end* register is set to one less than the number of characters to be displayed on a line, and the *offset* register determines the line width of the display. The *start and stop horizontal blanking* registers dictate the timing of the horizontal blank interval in the display sequence. The *vertical retrace start and end* registers create the vertical retrace period, while the *vertical total* register contains the total number of scan lines that are in a full display cycle, minus two. The *vertical display enable end* register defines the end of the displayed image, and the *start* and *end vertical blanking* registers define the vertical blanking period. The *CRT controller overflow* register is used to contain the excess bits of several registers, such as the vertical retrace start, and the vertical display end registers. The start address for the start of the video image after the vertical retrace period is contained in the *start address high* and *start address low* registers. The *preset row scan* register can be used to program the number of the first scan row to be displayed after the vertical retrace period.

The *maximum scan* register creates the 400-line display from the 200-line long image. It also creates the number of scan lines that are in each row of characters. The *cursor start* and *end* registers and the *cursor location high* and *low* registers position the cursor on the screen. The *underline location* register determines the position of the underline in the character. The *line compare* register can be used to scroll part of the screen image yet leave the rest, say a menu line, unaffected by the scroll.

The *mode control* register in the CRTC performs several controlling functions. This register can be used to reset or enable the horizontal and vertical retrace or to select doubleword, word, or byte addressing. Among its other functions, this register provides a bit for compatibility with the CGA graphics modes, which were designed originally to use a 6845 CRTC. The horizontal retrace select bit can divide the vertical timing clock by two and double the vertical resolution of the CRTC.

Sequencer Registers

There are six sequencer registers in the VGA. The *sequencer address* register is used to point to the data registers within the sequencer. The other registers are indexed from this address register. The *reset* register has two modes—synchronous and asynchronous. The synchronous reset stops the sequencer and clears its contents. This reset should be requested during an active display period, prior to the setting of the sequencer's clocking mode register and the miscellaneous output register, to ensure that no video data is lost from the video RAM. The asynchronous reset can be used to clear and halt the sequencer immediately, but there is a danger of losing some of the video RAM data. Setting both of these reset bits allows the sequencer to continue.

The *clocking mode* register has several functions. For updates to the video RAM that require extensive changes to its contents, the video screen can be turned off to allow fast loading of the RAM. The section of the sequencer that translates the video data into serial form for display can be loaded through the register with either 1, 2, or 4 bytes every character clock. The video modes that have a horizontal resolution of 320 or 360 pixels—modes 0H, 1H, 4H, 5H, DH, and 13H—use a dot clock that is half the rate of the master dot clock and is set in the clocking mode register. In order to accommodate the 9-pixel-wide characters in VGA modes 0 + , 1 + , 2 + , 3 + , and 7 + , the sequencer needs to be set to generate 9 pixels per character clock; this is done through the clocking mode register.

The *map mask* register, when used in association with the *memory mode* register, allows each of the video bit-maps to be enabled or disabled, so they can be written to by the CPU. Any or all of the maps can be enabled. It is possible to modify this memory 32 bits at a time, producing a method of fast video clearing. The memory mode register also allows over 64 KB of video memory to be used, thus providing access to the 256 KB of video memory on the system board. This is necessary to allow different character maps to be accessed via the *character map select* register.

Graphics Registers

The nine graphics registers are used in the graphics controller section of the VGA. In a fashion similar to the other VGA registers, the *graphics address* register is used as a pointer to the other graphics controller registers. The *set/reset* register is used in conjunction with the *enable set/reset* register to write to the memory maps. Individual memory maps can be selected by using these registers. The *read map select* register indicates the memory map that is to be read by the CPU. The *color compare* register can return data to the CPU, showing all of the pixels in the video memory map that are a specified color. The *data rotate* register performs a logical AND, OR, or XOR on the data in the video map, with the data that is presented in the CPU memory latched. This register can also rotate the data that is presented to the VGA by the CPU to the right, between 0 and 7 bits.

The *graphics mode* register loads the shift registers in the format required for the 256-color mode. This register also controls the formatting of the shift registers, to accommodate CGA-compatible modes 4 and 5. These graphics modes store their video information with the even scan lines in one area of memory and the odd scan lines in another. The CGA compatibility extends further in this register, with a bit being available to select odd or even addressing for the CGA video modes. Another function of this register is to select the read type; the CPU can either read the contents of the selected memory map or the result of the color compare function.

The selection of memory maps to be included in the color compare functions is also controlled by the *color don't care* register. The *bit mask* register can mask any bit so that it cannot be altered by the CPU. The write type of the VGA is selected in the graphics mode register. There are four different write operations that permit the data to be transferred in different forms from the CPU to the video subsystem. The miscellaneous register in the graphics controller controls the size of the memory maps for the video data that is mapped into the regenerative buffer for display. The video modes that store the video data as odd and even scan lines also need to set a bit in this register, as well as bits in the graphics mode register. This register is also used to change between text and graphics modes.

Chapter 5

General Purpose Registers

There are five general purpose registers in the VGA, with the *feature control* register present but currently reserved by IBM. The two *input status* registers are read only and provide information on the status of the VGA. Information provided in these registers includes the status of the CRT interrupt (whether a vertical retrace is due); the switch sense bit, which determines whether the system has a color or monochrome monitor attached; and if video information is currently displayed. The display enable bit, which was used on previous adapters to ensure that data was written to video RAM only during inactive video periods, is also present in this register, even though it is no longer necessary for the video system. Its presence provides compatibility with existing software. The *video subsystem enable* register allows video and I/O address decoding, regardless of the state of the sleep bit that may have been set while POS was turning off the video card. The *miscellaneous output* register selects the polarity for the vertical and horizontal sync pulses; this is used in video modes that require a different number of scan lines. The clock source is also selected through this register. Two clock rates are available within the VGA: one for the 640-pixel horizontal modes and the other for the 720-pixel horizontal video modes. A user-supplied clock is enabled or disabled by setting or resetting a bit in the miscellaneous output register. However, the clock itself does not go through this register. The I/O addresses for the compatibility modes of the CGA and the monochrome monitor are selected via this register, and the video RAM can also be enabled or disabled.

Summary

Although the video capabilities of the previous display adapters have been extensively documented, some applications that function on the original equipment may not run on the Personal System/2 machines without modification. The EGA in particular was programmed at a hardware level by many applications; care should be taken to ensure that the expected colors are set on the Personal System/2 machines. The video capabilities are expanded on the new machines, allowing growth (however small) in the video standard. The move towards multitasking machines is accommodated by the ability to preserve and restore the current video state on the machines.

6

The Model 50

The Personal System/2 family is the second generation of the IBM PC family. Just as Models 25 and 30 trace their roots to the PC and XT, the Personal System/2 Model 50 is a direct descendant of the IBM PC AT. The Model 50 is a desktop computer that provides performance comparable with that of the AT, along with more standard features, a smaller size, and a lower price. The Model 50 is well suited to serve as a medium-performance personal productivity workstation, particularly in a networked environment.

Features

The Model 50 consists of a system unit, keyboard, and display designed for desktop use (see Figure 6-1), and features an Intel 80286 microprocessor, as did the AT and XT 286. The processor and system bus run at 10 MHz (25 percent faster than the 8 MHz speed of the AT). The optional 80287 numeric coprocessor also runs at 10 MHz (88 percent faster than the 5.33 MHz of the AT's 80287). As might be expected, this descendant of the AT provides 16 MB of memory address space, an eight-channel direct memory access (DMA) controller, sixteen-level interrupt system, system clock and timers, real-time clock with complementary metal oxide silicon (RT/CMOS) RAM and battery backup, and an audio subsystem with speaker.

The Model 50 comes standard with 1 MB of memory installed on the system board (640 KB base memory and 384 KB extended memory). Like all other members of the Personal System/2 family, it features the IBM 101-key Enhanced Keyboard. This is the same keyboard that is available with late model ATs and XTs; however, the connector on the system unit end of its cord is different. A connector is also standard on the system board for the connection of an IBM

Chapter 6

Figure 6-1 The IBM Personal System/2 Model 50. *The Model 50 is a desktop computer that features a 80286 and the Micro Channel architecture.*
Photo courtesy of International Business Machines Corporation.

mouse or other pointing device. A keylock is provided, but it only secures the cover; it does not disable the keyboard as is the practice on the AT. Keyboard security is provided via a password scheme. One parallel and one serial port are provided on the system board. The Model 50 system board improves significantly on that of the AT, providing more memory, plus the pointing device port, parallel port, and serial port.

Another standard feature of the Model 50 is the Video Graphics Array (VGA) video subsystem described in Chapter 5. This subsystem, when used with the Personal System/2 family of analog color and monochrome displays, supports several graphics modes (640 by 480 pixels in 2 and 16 colors, and 320 by 200 pixels in 256 colors) and high resolution text mode (720 by 400 pixels in monochrome and 16 colors), while maintaining compatibility with CGA and EGA graphics and text modes. The VGA drives any member of the Personal System/2 family of displays through a connector on the system board. Neither the connector nor the VGA interface to the display is compatible with any IBM PC displays; however, some third-party adaptable sync monitors can be used with the VGA.

The Model 50 features a 16-bit implementation of IBM's Micro Channel architecture, as described in Chapter 4. This architecture, which is designed to support multi-tasking operating systems, is not compatible with the bus architecture used with the IBM PC family and the Personal System/2 Models 25 and 30. The Model 50 has three 16-bit connectors and one 16-bit slot with auxiliary video extension. Micro Channel connectors accept only expansion boards that are specifically designed for use with the Micro Channel, usually indicated by a suffix of /A in the name of the board. These boards are physically and electrically different from, and may not be interchanged with, IBM PC family or Personal System/2 Model 25 and 30 adapter boards.

A single 1.44 MB, 3.5-inch diskette drive is provided as standard equipment on the Model 50; a second 1.44 MB drive can be added as an option. These drives can also read and write diskettes in a 720 KB mode, which is compatible with that used by the drives on Models 25 and 30, the IBM PC Convertible, and several other portable and desktop PCs. A hard disk is also provided as standard equipment. The hard disk is a compact, 20 MB unit.

Overall, the Model 50, with its compact system unit and sleek new displays, has a modern look that the AT definitely lacks. The only common element is the 101-key keyboard, which is itself a modernized version of the original AT 84-key keyboard.

Physical Description

With dimensions of 14.1 inches wide by 16.5 inches deep by 5.5 inches high, and a weight of 21 pounds, the Model 50 system unit is smaller and considerably lighter than that of the AT, which measures 21.25 inches wide by 17.28 inches deep by 6.38 inches high and weighs 43 pounds. In fact, the system unit looks rather narrow sitting behind its IBM Enhanced Keyboard. In addition to weighing less and requiring less space, the Model 50 uses less electricity and generates considerably less heat (494 BTU/hour versus 1229 BTU/hour). It is also noticeably quieter in operation (46 dB average when operating and 40 dB average when idle, measured from a distance of 1 meter).

In a break from past IBM PC practice, the power switch is on the right front of the system unit. It is convenient to operate and, at the same time, adequately recessed to prevent accidental operation. Nearby green and yellow lights indicate power on and hard disk activity, respectively. The standard 1.44 MB, 3.5-inch diskette drive is located on the left front of the system unit, with either a second 1.44 MB drive or a matching blank cover plate to its right. As with other members of the family, the Model 50's model and serial numbers are displayed on the front (bottom right), as well as the back of the system case. This feature is extremely useful for organizations that deal with large equipment inventories.

Rear Panel Connectors

The system unit has a keylock for the system cover, but it is located on the rear of the unit, as shown in Figure 6-2, rather than the front. The system power connector is located on the left rear of the system unit; however, an auxiliary outlet to provide power for a display is not supplied, as is the case on the AT. The keyboard, mouse, parallel interface, serial interface, and video display connectors can be seen to the right of the power connector. These same connectors are standard on all Personal System/2 models.

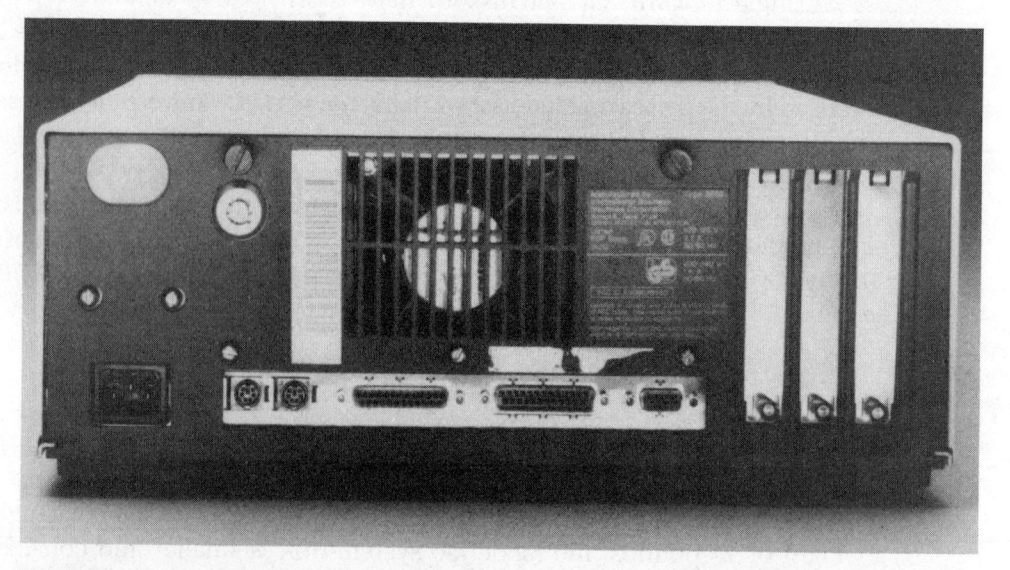

Figure 6-2 Rear of the Model 50 System Unit. *The keylock on the rear panel secures the cover, but does not disable the keyboard.*
Photo by Bill Schilling.

The keyboard and mouse use identical 6-pin miniature DIN connectors. These connectors are different from the keyboard connector used on PCs. The parallel and serial interfaces both use standard 25-pin D-shell connectors—the connector for the parallel port being female and the serial port male. The use of the 25-pin connector, rather than the less common 9-pin connector used on the AT, is a beneficial side effect of mounting these connectors on the system board rather than on the bracket of an expansion board as they are on the AT—an example of IBM's preference for using the larger 25-pin connector when sufficient space is available. The video display connector is a 15-pin D-shell connector

The Model 50

that is not compatible with the 9-pin video display connector used on PC family video controller boards.

Covers for the Model 50's three available Micro Channel expansion slots are located on the right rear of the system unit. These covers/end brackets are isolated from the adapter boards to which they are attached and isolated from the system ground, so they act as a shield against electromagnetic interference (EMI).

Inside the System Unit

Removing the system unit cover is simply a matter of unlocking the keylock, loosening the two knurled head screws on the top rear of the system unit, pulling the cover forward slightly, and then lifting it up. These screws and the knurled screws that secure the Micro Channel expansion slot covers are slotted and can be loosened by hand, using the cardboard disk that comes with the system's cover lock keys or a flat-blade screwdriver.

Removing the cover reveals that the inside of the Model 50 is like a split-level house, with the mass storage devices on the upper level, the system board on the lower level, and the Micro Channel adapters spanning the two on the left side. (See Figure 6-3.) In a marked departure from the construction of the AT, the system case is made of a composite plastic material rather than metal, except for the top and sides of the removable cover. The plastic used is lightweight, yet sturdy and durable.

The speaker and battery assembly is mounted in the front of the unit to the

Figure 6-3 Front of the Model 50 Uncovered. *The open position for the second 3.5-inch diskette drive can be seen to the left of the power supply.* Photo by Bill Schilling.

Chapter 6

left of the 1.44 MB diskette drive. The open bay to the right of the diskette drive is available for either a second 1.44 MB drive or the adapter assembly used to connect the Model 4869 external 360 KB, 5.25-inch diskette drive (described in Chapter 9) to the system. Each bay contains an interface connector; the two connectors are attached to a board that connects to the diskette drive controller contained on the system board. The diskette drive connectors provide power as well as signal lines for the diskette drives.

The Model 50 system unit is assembled with no connecting cables. This not only simplifies assembly of the system, but also makes it easier to maintain and upgrade. All system components—power supply, system board, and speaker/battery module—connect together without cables. (See Figure 6-4.) Furthermore,

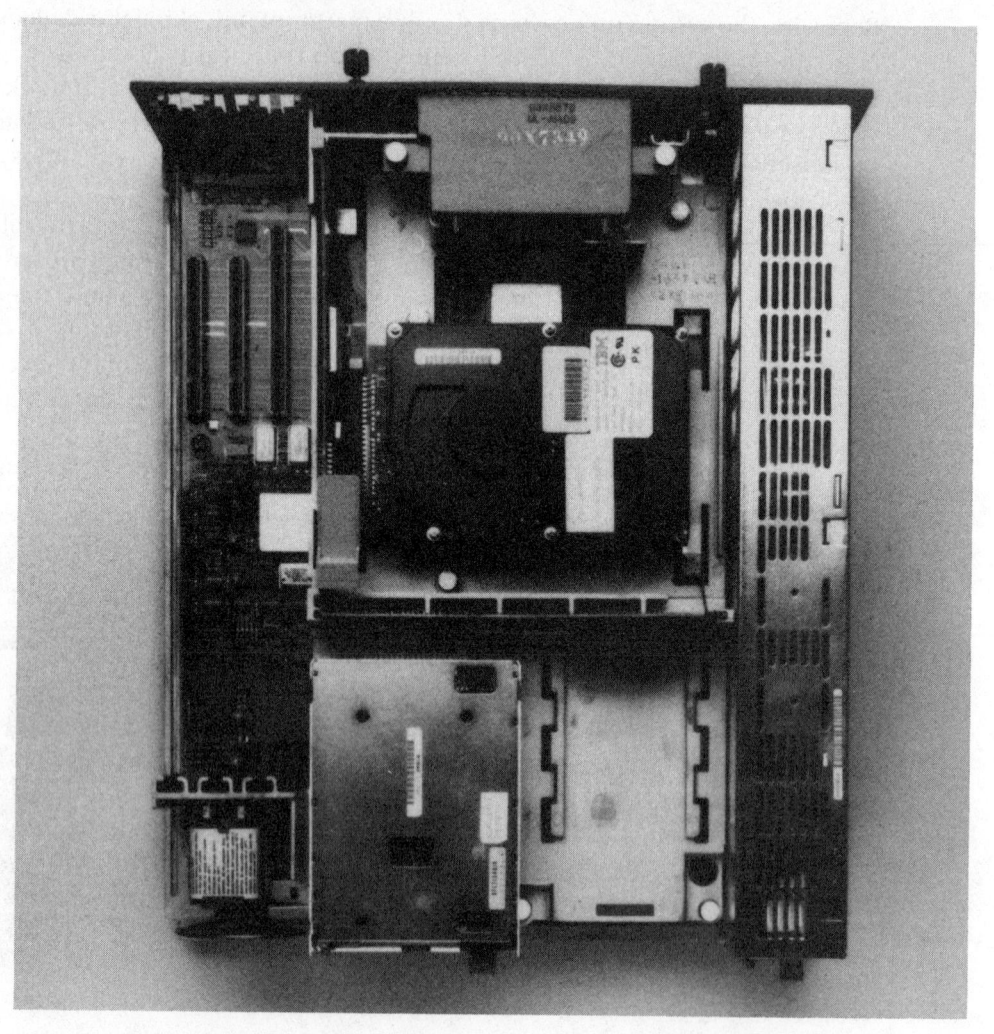

Figure 6-4 Inside the Model 50 System Unit. *The hard disk is located on the top level of the unit, behind the 3.5-inch diskette drive bays.*
Photo by Bill Schilling.

the only screws used are the three 6 mm slotted hexagonal screws that secure the power supply to the case and the six screws of the same type that secure the system board. The power supply is a narrow unit mounted on the right side of the system unit.

The system cooling fan and 3.5-inch hard disk drive are located to the left of the power supply. The hard disk drive connects directly to its controller, which is plugged into a dedicated Micro Channel slot in the system board.

The blue plastic tabs on each end of the hard disk adapter can be seen to the left of the disk drive at the rear and near the center of the system unit. These tabs act as handles, making the insertion and removal of Micro Channel boards easy on the fingers. The Model 50's three available Micro Channel connectors are to the left of the hard disk adapter. Micro Channel adapters are secured in the rear by a knurled head screw on the outside of the system unit and in the front by a plastic card guide inside the system unit. The card guides are somewhat wider than Micro Channel boards, to allow boards to be installed more easily. The wide handles on the front end of adapters serve to provide adequate spacing between adjacent boards, but the board nearest the outside cover is free to wobble a bit. Micro Channel expansion boards from IBM come with stick-on labels that can be attached to the front handle of the board, for ease of identification.

Major system components are held together by built-in tabs and slots and white plastic pop-up connectors, like the ones to the left of each end of the power supply. The plastic tool supplied by IBM for loosening the pop-up connectors is stored under the system speaker. Removing a diskette drive is merely a matter of lifting a tab underneath its front and pulling the drive forward, to disconnect it from the board that connects it to the system board. The hard disk is similarly removed by depressing two plastic tabs to the left of the power supply and sliding the drive to the right to detach it from its controller.

The Model 50 hard disk controller, shown in Figure 6-5, is a low power version of the ST-506/412 interface, but supports only one drive. It features surface mount technology (SMT) and IBM very large scale integration (VLSI) integrated circuits, and one potted device covered with a ceramic shield. The controller supports automatic head and track switching, features a 2 KB track buffer, and provides power as well as signals to the hard disk. Because it is connected directly to the hard disk, it must always occupy Micro Channel expansion slot 4.

Once the hard disk is detached, it is removed by lifting it upward. Full access to the system board can be provided by removing any other Micro Channel boards and loosening the appropriate pop-up connectors that secure several system components. These include the system fan, the mass storage devices, the battery and speaker assembly, and the composite plastic structure that supports them.

The system board can be removed after removing the three interior screws that secure it to the bottom of the case and the three exterior screws that secure it to the back. This can be done without removing the power supply, because the

Chapter 6

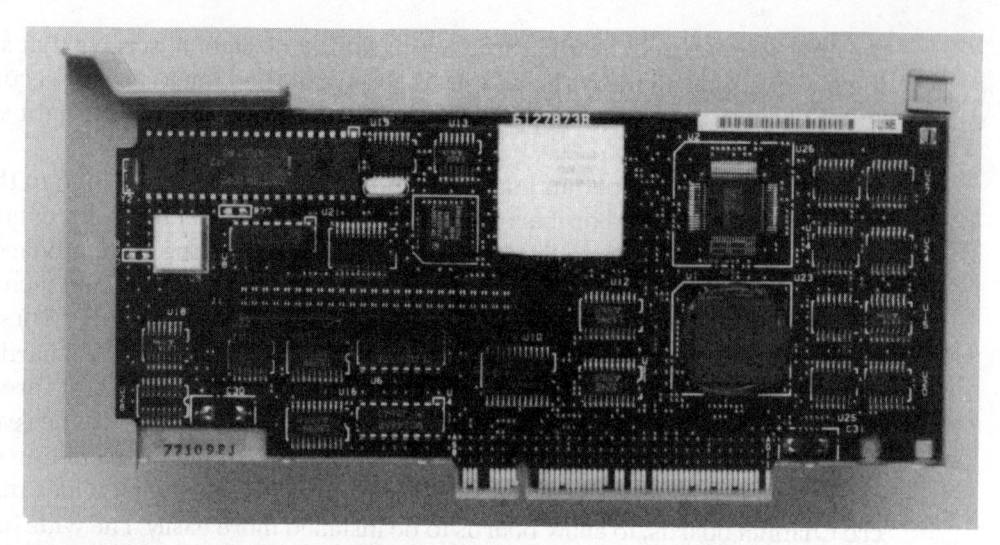

Figure 6-5 Model 50 Hard Disk Controller. *The Model 50 hard disk controller, which supports one drive, is a low power requirement version of the ST-506/412 interface.* Photo by Bill Schilling.

system board does not extend under the power supply. In keeping with the cableless interior of the system unit, the system board receives power from a connector built into the side of the power supply.

The system board contains a socket-mounted, 10 MHz 80286 with a nearby socket for an optional 10 MHz 80287 math coprocessor, as shown in Figure 6-6. The Model 50's 128 KB of read only memory (ROM) is contained in four modules in the front center of the system board. The connector for the system's diskette drives is located behind the ROM modules, and the integrated diskette controller is located to its right. The system board RAM is located to the right of the diskette controller connector. It consists of two memory module packages, each of which comprises six 1-megabit chips mounted in a single-in-line package (SIP). Each package provides 512 KB of parity-checked memory, for a total of 1 MB of system board memory. Each SIP is mounted in a 30-pin connector. SIPs are removed by spreading their retaining brackets and rotating their tops away from the power supply.

The system board makes extensive use of surface mount technology and VLSI chips. The large metal-covered chip near the back of the system unit is the Video Graphics Array. Behind it are the system's video, serial, parallel, and fuse-protected pointing device and keyboard connectors. To the center left are the system's 14.3818 MHz, 32 MHz, and 40 MHz timing crystals. Behind the crystals are the system's four 16-bit Micro Channel connectors. The rightmost one is reserved for the hard disk controller; next is a 16-bit slot with auxiliary video extension, which allows access to the system video output port; the remaining two are normal 16-bit Micro Channel connectors.

The 80286 memory expansion board described in Chapter 9 can be in-

Figure 6-6 Model 50 System Board. *The Model 50 has four Micro Channel expansion slots. One is reserved for the hard disk controller.*
Photo by Bill Schilling.

stalled in a Micro Channel slot to expand system RAM. This board comes with a minimum of 512 KB of parity checked RAM; when fully populated, this board holds 2 MB of RAM. The board is populated using 256 KB SIPs rather than the 512 KB SIPs used on the Model 50 system board. Since only three Micro Channel connectors are available on the Model 50, this limits system RAM to 7 MB using IBM-supplied memory options.

The Model 50 is designed for use in a wide variety of operating environments, as shown below.

Air temperature
  System on: 15.6 to 32.2 degrees C (60 to 90 degrees F)
  System off: 10.0 to 43.0 degrees C (50 to 110 degrees F)
Humidity
  System on: 8 percent to 80 percent
  System off: 20 percent to 80 percent
Altitude
  Maximum altitude 2133.6 meters (7000 feet)
Heat output
  494 BTU/hour

For electromagnetic compatibility purposes, the Model 50 is certified as an FCC Class B device. The current leakage of the Model 50 does not exceed 500 microamperes. Like Models 60 and 80, it is designed to meet current leakage requirements established by the National Fire Protection Code NFPA 76B for data processing equipment used in hospital environments. IBM advises users to consult local ordinances, National Fire Protection codes, and UL codes for specific details on the use of data processing equipment in a hospital environment.

Functional Description

The Model 50 features a well-matched group of key components that work together as a computer system. Many of these are integrated into a single system component—the system board.

The Model 50 system board provides the following integrated features:

Intel 80286 system microprocessor (CPU)

Intel 80287 math coprocessor

1 MB random access memory (RAM)

128 KB read only memory (ROM)

Real-time clock with 64-byte CMOS RAM and battery backup

16-bit Micro Channel with five expansion connectors

Eight-channel DMA controller

Sixteen-level interrupt system

Three system timers

64K possible I/O ports

Three system control ports

Audio subsystem with speaker

Keyboard and pointing device controller

Integrated video graphics system

Serial port controller

Parallel port controller

Diskette drive controller

Figure 6-7 shows a functional diagram of the system board.

The central component of any computer is its central processing unit (CPU). The Model 50's CPU is an 80286 microprocessor; since it is located on the system board, it is the most important feature of the system board.

80286 Microprocessor

The Model 50's 80286 microprocessor provides the following features:

10 MHz operation

24-bit addressing

16-bit data interface

8086-compatible real-address mode

Protected virtual-address mode

16 MB of physical-address space

1 gigabyte of virtual-address space

The 80286's instruction set is extensive, including instructions for string input/output as well as a comprehensive set of conventional arithmetic and logical instructions. A list of the available instructions is shown in Appendix C.

The Model 50's 80286 operates at 10 MHz, compared to the 8 MHz operation of the AT's 80286. The 10 MHz operation provides a clock cycle time of 100 nanoseconds. The CPU can perform memory access and I/O operations in as little as two clock cycles (200 nanoseconds). Because of the slower speed of the system board RAM, ROM, and I/O devices, the CPU must wait at least one additional clock cycle (one wait state) for each operation to complete. This results in a 300-nanosecond cycle time for system board RAM and ROM, and a minimum system board I/O cycle time of 300 nanoseconds.

The 80286 is compatible with the 8086 when operating in real-address mode, and provides enhanced functionality when operating in the protected virtual-address mode. When operating in real mode, the processor can access up to 1 MB of memory, using 20-bit physical addresses. As with the 8088 used in the PC and the 8086 used in Models 25 and 30, these addresses are generated from a 16-bit segment pointer and a 16-bit offset. Segment pointers address

Chapter 6

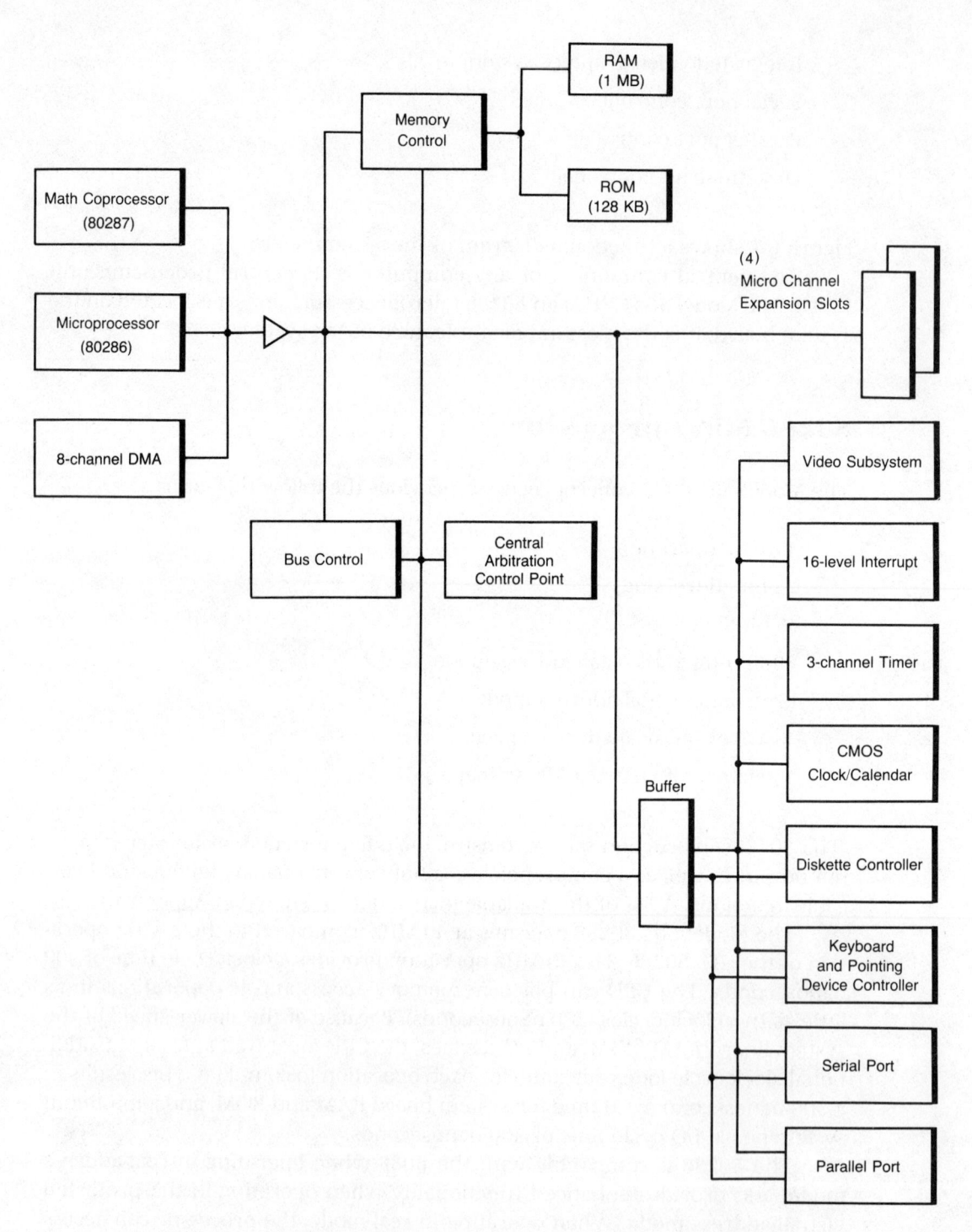

Figure 6-7 Model 50 System Block Diagram. *The system board features standard I/O controllers as well as the system microprocessor and memory.*

memory in units of 16 bytes. The 20-bit physical address of an item is determined by shifting the segment pointer left 4 bits (effectively multiplying it by 16) and adding the offset to it.

Protected mode supports multi-tasked operation and the execution of tasks whose memory size exceeds the size of physical memory. In protected mode, the processor can address up to 16 MB of physical memory using 24-bit addresses, and each task can have a virtual-address space of up to 1,073,741,824 bytes (1 gigabyte) mapped into physical memory. Protected mode uses a 16-bit selector and a 16-bit offset to generate 24-bit memory addresses. The selector is used as the index of one or two memory-resident tables (descriptor tables) of 24-bit segment base addresses. The 16-bit offset is added to the selected segment base address to form the physical address.

The Math Coprocessor

The optional 80287 math coprocessor operates at 10 MHz, compared to the 5.33 MHz operation of the AT's 80287. The 80287 performs high speed arithmetic, logarithmic, and trigonometric operations. The coprocessor has eight 80-bit registers and can work with seven numeric data types. See Table 6-1 for a description of the data types supported.

Table 6-1 Data Types Supported by the 80287. *An application must be written to make use of the coprocessor for any change in performance to be observed.*

Data Type	Length (Bits)	Significant Digits	Approximate Range
Word integer	16	4	32,768
Short integer	32	9	2×10^9
Long integer	64	19	9×10^{18}
Packed decimal	80	18	10^{18}
Single precision (short real)	32	6–7	$10^{\pm 38}$
Double precision (long real)	64	15–16	$10^{\pm 308}$
Extended precision (temporary real)	80	19	$10^{\pm 4392}$

The coprocessor works in parallel with the 80286. It functions as an I/O device through I/O ports 00F8H, 00FAH, and 00FCH. The 80286 sends operation codes and operands, and receives results through these I/O ports. The 80287 can generate a hardware interrupt if any of six different exception conditions occurs during instruction execution. An exception mask may be set to keep specific exceptions from causing the interrupt to be generated.

The 80287, like the 80286, operates in both real and protected virtual-address mode. It operates in real-address mode after a power-on reset, a system reset, or an I/O write to port 00F1H. It is placed in protected virtual-address mode by the execution of a SETPM ESC instruction.

When operating in real mode, the 80287 is compatible with the 8087 math coprocessor used in IBM PC's and the Personal System/2 Models 25 and 30. Software designed for use with the 8087 can usually operate on the Model 50 without major modifications. However, certain exception handling routines may need to be changed because of differences in the way the 80287 and 8087 handle numeric exceptions. See Chapter 4 for additional information on math coprocessor compatibility.

Random Access Memory and Read Only Memory

The Model 50 has three types of memory: read/write random access memory (RAM), read only memory (ROM), and real-time clock/complementary metal oxide silicon (RT/CMOS) RAM. ROM and RAM coexist together in the 80286's 16 MB physical address space, as shown in Figure 6-8. CMOS RAM is accessed via a system control port. See *System Control Ports*.

The Model 50 comes standard with 1 MB of RAM installed on the system board. Of this, 640 KB is base, also called conventional, memory beginning at address 0H, and 384 KB is extended memory beginning at address 100000H (1 MB). The 1 MB of RAM is made up of two 512K by 9 bit (1 parity bit for each 8 data

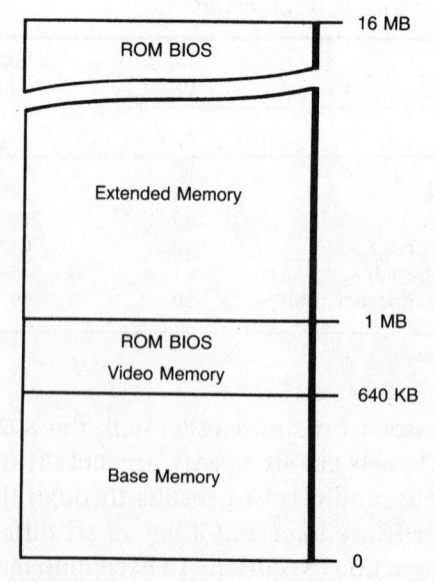

Figure 6-8 Model 50 Physical Address Space. *The 80286 can address up to 16 MB, whereas the 8086 can only address 1 MB.*

bits) memory module packages. Timing of the memory signals is specified by IBM as similar to that of Hitachi's HB61009BR-15 256 by 9 bit, 150-nanosecond dynamic RAM. RAM operates at 10 MHz with one wait state inserted, for a cycle time of 300 nanoseconds. Memory refresh requests are generated every 15 microseconds, and the memory must be accessed or refreshed 512 times before it can be used. When the system board is in setup mode (see Chapter 4) the system board memory can be disabled by a command to bit 0 of I/O port address 0103H. If the bit is set to 1, the memory is enabled; setting it to 0 disables the entire 1 MB of memory.

The Model 50 system ROM consists of four 32K by 8 bit modules configured in a 64K by 16 bit arrangement (twice the amount provided on the AT). ROM operates with one wait state, and it is not parity checked. The same system ROM occupies two sets of addresses—the top of the first and last megabyte of the 80286's 16 MB physical-address space (beginning at 0E0000H and FE0000H). System ROM contains the power-on self test (POST), a two-level basic input/output system (BIOS), and the BASIC language interpreter. The compatibility BIOS (CBIOS), which supports many current applications, supports 1 MB of memory-address space. An additional advanced BIOS (ABIOS), which provides support for IBM's OS/2 and other multi-tasking operating systems, supports a 16 MB address space. The ABIOS is designed for access by the device driver layer of an operating system such as OS/2, not applications. In order to achieve compatibility across systems, applications should interface with the operating system application program interface (API), not the system BIOS. Chapter 4 discusses the BIOS in Personal System/2 computers.

Real-Time Clock and Battery-Backed RAM

The Model 50's Motorola MC146818A real-time clock and CMOS RAM chip contains the system's real-time clock and 64 bytes of nonvolatile RAM. The first 14 bytes of RAM are used by the clock circuitry and the remainder is used for configuration and system status information. (See Table 6-2.) The system setup program on the Reference Diskette initializes status registers A, B, C, and D when the time and date are set. The 6-volt battery in the speaker and battery assembly maintains power to the system's configuration RAM when the system power supply is not in operation.

The CMOS RAM can be accessed using I/O ports 70H and 71H. Data is written to the RAM by executing an OUT to port 0070H with the RAM address to be written to, followed by an OUT to port 0071H with the data to be written. (Note that bit 7 at port 0070H stores the NMI mask bit.) Application programs should not normally write to this RAM. The stored information is only required for system initialization purposes, and should be updated using programs contained on the Reference Diskette provided with the system.

Reading is accomplished by performing an OUT to port 0070H with the

Table 6-2 Real-Time Clock RAM Address Map. *The configuration information for the system, in addition to the clock information, is stored in battery-backed RAM.*

Address (Hex)	Function
000	Seconds
001	Second alarm
002	Minutes
003	Minute alarm
004	Hours
005	Hour alarm
006	Day of week
007	Day of month
008	Month
009	Year
00A	Status register A
00B	Status register B
00C	Status register C
00D	Status register D
00E	Diagnostic status byte
00F	Shut-down status byte
010	Diskette drive byte
011	First hard disk type byte
012	Second hard disk type byte
013	Reserved
014	Equipment byte
015–016	Low and high base memory bytes
017–018	Low and high memory expansion bytes
019–031	Reserved
032–033	Cyclic redundancy check (CRC) bytes
034–036	Reserved
037	Date century byte
038–03F	Reserved

RAM address to be read, followed by an IN from port 0071H, with the data read being returned in the AL register. I/O operations to the CMOS RAM should be performed with interrupts inhibited. Furthermore, the read from port 0071H must be performed immediately after the write or else unreliable operation of the CMOS RAM may result.

Micro Channel

The Model 50 features a 16-bit implementation of the Micro Channel architecture. Micro Channel supports the asynchronous transfer of data between memory, I/O devices, and the CPU. It supports 16-bit addresses allowing 8- or 16-bit I/O transfers within a 64 KB range, or 24-bit addresses allowing 8- or 16-bit memory transfers within a 16 MB range. The Model 50's Micro Channel uses three 16-

bit connectors and one 16-bit connector with auxiliary video extension. Minimum default cycle time for the Model 50's Micro Channel is 200 ns.

The Micro Channel uses level-sensitive interrupts with interrupt sharing on all levels. Eight direct memory access (DMA) channels are supported for 8- and 16-bit DMA transfers. A central arbitration control point allows up to fifteen devices to share and control the channel. See Chapter 4 for additional information.

DMA Controller

The direct memory access (DMA) controller allows I/O devices to access memory directly, rather than via the CPU. The Model 50's DMA controller provides the following features:

- Register/program compatibility with IBM AT DMA channels
- 24-bit memory and 16-bit I/O port address capability
- Eight independent DMA channels
- Each channel programmable for byte or word transfer

The DMA is software programmable. The CPU can access the DMA controller's internal registers to set up and initiate a DMA transfer.

The DMA controller performs data transfers between memory and I/O devices, memory read operations, and memory refresh cycles. Serial data transfers are performed with a minimum of three 100-nanosecond clock cycles for I/O operations, and a minimum of two 100-nanosecond clock cycles for memory read/write operations. Read/write operations to system board memory require three clock cycles (300 nanoseconds).

Interrupts

The Model 50 provides sixteen levels of interrupts. Interrupts are controlled using two Intel 8259A interrupt controllers. These controllers are also used on the PC (which uses one 8259A) and the AT (which uses two); however, on the Personal System/2 Models 50 and above, they are initialized to use level-sensitive mode rather than the edge-triggered mode used on the PC and AT. Any of the interrupt levels may be masked except for certain nonmaskable interrupt (NMI) functions.

As on the PC and AT, each interrupt request line attached to the interrupt controller has a priority associated with it. Upon receipt of requests from these lines, the interrupt controller accepts the request from the line with the highest priority. It then issues an interrupt to the CPU if the request accepted is of higher priority than the interrupt (if any) currently being serviced. Once the

CPU acknowledges receipt of the interrupt, the controller directs it to the service routine that must be executed to service the interrupt. Interrupt request level assignments are shown in Table 6-3 in decreasing order of priority.

Table 6-3 Interrupt Assignments. *The NMI and system timer have the highest interrupt priority, as they are the most critical to the function of the system.*

Level	Function	Level	Function
NMI	Parity, watchdog timer, arbitration timeout, channel check		
IRQ0	Timer		
IRQ1	Keyboard		
IRQ2	Cascade interrupt control:	IRQ8	Real-time clock
		IRQ9	Redirect cascade
		IRQ10	Reserved
		IRQ11	Reserved
		IRQ12	Mouse
		IRQ13	Coprocessor error
		IRQ14	Hard disk
		IRQ15	Reserved
IRQ3	Serial alternate		
IRQ4	Serial primary		
IRQ5	Reserved		
IRQ6	Diskette		
IRQ7	Parallel port		

The highest priority is the nonmaskable interrupt, and the lowest is IRQ7. An NMI indicates that a parity error, channel check, channel timeout, or watchdog timer timeout has occurred. NMI interrupt requests caused by system board parity errors and channel checks may be ignored by setting the NMI mask bit at I/O port address 0070H. Watchdog timer and system channel timeout exceptions are not masked by this bit.

Interrupt levels 8 through 15 are handled by the second 8259A interrupt controller. These requests are serviced through IRQ2 of the first controller. In order to maintain compatibility with the PC and its eight levels of interrupts, IRQ9 is defined as the replacement interrupt level for those devices that use IRQ2.

System Timers

The Model 50 has three programmable timer/counters—channel 0, channel 2, and channel 3. The first two are similar to the PC and AT's channel 0 and channel 2 timers. Channel 3 is a Personal System/2 feature, not provided for members of the IBM PC family (or Models 25 and 30). Each channel has a counter associated with it that counts down from a preset value. Counters 0 and 2 are 16-bit counters that can count down in binary or binary coded decimal (BCD). Counter 3 is an 8-bit counter that counts down in binary only.

Channel 0 drives hardware IRQ0. Channel 2 provides tone generation for the audio subsystem. Both timers are driven by a 1.190 MHz clock signal. Channel 3, the watchdog timer, may be used to detect error conditions wherein IRQ0 is not being serviced. If IRQ0 is active for more than one period of channel 0's clock output signal, the watchdog timer's counter is decremented. When the count reaches zero, a nonmaskable interrupt is generated. Watchdog timer operation is defined only for mode 2 (rate generator) and mode 3 (square wave) operation of channel 0. It is enabled and disabled through BIOS interfaces.

I/O Ports

System input/output is performed via 64K possible I/O ports. Each input/output device has one or more ports associated with it. These ports are accessed directly by the CPU using IN and OUT instructions or through DMA transfer operations. The Micro Channel architecture and VGA provide a number of new and enhanced features. To support these features, IBM is now using previously reserved parts of the I/O port address map. The I/O addresses used by Model 50 system board devices are shown in Table 6-4.

System Control Ports

I/O ports 70H, 92H, and 61H are used for system control. Port 70H is used to enable/mask the nonmaskable interrupt signal, and also for real-time clock CMOS RAM operations. Bit 7 of port 0070H stores the NMI interrupt mask. When this bit is set to 1, the nonmaskable interrupt is enabled; when set to 0, the NMI is masked off. This bit is write only; it is set to 0 by a power-on reset of the system. Bit 6 is reserved, and bits 5 through 0 receive the address of the CMOS RAM location to be read or written as described previously.

System control port A (92H) supports several functions unique to the Personal System/2. Functions supported by each bit located at this port are shown in Table 6-5. All nonreserved bits may be read or written to unless otherwise indicated. Bits 6 and 7 control the hard disk activity light mounted in the system power supply unit. Setting either bit to 1 turns the light on; setting both bits to 0 turns the light off. Both bits are set to 0 at power-on reset. Bits 2 and 5 are reserved. Bit 4 is set to 1 when a watchdog timeout occurs. Bit 3 locks the 8-byte password area of RT/CMOS RAM when it is set to 1. Once set by POST, it can only be cleared by turning off the system. Bit 1 controls address bit A20 when the 80286 is in real mode. The A20 signal is active when this bit is set to 1, inactive when it is set to 0. The bit is set to 0 during a system reset. Bit 0 provides an alternate means of resetting the 80286.

Chapter 6

Table 6-4 System Board I/O Address Map. *The Micro Channel architecture and integrated video subsystem are implemented using some previously reserved I/O port addresses.*

Addresses (Hex)	Device
0000–001F	DMA controller
0020, 0021	Interrupt controller 1
0040, 0042, 0043, 0044, 0047	System timers
0060	Keyboard, mouse
0061	System control port B
0064	Keyboard, mouse
0070, 0071	RT/CMOS RAM and NMI mask
0074–0076	Reserved
0081, 0082, 0083, 0087	DMA page registers (0–3)
0089, 008A, 008B, 008F	DMA page registers (4–7)
0090	Central arbitration control port
0091	Card-selected feedback
0092	System control port A
0093	Reserved
0094	System board setup
0096, 0097	POS, channel connector select
00A0, 00A1	Interrupt controller 2
00C0–00DF	DMA controller
00F0–00FF	Math coprocessor
0100–0107	Programmable option select
0278–027B	Parallel port 3
02F8–02FF	Serial port 2 (RS232C)
0378–037B	Parallel port 2
03BC–03BF	Parallel port 1
03B4, 03B5, 03BA, 03C0–03C5	Video subsystem
03CE, 03CF, 03D4, 03D5, 03DA	Video subsystem
03C6–03C9	Video DAC
03F0–03F7	Diskette drive controller
03F8–03FF	Serial port 1 (RS232C)

Table 6-5 System Control Port A Functions. *The system control port A provides additional functions not available on the PC family of computers.*

Bit	Function
7	Hard disk activity light bit A
6	Hard disk activity light bit B
5	Reserved (0)
4	Watchdog timer status (read only)
3	Security lock latch
2	Reserved (0)
1	Alternate gate A20
0	Alternate system microprocessor reset

The 80286 operates in real-address mode when it is powered up. An instruction may be executed to switch it into protected virtual-address mode; however, an instruction is not available to switch it back. The ability to switch back is necessary for operating systems such as OS/2 that must time share the processor between a real-mode DOS application and protected-mode OS/2 applications. A switch from protected mode to real mode is accomplished on the AT by setting a flag in location 0FH of the system's battery-backed configuration RAM, and then sending a command to the 8042 keyboard controller directing it to reset the 80286. Once reset, the 80286 begins executing the system BIOS power-on self test (POST), which checks the flag in the configuration RAM to see if a soft reset or a power-on reset has occurred, and then proceeds accordingly. For compatibility with the AT, this procedure is also supported on the Model 50; however, the alternate system microprocessor reset is much faster, requiring only 13.4 microseconds.

The alternate system microprocessor reset bit is set to 0 by a system reset or by a write operation. When a write sets the bit from 0 to 1, the alternate processor reset signal (which is logically OR'd with the reset output of the 8042) is pulsed for between 100 and 125 nanoseconds. It takes at least 6.72 microseconds for the reset of the 80286 to occur. Once toggled, the bit remains set to 1 so that the system power-on self test (POST) can read it and determine that a switch from protected mode to real mode has occurred. If the bit is 0, POST assumes that the system was just powered on and begins testing the system.

Functions provided by system control port B (61H) are shown in Table 6-6.

For a write operation, setting bit 7 to 1 resets IRQ0. Bits 6–4 are reserved for write operations. Setting bit 3 to 0 enables channel check; setting bit 2 to 0 enables parity check. Bits 3 and 2 are set to 1 during power-on reset. Setting bit 1 to 1 enables speaker data. The output of system timer 2 is logically ANDed with this bit, and the result drives the Micro Channel audio sum node signal. Setting bit 0 to 1 enables the timer 2 gate to the system speaker; setting it to 0 disables the gate.

For a read operation, if bit 7 is set to 1, a parity check NMI has been detected. If bit 6 is equal to 1, a channel check NMI has been detected. Bit 5 indicates the state of the timer 2 output. For a read operation, bit 4 changes state with each refresh request. A read operation returns the result of the last write to bits 3–0.

Audio Subsystem

The audio subsystem consists of a 2.6-inch speaker and the linear amplifier that drives it. The linear amplifier is driven using system control port 61H to control the output from system timer channel 2. See Chapter 4 for details.

Table 6-6 System Control Port B Functions. *The system control port B is used in different ways, depending on whether it is being written to or read from.*

Bit	Function
Write Operations	
7	Reset timer 0 output latch (IRQ0)
6	Reserved
5	Reserved
4	Reserved
3	Enable channel check
2	Enable parity check
1	Speaker data enable
0	Timer 2 gate to speaker
Read Operations	
7	Parity check
6	Channel check
5	Timer 2 output
4	Toggles with each refresh request
3	Enable channel check
2	Enable parity check
1	Speaker data enable
0	Timer 2 gate to speaker

Keyboard Controller

The keyboard port and the pointing device (mouse) port are controlled by an Intel 8042 chip on the system board. This controller receives data from the keyboard as serial data and presents it to the system as a byte of data. The 8042 can be programmed by the system CPU to not send keyboard data to the system until a previously stored password is entered from the keyboard. This mechanism allows several modes of system security not available with the AT's keylock. See the *Security* section in this chapter for additional information.

Video Graphics Array

The Model 50 features a full implementation of the Personal System/2 VGA video subsystem. The VGA drives any member of the Personal System/2 family of analog color and monochrome displays through its 15-pin D-shell connecter mounted on the system board. It supports video modes provided by the IBM monochrome, CGA, and EGA adapters, as well as providing several new modes. The new modes include several graphics modes (640 by 480 pixels in 2 and 16 colors, and 320 by 200 pixels in 256 colors), along with high resolution text mode (720 by 400 pixels in monochrome and 16 colors). See Chapter 5 for additional information.

Serial Port Controller

The standard RS232 serial port is also integrated into the system board. It has a 25-pin male D-shell connector and is controlled by a National Semiconductor NS16550 programmable asynchronous communications controller. The NS16550 is compatible with the serial portion of the AT serial/parallel adapter. The controller automatically adds and removes start, stop, and parity bits. Characters with 5, 6, 7, and 8 data bits and 1, 1.5, or 2 stop bits are supported. Data rates from 50 to 19,200 bps are supported. Double buffering is provided as well as false start bit detection and line break generation and detection functions. The following modem control functions are provided:

- Clear to send (CTS)
- Request to send (RTS)
- Data set ready (DSR)
- Data terminal ready (DTR)
- Ring indicator (RI)
- Data carrier detect (DCD)

The serial port can be configured for use as COM1 or COM2 using the system configuration software on the Model 50/60 Reference Diskette provided with the system. It can be programmed to use interrupt level 4 (for COM1) or interrupt level 3 (for COM2). The serial port controller provides full compatibility with the asynchronous communications boards provided for use on the PC and AT, with a communications rate limit of 19,200 bps rather than 9,600 bps.

Parallel Port Controller

The standard parallel port is integrated into the system board. It has a 25-pin D-shell connector. The parallel port can be used to drive various IBM PC and compatible printers. It also has a bi-directional mode that allows the transfer of data to a Personal System/2 from another PC, but not vice versa. See Chapter 4 for additional information. The parallel port may be configured as LPT1, LPT2, or LPT3 using the system configuration software on the Model 50/60 Reference Diskette provided with the system.

Mass Storage

The Model 50 comes standard with a 3.5-inch, 1.44 MB diskette drive and a 20 MB hard disk. A second 3.5-inch, 1.44 MB diskette drive is available as an option. The characteristics of the drive are summarized below.

Number of heads	2
Number of tracks	80
Sectors per track	
High density	18
Low density	9
Bytes per sector	512
Media capacity	
High density	1.44 MB
Low density	720 KB
Media type	3.5-inch
Access time	
Track to track	6 ms
Settle time	15 ms
Motor start time	500 ms
Rotation rate	300 rpm
Transfer rate	
High density	500,000 bps
Low density	250,000 bps

Unlike the 3.5-inch diskette drives used in Models 25 and 30, this unit operates in high as well as low density mode, storing 1.44 MB of formatted data in high density mode and 720 KB in low density mode. When used in low density mode, the drive is fully compatible with the 720 KB, 3.5-inch diskette drives used on the IBM Convertible and the Personal System/2 Models 25 and 30, and optionally available for other members of the PC family. The blue eject button on the drive is marked 1.44 to distinguish it from the 720 KB diskette drive used on Models 25 and 30, which is similar in appearance.

The diskette drive controller is integrated into the system board. It performs read/write operations on diskettes without attempting to determine if they are high or low density. IBM recommends that the 1.44 MB diskette drive not be used to format a low density (1 MB unformatted) diskette in high density mode (1.44 MB formatted); and likewise that it not be used to format a high density diskette (2 MB unformatted) in low density (720 KB formatted) mode. This is a wise precaution because the actual magnetic media used in the two types of diskettes are different.

An external 360 KB, 5.25-inch diskette drive may be installed on the Model 50 instead of the second 1.44 MB drive. This external drive requires one Micro Channel expansion slot. It reads and writes 360 KB diskettes, but it can neither read nor write 1.2 MB diskettes. This makes data migration and data exchange between these machines and the AT, with its standard 1.2 MB drive, something of a problem.

Data can be transferred directly from a member of the IBM PC family to any member of the Personal System/2 family via the parallel port, using a cable connector and software supplied with the IBM Data Migration Facility. However, this is a one-way street. PCs cannot receive data back from a Personal System/2 using this facility, because they do not have bidirectional parallel ports as the Personal System/2 computers do. See Chapter 14 for additional information.

The hard disk on the Model 50 is a compact unit with 3.5-inch diameter disk platters. Technical data for this disk is shown in the list below. As can be seen from the specifications, this disk is more similar in capacity and performance to that of the XT and XT/286 than that of the AT.

Number of data heads	4
Number of cylinders	612
Sectors per track	17
Bytes per sector	512
Formatted capacity	
(million bytes)	20
Access time	
Track to track	15 ms
Average	80 ms
Maximum	180 ms
Rotation rate	3,600 rpm
Transfer rate	
(million bits per second)	5

The 20 MB hard disk provided as standard equipment on the Model 50 is the only disk currently available for it from IBM. However, the Model 50 provides BIOS support for the same thirty-two types of hard disks that Models 60 and 80 do. (See Chapter 7 for a list of the types of disks supported.) Thus, IBM (or some third party) may offer higher capacity and performance hard disks for the Model 50 in due time.

Power Supply

The system's 94-watt power supply provides power for the system board, channel adapters, diskette drives, hard disk, keyboard, and pointing device. It operates with two ranges of input power and automatically selects the appropriate range. The ranges are 90 to 137 V_{ac} at 5 amperes or 180 to 265 V_{ac} at 3 amperes, 50 to 60 ± 3 Hz. The power supply outputs + 5, + 12, and − 12 V_{DC} to the system board. Its switch is located on the front of the unit along with two light-emitting diodes (LEDs). The green LED indicates that the power supply is on, and the yellow LED indicates hard disk activity. The yellow LED is powered by the power supply's system status line, which is controlled using bits 6 and 7 of system control port A (0092H).

The power supply is capable of no-load operation. Conversely, if a DC output is shorted, the power supply shuts down with no damage to the power supply. If input power to the supply is interrupted, it automatically restarts when input power is restored. An internal fuse provides protection against input power overload. A power-good signal indicates proper operation of the power supply. At power off, the power-good goes inactive before output voltage falls below regulation limits.

The system board receives power directly from the power supply via a 50-pin edge connector mounted on the side of the power supply. Seventeen +5, six +12, and one −12 V_{DC} lines are provided, all with matching ground connections. The remaining two lines are for the power-good and system status signals.

Keyboard

The Model 50 uses the 101-key Enhanced Keyboard available on all Personal System/2 models and late model XTs and ATs. This keyboard is described in detail in Chapter 4. It is the best PC keyboard offered by IBM to date (once one becomes accustomed to the location of the function keys). The keys provide tactile and audible feedback without the clicky noise characteristic of the IBM PC keyboard. Both it and the Model 50 system unit are sufficiently deep, however, to make it impractical to use a Model 50 on a 24-inch deep work surface without a separate support for the keyboard.

Security

Three kinds of password protection are available on the Model 50: power-on password, keyboard password, and server mode. Power-on password protection enables users to set a password that must be entered whenever the computer is turned on. When the power is turned on, a small key appears on the screen. The user has three chances to enter the correct password, or the computer must be turned off and turned on again. When a correct password is entered, the computer starts its boot-up process.

The power-on password is set by initializing the system from the Reference Diskette and choosing *Set Features* and *Set Passwords* from the main menu. When the password is entered, it is stored in the machine's battery-backed CMOS memory. If users forget their passwords, they can wipe the memory by removing the battery for at least 20 minutes (of course, unauthorized users could do this too, unless the system cover is locked). For hardware maintenance purposes, password checking can be circumvented by shorting the two pins on the side of the speaker and battery assembly.

Network server mode, a variation on the power-on password, is intended for computers used as network servers. This mode is also chosen from the menu on the Reference Diskette. When server mode is in effect, the system can be restarted without entering the power-on password, but the keyboard remains locked until the correct power-on password is entered. This allows server and other remote systems to restart automatically after a power failure, even though the keyboard is locked. This is a marked improvement over the AT, which will not allow the system to be reinitialized if the keylock is locked.

The keyboard password enables a user to lock the keyboard without turning off the system. The KP program on the Reference Diskette is used to set the keyboard password, which can be different from the power-on password. Once the password has been selected, typing KP locks the keyboard and typing the password unlocks it. A disadvantage of the password system, compared to the AT keylock, is that the keyboard cannot be locked with an application running.

Installation and Use

Thanks to conveniences provided with the Micro Channel architecture and the number of functions and capabilities built into the system board, the Model 50 is much easier to set up and install than an AT. A setup sheet provides instructions on how to unpack and set up the computer. Installation consists of little more than connecting the system power cord and the keyboard cord, connecting the display to the system unit, and connecting the power cords for the system unit and display to electrical outlets.

The Model 50 Quick Reference manual provides simple, well-illustrated directions on system setup and option installation. Initial installation consists of booting the system using the Model 50/60 Reference (setup) Diskette packaged with the Quick Reference manual, and using the menu-driven program that is loaded automatically. IBM has wisely write protected the Reference Diskette, so that its contents cannot be deleted or overwritten.

Once the computer is powered on, POST begins. The Model 50's POST displays a running total of the amount of operational memory (in KB) it finds. Then, if all tests are completed successfully, it sounds one beep on the system speaker and attempts to initialize the system. System initialization is attempted first from diskette drive A, then, if no diskette is found there, from hard disk C. If drive C contains no operating system, a screen is displayed that graphically prompts the user to insert a system diskette into drive A and press function key F1. Figure 6-9 shows this prompt screen.

Chapter 6

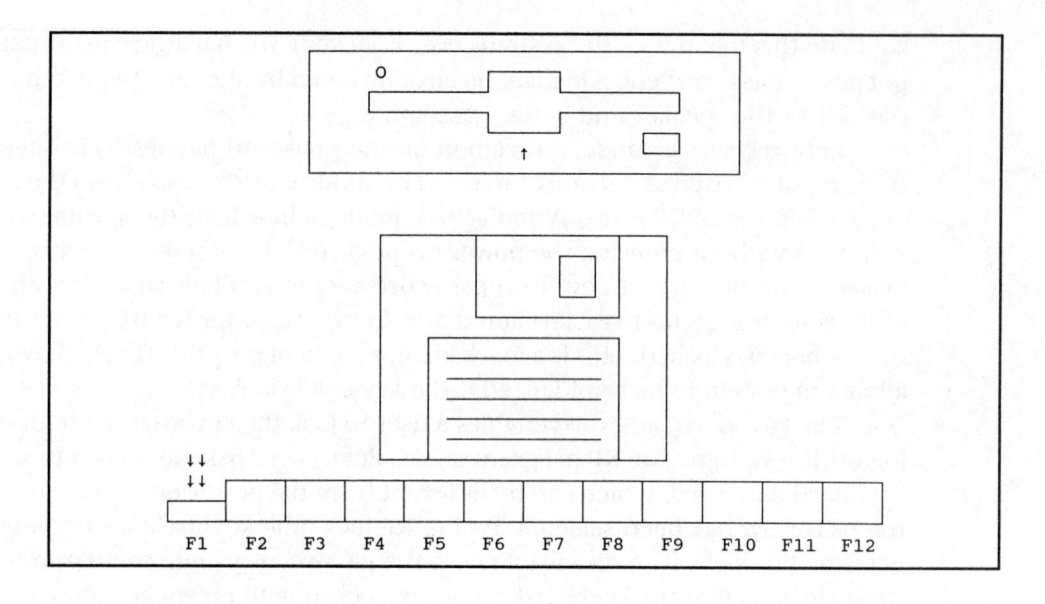

Figure 6-9 Load System Diskette Prompt. *This screen is displayed if POST is unable to find a system initialization program.*

When the Reference Diskette is used to initialize the system, the special version of COMMAND.COM that it contains displays an IBM logo screen. When Enter is pressed, the main menu is displayed (see Figure 6-10).

Before initializing the system from the Reference Diskette, POST sounds two beeps and displays 162 or 165 if the configuration stored in the system's battery-backed configuration RAM does not match the actual configuration of the system. It displays 163 if the time and date are not stored in the configuration RAM. If numeric codes are displayed, screens displayed after the IBM logo explain them. Additional information about error codes is provided in the *Problem Diagnosis* section of Chapter 7. The last screen gives the user the option of requesting the automatic update of the configuration RAM and reinitializing the system once the configuration RAM is updated. This is usually the only action the user needs to take. See the *Installing Options* section for additional information.

If POST sounds two beeps and displays 301, a keyboard error has been detected. This usually means that a key is stuck down, the keyboard is not connected to the system unit, or the keyboard has been connected to the pointing device connector by mistake. If the system is initialized from the Reference Diskette, this code is only displayed between the time that POST concludes and the IBM logo screen appears. Upon receipt of an Enter keycode, the system displays a message explaining that the 301 code indicates a keyboard error; however, the system cannot receive the Enter keycode until the keyboard is properly connected. Therefore, when a code 301 is displayed and/or the system does not re-

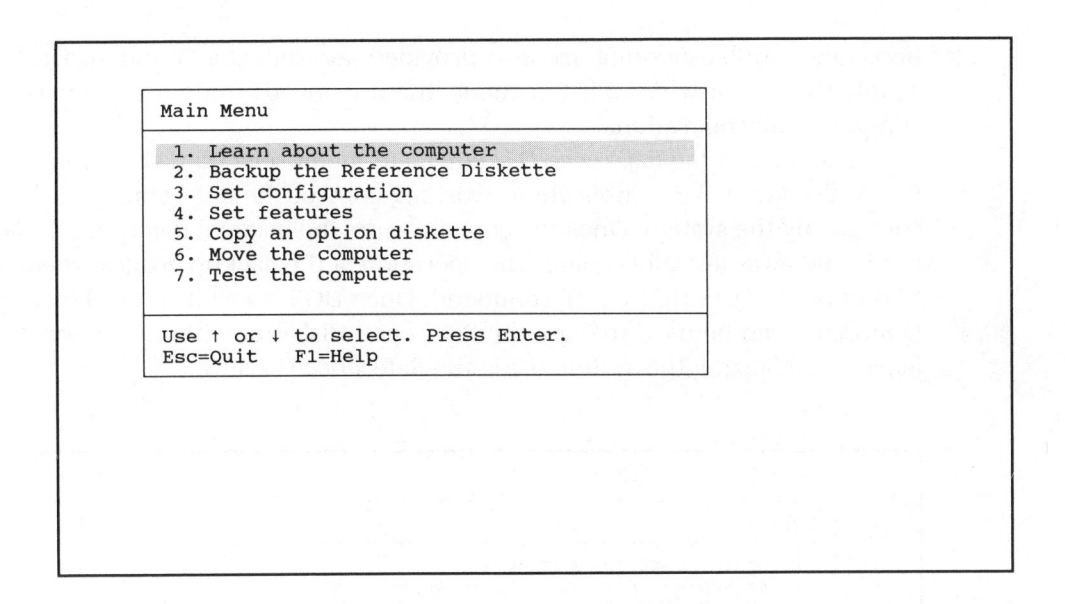

```
Main Menu

1. Learn about the computer
2. Backup the Reference Diskette
3. Set configuration
4. Set features
5. Copy an option diskette
6. Move the computer
7. Test the computer


Use ↑ or ↓ to select. Press Enter.
Esc=Quit    F1=Help
```

Figure 6-10 Reference Diskette Main Menu. *The setup procedure for the Model 50 is far more user friendly than that for the AT.*

spond to Enter being pressed, a check for stuck keys or an improperly connected keyboard is in order.

From the Reference Diskette main menu the user can learn about the computer, back up the Reference Diskette, set or view the system configuration, set system features such as date and time, copy the option diskette (which contains several hidden files), park the heads on the system's hard disks, and run diagnostic tests on the computer. An entry is selected from the menu either by entering its number or by moving the highlight bar to the entry, using the cursor keys, then pressing Enter. Pressing Esc while using a menu causes the display of the previous menu. Pressing Esc and then Enter while using the main menu causes the system to be initialized.

Selecting *Learn about the computer* from the Reference Diskette main menu initiates a slide-show-type program that provides a well-illustrated tutorial on the Model 50 and its hardware and software capabilities. The tutorial is organized into six files or chapters, including one on how to use the program, plus an index. The index is useful because it can be summoned from within a chapter by pressing function key F8. The user may then select from an alphabetical list of key terms used in the tutorial. Once a selection is made, the primary page containing the entry is displayed.

The tutorial describes the Model 50's hardware capabilities, complete with images of the items being described. The images displayed use the VGA's 320 by 200 pixel, 256-color mode, and provide the user with an accurate pictorial representation of the devices being discussed. Chapters on software, communications, and troubleshooting are also provided, as well as a "What To Do Next" chapter, which provides a list of things the user needs to do next to make the computer fully operational.

The *Backup the Reference Diskette* main menu selection, shown in Figure 6-11, is used to make a duplicate or working copy of the Reference Diskette for configuring the system. Once the copy is made, the original Reference Diskette can be put away for safekeeping. The operation of the backup program is similar to that of the DOS DISKCOPY command. Once DOS is installed on the system, DISKCOPY can be used to copy the Reference Diskette, without the inconvenience of initializing the system using the Reference Diskette.

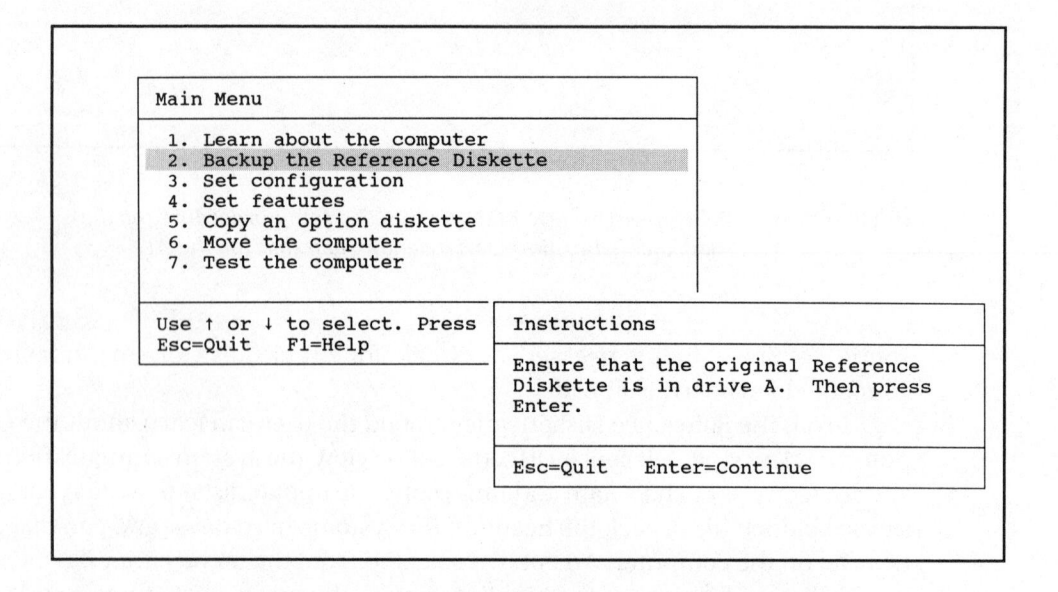

```
Main Menu
    1. Learn about the computer
    2. Backup the Reference Diskette
    3. Set configuration
    4. Set features
    5. Copy an option diskette
    6. Move the computer
    7. Test the computer

Use ↑ or ↓ to select. Press      Instructions
Esc=Quit    F1=Help
                                  Ensure that the original Reference
                                  Diskette is in drive A.  Then press
                                  Enter.

                                  Esc=Quit    Enter=Continue
```

Figure 6-11 Backup the Reference Diskette Menu. *A backup of the Reference Diskette is required to store system configuration information.*

On systems with only one diskette drive, it is necessary to shuffle the Reference Diskette and the target (backup copy) diskette in and out of the drive as the program alternately reads information from the source and writes it to the target diskette. If the target diskette is not formatted, the program formats it before it begins the copy operation. Disk shuffling is not required on systems with two 1.44 MB diskette drives, but because the Reference Diskette contains several hidden files, either the menu program or DISKCOPY (not COPY *.*) must be used to completely copy the contents of the Reference Diskette.

The *Set configuration* main menu selection references one of the most powerful items on the Reference Diskette. This feature allows the user to view the current configuration as stored in the system's CMOS RAM; change, back up, or

The Model 50

restore the system configuration; or perform an automatic update of the system configuration based on information determined from the installed hardware. The *Set configuration* menu selection is shown in Figure 6-12.

```
Main Menu
   Set Configuration
     1. View configuration
     2. Change configuration
     3. Backup configuration
     4. Restore configuration
     5. Run automatic configuration

  U
  E   Press a number to select.
      Esc=Quit    F1=Help
```

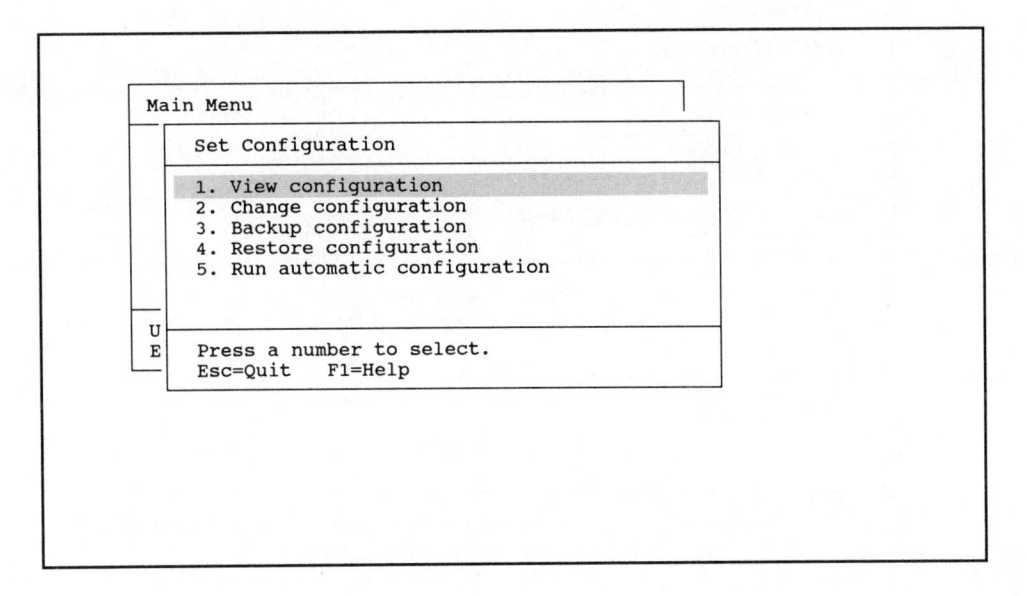

Figure 6-12 Set Configuration Menu. *Information about installed options and adapters is stored in the battery-backed CMOS RAM via this menu.*

Selecting *View configuration* from the menu produces a display of the system configuration information stored in the system's CMOS RAM. (See Figure 6-13.) Information displayed includes the amount of memory installed, the amount of memory usable upon completion of POST, a description of the devices installed on or controlled directly from the system board (*Built in Features*), and a description and configuration of the boards installed in the system's Micro Channel expansion slots.

The configuration display can be scrolled by using PgUp, PgDn, and the up and down cursor keys. The second screen of the display is shown in Figure 6-14. On the Model 50, slot 4 is always occupied by the hard disk adapter. The configuration display tells that it is a type 30 drive, with a Micro Channel bus arbitration level of 3.

The *Change configuration* menu's display is similar to that of *View configuration*, differing only in the title bar and the fact that items that can be changed are contained in brackets. The display is shown in Figure 6-15. Items to be changed are selected by stepping up or down through the items using the cursor keys. The display scrolls in the same way as the *View configuration* display. Once an item to be changed is selected, the next or previous acceptable value for the item

Chapter 6

```
┌─────────────────────────────────────────────────────────────────┐
│ View Configuration                                              │
│                                                                 │
│   Total System Memory                                           │
│     Installed Memory ...................... 2048 KB (2.0 MB)   │
│     Usable Memory ......................... 2048 KB (2.0 MB)   │
│                                                                 │
│   Built In Features                                             │
│     Installed Memory ...................... 1024 KB (1.0 MB)   │
│     Diskette Drive A Type ................. 1.44MB 3.5"        │
│     Diskette Drive B Type ................. Not Installed      │
│     Math Coprocessor ...................... Installed          │
│     Serial Port ........................... SERIAL_1           │
│     Parallel Port ......................... PARALLEL_1         │
│                                                                 │
│   Slot1 - IBM Dual Async Adapter                                │
│     Connector 1 ........................... SERIAL_2           │
│     Connector 2 ........................... SERIAL_3           │
│                                                                 │
│   Slot2 - IBM 2 MB 16-bit Memory Adapter                        │
│                                                                 │
│   Esc=Quit                                                      │
│   F1=Help                           ↓    End     PageDown       │
│                                                                 │
└─────────────────────────────────────────────────────────────────┘
```

Figure 6-13 First View Configuration Screen. *The currently stored configuration information shows the amount of memory available, as well as the installed devices.*

```
┌─────────────────────────────────────────────────────────────────┐
│ View Configuration                                              │
│                                                                 │
│     Diskette Drive A Type ................. 1.44MB 3.5"        │
│     Diskette Drive B Type ................. Not Installed      │
│     Math Coprocessor ...................... Installed          │
│     Serial Port ........................... SERIAL_1           │
│     Parallel Port ......................... PARALLEL_1         │
│                                                                 │
│   Slot1 - IBM Dual Async Adapter                                │
│     Connector 1 ........................... SERIAL_2           │
│     Connector 2 ........................... SERIAL_3           │
│                                                                 │
│   Slot2 - IBM 2 MB 16-bit Memory Adapter                        │
│     Installed Memory ...................... 1024 KB (1.0 MB)   │
│                                                                 │
│   Slot3 - Empty                                                 │
│                                                                 │
│   Slot4 - IBM Fixed Disk Adapter                                │
│     Type of drive ......................... 30                 │
│     Arbitration Level ..................... Level_3            │
│                                                                 │
│   Esc=Quit                          ↑    Home    PageUp         │
│   F1=Help                           ↓    End                    │
│                                                                 │
└─────────────────────────────────────────────────────────────────┘
```

Figure 6-14 Second View Configuration Screen. *This screen shows information about the hard disk and other adapters installed.*

is displayed by pressing the F5 or F6 key. For instance, the system board serial port can be set to be parallel port 1 or 2 by selecting the item, using F5 or F6 to display the desired value, and then pressing F10 to save the configuration in the CMOS RAM. If the user selects a value that conflicts with one already selected for another system device, a message indicates that a conflict exists.

Backup configuration, shown in Figure 6-16, saves the system configuration in a Reference Diskette file (SYSCONF). This feature can be used only with a backup copy of the Reference Diskette, since the original diskette is write protected. *Restore configuration* provides the ability to quickly restore the previously saved configuration to a system whose battery (which maintains information in CMOS RAM) has been disconnected or replaced.

Run automatic configuration is one of the most powerful and useful features available from the *Set configuration* menu. The display for this selection is shown in Figure 6-17. This utility automatically updates the system configuration to completely reflect the addition or removal of system devices. *Change configuration* is usually only needed to set user preferences, such as which port number a serial or parallel port on a particular board is to use.

If configuration changes are made and saved to the CMOS RAM using either *Change configuration* or *Run automatic configuration*, after pressing Esc to leave the *Set configuration* menu the user is instructed to press Enter to reinitialize the system and activate the configuration changes. (See Figure 6-18.)

```
Change Configuration                                           * Conflicts

    Total System Memory
        Installed Memory ..................... 2048 KB (2.0 MB)
        Usable Memory ........................ 2048 KB (2.0 MB)

    Built In Features
        Installed Memory ..................... 1024 KB (1.0 MB)
        Diskette Drive A Type ................ [1.44MB 3.5"    ]
        Diskette Drive B Type ................ [Not Installed       ]
        Math Coprocessor ..................... Installed
        Serial Port .......................... [SERIAL_2] *
        Parallel Port ........................ [PARALLEL_1]

    Slot1 - IBM Dual Async Adapter
        Connector 1 .......................... [SERIAL_2]
        Connector 2 .......................... [SERIAL_3]

    Slot2 - IBM 2 MB 16-bit Memory Adapter

    Esc=Quit    F5=Previous    F10=Save    ↑    Home
    F1=Help     F6=Next                    ↓    End     PageDown
```

Figure 6-15 Change Configuration Screen. *Configuration items that can be changed are shown in brackets.*

Chapter 6

```
Main Menu
  ┌─Set Configuration──────────────────┐
  │                                    │
  │  1. View configuration             │
  │  2. Change configuration           │
  │  3. Backup configuration           │
  │  4. Restore configuration          │
  │  5. Run automatic configuration    │
  │                                    │
U │                                    │
E │  Press a number to select.   ┌─Information──────────────────┐
  │  Esc=Quit    F1=Help         │                              │
  │                              │  Backup complete.            │
  └──────────────────────────────│                              │
                                 │                              │
                                 │  Press Enter to continue.    │
                                 └──────────────────────────────┘
```

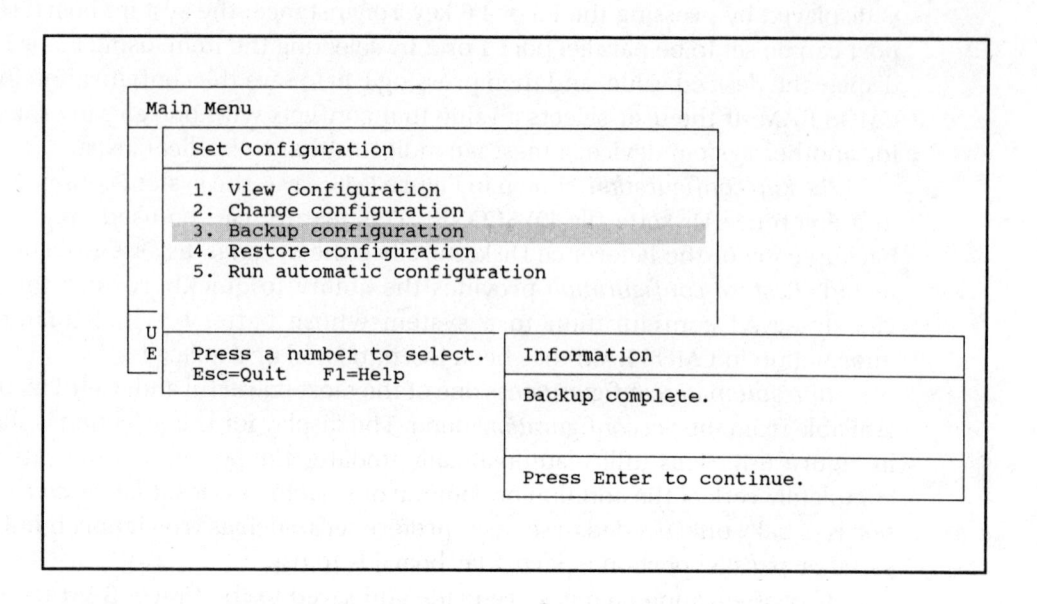

Figure 6-16   Backup Configuration Menu. *Existing configuration information can be restored to the battery-backed system configuration RAM if this feature has been used.*

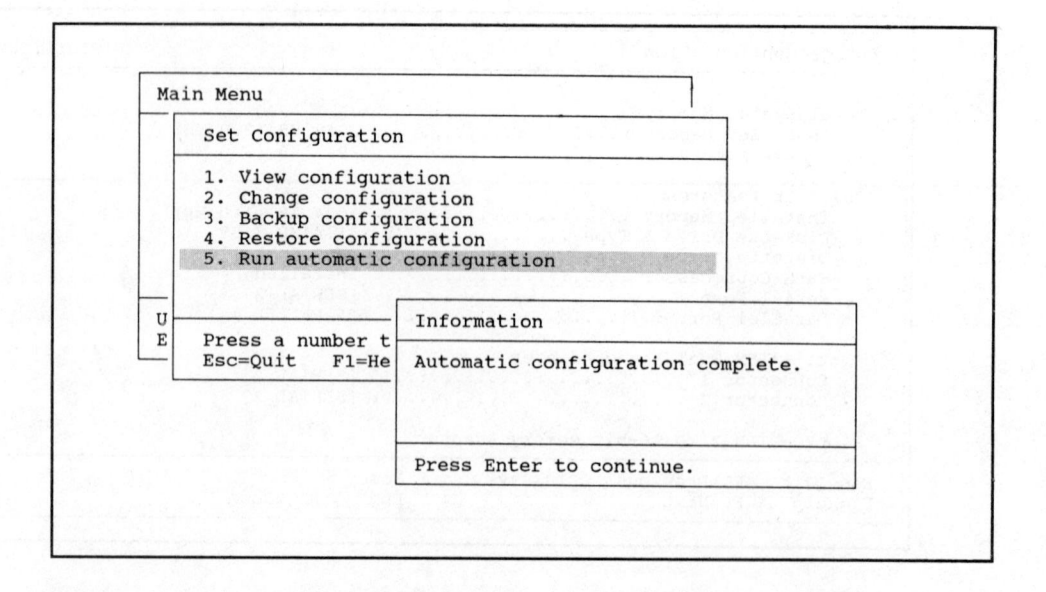

```
Main Menu
  ┌─Set Configuration──────────────────┐
  │                                    │
  │  1. View configuration             │
  │  2. Change configuration           │
  │  3. Backup configuration           │
  │  4. Restore configuration          │
  │  5. Run automatic configuration    │
  │                                    │
U │                              ┌─Information──────────────────┐
E │  Press a number t            │                              │
  │  Esc=Quit    F1=He           │  Automatic configuration complete. │
  │                              │                              │
  └──────────────────────────────│                              │
                                 │  Press Enter to continue.    │
                                 └──────────────────────────────┘
```

Figure 6-17   Run Automatic Configuration Menu. *For many situations, the automatic configuration program is adequate.*

186

```
┌─────────────────────────────────────────────────────────────┐
│   ┌─────────────────────────────────────┐                   │
│   │ Main Menu                           │                   │
│   │   ┌─────────────────────────────────┴──┐                │
│   │   │ Set Configuration                  │                │
│   │   │   ┌────────────────────────────────┴───────┐        │
│   │   │   │ 1. View configuration                  │        │
│   │   │   │ 2. Change configuration                │        │
│   │   │   │ 3. Backup configuration                │        │
│   │   │   │ 4. Restore configuration               │        │
│   │   │   │ 5. Run automatic configuration         │        │
│   │   │   │         ┌──────────────────────────────┴─────┐  │
│   │   │   │         │ Information                        │  │
│   │   │ U ├─────────┤────────────────────────────────────┤  │
│   │   │ E │ Press a number to select.│ Configuration changes have been │
│   │   └───┤ Esc=Quit    F1=Help      │ made. Press Enter to restart the│
│           └──────────────────────────┤ computer and activate the changes.│
│                                      │ Press Esc to return to the Main │
│                                      │ Menu.                           │
│                                      │                                 │
│                                      │ Esc=Quit                        │
│                                      └─────────────────────────────────┘
└─────────────────────────────────────────────────────────────┘
```

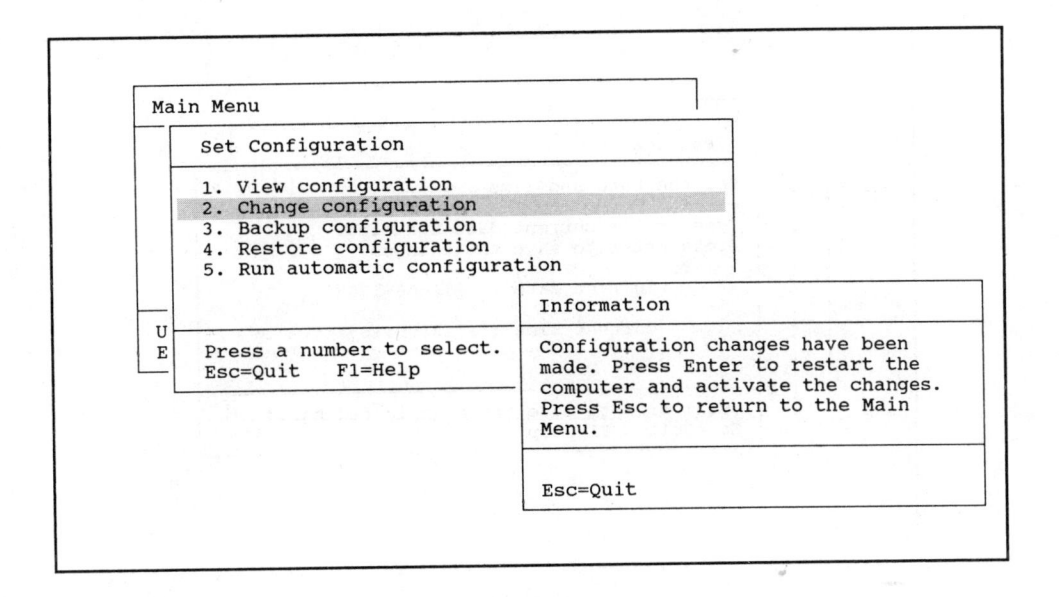

Figure 6-18   Configuration Change Information Display. *The system must be reset for configuration changes to be activated.*

The *Set Features* main menu selection, shown in Figure 6-19, provides the ability to set the date and time maintained in CMOS RAM, set system passwords, and select the keyboard key repeat speed. The time and date is set by keying numbers into delimited fields; the cursor keys can be used to position the cursor on particular fields or portions of fields to be changed.

The *Set passwords* menu, shown in Figure 6-20, is used to set the system power-on password, set the keyboard password, and set network server mode. The power-on password may not be changed or removed using this menu; selecting *Change power-on password* or *Remove power-on password* merely produces displays telling the user that these actions can be performed when the password prompt (a small key) is displayed at system power-up.

The *Set keyboard speed* menu allows the key repeat rate for the keyboard to be set to normal (15 characters per second) or fast (30 characters per second). Normal is the default setting. Figure 6-21 demonstrates the use of the F1 key to provide help information, which is sometimes (as in this case) more specific than the information provided in the Model 50 Quick Reference manual.

The *Copy an option diskette* main menu selection, shown in Figure 6-22, allows the adapter description files (ADF) and diagnostic programs, provided on diskette along with a hardware option, to be copied to the backup copy of the Model 50/60 Reference Diskette. This is required in order for the Reference Diskette to contain all of the description files and diagnostic routines needed to test all devices installed in the system. DOS users with one diskette drive can perform the same operation by copying the files on the option diskette to the hard disk and then to the backup copy of the Reference Diskette (as long as there are

Chapter 6

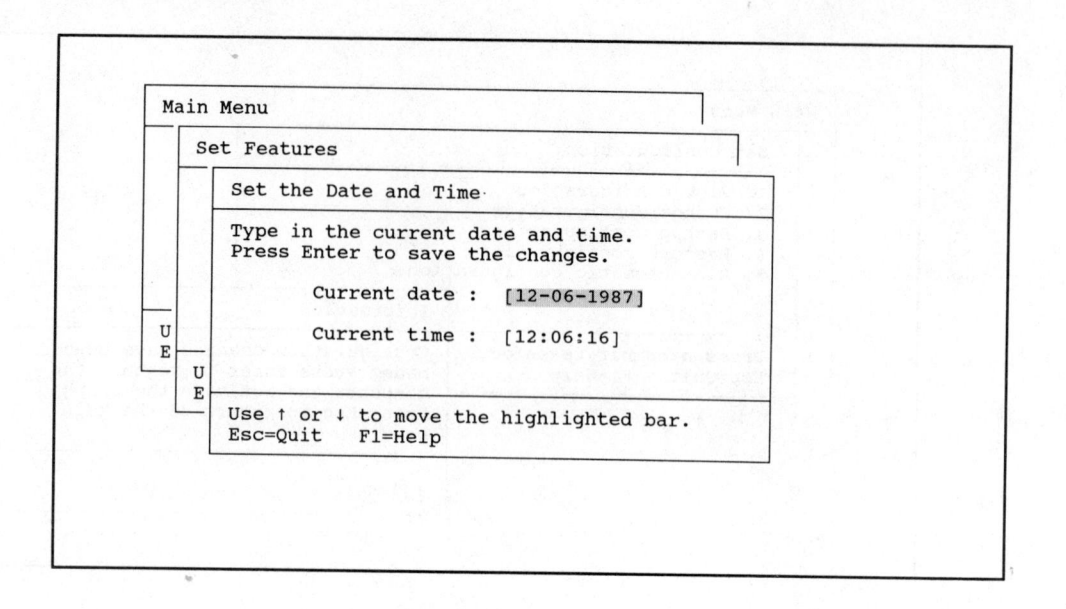

```
 Main Menu
  ┌─────────────────────────────────────────────┐
  │ Set Features                                │
  │  ┌────────────────────────────────────────┐ │
  │  │ Set the Date and Time                  │ │
  │  │                                        │ │
  │  │ Type in the current date and time.     │ │
  │  │ Press Enter to save the changes.       │ │
  │  │                                        │ │
  │  │       Current date :   [12-06-1987]    │ │
  │  │                                        │ │
  │U │       Current time :   [12:06:16]      │ │
  │E │                                        │ │
  │  │U                                       │ │
  │  │E                                       │ │
  │  │ Use ↑ or ↓ to move the highlighted bar.│ │
  │  │ Esc=Quit    F1=Help                    │ │
  │  └────────────────────────────────────────┘ │
  └─────────────────────────────────────────────┘
```

Figure 6-19 Set the Date and Time. *For normal situations, the date and time are preserved by the battery-backed CMOS RAM; this feature allows them to be reset.*

```
 Main Menu
  ┌─────────────────────────────────────────────┐
  │ Set Features                                │
  │  ┌────────────────────────────────────────┐ │
  │  │ Set Passwords                          │ │
  │  │                                        │ │
  │  │  1. Set power-on password              │ │
  │  │  2. Change power-on password           │ │
  │  │  3. Remove power-on password           │ │
  │  │  4. Set keyboard password              │ │
  │  │  5. Set network server mode            │ │
  │U │                                        │ │
  │E │                                        │ │
  │  │U                                       │ │
  │  │E                                       │ │
  │  │ Use ↑ or ↓ to select. Press Enter.     │ │
  │  │ Esc=Quit    F1=Help                    │ │
  │  └────────────────────────────────────────┘ │
  └─────────────────────────────────────────────┘
```

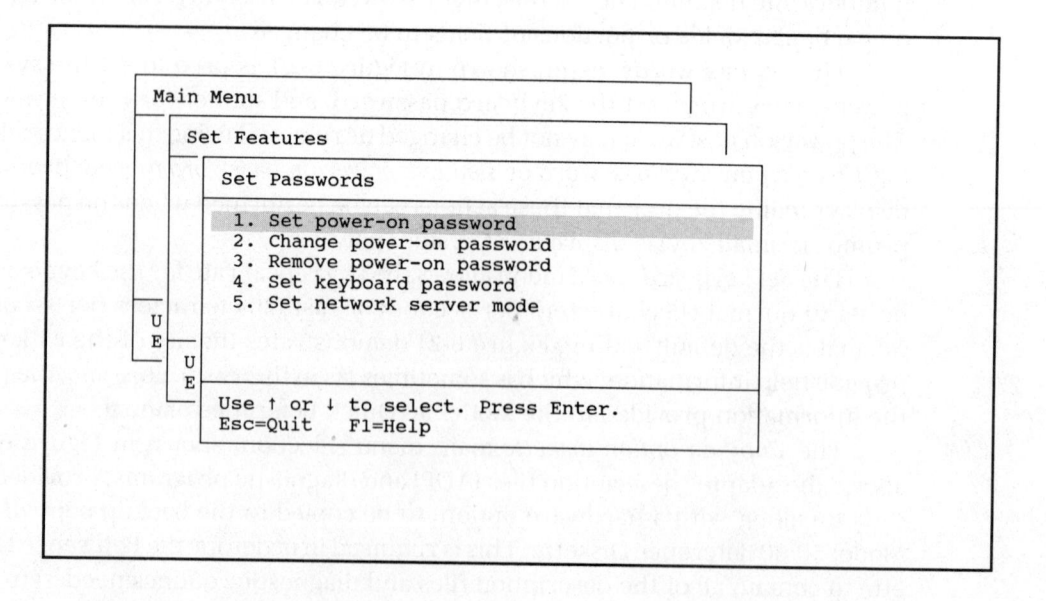

Figure 6-20 Set Passwords Menu. *The power-on and network server mode passwords can be set using this menu. The KP program is used to set the keyboard password.*

The Model 50

```
┌─────────────────────────────────────────────────────────────┐
│  Main Menu                                                  │
│  ┌────────────────────────────────────────────────┐         │
│  │ Set Features                                   │         │
│  │ ┌──────────────────────────────────────┐       │         │
│  │ │ Set Keyboard Speed                   │       │         │
│  │ │                                      │       │         │
│  │ │ Press Enter to save the keyboard speed.      │         │
│  │ │                                      │       │         │
│  │ │     Keyboard Speed: [Normal]         │       │         │
│  │ │                      Fast            │       │         │
│  │U│                                      │       │         │
│  │E│  ┌──────────────────────────────────────────┐│         │
│  │ │U │ Help                    Page  1 of  1    ││         │
│  │ │E │                                          ││         │
│  │ │  │ The normal keyboard speed puts           ││         │
│  │ │ Use ↑ or ↓ to move the h│ characters on the screen at a rate │
│  │ │ Esc=Quit    F1=Help     │ 15 characters per second.          │
│  │ │                         │                                    │
│  │ │                         │                                    │
│  │ │                         │ Esc=Quit                           │
│  │ │                         └──────────────────────────────────┘ │
│  └────────────────────────────────────────────────┘         │
└─────────────────────────────────────────────────────────────┘
```

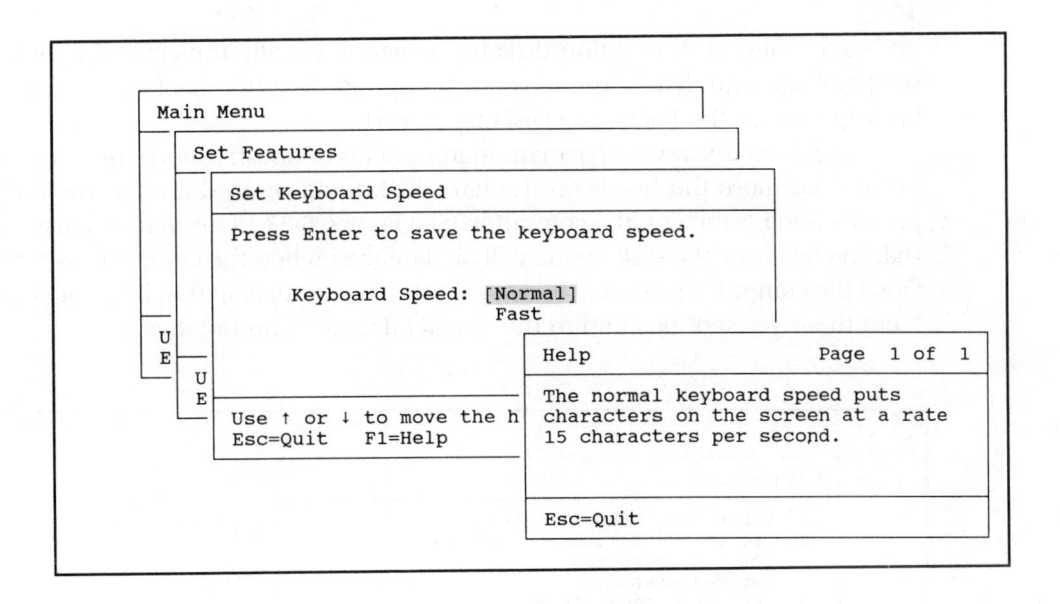

Figure 6-21   Set Keyboard Speed Menu and Help Display. *The repeat rate for the keyboard can be adjusted using this menu.*

```
┌─────────────────────────────────────────────────────────────┐
│  ┌──────────────────────────────────────────┐               │
│  │ Main Menu                                │               │
│  │                                          │               │
│  │ 1. Learn about the computer              │               │
│  │ 2. Backup the Reference Diskette         │               │
│  │ 3. Set configuration                     │               │
│  │ 4. Set features                          │               │
│  │ 5. Copy an option diskette               │               │
│  │ 6. Move the computer                     │               │
│  │ 7. Test the computer                     │               │
│  │                                          │               │
│  ├──────────────────────────┐               │               │
│  │ Use ↑ or ↓ to select     │               │               │
│  │ Esc=Quit    F1=Help      ┌──────────────────────────────┐│
│  │                          │ Information                  ││
│  └──────────────────────────┤                              ││
│                             │ Insert your New Option Diskette in drive │
│                             │ A:                           ││
│                             │                              ││
│                             │ Press Enter to continue.     ││
│                             │ Esc=Quit                     ││
│                             └──────────────────────────────┘│
└─────────────────────────────────────────────────────────────┘
```

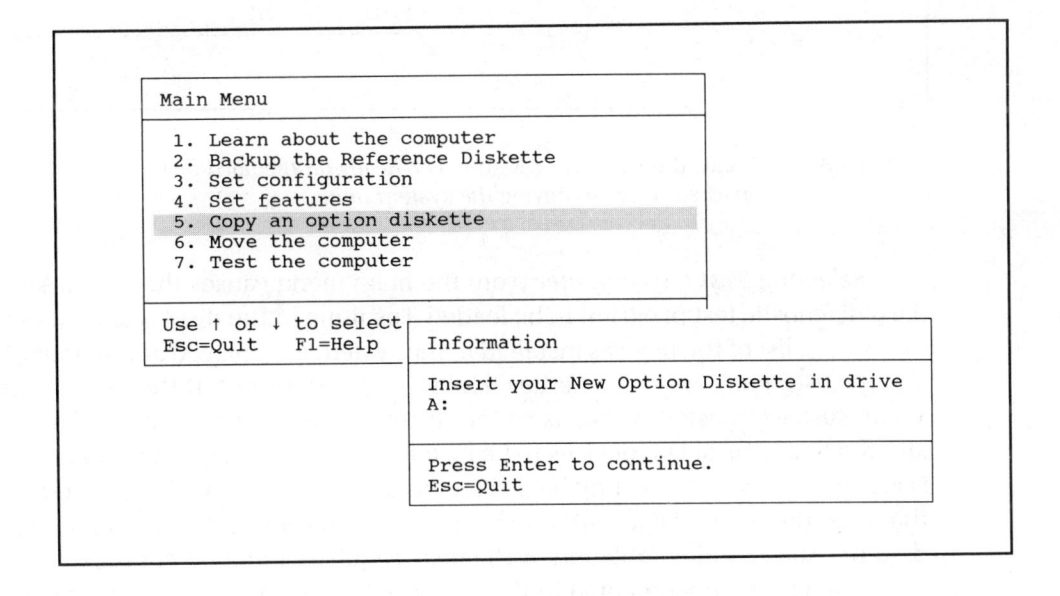

Figure 6-22   Copy an Option Diskette Menu. *Adapters that are installed in the Micro Channel may have configuration and diagnostic software supplied with them.*

189

Chapter 6

no hidden files on the option diskette, which is usually the case). On systems with two diskette drives, the files on the option diskette may be copied to the backup copy of the Reference Diskette directly.

The *Move the computer* main menu selection doesn't move the computer, but it does move the heads on the hard disk to an unused area of the disk in preparation for moving the computer. (See Figure 6-23.) This makes it less likely that the heads or the disk media will be damaged when the computer is moved. Once the computer is moved, it can be restarted as normal; the disk heads move from their "parked" position to the normal data area automatically.

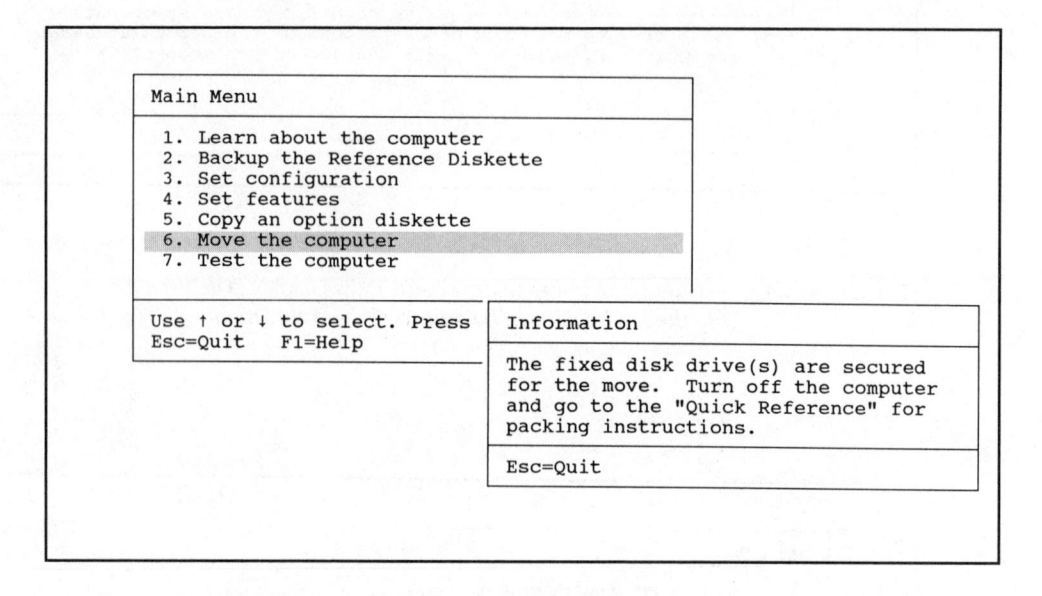

```
Main Menu

  1. Learn about the computer
  2. Backup the Reference Diskette
  3. Set configuration
  4. Set features
  5. Copy an option diskette
  6. Move the computer
  7. Test the computer

Use ↑ or ↓ to select.  Press      Information
Esc=Quit    F1=Help
                                  The fixed disk drive(s) are secured
                                  for the move.  Turn off the computer
                                  and go to the "Quick Reference" for
                                  packing instructions.

                                  Esc=Quit
```

Figure 6-23   Move the Computer Menu. *The heads on the hard disk can be parked prior to moving the system unit.*

Selecting *Test the computer* from the main menu causes the system's standard diagnostic test program to be loaded, resulting in the display shown in Figure 6-24. A list of the devices installed in the system is displayed at the upper left and the user must answer whether the list is correct or not. If the user answers Y (for yes), each listed device is tested in turn. A user answering N (for no) is allowed to add or delete devices to be tested from the list. Once the updated list is accepted as correct, testing begins on the devices listed. During the course of the tests, the user is required to press keys and to remove and insert a scratch diskette. Successful completion of the tests requires about 10 minutes.

Though not documented in the Quick Reference manual, the Model 50/60 Reference Diskette contains an advanced diagnostic program in addition to the standard diagnostic program. This program is invoked by entering Ctrl-A from the Reference Diskette main menu. These diagnostics hardware format a hard disk and provide the selective diagnostic services provided by the IBM PC family advanced diagnostics. See Chapter 7 for additional information.

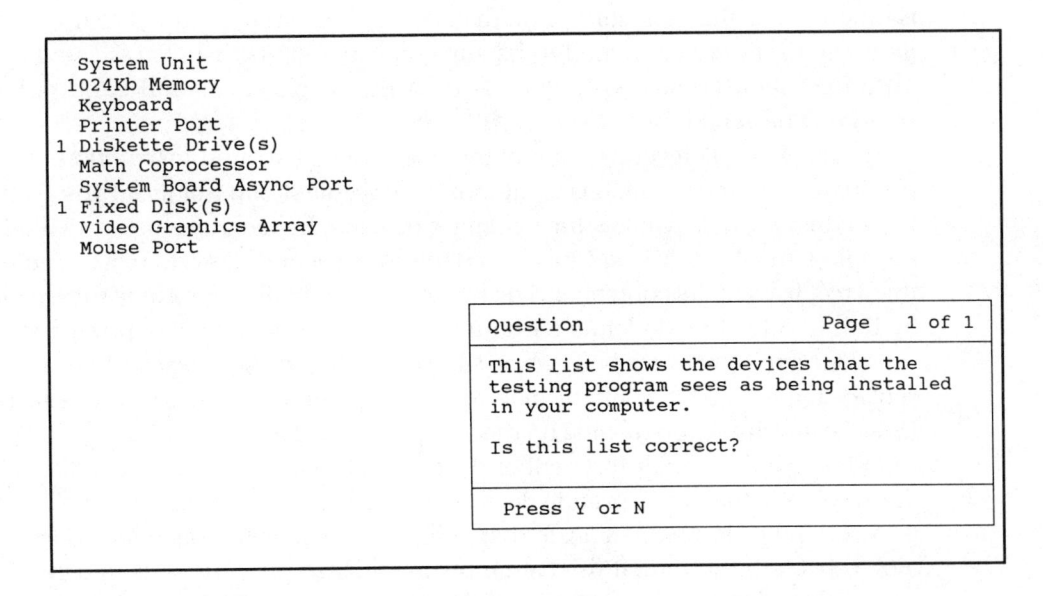

```
System Unit
 1024Kb Memory
 Keyboard
 Printer Port
1 Diskette Drive(s)
 Math Coprocessor
 System Board Async Port
1 Fixed Disk(s)
 Video Graphics Array
 Mouse Port
```

Question	Page 1 of 1
This list shows the devices that the testing program sees as being installed in your computer. Is this list correct?	
Press Y or N	

Figure 6-24 Standard Diagnostics Display. *Basic diagnostics for the system can be accessed directly from the Reference Diskette.*

Installing Options

Installing options on the Model 50 is generally quick and easy. If the option includes a diskette that contains Micro Channel device configuration information, that information should be copied to the working copy of the Model 50 Reference Diskette, using the DOS or OS/2 COPY command, or the Reference Diskette's *Copy an option diskette* program, before the physical installation process begins.

The physical installation process generally requires no tools. For safety, the power cord should be removed before the cover is removed. Once the two thumb screws on the rear of the system unit are loosened, the cover is pulled forward and lifted up and off. After this is done, the system's Micro Channel expansion slots are easily accessible. Physically installing a Micro Channel adapter is a matter of removing an expansion slot cover from an unused slot, inserting the board, and tightening the thumb screw on the rear of the system unit to hold the adapter firmly in place. Most adapters can be installed in any available slot; the 8514/A Display Adapter or any other video adapter must be installed in expansion slot 3, which contains the video extension connector (and is thus longer than the other sockets).

Installation of an Intel 80287 math coprocessor is equally straightforward, only requiring the insertion of the coprocessor into its socket, with the notched

Chapter 6

(pin 1) end toward the back of the system unit. The socket is directly accessible unless there is an adapter board installed in the Micro Channel expansion slot closest to the edge of the unit; if there is, removal takes only a few seconds. This is much easier than installing a math coprocessor in the cramped rear quarters of an AT. Users are cautioned to be sure that the 80287 installed is rated for 10 MHz operation (indicated by 80287-10 on the IC package), and is not a lower rated one designed for use on an AT.

The Quick Reference manual includes well-illustrated instructions for replacing the battery (which is located in a convenient clip holder at the front of the system unit), as well as for installing options. Instructions are provided for installing an 80287, Micro Channel adapters, a second diskette drive, and even the hard disk and its controller. The latter is an indication of past or future plans by IBM to offer the Model 50 without a hard disk as standard equipment.

Once an option has been physically installed and the cover and power cord replaced, the system's configuration must be updated by inserting the Reference Diskette in drive A and turning on the system power. The power-on self test (POST) sounds two beeps and displays either the code 162 (see Figure 6-25), to indicate that the system's actual configuration does not match the configuration stored in CMOS RAM, or the code 165 to indicate that an option adapter has been added to or removed from the system. It then attempts to initialize the system from drive A.

If no diskette is found in drive A, the system waits, displaying only the error code. Veteran AT users will likely guess that this is the cue to press function key F1 to continue; however, all Personal System/2 users may not be this well in-

```
02048 KB OK
162
```

Figure 6-25 POST Error 162. *The Reference Diskette must be used to change system configuration information when an option is added to the system.*

formed. Inserting the Model 50/60 Reference Diskette in drive A before pressing F1 provides the informative display shown in Figure 6-26, once the system is initialized (and the obligatory IBM logo is displayed and acknowledged by pressing Enter). Page 2 of the display is accessed by pressing PgDn. From here the user can request an automatic update of the system's CMOS RAM and then reinitialize the system, after the configuration is updated. (See Figure 6-27.)

```
Adapter Configuration Error - 00165      Page 1 of 2

The computer's internal self-tests found an option
adapter that is different from the option adapters
indicated in the computer's configuration.

This error occurs if option adapters are added,
removed, or are not working properly.

If you have added or removed an adapter, run
automatic configuration.  To view or change the
results of automatic configuration, go to the Main
Menu of this diskette and select "Set configuration."


PageDown
```

Figure 6-26 Configuration Error Screen. *The Reference Diskette is used to display the error message corresponding to a POST error code.*

```
Adapter Configuration Error - 00165      Page 2 of 2

Select "View configuration" or "Change configuration"
from the Set Configuration menu.

If you have not added or removed an adapter, do not
run automatic configuration.  Go to the Main Menu of
this diskette.  Select "Test the computer" to
determine the cause of the error and what action to
take.



Automatically configure the system? (Y/N)
         PageUp
```

Figure 6-27 Configuration Error Screen, Page 2. *The system can be configured using the automatic configuration option in most instances.*

Chapter 6

This awkward error identification process is required because IBM stores the error message text outside the machine, so that it can be in any language the user requires, worldwide. While this may be convenient for IBM, it isn't particularly convenient for users, since even holding down a key while POST is running causes the system to cryptically display a 301 code and then do nothing until the user presses F1.

As noted in the error display, a 162 error can also occur if certain external devices are not powered on. The 5.25-inch external diskette drive is such a device. To avoid this error, the diskette drive should be powered on before the system unit. If the user forgets to do so, pressing F1 after the 162 code is displayed causes the system to initialize from either the 3.5-inch diskette in drive A or from the hard disk.

Warranty and Service

IBM provides a one-year limited warranty on the Model 50. Customer carry-in is required for warranty repairs; on-site warranty repair is available as an option. Reflecting expected reliability and ease of repair, the minimum cost of IBM on-site maintenance for the Model 50 is $180. This is significantly less than the $546 minimum maintenance charge announced with the 8 MHz IBM AT, Model 339 in 1986.

Documentation

The Model 50 Quick Reference manual provides well-illustrated directions on system setup and option installation, but very little information about how the system works. This valuable information is available only in supplementary manuals at additional cost. These manuals include individual technical reference manuals for system devices and adapters, the Model 50/60 Technical Reference, the Personal System/2 and Personal Computer BIOS Interface Technical Reference, the Hardware Maintenance Reference, and the Personal System/2 Hardware Maintenance Service manuals.

The Model 50/60 Technical Reference provides in-depth technical information about the two systems. The level of detail varies from chapter to chapter, but new features such as the Micro Channel architecture and the Video Graphics Array are described at length. It is a much more organized description of the system than the AT Technical Reference, which seems to be a collection of information not provided elsewhere. However, it does not provide schematics, as the AT Technical Reference does.

The BIOS Interface Technical Reference provides information on the basic input/output system (BIOS) of all IBM PC and Personal System/2 computers. As one might expect, this involves a lot of information and a fair number of footnotes and explanations to differentiate between the services provided on different models. Although BIOS listings are not provided in the technical reference of each computer, as in the past, the functional operation of the BIOS is described in detail in an organized manner.

The Hardware Maintenance Reference manual provides general maintenance information, including product descriptions, disassembly procedures, and an introduction to diagnostics. The Hardware Maintenance Service manual consists of a two-part diagnostic manual and a copy of the Model 50/60 Reference Diskette. The manual provides detailed specifications and part numbers for the various assemblies that comprise a Model 50 or 60. It also explains how to diagnose system problems, using the advanced diagnostics contained on the Reference Diskette and an electrical multimeter. Maintenance personnel are led through a series of step-by-step procedures that allow problems to be isolated to a particular system device. Once a problem is isolated, the manual recommends corrective action, which is all too often "Replace the system board."

Disk Cache Program

IBM provides a hard disk cache program, called IBMCACHE, with Models 50, 60, and 80. This program uses a set of sector buffers to improve disk read performance. The sector buffers may be located in either base or extended memory. The cache program runs in base memory. For each disk read initiated by the system, the cache program determines if the requested sector is stored in one of its sector buffers. If it is, the information is provided directly, without accessing the disk. If the requested sector is not in the sector buffers, it is read from the disk, along with additional sectors, to satisfy the current request and in anticipation of subsequent sequential read requests.

The minimum size of the cache of sector buffers is 16 KB. The cache program is implemented as a DOS device driver (IBMCACHE.SYS). The device driver, when installed, attaches itself to BIOS interrupts 13H and 15H (disk/diskette and system services, respectively). The device driver intercepts interrupt 13H requests, and immediately passes control to the interrupt 13H BIOS routine if the request is not for hard disk services. If the request is for hard disk services, the cache program checks to determine if the function requested is one of the following:

Read Sectors
Write Sectors
Format Track

Chapter 6

Format Unit
Write Long

Treatment of *Read Sectors* requests depends on the number of sectors requested in the read command. If the number of sectors specified in the read request is less than two times the cache page size, the requested sectors are read from the sector buffers if they are contained there, or read from the disk, stored in the sector buffers, and furnished to the requestor. If the number of sectors to be read is greater than or equal to twice the cache page size (the number of sectors read each time the disk is accessed), the cache program passes the request to the BIOS routine for processing. Requests for a large number of sectors are not processed, because such processing would likely fill all the cache buffers with new information and because breaking large reads up into a series of cache page size reads results in disk rotational delays that slow down the read process.

When a *Write Sectors* function is requested, the cache program determines what pages in the cache are affected and updates them, using the data from the requestor's output buffer. It then issues the original write request to the BIOS routine. Thus, the appropriate pages in the cache get updated and the write proceeds at full speed once initiated.

The cache program provides two additional functions for interrupt 13H. These functions allow an application to get cache statistics or flush the cache. The cache statistics consist of page number, size of cache page, and performance data. The calling sequence for *get cache* statistics is:

(AH) = 1DH
(AL) = 01H

On return:
CF = 1 indicates an error.
CF = 0 indicates no error.
(AH) = error code (0 if CF = 0).
(ES:BX) = pointer to statistics area (if CF = 0).

The *flush cache* function allows all entries in the cache to be invalidated. This is useful when a format is to be performed or the media is changed in a removable media drive. The calling sequence for the flush cache function is:

(AH) = 1DH
(AL) = 02H

On return:

    CF = 1 indicates an error.

    CF = 0 indicates no error.

    (AH) = error code (0 if CF = 0).

The cache sector buffers may be either in conventional or extended memory, as specified on the `DEVICE=` statement in the CONFIG.SYS file. To keep internal operation as much the same as possible, irrespective of whether the sector buffers are located in conventional or extended memory, a one-page intermediate buffer is maintained in conventional memory, regardless of where the sector buffers are stored.

The cache program is designed to provide minimal interference with other programs that use extended memory, when its sector buffers are stored in extended memory. The program allocates the space for sector buffers starting at the high end of extended memory and working downward, whereas most programs start with low addresses and work upward. It intercepts requests to the BIOS interrupt 15H, subfunction 88H (Determine Memory Size), and returns the amount of extended memory that is not being used by the cache, rather than the total amount of extended memory in the system. It also coexists with VDISK software. If VDISK device drivers are installed, the cache program determines if enough extended memory is available over and above that used by the VDISK drivers, for the cache to be installed.

If the function requested is *Format Track* or *Format Unit*, the cache program marks all sector buffers invalid. This is done because the format operations erase data on the disk, thus invalidating the contents of the cache sector buffers.

The cache program invalidates all cache pages containing sectors referenced in a *Write Long* request. This is done because a *Write Long* can be used when DOS is performing error recovery on the hard disk, and thus can invalidate data stored in the cache.

IBMCACHE.COM, a program for installing the cache program on the hard disk as a DOS device driver, is provided on the Model 50/60 Reference Diskette along with the cache program. Both files are hidden, most likely to keep them from easily migrating to computers other than the Personal System/2.

Although the cache program significantly increases the performance of many disk-intensive applications (particularly on the Model 50, with its slow disk), its inclusion with Models 50, 60, and 80 seems to have been something of an afterthought by IBM. It is not mentioned or described in the Quick Reference manual, its only documentation being a two-page insert included with the Quick Reference manual and the online help available when the cache installation program is run.

Chapter 6

Accessing the Reference Diskette and entering the command IBMCACHE causes the display of the menu shown in Figure 6-28. From this menu the user can install the disk cache onto the system's drive C, view or change the cache's configuration, or remove the cache from drive C. An entry is selected from the menu either by entering its number or by moving the highlight bar to it, using the cursor keys, and pressing Enter.

```
Disk Cache Main Menu
  1. Install disk cache onto drive C:
  2. View disk cache settings
  3. Change disk cache settings
  4. Remove disk cache from drive C:


Use ↑ or ↓ to select. Then press Enter.
Esc=Quit    F1=Help
```

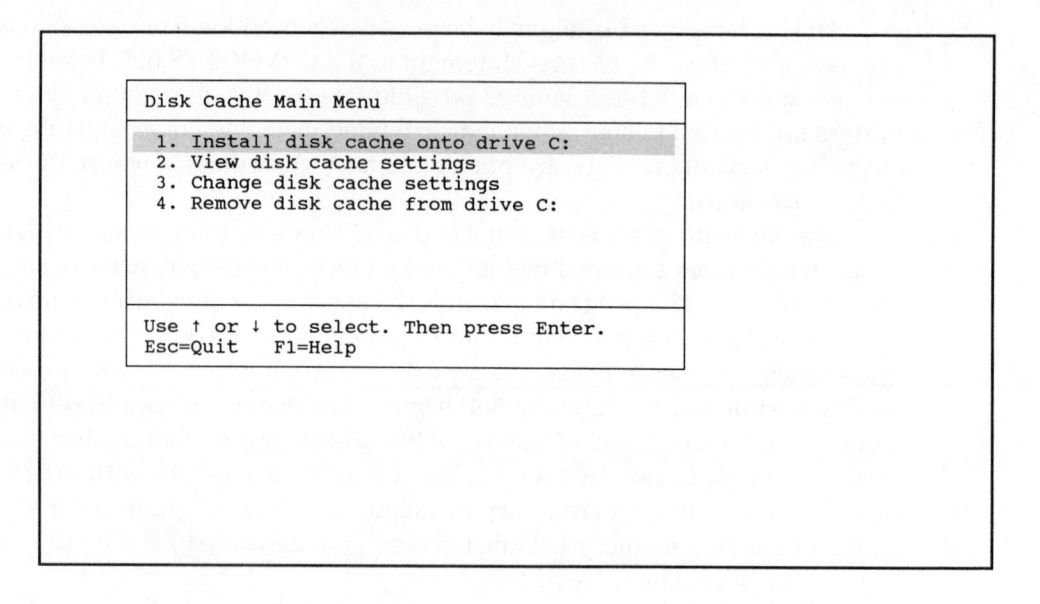

Figure 6-28 Disk Cache Main Menu. *The cache program supplied with the Model 50 provides an increase in disk performance as seen by the user.*

If the user selects the *Install disk cache into drive C:* option, the display shown in Figure 6-29 results. If the user answers yes, the device driver IBMCACHE.SYS is copied to the root directory of drive C. The installation program creates a CONFIG.SYS file consisting of the statement

```
device=\ibmcache.sys    64 /NE /P4
```

followed by the contents of any existing CONFIG.SYS file. The existing CONFIG.SYS file (if any) is renamed CONFIG.BAK.

The device statement above indicates that the root directory file IBMCACHE.SYS is to be loaded as a device driver. The parameters 64, /NE, and /P4 define the size and operational characteristics of the cache: 64 indicates that 64 KB is to be used for sector buffers; /NE indicates that the buffers are to be located in conventional (not extended) memory; and /P4 indicates that the cache page size (the number of sectors to be read each time the disk is accessed) is to be 4 sectors. Once the installation program has modified the CONFIG.SYS file, the system must be reinitialized in order for the IBMCACHE.SYS device driver to be loaded. After that, operation is automatic.

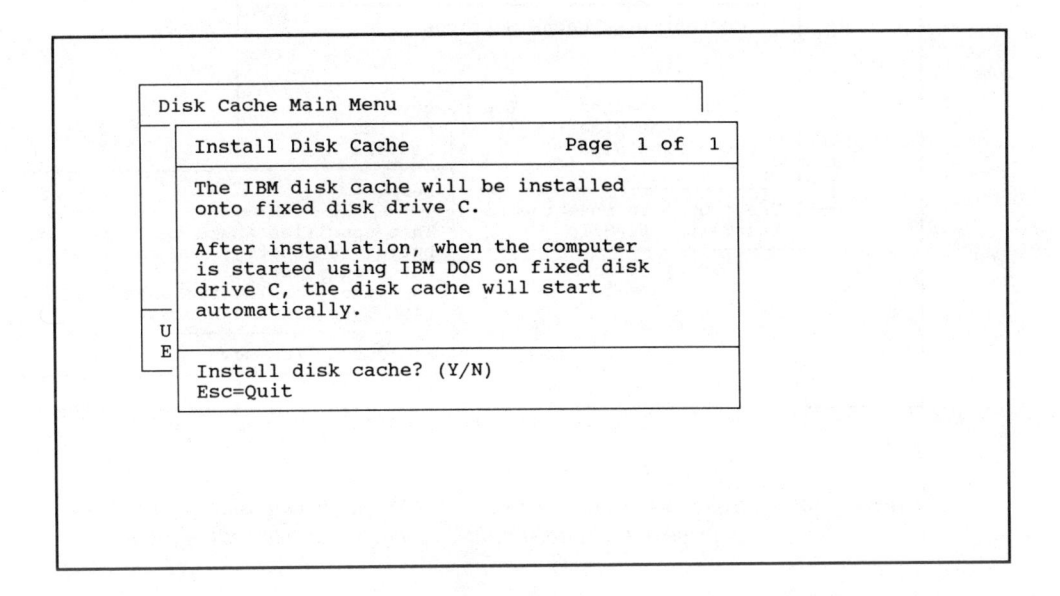

```
Disk Cache Main Menu

 Install Disk Cache              Page  1 of  1

 The IBM disk cache will be installed
 onto fixed disk drive C.

 After installation, when the computer
 is started using IBM DOS on fixed disk
 drive C, the disk cache will start
 automatically.

 Install disk cache? (Y/N)
 Esc=Quit
```

Figure 6-29 Install Disk Cache Menu. *The cache buffer can be located in base or extended memory.*

Once the cache is installed, the IBMCACHE program may be run from the Reference Diskette to view or change the cache's configuration or to remove the cache. Selecting *View disk cache settings* from the Disk Cache main menu results in a display that tells whether the cache is in conventional or extended memory, and gives its size and the cache page size. Online help about any of the three items displayed may be obtained by pressing F1 while the item is highlighted. (See Figure 6-30.)

Upon returning to the main menu, the user can select any or all of the three items to be changed, and step forward and backward through acceptable values using function keys F5 and F6. (See Figure 6-31.) The values shown in the display configure the cache to use the 384 KB of extended memory (provided standard on the Model 50) for its sector buffers. This configuration doesn't perform quite as well as it would with the same amount of conventional memory for sector buffers. However, it provides a large disk cache with only 16 KB of conventional memory used by the device driver. Some communications data may be lost if high speed communications are performed while the cache is operating in extended memory, because interrupts are inhibited during the block move operations used to move data between extended and conventional memory.

The cache may be removed from the system by selecting *Remove disk cache*

Chapter 6

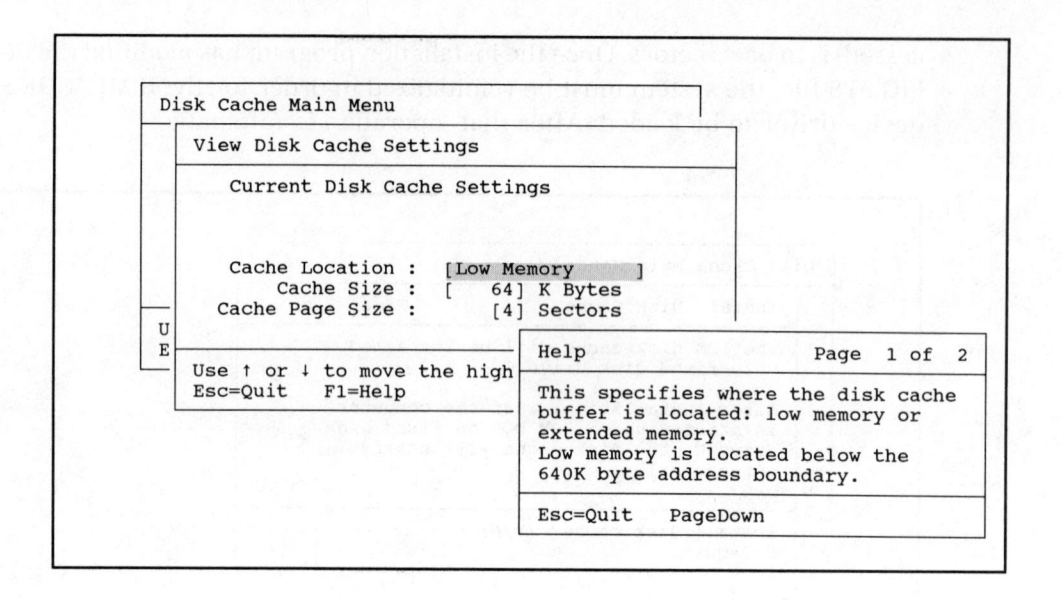

```
┌─────────────────────────────────────────────────────────────┐
│  ┌──────────────────────────────────┐                        │
│  │ Disk Cache Main Menu             │                        │
│  │ ┌──────────────────────────────┐ │                        │
│  │ │ View Disk Cache Settings     │ │                        │
│  │ │                              │ │                        │
│  │ │   Current Disk Cache Settings│ │                        │
│  │ │                              │ │                        │
│  │ │                              │ │                        │
│  │ │   Cache Location  : [Low Memory    ]                    │
│  │ │      Cache Size   : [   64] K Bytes │                   │
│  │ │   Cache Page Size :      [4] Sectors│                   │
│U │ │                              │ │                        │
│E │ │                         ┌────┴─┴──────────────────────┐ │
│  │ │ Use ↑ or ↓ to move the high│ Help        Page 1 of 2  │ │
│  │ │ Esc=Quit   F1=Help        │                           │ │
│  │ └───────────────────────────┤ This specifies where the disk cache │
│  │                             │ buffer is located: low memory or    │
│  │                             │ extended memory.                    │
│  │                             │ Low memory is located below the     │
│  │                             │ 640K byte address boundary.         │
│  │                             │                                     │
│  │                             │ Esc=Quit     PageDown               │
│  │                             └─────────────────────────────────────┘
└─────────────────────────────────────────────────────────────┘
```

Figure 6-30   View Disk Cache Settings. *The menu-driven utility on the Reference Diskette simplifies cache parameter adjustment.*

```
┌─────────────────────────────────────────────────────────────┐
│  ┌──────────────────────────────────┐                        │
│  │ Disk Cache Main Menu             │                        │
│  │ ┌──────────────────────────────┐ │                        │
│  │ │ Change Disk Cache Settings   │ │                        │
│  │ │                              │ │                        │
│  │ │ Type or select values in the highlighted                │
│  │ │ bar.  Press Enter when done. │ │                        │
│  │ │                              │ │                        │
│  │ │                              │ │                        │
│  │ │   Cache Location  : [Extended Memory]                   │
│  │ │      Cache Size   : [  384] K Bytes                     │
│  │ │   Cache Page Size :      [4] Sectors                    │
│U │ │                              │ │                        │
│E │ └──────────────────────────────┘ │                        │
│  │ Use ↑ or ↓ to move the highlighted bar.                   │
│  │ Esc=Quit   F1=Help   F5/F6=Select                         │
│  └──────────────────────────────────┘                        │
└─────────────────────────────────────────────────────────────┘
```

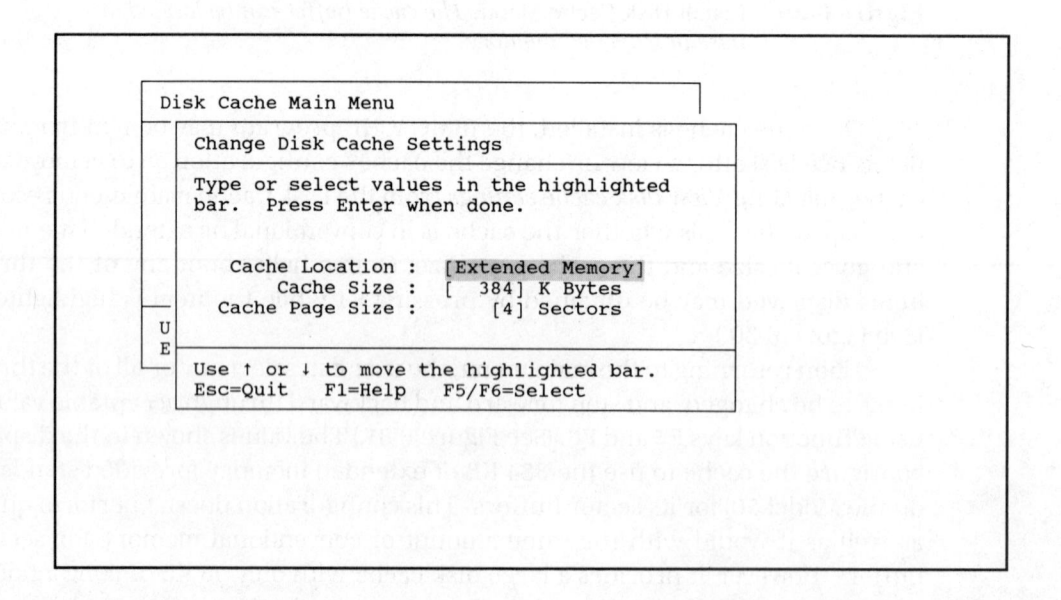

Figure 6-31   Change Disk Cache Settings. *The page size, total size, and location (base or extended memory) of the cache can be specified.*

from drive C: from the Disk Cache main menu. Upon selection of this item, IBMCACHE.SYS is removed from the root directory of drive C and the IBMCACHE.SYS device driver line is removed from the CONFIG.SYS file. The system must then be reinitialized to purge the device driver and sector buffers from memory. Users can, of course, achieve the same effect by manually deleting the driver and changing the CONFIG.SYS file using a text editor. Configuration changes can likewise be made by editing CONFIG.SYS. Minimum, maximum, and default values for cache size are 16, 512, and 64 KB of conventional memory. If extended memory is used for the sector buffers (indicated by the /E option), the minimum, maximum, and default cache sizes are 16, 15,360, and 128 KB. Valid values for cache page size are 2, 4, and 8 sectors; the default value is 4. Users who appreciate uncluttered root directories can move the IBMCACHE.SYS file to a utility directory, and change the CONFIG.SYS DEVICE= statement to indicate its new location.

Performance and Compatibility

As a successor to the AT, one might expect the Model 50's performance to be uniformly better than that of the AT. While this is true for many aspects of its performance, it is not true overall, at least not without software assists such as the disk cache (which can help the AT as well). While the Model 50's 80286 CPU operates 25 percent faster than that of the AT (10 MHz versus 8 MHz) and the 80287 operates 88 percent faster (10 MHz versus 5.33 MHz), its hard disk is twice as slow as that of the AT.

More important to more users, however, is how quickly and how well the Model 50 performs day-to-day applications. Typical applications include word processing, data management, graphics, and program development. Comparing the performance of the Model 50 with other machines in these four areas provides a better representation of its actual performance than the various clock rates and memory speeds alone.

Table 6-7 shows the performance of the Model 50 in non-disk-dependent tests using the WordPerfect word processing application and the Lotus 1-2-3 spreadsheet application. In the WordPerfect test, a large text file was loaded into memory and the times to move from the top to the bottom of the file, and from the bottom to the top, were measured. As expected, the Model 50 with its 10 MHz 80286 performed the test about 25 percent faster than the AT or AT-compatible Compaq DeskPro 286. Similar, but less striking results were observed when using WordPerfect to replace all occurrences of a "j" in the file with a "k".

To test the Model 50's computational speed, calculations were performed on three different spreadsheets, using Lotus 1-2-3. (See Appendix A for a detailed description of all tests that were performed.) The use of the math coprocessor on the Model 50 had little effect on the calculation time for the first

Chapter 6

Table 6-7. Non-Disk-Dependent Performance. *The Model 50 was faster overall than the AT in non-disk-dependent tests.*

Test Machine	Equipment	WordPerfect Move Top to Bottom	WordPerfect Move Bottom to Top	WordPerfect Search and Replace	Lotus 1-2-3 Recalc— Add	Lotus 1-2-3 Recalc— Multiply	Lotus 1-2-3 Recalc— Mix
Model 30	No math	49	38	58	8	14	91
	Math	49	38	58	8	8	9
AT 339	No math	28	21	32	5	8	45
	Math	28	21	32	5	5	7
Compaq 286	No math	28	21	34	5	7	45
	Math	28	21	34	5	5	7
Model 50	No Math	22	16	27	4	6	36
	Math	22	16	27	4	4	5
Model 60	No math	22	16	27	4	6	36
	Math	22	16	27	4	4	5
AST 10 MHz	No math	16	12	20	3	4	28
	Math	16	12	20	3	3	4

All results are in seconds.

spreadsheet, which used addition; some effect on the second, which used multiplication and division; and a dramatic effect on the third, which used square root and natural logarithm functions. The Model 50 performed all spreadsheet tests 20 to 25 percent faster than the AT, both with and without the math coprocessor on each machine. Even though the Model 50's math coprocessor is much faster than the AT's, the spreadsheet recalculation time was obviously dependent on the speed of the memory and the 80286.

The Model 50 provided significantly increased performance over that of the Model 30, in addition to providing the capability of running protected mode operating systems such as OS/2. It uniformly performed the tests in half the time required by the Model 30.

The Model 60, with its functionally identical architecture, performed the tests in the same time as the Model 50. The AST Premium/286—an AT-compatible with a 10 MHz 80286, 8 MHz 80287, and zero-wait-state memory—outperformed Models 50 and 60 in the word processing and spreadsheet calculation tests. Reflecting the performance of its zero wait state memory, it performed the tests 20 to 25 percent faster than Models 50 and 60, making it approximately 40 percent faster overall than the AT.

To test the Model 50's file processing capabilities, WordPerfect was used to load a document file, Lotus 1-2-3 was used to load a spreadsheet, and dBASE III PLUS was used to perform two sorts on a file. PC-KEY-DRAW was used to display a graphics file and MASM 5.0 was used to assemble a file. All tests were performed from RAM disk, from the hard disk, and from the 3.5-inch diskette drive using 1.44 MB and 720 KB diskettes. Results of the tests with the files stored on an extended memory RAM disk are shown in Table 6-8.

Table 6-8 RAM-Disk-Dependent Performance. *The Model 50 performed file functions from RAM disk 20 to 25 percent faster than the AT.*

Test Machine	WordPerfect Load File	Lotus 1-2-3 Load File	dBASE III PLUS Sort on Title	dBASE III PLUS Sort on Disk ID	PC-KEY-DRAW Run Sample Macro	MASM Assemble File
AT 339	11	22	9	10	520	8
Compaq 286	9	22	10	11	473	8
Model 50	9	17	7	8	406	6
Model 60	9	17	8	8	406	6
AST 10 MHz	10	16	7	8	384	5

All results are in seconds.

The Model 50 performed operations that used RAM disk files 20 to 25 percent faster than the AT. The Model 60 did likewise. The AT-compatible AST Premium/286, with its 10 MHz 80286 and zero wait state memory, performed all tests in about the same time as the Model 50. This is indicative of the processor-intensive nature of these particular tests, since the AST's memory is faster but its processor is the same speed as that of the Model 50.

The WordPerfect document and Lotus 1-2-3 spreadsheet were loaded from the Model 50's hard disk, and the dBASE III PLUS sorts were performed there also. PC-KEY-DRAW was used to display the same graphics file and MASM 5.0 was used to assemble the same source file, with the files stored on the hard disk instead of RAM disk. Results of the tests with the files stored on the hard disk, with and without a 384 KB extended memory disk cache in use, are shown in Table 6-9.

Even though the rated access time of the Model 50's hard disk was much slower than that of the AT, it performed faster in operations where the same data was used a number of times and was thus retained in the disk cache. When the cache software was used, the dBASE III PLUS sorts were performed 20 percent faster than on an AT (without disk cache software). Using the same test without the disk cache software in use, the Model 50 performed 45 percent slower, making it 15 percent slower than the AT and nearly as slow as the Model 30.

The Model 60, with its faster hard disk, was about as fast without the cache software installed as the Model 50 was with it installed. The AST Premium/286, which also has a fast hard disk, was faster than either the Model 50 or Model 60 on file loads, and offered performance comparable to that of the Model 60 on the file sorts. Both it and the AT can benefit from third-party disk cache software in the same way that the Model 60 benefited from it in several of the tests.

Chapter 6

Table 6-9 Hard-Disk-Dependent Performance. *The Model 50 performed hard-disk-dependent functions faster than the AT when the disk cache software was used.*

Test Machine	Equipment	WordPerfect Load File	Lotus 1-2-3 Load File	dBASE III PLUS Sort on Title	dBASE III PLUS Sort on Disk ID	PC-KEY-DRAW Run Sample Macro	MASM Assemble File
Model 30	No cache	25	44	32	34	655	16
AT 339	No cache	12	29	25	26	526	11
Compaq 286	No cache	11	28	21	22	477	9
Model 50	No cache	10	27	29	29	409	7
	Cache	10	20	20	21	405	7
Model 60	No cache	10	26	24	21	406	7
	Cache	10	19	22	21	406	7
AST 10 MHz	No cache	8	17	25	24	388	9

All results are in seconds.

Although speed of the diskette drives is less critical on the Model 50 than on the Model 25 or an IBM PC, the disk-dependent tests were run on the Model 50's 3.5-inch diskette drive, using diskettes formatted at 1.44 MB. See Table 6-10 for results. When using high density diskettes, the performance of the Model 50 was the same as that of the Model 60 and more or less comparable to that of the AT. The AT was about 10 percent faster on the spreadsheet load and on the file assembly, which can be attributed to its high density drive's faster track-to-track access time. With the processing-intensive PC-KEY-DRAW macro, the Model 50 was its usual 20 percent faster.

Table 6-10 High Density Diskette-Dependent Performance. *When using high density diskettes, the performance of the Model 50 was comparable to that of the AT.*

Test Machine	WordPerfect Load File	Lotus 1-2-3 Load File	dBASE III PLUS Sort on Title	dBASE III PLUS Sort on Disk ID	PC-KEY-DRAW Run Sample Macro	MASM Assemble File
AT 339	29	112	204	223	583	19
Compaq 286	33	112	171	169	543	21
Model 50	29	128	204	207	469	20
Model 60	29	128	204	204	468	20
AST 10 MHz	20	108	199	223	465	16

All results are in seconds.

The same tests were run using low density diskettes. Operations such as the WordPerfect file load, which were affected by cutting the transfer rate in half, were much slower, as shown in Table 6-11. This series of tests was so dependent on diskette drive performance that the performance of the Model 50 was little better than that of the Model 30, except for the more processor-intensive PC-KEY-DRAW graphics application.

Table 6-11 Low Density Diskette Performance. *The Model 30 performed about as fast as the Model 50 when using low density diskettes.*

Test Machine	WordPerfect Load File	Lotus 1-2-3 Load File	dBASE III PLUS Sort on Title	dBASE III PLUS Sort on Disk ID	PC-KEY-DRAW Run Sample Macro	MASM Assemble File
Model 30	54	140	219	224	704	36
AT 339	43	115	193	184	572	30
Compaq 286	43	115	181	182	546	27
Model 50	45	134	209	211	480	22
Model 60	45	134	209	211	465	23
AST 10 MHz	40	112	176	180	468	26

The Model 50 ran all test applications without difficulty, attesting to its PC software compatibility. On the non-disk-dependent tests it performed the same as the Model 60, as expected. It offered more performance than the Model 30 and AT 339, but somewhat less than the AST Premium/286. In tests that involved extensive hard disk access, the Model 50 did well in situations where the same data was accessed frequently. In those situations, the disk cache made the Model 50's slow disk competitive with the faster disks used on the AT, AT compatibles, and Model 60. In all other disk-intensive situations, the other machines came out ahead.

Summary

The Model 50 is a capable desktop replacement for the AT. It is slightly faster, smaller, quieter, and less expensive to purchase, operate, and maintain. It offers improved performance over the AT and is very software compatible. Aside from the problem of diskette incompatibility, the vast majority of PC and AT applications will run on the Model 50 the first time and every time.

Some applications require more expansion slots and more mass storage than the Model 50 provides. The Model 60, with its eight expansion slots and its ability to support large hard disks, is available to fill this need. Either it or the Model 80 is a better solution for a server or multi-user system. Of course, along with that solution comes larger size, greater weight, and higher cost.

For many, the Model 50 will be the most cost-effective personal productivity workstation offered in the Personal System/2 family. In a stand-alone environment it is limited somewhat by the capacity of its hard disk; however, alternative hard disks are available from third parties and will likely be available from IBM as the cost of hard disks continues to drop. The Model 50 is a very good workstation for the LAN environment. A version of the Model 50 without a hard disk is available from IBM as a special bid offering. This configuration gives the user the processing power of an 80286, the ability to run OS/2, and the file system and communications capabilities provided by a network, without the extra cost of a slow hard disk.

7

The Model 60

Whereas the Model 50 is simply a replacement for the AT or XT 286, the Model 60 is a larger, more versatile machine that improves upon the AT in almost every way. In addition to being faster, it offers more standard features, larger storage capacity, more expansion capability, and quieter operation, and it requires less space on the desktop. The Model 60 is well suited for use as a personal productivity workstation using DOS 3.3 or OS/2, and it is also well suited for use as a small to medium size network file server.

Features

The Model 60 offers all the capabilities of the Model 50, plus additional performance and capacity, particularly in the areas of hard disk storage and system expansion slots. The Model 60 comes standard with a larger hard disk and can support a second hard disk. It has four additional Micro Channel expansion connectors and a 2 KB CMOS RAM extension not provided on the Model 50.

The Model 60 consists of a separate display and keyboard designed for desktop use and a floor-standing system unit. (See Figure 7-1.) The display and keyboard are the same as those used on the Model 50 and other members of the Personal System/2 family; however, the system unit is quite different from that of the Model 50.

The Model 60 features an Intel 80286 microprocessor, as did the AT. The processor and system bus run at 10 MHz (25 percent faster than the 8 MHz speed of the AT). The optional 80287 numeric coprocessor also runs at 10 MHz (88 percent faster than the 5.33 MHz of the AT's 80287). As might be expected, this descendant of the AT provides 16 MB of memory address space, an eight-channel direct memory access (DMA) controller, sixteen-level interrupt system,

Chapter 7

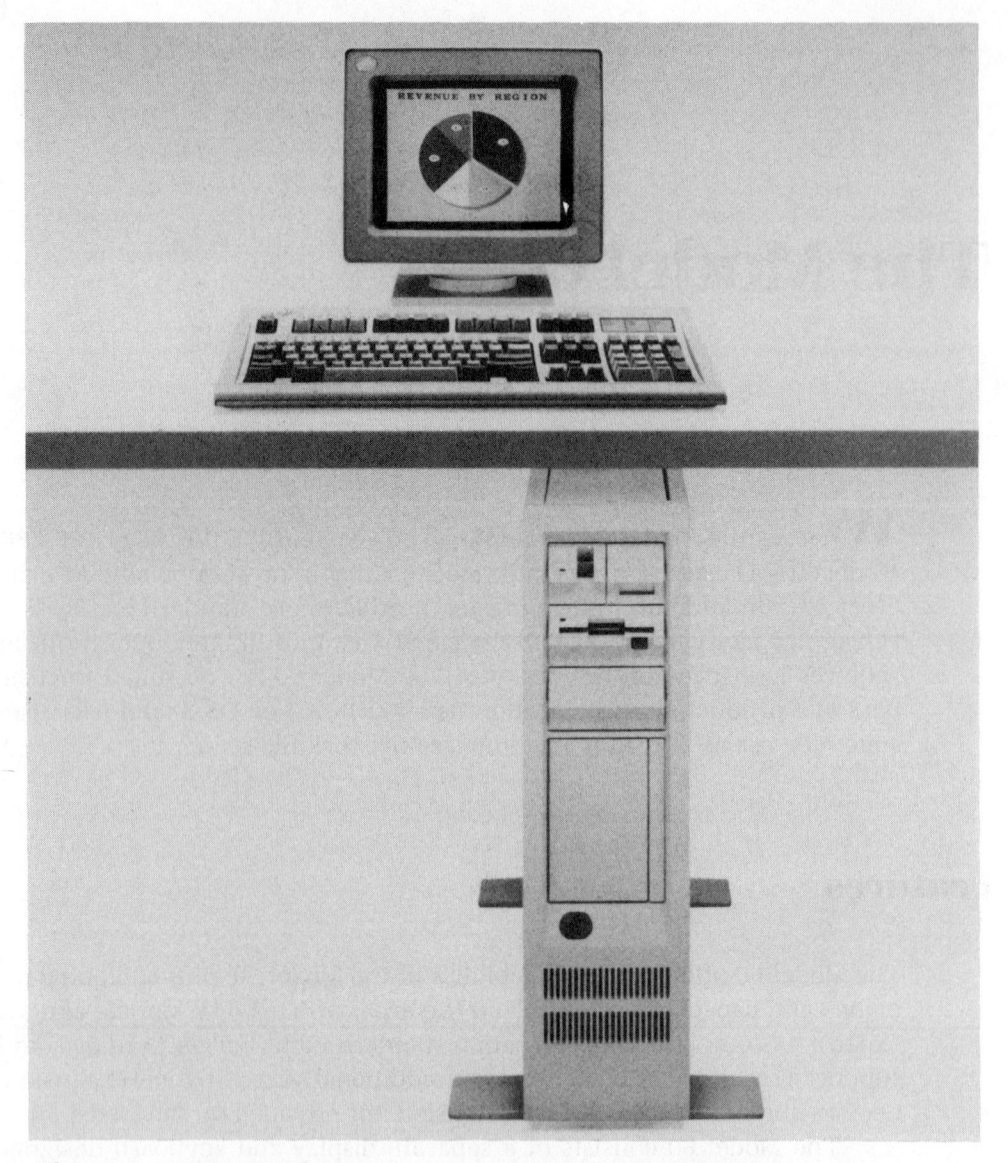

Figure 7-1   The IBM Personal System/2 Model 60. *The Personal System/2 Model 60 is a floor-standing 80286-based computer.*
Photo courtesy of International Business Machines Corporation.

system clock and timers, real-time clock with CMOS RAM and battery backup, and an audio subsystem with speaker.

The Model 60 comes standard with 1 MB of memory installed on the system board (640 KB base memory and 384 KB extended memory). Like all other members of the Personal System/2 family, it features the IBM 101-key Enhanced Keyboard. This is the same keyboard that is available with late model ATs and XTs; however, the connector on the system unit end of its cord is different. A connector is also provided standard on the system board for the connection of an IBM mouse or other pointing device. A keylock is provided, but it only secures the cover; it does not disable the keyboard as on the AT. Keyboard security is provided via a password scheme. One parallel and one serial port are provided on the system board. The Model 60 system board improves significantly on that of the AT, providing more memory, plus the pointing device port, parallel port, and serial port.

Another standard feature of the Model 60 is the Video Graphics Array (VGA) video subsystem described in Chapter 5. This subsystem, when used with the Personal System/2 family of analog color and monochrome displays, supports several graphics modes (640 by 480 pixels in 2 and 16 colors, 320 by 200 pixels in 256 colors) and high resolution text mode (720 by 400 pixels in monochrome and 16 colors), while maintaining compatibility with CGA and EGA graphics and text modes. The VGA drives any member of the Personal System/2 family of displays, through a connector on the system board. Neither the connector nor the VGA is compatible with any IBM PC display; however, some third-party adaptable sync monitors can be used with the VGA.

The Model 60 features a 16-bit implementation of IBM's Micro Channel architecture, as described in Chapter 4. This architecture, which is designed to support multi-tasking operating systems, is not compatible with the bus architecture used with the IBM PC family and Personal System/2 Models 25 and 30. The Model 60 uses seven 16-bit connectors and one 16-bit slot with auxiliary video extension. Micro Channel connectors only accept expansion boards that are specifically designed for use with the Micro Channel, usually indicated by a suffix of /A in the name of boards provided by IBM. These boards are physically and electrically different from, and may not be interchanged with, IBM PC family (or Personal System/2 Model 25 and 30) adapter boards.

A single 1.44 MB, 3.5-inch diskette drive is provided as standard equipment on the Model 60; a second 1.44 MB drive can be added as an option. These drives can also read and write diskettes in a 720 KB mode, which is compatible with that used by the drives on Models 25 and 30, the IBM PC Convertible, and several other portable and desktop PCs.

Two different configurations of the Model 60 are offered, distinguished only by the hard disk provided as standard equipment. The 8560-041 has a 44 MB, 5.25-inch hard disk; the 8560-071 comes standard with a 70 MB, 5.25-inch hard disk. The 041 can support a second 44 MB hard disk; the 071 can support a second 70 MB or 115 MB hard disk.

The Model 60, with its floor-standing system unit and sleek new displays, has a modern look that the AT definitely lacks. Overall it is quieter, less bulky, and generally easier to set up and operate. The only major element common to it and the AT is the 101-key keyboard, which is itself a modernized version of the original AT 84-key keyboard.

Chapter 7

Physical Description

Unlike the Model 50, the Model 60 system unit is actually larger than that of the AT. Its composite plastic case measures 23.5 inches high by 19 inches deep by 5.5 inches wide, compared to the AT's 21.25 inches wide by 17.28 inches deep by 6.38 inches high. The system unit stands on the floor, usually under a table or beside a desk, stabilized by two 12.5-inch feet that are rotated inward for shipment.

Because the system unit is no longer on the desktop for the system display to rest on, users accustomed to desktop ATs may find the Model 60's display too low at first. In truth, when mounted on their stands, the Personal System/2 family displays are about the same height above the table as modern ergonomically designed video display terminals. Most users will probably either make the adjustment or invest in one of the height adjustable display stands available for video display terminals.

At 41.8 pounds, the Model 60 system unit weighs almost as much as the AT's 43 pounds; and it actually generates slightly more heat—1240 BTU/hour compared to the AT's 1229 BTU/hour. It generates the same low noise level as the Model 50 (46 dB average, operating; 40 dB average, idle, measured at a distance of 1 meter). Since it is a floor-standing, rather than a desktop unit it seems much quieter to the user than an AT.

The fold-down handle built into the top and the hand holds built into the sides of the system case are handy for positioning the system, both during installation and when a system needs to be moved or upgraded. The system unit is a bit heavy to be carried far by the handle, but it can be pulled along across smooth surfaces easily. The system unit fits nicely under or beside a desk; 2 to 3 inches of clearance should be provided around the system unit to allow proper ventilation and cooling.

The system power switch is located at the top front of the unit, permitting easy access. It is recessed to prevent accidental operation, but not as well as the Model 50's power switch. As on the Model 50, nearby green and yellow lights indicate power on and hard disk activity, respectively. As with other members of the family, the Model 60's model and serial numbers are displayed on the front (below and to the right of the power switch), as well as on the back of the system case. This feature, which others would do well to copy, is extremely useful, particularly in organizations with large amounts of equipment. The standard 1.44 MB, 3.5-inch diskette drive is located below the power switch, and either a second 1.44 MB diskette drive or a matching cover is located immediately below it.

Rear Panel Connectors

The system power connector is located on the top rear of the system unit, as shown in Figure 7-2. An auxiliary outlet to provide power for a display is not provided, as it is on the AT. The keyboard, mouse, parallel interface, serial interface, and video display connectors are mounted vertically beneath the power connector. These same connectors are standard on all Personal System/2 models.

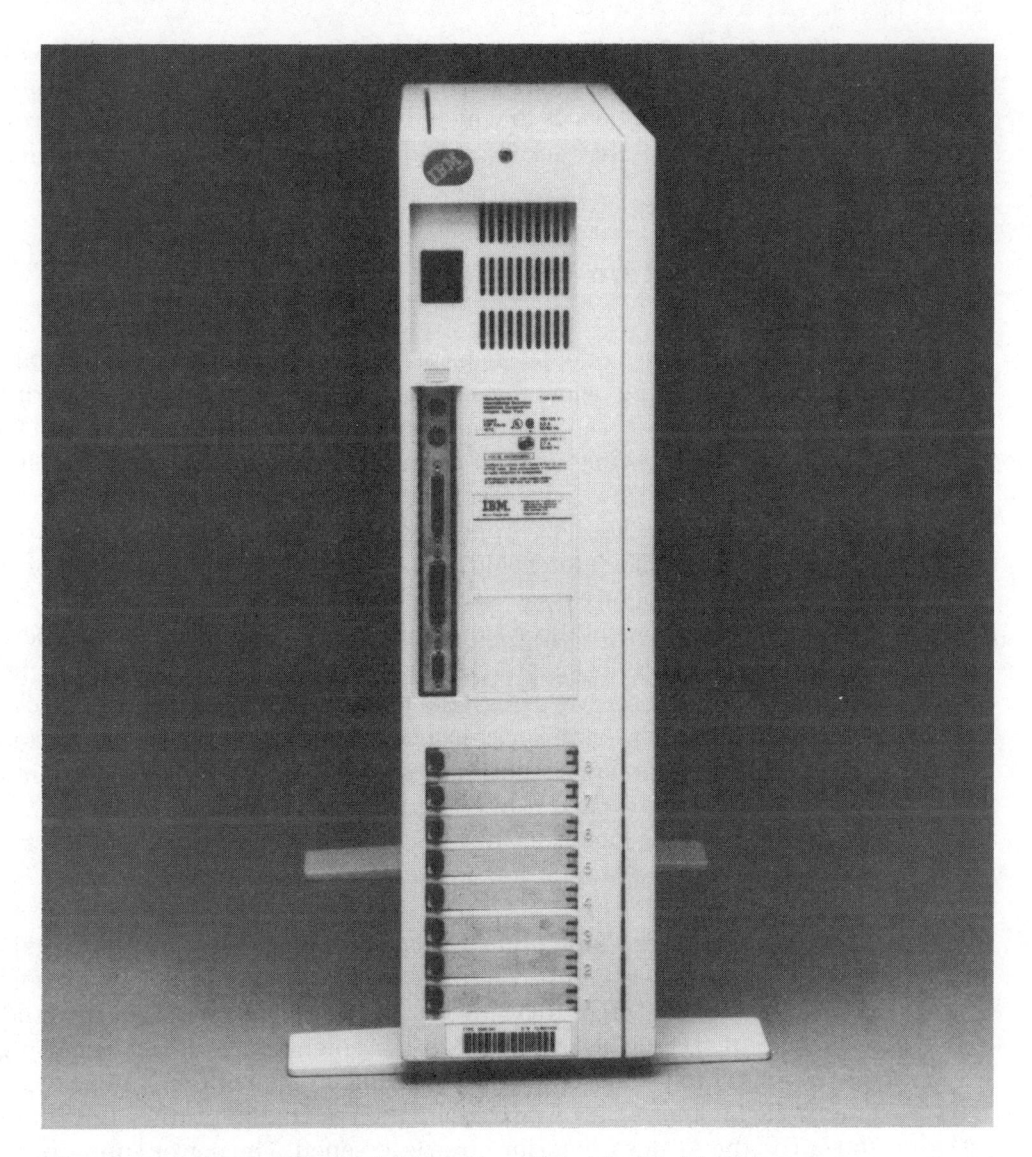

Figure 7-2 Rear of the Model 60 System Unit. *The Model 60 does not include an auxiliary power connector for the monitor, as does the AT.* Photo by Bill Schilling.

The keyboard and mouse connectors use identical 6-pin miniature DIN connectors. These connectors are different from the keyboard connector used on PCs. The parallel and serial interfaces both use standard 25-pin D-shell con-

211

nectors; the connector for the parallel port is female and for the serial port is male. The video display connector is a 15-pin D-shell connector that is not compatible with the 9-pin video display connector used on PC family video controller boards.

Covers for the Model 60's seven available Micro Channel expansion slots are located below the interface connectors. These covers or end brackets are isolated from the adapter boards to which they are attached and isolated from the system ground, so that they act as a shield against electromagnetic interference.

Inside the System Unit

The front cover of the Model 60 snaps off easily for maintenance or installation of options. The opening below the standard 3.5-inch diskette drive is a diskette drive bay that is already cabled for a second 3.5-inch diskette drive. (See Figure 7-3.) The adapter assembly used to connect the Model 4869 external 360 KB, 5.25-inch external diskette drive (described in Chapter 9) to the system can also be installed in this bay. The larger opening below the second diskette drive bay is to accommodate the 3363 Internal Optical Disk Drive (see Chapter 9) or other storage device that requires user access for its operation.

Access to the inside of the system unit is obtained by removing the left side cover. This is done by unlocking the keylock located at the top; loosening the two large slotted retaining screws using a flat blade screw driver, the cardboard disk that comes with the system's keys, or a coin; and lifting the cover out and up. Since this is a floor-standing rather than a desktop unit, the drudgery of moving the display is no longer part of opening up the system unit. In fact, if care is taken to see that adequate clearance is available on the left side of the system unit, access to the system unit for upgrade and maintenance purposes is very simple indeed.

A look inside the system unit reveals that the Model 60 was designed with expansion in mind. (See Figure 7-4.) The 207-watt power supply is mounted at the top of the case. It is secured by three 6 mm slotted hexagonal headed screws; a plastic wire-tie is attached for use as a convenient handle when removing the power supply. The system cooling fan is located in the bottom of the power supply. Since the fan blows downward inside the system unit, rather than outward as the fan on the AT does, exterior noise is lessened. The power supply operates with either 110 V_{ac} or 220 V_{ac}. A multi-wire cable with a large 15-pin connector furnishes power to the system board. Two standard disk drive power connectors are mounted in the bottom of the power supply.

The standard 1.44 MB diskette drive is mounted in a 3.5-inch drive bay below the power supply. This bay and the one immediately below it house connectors that are attached to the system board by a ribbon cable. These connectors, which provide both power and signals for the diskette drives, are the same as those on

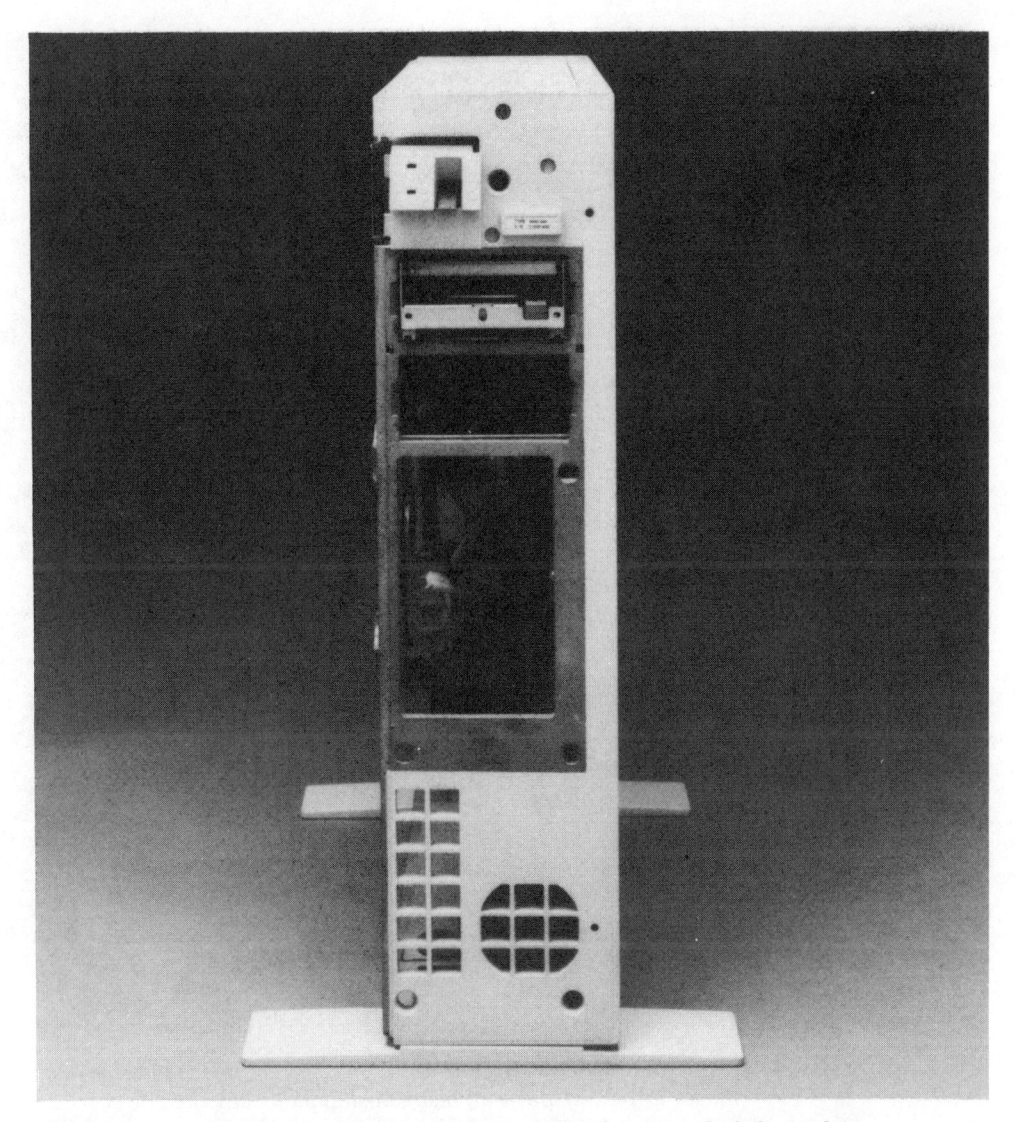

Figure 7-3 Front of the Model 60, Uncovered. *The 3.5-inch diskette drive bays, the position for the second hard disk, and the system speaker are visible from the front.* Photo by Bill Schilling.

the Model 50. Diskette drives may be easily exchanged between the Model 50 and Model 60 for upgrade or troubleshooting purposes. Drives are removed and installed using a procedure similar to that used on the Model 50. On the Model 60, the diskette bays are accessed by removing the front cover, pulling it out and up from the bottom. To remove a diskette drive it is also necessary to remove the side cover, so that the plastic release tabs on the rear of the drive bay can be pressed while lifting the tab under the front of the diskette drive and pulling.

Below the diskette drive bays is a frame for vertically mounting 5.25-inch

Chapter 7

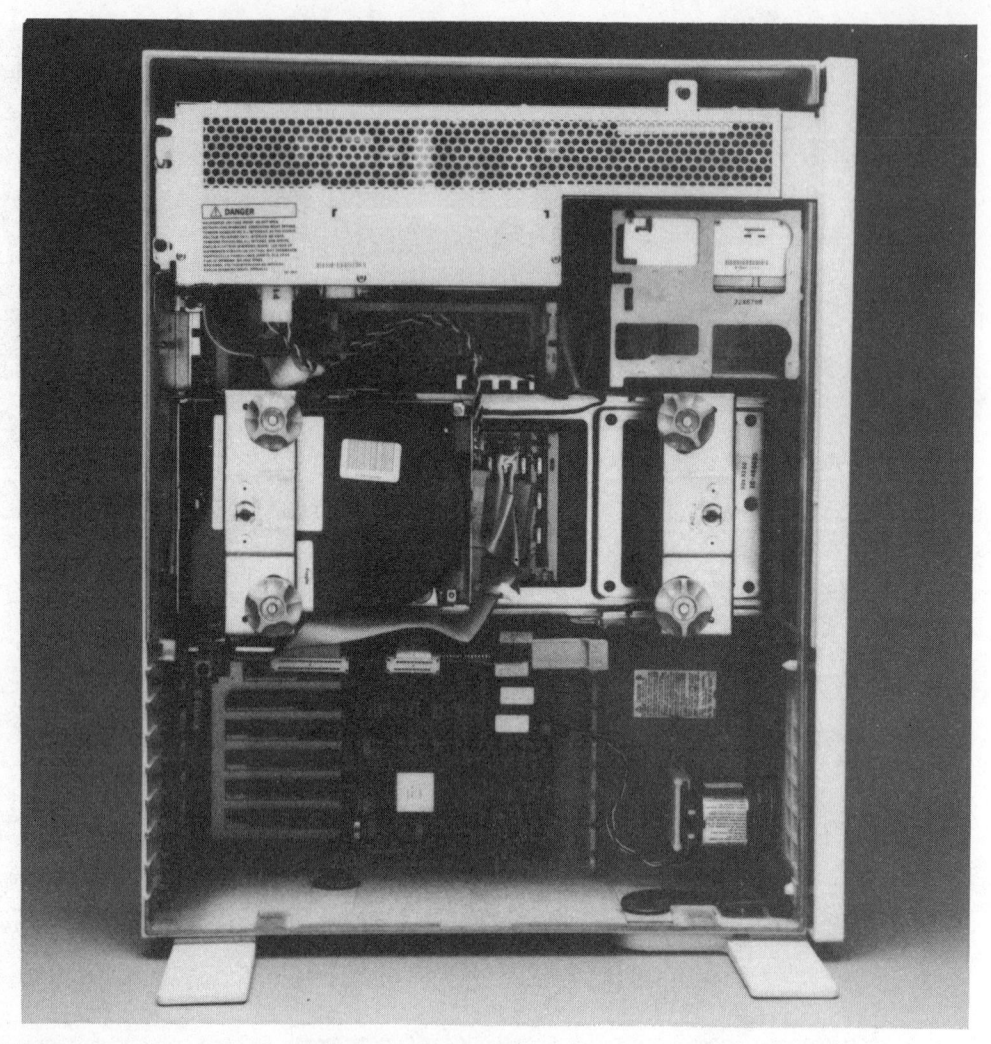

Figure 7-4 Inside the Model 60 System Unit. *The system board occupies the lower part of the unit, being positioned below the power supply.* Photo by Bill Schilling.

mass storage devices. Devices mounted in this frame are held by a metal mechanism that is adjusted by turning two large blue knobs. This mechanism tightens against the device's side-mounted rails, to hold the device securely. The Model 8560-041 contains a 44 MB hard disk. The Model 8560-071 comes standard with a 70 MB hard disk. The 041 supports an additional 44 MB hard disk drive; the 071 supports an additional 70 MB or 115 MB hard disk drive. The 041 uses an ST-506 disk controller, manufactured by IBM and shown in Figure 7-5. The controller provides the same features as that used on the Model 50, but supports two drives instead of one. The 071 features an Enhanced Small Device Interface (ESDI) disk adapter, which is also available on the Model 80.

Figure 7-5 Model 60 ST-506 Hard Disk Controller. *The Model 60 hard disk controller is an ST-506/412 interface that supports two drives.*
Photo by Bill Schilling.

The system speaker/battery assembly is located near the bottom front of the system unit and is connected to the system board by a small twisted wire cable. Between it and the system board are grooves molded into the system case, which somewhat control the movement of installed Micro Channel boards. The boards are free to wobble about when the system cover is off. They are stabilized when the cover is in place by a block of foam plastic material on the inside of the cover, which presses against them.

In order to remove the system board for repair or replacement, one must first remove the 5.25-inch mass storage device frame. This frame is secured to the system case using four 6 mm slotted hexagonal screws. These screws, like the ones that secure the power supply and the system board, mate with metal fittings in the composite plastic case. Removing the system board is accomplished by removing all Micro Channel boards, disconnecting the system board power connector and diskette drive connector, and removing the seven 6 mm slotted hexagonal screws that secure the system board to the case, and the one screw of the same type that connects it to the system ground.

It is not necessary to remove the power supply in order to remove the system board, because the system board does not extend under it. Once the system board is removed, a metal shield that makes contact with the power supply and various grounding pads on the system board is visible. The irregular shape of this shield indicates that it is installed to extend the system board grounding plane and to control electromagnetic emissions.

The Model 60 system board is functionally the same as that of the Model 50, but with certain enhancements. The most obvious enhancement is that the Model 60 system board (which is slightly bigger than that of the AT) has seven available Micro Channel connectors, compared to the three available on the Model 50. (This is one less than the total number of connectors on each system board, because the

Chapter 7

hard disk controller that is standard on each uses one Micro Channel connector.) The Model 60's Micro Chanel connectors are shown in Figure 7-6.

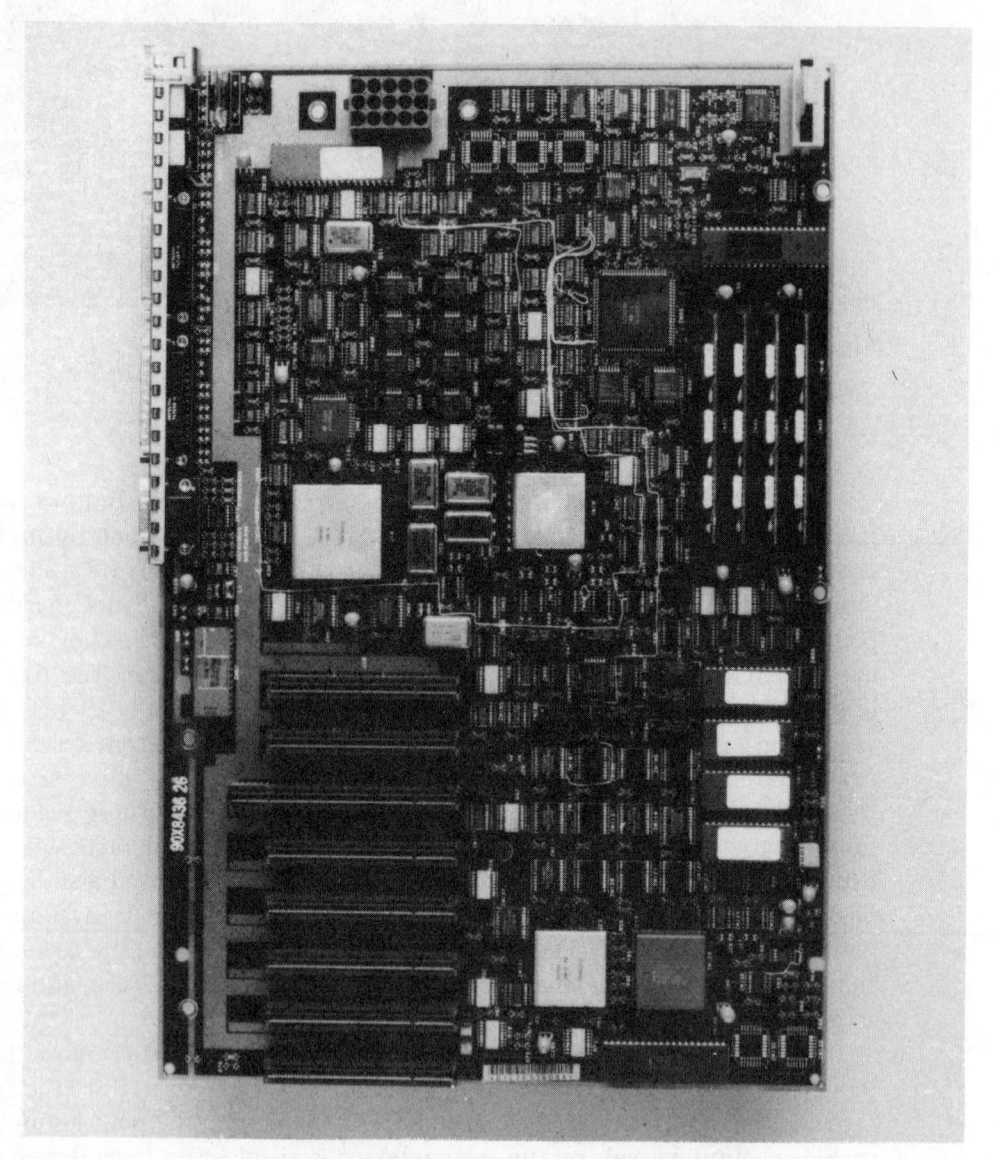

Figure 7-6 Model 60 System Board. *The Model 60 system board has eight Micro Channel expansion slots, compared to four on the Model 50.* Photo by Bill Schilling.

The Model 60 contains 2 KB of CMOS RAM in addition to the 64 bytes of battery-backed RAM contained on the Motorola MC146818A real-time clock with complementary metal oxide silicon (RT/CMOS) RAM device (used on both it and the Model 50). This RAM extension is reserved for storage of system information and for diagnostic use. The 6-volt battery in the speaker and battery assembly, which maintains information in the RT/CMOS RAM, also maintains the information in the RAM extension.

The system's socket-mounted 10 MHz 80286 microprocessor is located on the lower right of the system board, with the socket for an optional 10 MHz 80287 math coprocessor just below it. The Model 50's 128 KB of read only memory (ROM) is contained in four modules located above the 80286.

The system board RAM is located above the ROM. It is made up of four single-in-line packages (SIPs), each of which provides 256 KB of parity-checked memory. These SIPs may be removed by carefully spreading their retaining brackets and then rotating the top of the SIP toward the front of the system board. The 512 KB modules used on the Model 50 are physically the same size as the SIPs used on the Model 60. It seems reasonable that IBM might offer a redesigned Model 60 with 2 MB of system board memory, when the OS/2 operating system with its large extended memory requirements is more widely used.

As with the Model 50, expansion memory is added to the Model 60 using the 80286 memory expansion boards. (See Chapter 9 for additional information.) This board uses the same SIPs that are used on the system board; expansion boards may contain between 512 KB and 2 MB of memory. Using this memory expansion option, the Model 60 with its seven available Micro Channel connectors may contain a maximum of 15 MB of RAM.

The integrated diskette controller and the blue connector that connects it to the diskette drive bays is above the system RAM. The system's 14.3818 MHz, 32 MHz, and 40 MHz timing crystals are situated near the center of the system board. To their left is the Video Graphics Array (VGA). To the left of the VGA are the system's video, serial, parallel, and fuse-protected pointing device and keyboard connectors.

The Model 60 is designed for use in a variety of operating environments. It can be operated in a wide range of air temperatures and humidities, as shown below.

Air temperature
    System on: 15.6 to 32.2 degrees C (60 to 90 degrees F)
    System off: 10.0 to 43.0 degrees C (50 to 110 degrees F)

Humidity
    System on: 8 percent to 80 percent
    System off: 20 percent to 80 percent

Altitude
    Maximum altitude 2133.6 meters (7000 feet)

Heat output
    1240 BTU/hour

Chapter 7

For electromagnetic compatibility purposes, the Model 60 is certified as an FCC Class B device. The current leakage of the Model 50 does not exceed 500 microamperes. Like Models 50 and 80, the Model 60 is designed to meet current leakage requirements established by the National Fire Protection Code NFPA 76B for data processing equipment used in hospital environments. IBM advises users to consult local ordinances and National Fire Protection and UL codes for specific details on the use of data processing equipment in a hospital environment.

Functional Description

The Model 60 features a well-matched group of key components that work together as a computer system. Many of these are integrated into a single system component—the system board. Figure 7-7 is a functional diagram of the system board.

The Model 60 system board provides the same functional characteristics as that of the Model 50, plus several enhancements. It has four additional Micro Channel expansion connectors, and a 2 KB CMOS RAM extension not provided on the Model 50. Features provided on the Model 60 include:

- Intel 80286 system microprocessor (CPU)
- Intel 80287 math coprocessor
- 1 MB random access memory (RAM)
- 128 KB read only memory (ROM)
- Real-time clock with 64-byte CMOS RAM and battery backup
- 2 KB CMOS RAM extension, also backed up by battery
- 16-bit Micro Channel with eight expansion connectors
- Eight-channel DMA controller
- Sixteen-level interrupt system
- Three system timers
- 64K possible I/O ports
- Three system control ports
- Audio subsystem with speaker
- Keyboard and pointing device controller
- Integrated video graphics system
- Serial port controller
- Parallel port controller
- Diskette drive controller

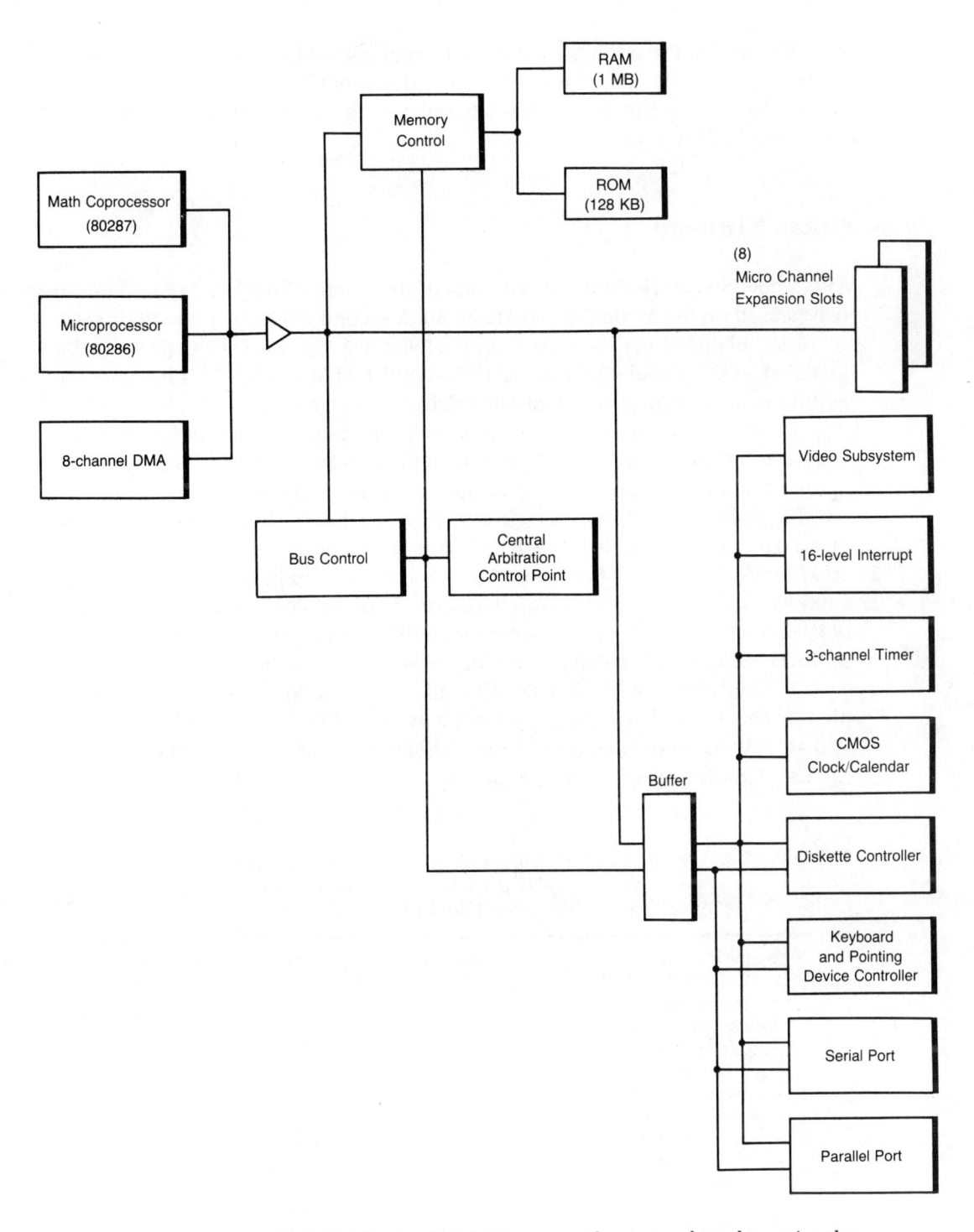

Figure 7-7 Model 60 System Block Diagram. *The system board contains the system processor and memory, standard I/O interfaces, and eight Micro Channel expansion slots.*

Chapter 7

Except for the additional Micro Channel connectors and the 2 KB CMOS RAM extension, the Model 60 system board is functionally the same as that of the Model 50. The functional characteristics of the Model 50 system board are described in Chapter 6.

Mass Storage

The Model 60 comes standard with one of the same 3.5-inch, 1.44 MB diskette drives used on the Model 50 and Model 80. A second 3.5-inch, 1.44 MB diskette drive is available as an option. An external 360 KB, 5.25-inch diskette drive can be installed on the Model 60 instead of the second 1.44 MB drive. See Chapter 6 for additional information on the diskette drives.

Hard disk support is the area in which the Model 60 most improves upon the AT and the Model 50. The Model 50, with its standard low-performance 20 MB hard disk and no provision for an additional internal hard disk, is more comparable to the XT 286 than the AT. On the other hand, the smallest hard disk offered on the Model 60 is a 44 MB unit, and the 8560-071 comes standard with a 70 MB disk. The 8560-041 can support a second 44 MB hard disk; the 071 can support a second 70 MB or 115 MB hard disk. Thus, while the maximum internal hard disk storage available from IBM for an AT is 60 MB, an 8560-071 can contain up to 185 MB using hard disks currently available from IBM.

Technical data on the hard disks available for the Model 60 is shown in Table 7-1. The 44 MB disk drive is controlled using an ST-506 controller; the 70 MB and 115 MB disks are controlled using an ESDI controller. IBM does not support the use of both controllers in the same system.

Table 7-1 Model 60 Hard Disk Technical Data. *Hard disks offered for the Model 60 provide higher performance and more capacity than those offered for the AT or Model 50.*

Specification	44 MB	70 MB	115 MB
Number of data heads	7	7	7
Number of cylinders	733	583	915
Sectors per track	17	36	36
Bytes per sector	512	512	512
Formatted capacity (million bytes)	44	70	115
Access time (ms):			
Track to track	10	5	6
Average	40	30	28
Maximum	80	60	60
Rotation rate (rpm)	3,600	3,600	3,600
Transfer rate (million bits per second)	5	10	10

The Model 60

The ESDI hard disk controller used on the 071 controls two ESDI devices, using a relative byte address and absolute byte address architecture. It features a 16 KB sector buffer and a data transfer rate of up to 10 million bits per second. It uses direct memory access for the transfer of data between the disk drive and the system, and can be programmed for burst mode transfer. Its on-board ROM BIOS supports interrupt sharing and provides extensive error handling and diagnostic capabilities. The adapter can retry certain operations after detection of an error without system intervention.

Although IBM only offers three different disk drives for use with the Model 60, like all Personal System/2 models except for Models 25 and 30, it supports thirty-two types of hard disks. Types 33 through 255 are reserved. The characteristics of the disks supported are listed in Table 7-2. The original AT (BIOS dated 1/10/84) supports drive types 0 through 14; ATs with a BIOS dated 6/10/85 or later support types 0 through 23. Models 25 and 30 support types 0 through 26 (even though IBM does not offer a hard disk for the Model 25).

Table 7-2 Disk Drives Supported. *Even though IBM currently offers only three different disk drives for the Model 60, it supports thirty-two types of hard disks.*

Disk Type	Number of Cylinders	Number of Heads	Precompensation Cylinder	Landing Zone	Defect Map
0			No hard disk installed		
1	306	4	128	305	No
2	615	4	300	615	No
3	615	6	300	615	No
4	940	8	512	940	No
5	940	6	512	940	No
6	615	4	None	615	No
7	462	8	256	511	No
8	733	5	None	733	No
9	900	15	None	901	No
10	820	3	None	820	No
11	855	5	None	855	No
12	855	7	None	855	No
13	306	8	128	319	No
14	733	7	None	733	No
15			Reserved		
16	612	4	All	663	No
17	977	5	300	977	No
18	977	7	None	977	No
19	1,024	7	512	1,023	No
20	733	5	300	732	No
21	733	7	300	732	No
22	733	5	300	733	No
23	306	4	All	336	No
24	612	4	305	663	No
25	306	4	None	340	No
26	612	4	None	670	No
27	698	7	300	732	Yes

Table 7-2   (cont.)

Disk Type	Number of Cylinders	Number of Heads	Precompensation Cylinder	Landing Zone	Defect Map
28	976	5	488	977	Yes
29	306	4	All	340	No
30	611	4	306	663	Yes
31	732	7	300	732	Yes
32	1,023	5	None	1,023	Yes

Power Supply

The Model 60's 207-watt power supply provides power for the system board, channel adapters, diskette drives, hard disks, keyboard, and pointing device. It operates with two ranges of input power and automatically selects the appropriate range. The ranges are 90 to 137 V_{ac} at 5 amperes or 180 to 265 V_{ac} at 3 amperes, 50 to 60 ±3 Hz. The power supply outputs +5, +12, and −12 V_{DC}. Its switch is located on the front of the unit along with two light-emitting diodes (LEDs). The green LED indicates that the power supply is on, and the yellow LED indicates hard disk activity. The yellow LED is powered by the power supply's system status line, which is controlled using bits 6 and 7 of system control port A (0092H).

The power supply is capable of no-load operation. Conversely, if a DC output is shorted, the power supply shuts down with no damage to the power supply. If input power to the supply is interrupted, it automatically restarts when input power is restored. An internal fuse provides protection against input power overload. A power-good signal indicates proper operation of the power supply. At power off, the power-good goes inactive before output voltage falls below regulation limits.

The system board receives power directly from the power supply via a multi-wire cable with a 15-pin connector. Five +5, one +12, one −12 V_{DC} outputs, and six signal ground lines are provided. The remaining two lines are for the power-good and system status signals. Power for two hard disk drives is available from the two 4-pin connectors located on the bottom of the power supply. Each connector provides one +5 and one +12 output, and two signal ground lines. Four-wire cables are used to carry the power from the power supply connectors to the hard disk drives.

Keyboard

The Model 60 uses the 101-key Enhanced Keyboard available on all Personal System/2 models and late model XTs and ATs. This keyboard is described in detail in Chapter 4. It is the best PC keyboard offered by IBM to date (once one becomes accustomed to the location of the function keys). The keys provide tactile and audible feedback without the clicky noise characteristic of the IBM PC keyboard. Because the Model 60's system unit sits on the floor rather than the desktop, the connecting cord included with the keyboard is slightly over twice as long as the Model 50's 4-foot keyboard cord.

Security

The Model 60 provides the same password protection as the Model 50: power-on password, keyboard password, and server mode. Power-on password protection enables users to set a password that must be entered whenever the computer is turned on. When the power is turned on, a small key appears on the screen. The user has three chances to enter the correct password, or the computer must be turned off and turned on again. When a correct password is entered, the computer starts its boot-up process.

The power-on password is set by starting the system from the Reference Diskette and choosing *Set Features* and *Set Passwords* from the main menu. When the password is entered, it is stored in the machine's battery-backed CMOS memory. See Chapter 6 for additional information.

Installation and Use

The Model 60, like the Model 50, is easier to set up and install than an AT. A setup sheet shows how to unpack and set up the computer. Due to the number of functions provided on the system board and the convenience of the Micro Channel Architecture, often the only reason to open the cover is to unlatch the feet that swivel out to steady the machine.

Once this is done, installation is similar to that for the Model 50—installing the system power cord and the keyboard cord, connecting the display to the system unit, and connecting the two power cords to electrical outlets. Since the Model 60 stands beside or under a desk, rather than on a desktop like the Model 50, this usually involves a bit of bending and stooping; but once installed, the machine operates very quietly and unobtrusively.

After the connections are made, the computer is powered on and initialized using the same Reference Diskette that is provided with the Model 50. After pressing Enter to get past the IBM logo display, the user has access to the Reference Diskette main menu.

Before initializing the system from the Reference Diskette, the Model 60's power-on self test (POST) sounds two beeps and displays 162 or 165 if the configuration stored in the system's battery-backed configuration RAM does not match the actual configuration of the system. It displays 163 if the time and date are not stored in the configuration RAM. If numeric codes are displayed, screens displayed after the IBM logo explain them. The last screen gives the user the option of requesting the automatic update of the configuration RAM and reinitializing the system once the configuration is updated. This is usually the only action the user needs to take. See Chapter 6 for more information.

If POST sounds two beeps and displays 301, a keyboard error has been detected. This usually means that a key is stuck down, the keyboard is not connected to the system unit, or the keyboard has been connected to the pointing device connector by mistake. If the system is initialized from the Reference Diskette, this code will only be displayed between the time when POST concludes and the IBM logo screen appears. Upon receipt of an Enter keycode, the system will display a message explaining that the 301 code indicates a keyboard error; however, the system cannot receive the Enter until the keyboard is properly connected. Therefore, when a code 301 is displayed and/or the system does not respond to Enter, a check for stuck keys or an improperly connected keyboard is in order.

From the Reference Diskette main menu one can learn about the computer, back up the Reference Diskette, set or view the system configuration, set system features such as date and time, copy the option diskette (which contains several hidden files), park the heads on the system's hard disks, and run diagnostic tests on the computer. An entry is selected from the menu either by entering its number or by moving the highlight bar to the entry, using the cursor keys, and pressing Enter. Pressing Esc while using a menu causes the display of the previous menu. Pressing Esc and then Enter while using the main menu causes the system to be initialized.

Most requirements can be met simply by using the *Run automatic configuration* option. This option automatically updates the system configuration stored in CMOS RAM to reflect the devices it detects installed in the system. Use of the manual *Change configuration* option is usually only required to set preferences, such as which port number a serial or parallel port on a particular board is to use. See Chapter 6 for additional information.

Though not documented in the Quick Reference manual, the Model 50/60 Reference Diskette contains an advanced diagnostic program as well as the standard diagnostic program. This program is invoked by entering Ctrl-A from the Reference Diskette main menu. These diagnostics hardware-format a hard disk and provide the selective diagnostic services provided by the IBM PC family advanced diagnostics. See *Problem Diagnosis* in this chapter for additional information.

Installing Options

Installing options on the Model 60 is functionally the same as on the Model 50, but physically different because of the Model 60's vertical construction. Installing options is generally quick and easy. If a diskette is provided with the option that contains Micro Channel device configuration information, that information should be copied to the working copy of the Model 60 Reference Diskette, using either the DOS or the OS/2 COPY command or the Reference Diskette's *Copy an option diskette* program, before the physical installation process begins.

The physical installation process generally requires no tools. For safety, the power cord should be removed before the cover is removed. Once the two large slotted screws on the side of the system unit are loosened, the side cover is pulled outward from the top and lifted away. Although the system unit can be pulled out from under a desk to facilitate this process, it is best to situate the system unit so that full access to the side cover is always available. Once the side cover is removed, the system's Micro Channel expansion slots are easily accessible.

Physically installing a Micro Channel adapter is a matter of removing an expansion slot cover from an unused slot, inserting the board, and tightening the thumb screw on the rear of the system unit to hold the adapter firmly in place. Most adapters can be installed in any available slot; the 8514/A Display Adapter, or other video adapters, must be installed in expansion slot 6, which contains the video extension connector (and is thus longer than the other sockets).

Installation of an Intel 80287 math coprocessor is equally straightforward, only requiring the insertion of the coprocessor into its socket, with the notched (pin 1) end toward the back of the system unit. The socket is directly accessible unless there is an adapter board installed in the bottom Micro Channel expansion slot; if there is, removal takes only a few seconds. Users are cautioned to be sure that the 80287 installed is rated for 10 MHz operation (indicated 80287-10 on the IC package), and is not a lower rated one designed for use on an AT.

The Quick Reference manual includes well-illustrated instructions for replacing the battery (which is located in a convenient clip holder at the bottom of the system unit), as well as for installing options. Instructions are provided for installing an 80287, Micro Channel adapters, a second diskette drive, and a second hard disk.

Once an option has been physically installed and the cover and power cord replaced, the system's configuration must be updated by inserting the Reference Diskette in drive A and turning on the system power. After displaying the amount of operational memory found, the power-on self test (POST) sounds two beeps and displays either the code 162 to indicate that the system's actual configuration does not match the configuration stored in CMOS RAM, or 165 to indicate that an option adapter has been added to or removed from the system. It then attempts to initialize the system from drive A.

If no diskette is found in drive A, the system waits, displaying only the amount of operational memory found and the error code(s). Inserting the Model 50/60 Reference Diskette in drive A and pressing F1 provides a display describing the error code, once the system is initialized (and the obligatory IBM logo is displayed and acknowledged by pressing Enter). Page 2 of the display is accessed by pressing PgDn; from here the user can request the automatic update of the system's configuration RAM and then reinitialize the system after the configuration is updated. See Chapter 6 for additional information.

Chapter 7

As noted in the error display, a 162 error can also occur if certain external devices are not powered on. The 5.25-inch external diskette drive is such a device. To avoid this error, the diskette drive should be powered on before the system unit. If the user forgets to do so, pressing F1 after the 162 code is displayed causes the system to initialize from either the diskette in drive A or from the hard disk.

Problem Diagnosis

The Personal System/2 Micro Channel architecture and the Reference Diskette programs greatly simplify system setup and installation; however, they cannot keep system components from failing from time to time. Some failures can be detected by the system's power-on self test, but some must be isolated using diagnostic routines.

The POST resident in system BIOS on all PC and Personal System/2 computers initializes and tests system components when the system is powered up. If all tests are passed, POST sounds a single beep before attempting to initialize the system from a diskette or the hard disk.

Any action other than this indicates a problem. In many instances, a problem can only be localized to a specific device by selectively replacing suspected faulty devices. If POST sounds no beeps, continuous beeps, or repeated short beeps, the system board or power supply is usually at fault, although faulty adapter boards using excessive power can sometimes cause this problem as well. Incorrect installation of the math coprocessor can also result in a blank screen and no beeps sounded.

One long and two short beeps or one short beep and a blank or unreadable display also indicates problems with the system board or power supply. One long and one short beep indicates problems with the display option adapter, system board, or power supply. One short beep and a screen prompting the user to insert a diskette indicates problems with the diskette drive, diskette drive bus adapter or cable, system board, or power supply.

If one beep is sounded and the screen remains blank even though the system appears to initialize from diskette or the hard disk, the user should check for loose display power cords and interface cables. The display should also be checked to be sure that it is powered on and the brightness is turned up to a viewable level.

POST identifies many different fault conditions by sounding two beeps and displaying an error code. These codes can indicate conditions ranging from the inconsequential (system time and date not set) to the extremely serious (hard disk drive failure). Upon the display of one of these codes, the user should insert the Reference Diskette into drive A and press F1, so the system can be initialized from the diskette. (Unfortunately, no message to this effect is displayed, only the numeric code.)

Once the system is initialized from diskette (and the IBM logo is displayed and acknowledged by pressing Enter) a screen describing the error discovered by POST is displayed. A listing of these error codes and their probable causes is given in Table 7-3. Depending on the error, the user should either allow the system to be automatically configured by answering Y when asked, or proceed to the Reference Diskette main menu by answering N.

Table 7-3 Personal System/2 Error Codes. *POST can identify a variety of system errors; it is often necessary to run diagnostic tests to determine the exact nature of the problem.*

Error Code	Probable Failing Unit
No beep, continuous beeps, or repeating short beeps	Power supply, system board, defective option or adapter
One long and one short beep	System board, display option adapter, power supply
One long and two short beeps	System board, power supply
One short beep and a blank or unreadable display	System board, power supply
One short beep and diskette prompt	Diskette drive, system board, power supply, diskette drive cable
System Board Codes	
111 (memory adapter)	Any adapter, system board
112, 113 (arbitration error)	Any adapter, system board
166 (adapter timeout error)	Any adapter, system board
161 (battery error)	Battery
162 (configuration error)	Any device
163 (date and time error)	System board
164 (expansion memory error) or 165 (system options not set)	System board or configuration not correctly set
1XX (not listed above)	System board
Memory Codes	
20X (with slot 0 indicated) or XXXXXX XXXX 201	Memory module package on the system board, or system board
20X (with other than slot 0 indicated)	Memory adapter board or memory module package on it
Keyboard Codes	
301	Keyboard, keyboard cable
303	System board, keyboard, keyboard cable
304	System board, keyboard
305	Fuse on system board
System Board Parallel Port Code	
401	System board

Chapter 7

Table 7-3 (cont.)

Error Code	Probable Failing Unit
Diskette Drive Codes	
602	Reference diskette
6XX	Diskette drive, system board, power supply, diskette drive cable
Math Coprocessor Code	
701	Math coprocessor, system board
System Board Serial Port Codes	
1102, 1106, 1108, 1109	System board or serial port on adapter board
1112, 1118, 1119	System board
1107	Communications cable, system board
11XX	System board
Dual Async Adapter Codes	
1202, 1206, 1208, 1209	Dual async adapter or serial port on system board or other adapter board
1212, 1218, 1219, 1227, 1233, 1234	Dual async adapter, system board
1207	Communications cable, dual async adapter
12XX (not listed above)	Dual async adapter, system board
Hard Disk Drive Codes	
1780 (drive C)	Drive C, disk controller, system board
1781 (drive D)	Drive D, disk controller, system board
1782 (hard disk controller)	Disk controller, system board
1790 (drive C) or 1791 (drive D)	Disk drive, disk controller. Action: format drive
17XX (not listed above)	Disk drive, disk controller, system board, power supply, disk drive cables
System Board Video Code	
24XX	System board
Mouse or Other Pointing Device Codes	
8601 or 8602	Pointing device
8603	System board
8604	System board, pointing device
Multiprotocol Adapter Codes	
10002, 10006, 10008, 10009	Multiprotocol adapter or other serial port device, including system board
10012, 10018, 10019, 10042, 10056	Multiprotocol adapter, system board
10007	Communications cable, multiprotocol adapter
100XX (not listed above)	Multiprotocol adapter, system board
ESDI Hard Disk Codes	
10480 (drive C)	Drive C, ESDI disk controller, system board

Table 7-3 (cont.)

Error Code	Probable Failing Unit
10481 (drive D)	Drive D, ESDI disk controller, system board
10482 or 10483 (disk controller)	ESDI disk controller, system board
10490 (drive C) or 10491 (drive D)	Disk drive, ESDI disk controller. Action: format drive
104XX (not listed above)	Disk drive, ESDI disk controller, system board, power supply, disk drive cables

Once at the main menu, the user may select *Test the computer* to initiate the standard diagnostics that test various system components in turn. If it is apparent which components are likely to be at fault, or if the user wishes to test the reliability of certain system components, use of the advanced diagnostic program contained on the Reference Diskette is recommended. This program, which is not documented in the Quick Reference manual, allows the execution of individual or multiple diagnostic routines and errors to be logged to diskette or the printer.

The *Advanced Diagnostic* menu, shown in Figure 7-8, is invoked by entering Ctrl-A from the Reference Diskette main menu. From this menu the user can either initiate system checkout or perform a low level format on the hard disk (usually only necessary after disk faults have developed). Online help can be obtained for this menu and several others in the diagnostics package by pressing F1.

```
Advanced Diagnostic Menu

1  System checkout
2  Format fixed disk




Use ↑ or ↓ to select. Press Enter.
Esc=Quit      F1=Help
```

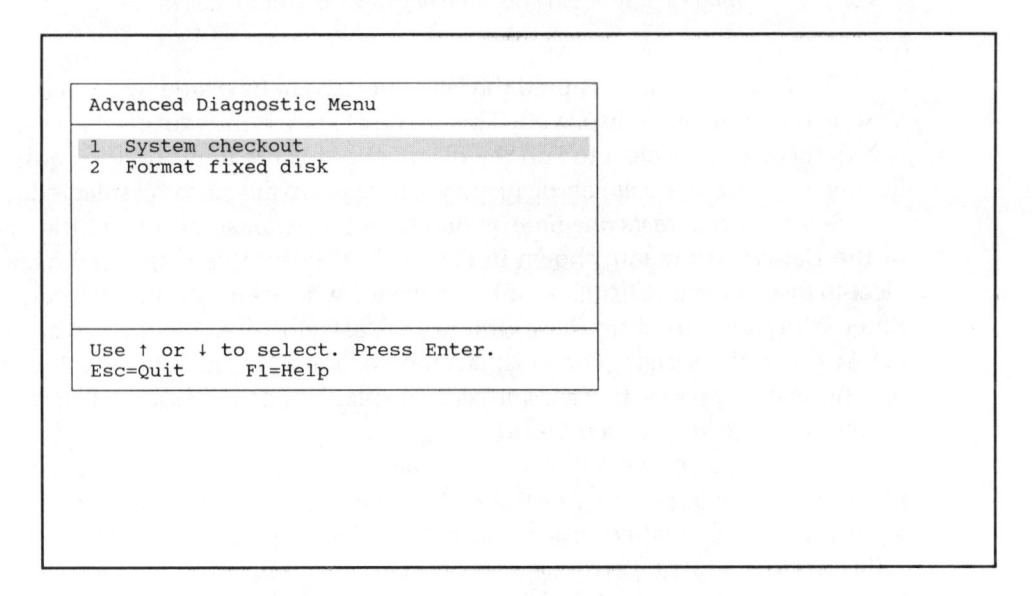

Figure 7-8 Advanced Diagnostic Menu. *The Advanced Diagnostic menu can be used to run the advanced diagnostic programs or to perform a low level format on a hard disk.*

Chapter 7

Selecting *System checkout* results in the display of a list of devices on the system, as is provided with the standard diagnostic program. (See Figure 7-9.) The user may either accept this list as correct and proceed to the next menu, or add devices to or remove them from the list of devices available for testing.

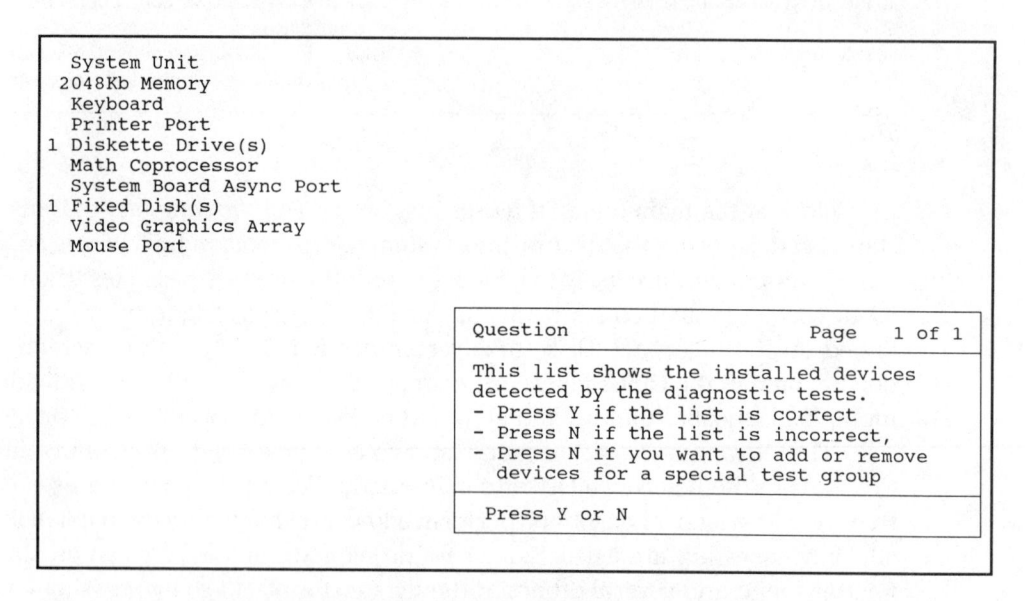

```
  System Unit
   2048Kb Memory
    Keyboard
    Printer Port
 1  Diskette Drive(s)
    Math Coprocessor
    System Board Async Port
 1  Fixed Disk(s)
    Video Graphics Array
    Mouse Port




                        ┌──────────────────────────────────────────────┐
                        │ Question                    Page    1 of 1   │
                        ├──────────────────────────────────────────────┤
                        │ This list shows the installed devices        │
                        │ detected by the diagnostic tests.            │
                        │ - Press Y if the list is correct             │
                        │ - Press N if the list is incorrect,          │
                        │ - Press N if you want to add or remove       │
                        │   devices for a special test group           │
                        ├──────────────────────────────────────────────┤
                        │ Press Y or N                                 │
                        └──────────────────────────────────────────────┘
```

Figure 7-9   System Device List. *The system device list can be accepted or modified by adding or removing devices from the list.*

Once the user has accepted the list of devices to be tested by entering Y, the *Test Selection* menu is displayed. This menu is shown in Figure 7-10. From this menu, the user may elect to run the diagnostics for the items on the equipment list once or continuously; error display and logging may also be selected.

Selecting *Run tests one time* or *Run tests continuously* results in the display of the *Device Test* menu, shown in Figure 7-11. From this menu, the user may elect to test all devices listed or only the device whose entry is highlighted when Enter is pressed. In either case, the user is asked if error messages are to be displayed on the screen. Each displayed error message must be acknowledged for the test to proceed; error message display is not necessary if errors are logged to the printer or a diskette.

Selecting *Log or display errors* from the *Test Selection* menu results in the display of the *Error Log* menu. (See Figure 7-12.) From this menu, the user can elect to log errors to a formatted diskette in drive A (log entries are written to the file ERROR.LOG), to LPT1. Error logging continues until *Stop error log* is selected from the menu. *View error log* is used to examine log entries saved on diskette. Error messages are displayed from the log as they appear when they occur. Messages consist of the time of day when the error occurred, the pass number of the test, the error number and message, and recommended action. Once an error logging option has been selected, the user is instructed to select *Run tests continuously* from the *Test Selection* menu. Control is returned to that menu when Enter is pressed.

```
┌─────────────────────────────────────────────────────────┐
│                                                         │
│   ┌─────────────────────────────────────────┐           │
│   │ Test Selection Menu                     │           │
│   │                                         │           │
│   │ 1 Run tests one time                    │           │
│   │ 2 Run tests continuously                │           │
│   │ 3 Log or display errors                 │           │
│   │                                         │           │
│   │                                         │           │
│   │                                         │           │
│   │ Use ↑ or ↓ to select. Press Enter.      │           │
│   │ Esc=Quit     F1=Help                    │           │
│   └─────────────────────────────────────────┘           │
│                                                         │
└─────────────────────────────────────────────────────────┘
```

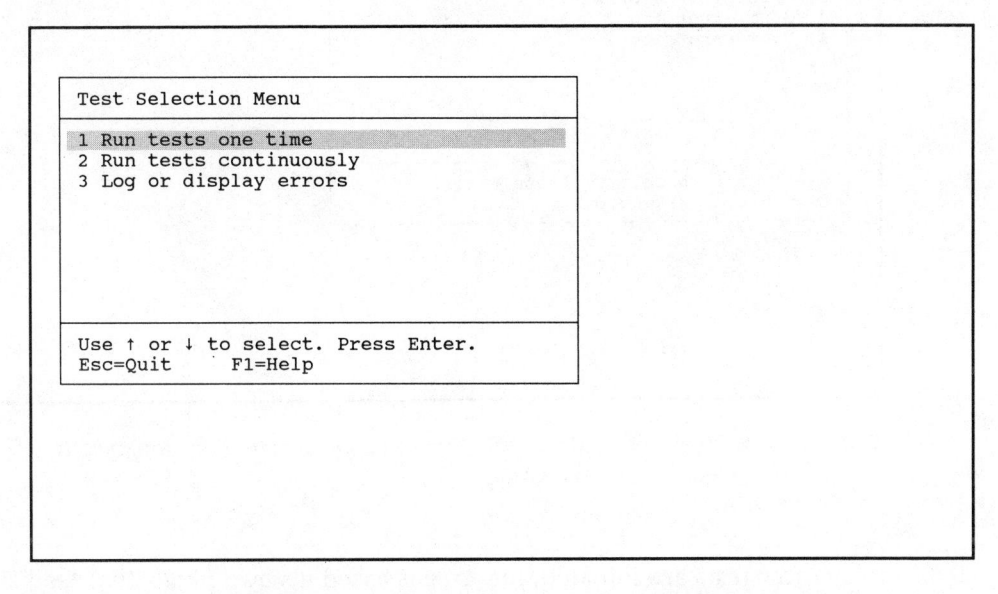

Figure 7-10 Test Selection Menu. *The Test Selection menu is used to run tests one or more times, and to log or display error messages.*

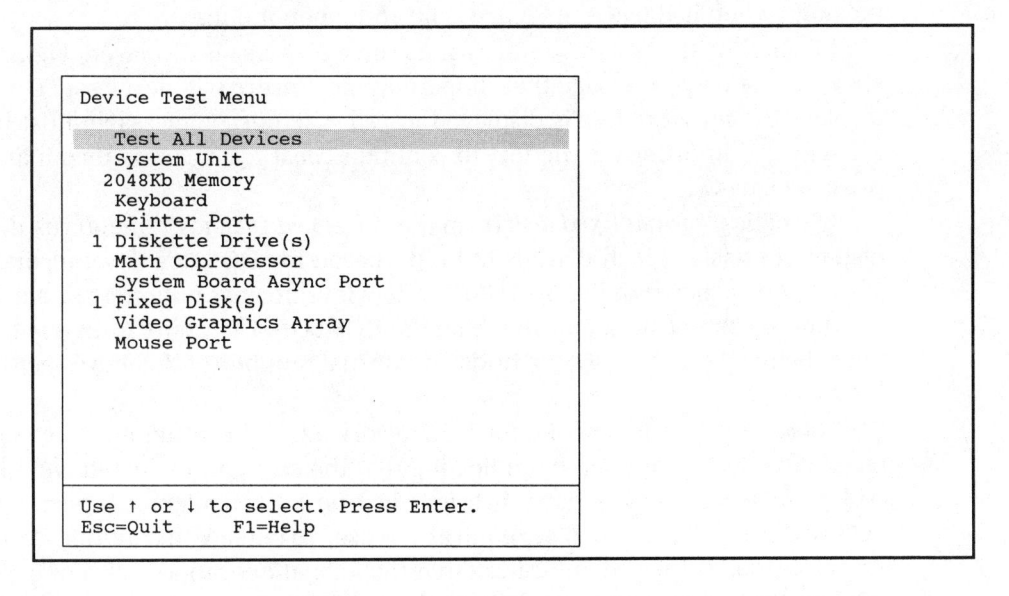

```
┌─────────────────────────────────────────────────────────┐
│                                                         │
│   ┌─────────────────────────────────────────┐           │
│   │ Device Test Menu                        │           │
│   │                                         │           │
│   │    Test All Devices                     │           │
│   │    System Unit                          │           │
│   │    2048Kb Memory                        │           │
│   │    Keyboard                             │           │
│   │    Printer Port                         │           │
│   │  1 Diskette Drive(s)                    │           │
│   │    Math Coprocessor                     │           │
│   │    System Board Async Port              │           │
│   │  1 Fixed Disk(s)                        │           │
│   │    Video Graphics Array                 │           │
│   │    Mouse Port                           │           │
│   │                                         │           │
│   │ Use ↑ or ↓ to select. Press Enter.      │           │
│   │ Esc=Quit     F1=Help                    │           │
│   └─────────────────────────────────────────┘           │
│                                                         │
└─────────────────────────────────────────────────────────┘
```

Figure 7-11 Device Test Menu. *Specific devices to be tested can be selected, or all devices listed can be tested.*

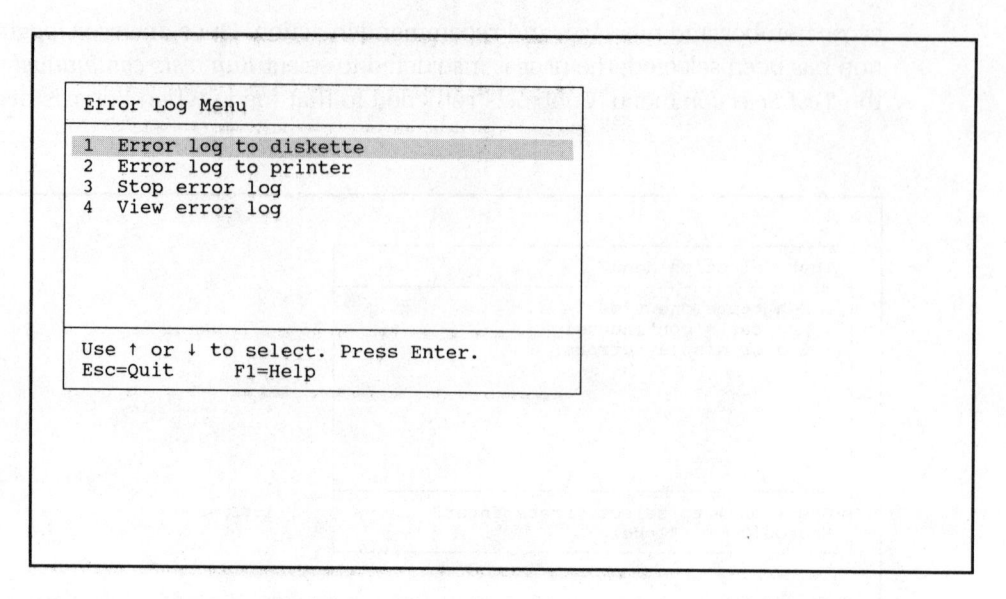

```
Error Log Menu

1  Error log to diskette
2  Error log to printer
3  Stop error log
4  View error log




Use ↑ or ↓ to select.  Press Enter.
Esc=Quit      F1=Help
```

Figure 7-12 Error Log Menu. *The Error Log menu is used to stop, direct, or view the error log.*

After tests are initiated, the user is asked if wrap plugs (that feed signals back into the port) are to be used on the ports to be tested, what video tests are to be run, and if a mouse is connected to the system. When these questions are answered, the selected tests are initiated and run in turn. Each test displays a message when it starts (see Figure 7-13) and when it ends.

Each time the test encounters an error, a message is displayed. Figure 7-14 shows a test error message that, hopefully, most users will never see. On color displays the message box is displayed in red. Each error message must be acknowledged in order for the test to continue. Most tests can be terminated by pressing Ctrl-C.

Selecting *Format fixed disk* from the *Advanced Diagnostic* menu results in a display that asks which drive is to be prepared for DOS (low level formatted). This menu is shown in Figure 7-15. The low level format option can be used to attempt to correct situations in which the DOS FORMAT utility is unable to prepare the hard disk for DOS or finds an unusual number of defective sectors on the disk.

Selecting *Prepare Drive C for DOS* (or drive D if the system has two drives) causes the display of a screen indicating that the operation selected will ERASE ALL information on the hard disk (not just on a particular disk partition, as FORMAT does). This selection requires the user to enter Y to continue. Another screen is then displayed, indicating that the format operation will require from 20 minutes to 2 hours, depending on the size of the disk . The user must then enter N (for No, do not STOP now) to indicate that the process is to proceed.

The Model 60

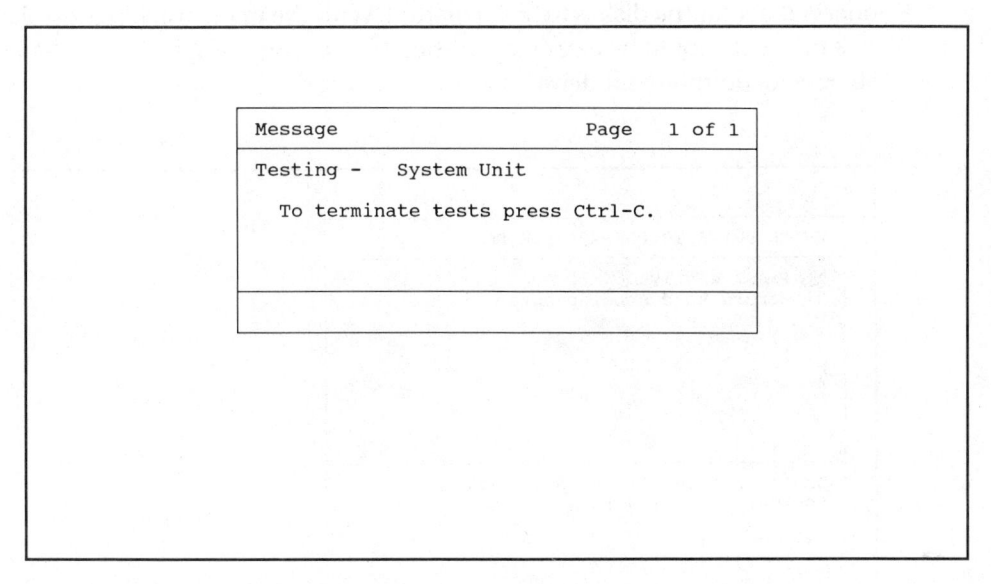

```
┌─────────────────────────────────────────────────┐
│                                                 │
│    ┌───────────────────────────────────────┐    │
│    │ Message                Page   1 of 1  │    │
│    │ Testing -   System Unit               │    │
│    │    To terminate tests press Ctrl-C.   │    │
│    │                                       │    │
│    │                                       │    │
│    ├───────────────────────────────────────┤    │
│    │                                       │    │
│    └───────────────────────────────────────┘    │
│                                                 │
└─────────────────────────────────────────────────┘
```

Figure 7-13 Test Start Message. *Each test displays a message when it starts; generally the display remains unchanged until the test is completed or an error is encountered.*

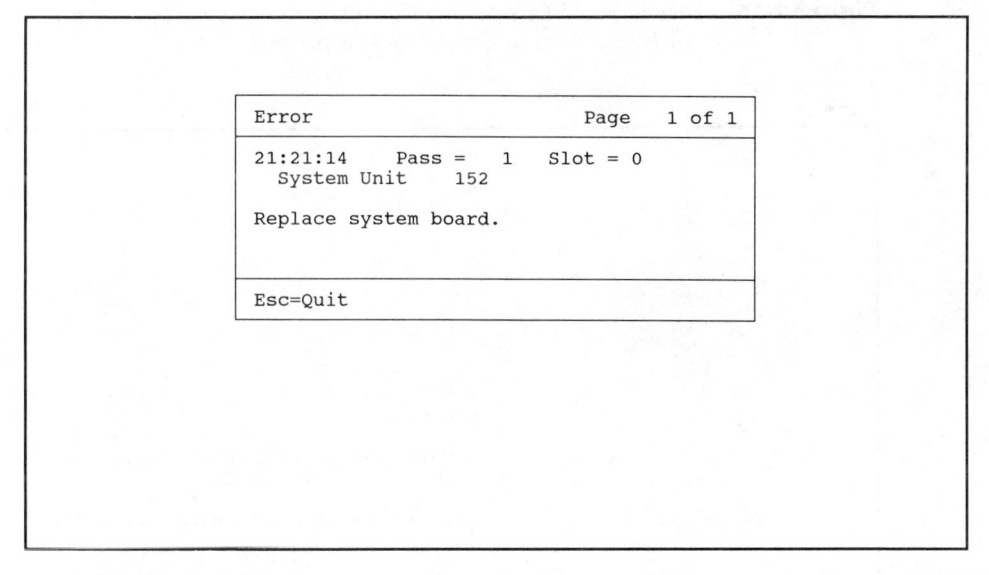

```
┌─────────────────────────────────────────────────┐
│                                                 │
│      ┌─────────────────────────────────────┐    │
│      │ Error                 Page   1 of 1 │    │
│      │ 21:21:14     Pass =  1   Slot = 0   │    │
│      │    System Unit     152              │    │
│      │ Replace system board.               │    │
│      │                                     │    │
│      ├─────────────────────────────────────┤    │
│      │ Esc=Quit                            │    │
│      └─────────────────────────────────────┘    │
│                                                 │
└─────────────────────────────────────────────────┘
```

Figure 7-14 Test Error Message. *Error messages must be acknowledged for the test to continue; on color displays the message box appears in red.*

The screen shown in Figure 7-16 is displayed next. The user is informed that all data on the hard disk drive will be destroyed, and is required to affirm

233

Chapter 7

once again that the disk is to be formatted. With the many warnings displayed by this utility, it should be extremely difficult for anyone using it to accidentally destroy data on the fixed drive.

```
WHICH DRIVE TO PREPARE FOR DOS?

1  Prepare Drive C for DOS
2  Return to control program




Use ↑ or ↓ to select.  Press Enter.
Esc=Quit        F1=Help
```

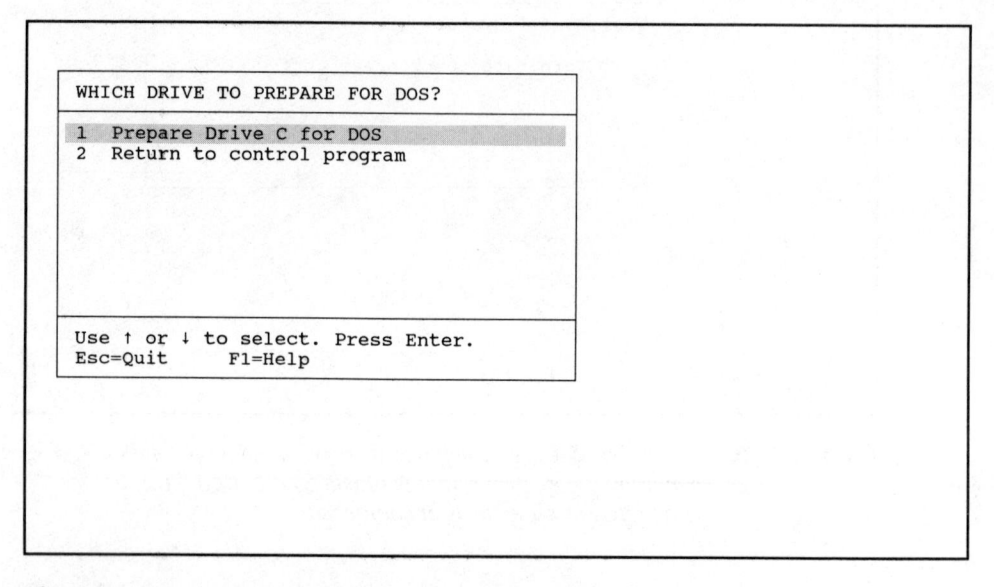

Figure 7-15   Format Fixed Disk Menu. *This menu can be used to initiate a low level format of drive C (or D if present).*

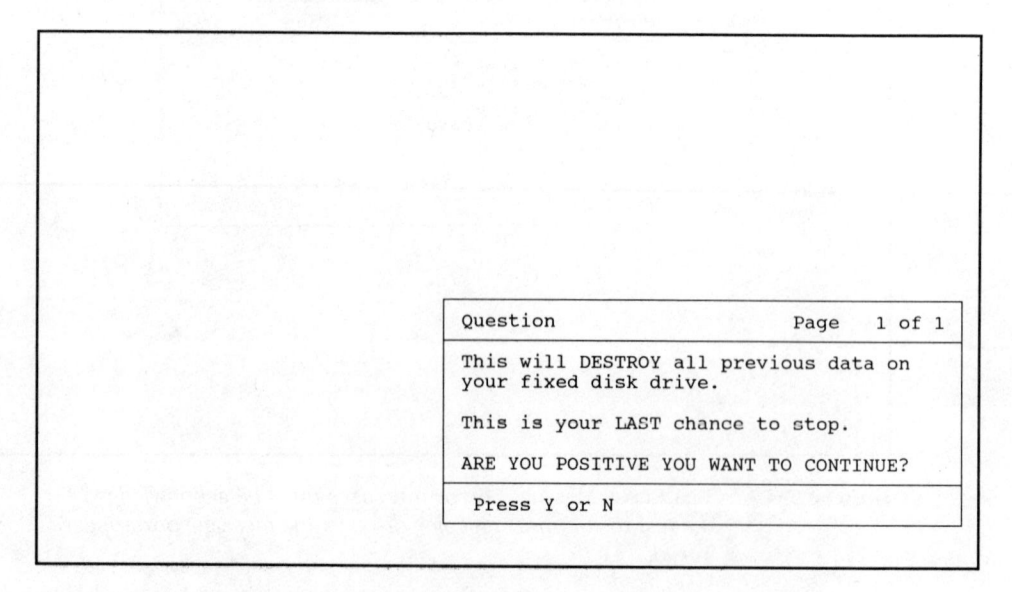

```
                              Question              Page    1 of 1
                              This will DESTROY all previous data on
                              your fixed disk drive.
                              This is your LAST chance to stop.
                              ARE YOU POSITIVE YOU WANT TO CONTINUE?
                              Press Y or N
```

Figure 7-16   Last Chance to Stop Display. *The low level format procedure makes it very clear to the user that the low level format will destroy all data on the hard disk.*

234

Once the low level formatting process begins, a screen is displayed and updated showing the cylinder being formatted and the number of defective sectors that have been encountered. Upon completion of the formatting process (which requires approximately 40 minutes for the 44 MB disk used on the Model 60 and correspondingly longer on larger size disks), a screen is displayed that describes the physical characteristics of the hard disk and the number of sectors that were encountered and marked bad during the formatting process. (See Figure 7-17.)

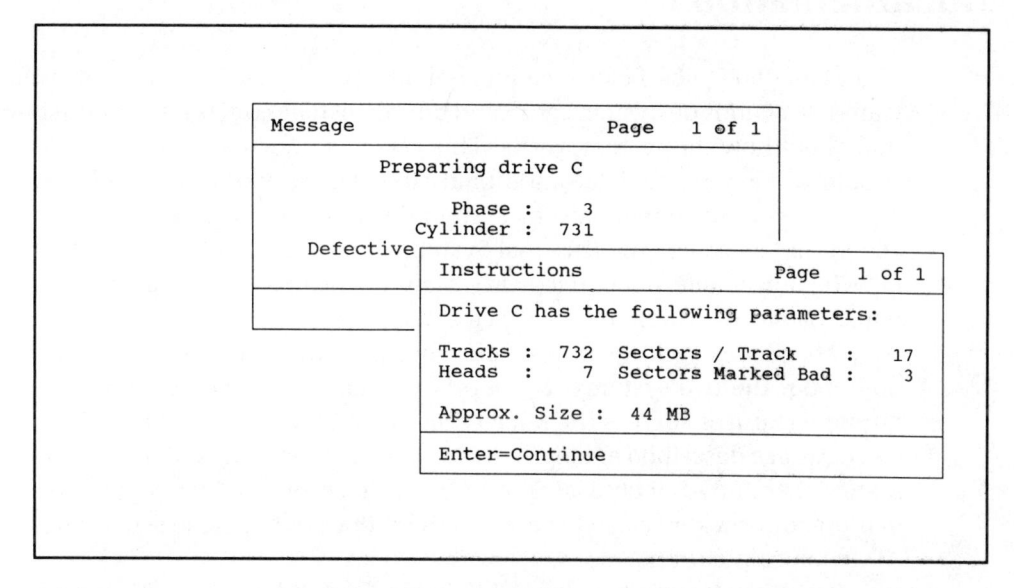

```
Message                          Page    1 of 1
           Preparing drive C
              Phase    :    3
              Cylinder :  731
    Defective
             ┌─────────────────────────────────────────────┐
             │ Instructions                  Page   1 of 1 │
             │ Drive C has the following parameters:       │
             │                                             │
             │ Tracks :  732   Sectors / Track     :  17   │
             │ Heads  :    7   Sectors Marked Bad  :   3   │
             │ Approx. Size :  44 MB                       │
             │ Enter=Continue                              │
             └─────────────────────────────────────────────┘
```

Figure 7-17 Format Fixed Disk Completion Screen. *The completion screen displays information about the formatted disk, including the number of sectors marked bad.*

Pressing Enter in response to the screen returns control to the *Which Drive to Prepare for DOS?* menu. Pressing Enter again returns control to the *Advanced Diagnostics* menu. From here pressing Esc and then Enter causes the system to be initialized. In this case, the system must be initialized from a DOS system diskette so that FDISK can be run to establish a DOS partition on the disk, and SELECT can be run to install DOS. Any attempt to use the disk before using FDISK will result in an Invalid Drive Specification message. See Chapter 10 for additional information.

Warranty and Service

IBM provides a one-year limited warranty on the Model 60. Customer carry-in is required for warranty repairs; on-site warranty repair is available as an op-

Chapter 7

tion. Reflecting expected reliability and ease of repair of these systems, the minimum cost of IBM on-site maintenance is $190 and $210 for the Model 60-041 and 071, respectively. These costs are significantly less than the $546 minimum maintenance charge announced with the 8 MHz IBM AT, Model 339 in 1986.

Documentation

The Model 60 Quick Reference manual, like the Model 50's, provides well-illustrated directions on system setup and option installation, but very little information about how the system works. This valuable information is available only in supplementary manuals at additional cost. These manuals include individual technical reference manuals for system devices and adapters, the Model 50/60 Technical Reference, the Personal System/2 and Personal Computer BIOS Interface Technical Reference, Hardware Maintenance Reference, and Personal System/2 Hardware Maintenance Service manuals.

The Model 50/60 Technical Reference provides in-depth technical information about the two systems. The level of detail varies from chapter to chapter, but new features such as the Micro Channel architecture and the Video Graphics Array are described at length. It is a much more organized description of the system than the AT Technical Reference, which seems to be a collection of information not provided elsewhere. However, it does not provide schematics, as the AT Technical Reference does.

The BIOS Interface Technical Reference provides information on the basic input/output system (BIOS) of all IBM PC and Personal System/2 computers. As one might expect, this involves a lot of information and a fair number of footnotes and explanations, to differentiate between the services provided on different models. Although BIOS listings are not provided in the technical reference of each computer, as in the past, the functional operation of the BIOS is described in detail in a fairly organized manner.

The Hardware Maintenance Reference manual provides general maintenance information such as product descriptions, disassembly procedures, and an introduction to diagnostics. The Hardware Maintenance Service manual consists of a two-part diagnostic manual and a copy of the Model 50/60 Reference Diskette. The manual provides detailed specifications and part numbers for the various assemblies that comprise a Model 50 or 60. It also explains how to diagnose system problems, using the diagnostics contained on the Reference Diskette and an electrical multimeter. Maintenance personnel are led through a series of step-by-step procedures that allow problems to be isolated to a particular system device. Once a problem is isolated, the manual recommends corrective action, which is all too often "Replace the system board."

Performance and Compatibility

The Model 60 is an improved AT in almost every way. It uniformly provides increased performance, capacity, and reliability. It also provides faster performance of day-to-day applications, which reflects the higher speed of its processor, memory, and disk systems.

Table 7-4 shows the Model 60's performance in the non-disk-dependent segments of word processing and spreadsheet tests. (See Appendix A for a detailed description of all tests that were performed.) The Model 60, with its functionally identical architecture, performed the tests in the same time as the Model 50. In the WordPerfect test, a large text file was loaded into memory; the times to move from the top to the bottom of the file and from the bottom to the top were then measured. As with the Model 50, the Model 60, with its 10 MHz 80286, performed the test about 25 percent faster than the AT or AT-compatible Compaq Deskpro 286. Similar but not as striking results were observed when using WordPerfect to replace all occurrences of a "j" in the file with a "k".

Table 7-4 Non-Disk-Dependent Performance. *The Model 60 was faster overall than the AT in non-disk-dependent tests.*

Test Machine	Equipment	WordPerfect Move Top to Bottom	Move Bottom to Top	Search and Replace	Lotus 1-2-3 Recalc— Add	Recalc— Multiply	Recalc— Mix
AT 339	No math	28	21	32	5	8	45
	Math	28	21	32	5	5	7
Compaq 286	No math	28	21	34	5	7	45
	Math	28	21	34	5	5	7
Model 50	No math	22	16	27	4	6	36
	Math	22	16	27	4	4	5
Model 60	No math	22	16	27	4	6	36
	Math	22	16	27	4	4	5
AST 10 MHz	No math	16	12	20	3	4	28
	Math	16	12	20	3	3	4
16 MHz Model 80	No math	12	9	14	2	3	20
	Math	12	9	14	2	2	3

All results are in seconds.

To test the Model 60's computational speed, calculations were performed on three different spreadsheets, using Lotus 1-2-3. The use of the math coprocessor on the Model 60 had little effect on the calculation time for the first spreadsheet, which used addition; some effect on the second, which used multiplication and division; and a dramatic effect on the third, which used

Chapter 7

square root and natural logarithm functions. The Model 60 performed all of the spreadsheet tests 20 to 25 percent faster than the AT as well, both when the math coprocessor was used on each machine and when it was not used on either machine. The Model 60's math coprocessor is much faster than the AT's, so the spreadsheet recalculation time was obviously dependent on the speed of the memory and the 80286.

The AST Premium/286—an AT-compatible with a 10 MHz 80286, 8 MHz 80287, and zero wait state memory—outperformed Models 50 and 60 in the word processing and spreadsheet calculation tests just described. Reflecting the performance of its zero wait state memory, it performed the tests 20 to 25 percent faster than Models 50 and 60, making it some 40 percent faster overall than the AT.

Even better performance, albeit at greater cost, was available from the 16 MHz Model 80. The Model 80 uniformly performed the tests just described some 50 percent faster than the Model 60, and about 60 percent faster than the AT. Even better performance is available from the Model 80 when using operating systems such as AIX Personal System/2 and applications that take advantage of its 32-bit architecture. (See Chapter 12.)

The Model 60's file processing capabilities were tested by using WordPerfect to load a document file, Lotus 1-2-3 to load a spreadsheet, and dBASE III PLUS to perform two sorts on a file. PC-KEY-DRAW was used to display a graphics file and MASM 5.0 was used to assemble a file. All tests were performed from RAM disk, the hard disk, and from the 3.5-inch diskette drive, using 1.44 MB and 720 KB diskettes. Results of the tests with the files stored on an extended memory RAM disk are shown in Table 7-5.

The Model 60, like the Model 50, performed operations that use RAM disk files 20 to 25 percent faster than the AT. The AT-compatible AST Premium/286,

Table 7-5 RAM-Disk-Dependent Performance. *The Model 60 performed file functions from RAM disk 20 to 25 percent faster than the AT.*

Test Machine	WordPerfect Load File	Lotus 1-2-3 Load File	dBASE III PLUS Sort on Title	Sort on Disk ID	PC-KEY-DRAW Run Sample Macro	MASM Assemble File
AT 339	11	22	9	10	520	8
Compaq 286	9	22	10	11	473	8
Model 50	9	17	7	8	406	6
Model 60	9	17	8	8	406	6
AST 10 MHz	10	16	7	8	384	5
16 MHz Model 80	5	9	4	5	277	3

All results are in seconds.

with its 10 MHz 80286 and zero wait state memory, performed all tests in about the same time as the Model 60. This is indicative of the processor-intensive nature of these particular tests, since the AST's memory is faster while its processor is the same speed as that of the Model 60. The Model 80 provided even better performance, completing the tests 30 to 50 percent faster than the Model 60 and about 60 percent faster than the AT.

The WordPerfect document and Lotus 1-2-3 spreadsheet were loaded from the Model 60's 44 MB hard disk, and the dBASE III PLUS sorts were performed there as well. PC-KEY-DRAW was used to display the same graphics file and MASM 5.0 was used to assemble the same source file, with the files stored on the hard disk instead of RAM disk. Results of the tests with the files stored on the hard disk, with and without a 384 KB extended memory disk cache in use, are shown in Table 7-6.

Table 7-6 Hard-Disk-Dependent Performance. *The Model 60 performed hard-disk-dependent functions faster than the AT.*

Test Machine	Equipment	WordPerfect Load File	Lotus 1-2-3 Load File	dBASE III PLUS Sort on Title	dBASE III PLUS Sort on Disk ID	PC-KEY-DRAW Run Sample Macro	MASM Assemble File
AT 339	No cache	12	29	25	26	526	11
Compaq 286	No cache	11	28	21	22	477	9
Model 50	No cache	10	27	29	29	409	7
	Cache	10	20	20	21	405	7
Model 60	No cache	10	26	24	21	406	7
	Cache	10	19	22	21	406	7
AST 10 MHz	No cache	8	17	25	24	388	9
16 MHz	No cache	7	13	26	26	283	4
Model 80	Cache	7	11	22	21	277	4

All results are in seconds.

The Model 60 offered about 25 percent faster performance than the AT on this series of tests. It was about as fast without the cache software installed as the Model 50 was with it installed. The AST Premium/286, which also has a fast hard disk, was faster than either the Model 50 or Model 60 on file loads, and offered performance comparable to that of the Model 60 on the file sorts. Both it and the AT can benefit from third-party disk cache software, in the same way that the Model 60 benefited from it in several of the tests.

The Model 80, with its ESDI 70 MB hard disk, performed the file loads up to 50 percent faster than the Model 60 (60 percent faster than the AT). On tests

Chapter 7

such as the sorts, where the hard drive's transfer rate was not as much of a factor, it offered performance comparable to that of the Model 60.

Except for loading software, the speed of the diskette drives was less critical on the Model 60 than on the Model 50; hard disks on Model 60s are large enough that they will probably be backed up using something other than the diskette drive. The disk-dependent tests were run on the Model 60's 3.5-inch diskette drive, using diskettes formatted at 1.44 MB. Table 7-7 shows the results. When using high density diskettes, the performance of the Model 60 was the same as that of the Model 50 and comparable to that of the AT. The AT was about 10 percent faster on the spreadsheet load and on the file assembly, which is attributable to the faster track-to-track access time of its high density drive. This series of tests is so I/O intensive that the Model 80 exhibited nearly the same performance as the Model 60; only in the more processor-intensive PC-KEY-DRAW graphics application was it appreciably faster.

Table 7-7 High Density Diskette-Dependent Performance. *When using high density diskettes, the performance of the Model 60 was comparable to that of the AT.*

Test Machine	WordPerfect Load File	Lotus 1-2-3 Load File	dBASE III PLUS Sort on Title	dBASE III PLUS Sort on Disk ID	PC-KEY-DRAW Run Sample Macro	MASM Assemble File
AT 339	29	112	204	223	583	19
Compaq 286	33	112	171	169	543	21
Model 50	29	128	204	207	469	20
Model 60	29	128	204	204	468	20
AST 10 MHz	20	108	199	223	465	16
16 MHz Model 80	28	125	196	194	339	18

All results are in seconds.

The same tests were run using low density diskettes. Operations such as the WordPerfect file load, which were affected by the transfer rate being cut in half, were much slower. (See Table 7-8.) Once again, the Model 80 did not perform appreciably better than the Model 60.

The Model 60, like all Personal System/2 machines, is very software compatible; it ran the test applications with no difficulty whatsoever. On the non-disk-dependent tests, the Model 60 performed the same as the Model 50, as expected. It offered 20 percent better performance than the AT 339, but somewhat less performance than the AST Premium/286 and less still than the Model 80. In situations that involved extensive hard disk access, the Model 60 provided better performance than the AT 339 and the Model 50. In such

Table 7-8 Low Density Diskette-Dependent Performance. *Operations such as file loads were 50 percent slower because of the slower transfer rate provided by low density diskettes.*

Test Machine	WordPerfect Load File	Lotus 1-2-3 Load File	dBASE III PLUS Sort on Title	dBASE III PLUS Sort on Disk ID	PC-KEY-DRAW Run Sample Macro	MASM Assemble File
AT 339	43	115	193	184	572	30
Compaq 286	43	115	181	182	546	27
Model 50	45	134	209	211	480	22
Model 60	45	134	209	211	465	23
AST 10 MHz	40	112	176	180	468	26
16 MHz Model 80	39	125	209	202	337	20

All results are in seconds.

situations, its performance was comparable to that of the AST Premium/286 and the Model 80. In the file load tests the Model 80, with its high transfer rate ESDI disk, came out on top.

Summary

The Model 60 improves upon the AT in almost every way. It is faster, quieter, and less expensive to purchase, operate, and maintain. It offers improved performance and capacity over the AT, and is very software compatible. Aside from the problem of diskette incompatibility, the vast majority of PC and AT applications will run on the Model 60 the first time and every time.

The Model 60 is a good next step up from the Model 50. It provides additional expansion slots and much greater mass storage than the Model 50. Many applications and functions can be performed on it without going to the extra performance and cost of the Model 80.

The Model 60 is a good medium level personal productivity workstation. It will operate quietly underneath and beside many desks for a long time to come. With its larger memory and disk storage capacity and its Micro Channel architecture, it is an ideal platform for OS/2. It is also well suited for service as a LAN server, and will be used as such with third-party network software as well as the OS/2 LAN Server. With its large disk capacity and numerous expansion slots, it will be a major player in the OS/2 extended edition data and communications management environment.

8

The Model 80

The 80386-based machine from IBM is the Personal System/2 model that has the most to offer for the future. It has good expansion capabilities and is based on the latest available microprocessor in the Intel 8086 family. Various versions of the Model 80 use leading edge processor and disk drives to produce their high performance. These limits are not currently stretched by operating systems such as DOS and OS/2, which do not make use of the additional modes of the 80386. The Model 80 is poised for the future in that it can support an operating system that exploits the other functions of the 80386 when, and if, one is ever written.

In addition to being the most technologically advanced machine in the current family, the Model 80 is generally superior to its siblings, because they are underpowered for their architecture. The Model 50 machine is overpriced for its performance and has little expansion capability. The Model 60's hard disk offers only slightly better performance than that of the AT, so the Model 80 may be a closer replacement for the AT than differences in architecture between the machines would indicate.

Originally, three versions of the Model 80 were announced: two 16 MHz models and one 20 MHz. Since that time, an additional 20 MHz version with larger mass storage has been announced. These machines are based on the same microprocessor that is used in the Compaq Deskpro 386, but the expansion bus is based on the Micro Channel architecture that is standard on Models 50 and 60. This architecture is described in detail in Chapter 4. It is well designed to accommodate bus expansion as microprocessors with greater addressing capabilities are required in new machines. There is no reason why the Micro Channel design could not be incorporated into a 64-bit bus or even larger bus.

Chapter 8

Features

The Model 80 is a floor-standing machine with looks and options that are very similar to the Model 60. Four models are being offered. Part number 8580-041 and part number 8580-071 are driven at 16 MHz. The other two models—the 8580-111 and the 8580-311—run at 20 MHz. The faster machines have greater capacity mass storage hard disks than the slower versions. But, the 20 MHz models offer more than just a faster processor. The memory arrangement in the 20 MHz machines is different from that used in the 16 MHz machines. The interaction with the bus is also different, due to the different microprocessor speeds and varying wait state requirements. Figure 8-1 shows a photograph of the 16 MHz Model 80.

The full capabilities of the Model 80 will not be realized until an operating system is available that takes advantage of the 80386 processor. OS/2 is only designed to take advantage of the protected mode of the 80286 processor. The protected mode of the 80386 is a superset of this mode and so is fully compatible with OS/2; the real mode of the 80386 is compatible with existing versions of DOS. Alternatively, control programs are available today that can provide most of the desired functionality with existing application software. These control programs use the virtual 8086 mode of the 80386 to run several programs at once. Some of the available control programs are discussed in Chapter 13.

All versions of the Model 80 consist of a system unit, which can be slid under a standard sized desk, and keyboard. The system unit and keyboard compose the Model 80 itself, but one of the analog Personal System/2 monitors is required for operation and should be purchased as an option. Monitors are discussed in detail in Chapter 9. Keyboard and monitors are supplied with long cables that permit under-desk installation of the system unit. All cables are attached via the rear panel of the unit; the diskette drives and power switches are on the front of the unit. Because the vents for cooling are on all sides of the unit, it should be positioned to avoid obstructing any side. IBM recommends a minimum of 2 inches on either side of the unit for adequate ventilation.

The keyboard is the standard 101-key Enhanced Keyboard that is supplied standard with all of the Personal System/2 machines except the Model 25. It has 12 function keys in a row along the top, with a separate cursor keypad and numeric keypad on the right hand side of the QWERTY keyboard. Any IBM Personal System/2 monitor can be used with the machine; the 8503 monochrome and the 8512, 8513, and 8514 color monitors all work, but the 8514 is the only monitor capable of displaying the full resolution of the optional 8514/A Display Adapter.

The machine is supplied with one 1.44 MB, 3.5-inch diskette drive as standard; an optional second drive can be added. The 16 MHz machines come with a single hard disk; the 8580-041 has a 44 MB hard disk and the 8580-071 has a 70 MB hard disk. A second 44 MB disk can be added to the 041, and a second 70 MB,

The Model 80

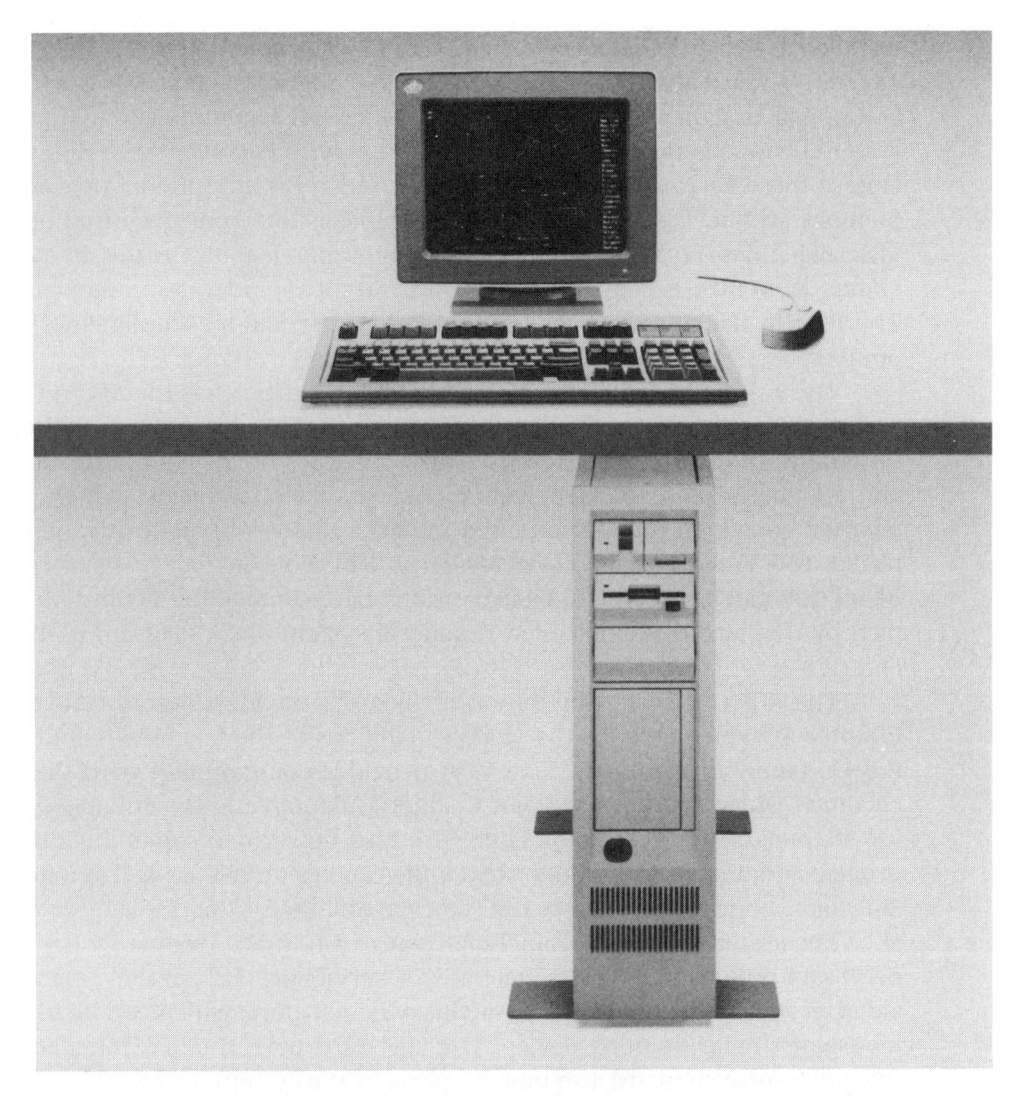

Figure 8-1 The Model 80. *The Model 80 is the 80386-based, floor-standing member of the Personal System/2 family.*
Photo courtesy of International Business Machines Corporation.

115 MB, or 314 MB hard disk can be added to the 071. The 20 MHz machines are also supplied with one hard disk as standard; the 8580-111 has a 115 MB hard disk and the 8580-311 has a 314 MB hard disk. An additional hard disk—either 70 MB, 115 MB, or 314 MB—can be added to either machine. For all machines, the second hard disk can be added without the addition of another controller card, because the standard disk controller expansion board that comes with the machines supports up to two disk drives.

The system board RAM that is supplied with the machine varies with the versions of the Model 80. On the 16 MHz machines, up to 2 MB of memory can be installed in the system board on two carriers, each containing 1 MB. On the 041, 1 MB is standard and the second is an optional upgrade; on the 071, the full 2 MB is supplied as standard. The 20 MHz machines can accommodate up to 4 MB of system board memory, again on two carriers with each carrier holding 2 MB. One of these carriers is standard and the second is an optional upgrade. The memory architecture of the 16 MHz machines is different from that of the 20 MHz machines, and the memory is not interchangeable. On the 20 MHz machines, ROM BIOS is remapped into RAM and the memory is arranged in pages. The 16 MHz machines have faster ROM that is not remapped and memory that is not paged.

The system board for all of the Model 80 versions has the Micro Channel architecture incorporated into its expansion bus. This architecture includes 32-bit slots as well as the 16-bit slots used on Models 50 and 60. Also, as on Models 50 and 60, one of the expansion slots includes a video extension, which permits adapter boards to make use of the integral video subsystem on the system board. There are three 32-bit slots and five 16-bit slots on the expansion bus. The 32-bit slots can be used by 16-bit expansion boards; one of the 16-bit slots is occupied by the hard disk controller. A general system block diagram is shown in Figure 8-2.

The video subsystem on the system board is the VGA (described in detail in Chapter 5), which extends the video capabilities of the machine from those of the PC family of machines. The VGA provides compatibility with the Monochrome Display Adapter, the Color Graphics Adapter, and the Enhanced Graphics Adapter of the PC family. There are also new video capabilities that can provide additional resolution for text and graphics modes, as well as a dramatic 256-color mode that enhances the color capabilities of the display.

The system board itself includes features that are frequently considered necessary options on PC class machines. A serial port and parallel port are provided as standard, along with an auxiliary device port, which can be used by a mouse or similar pointing device. The keyboard port, the disk interfaces, and the video subsystem are also integral parts of the system board. There is a battery-backed real-time clock that maintains both the date and time as well as the system configuration. As with the PC class machines, ROMs are used to contain BIOS, POST, and BASIC.

The Model 80 is designed to require minimum configuration and installation. The Reference Diskette that is supplied with the machine is capable of configuring the system, and there are no switches to be set inside the machine when additional options, such as memory, are installed.

The power supply on the Model 80 is an adequate 225-watt power supply with an auto restart feature that allows the unit to restart after a power loss. This feature is useful for unattended operation situations, such as when a machine is used as a network server; the machine reboots automatically, without having to be manually switched off and back on again. This feature was not available on the PC family of machines. Assuming that add-on boards used in the

The Model 80

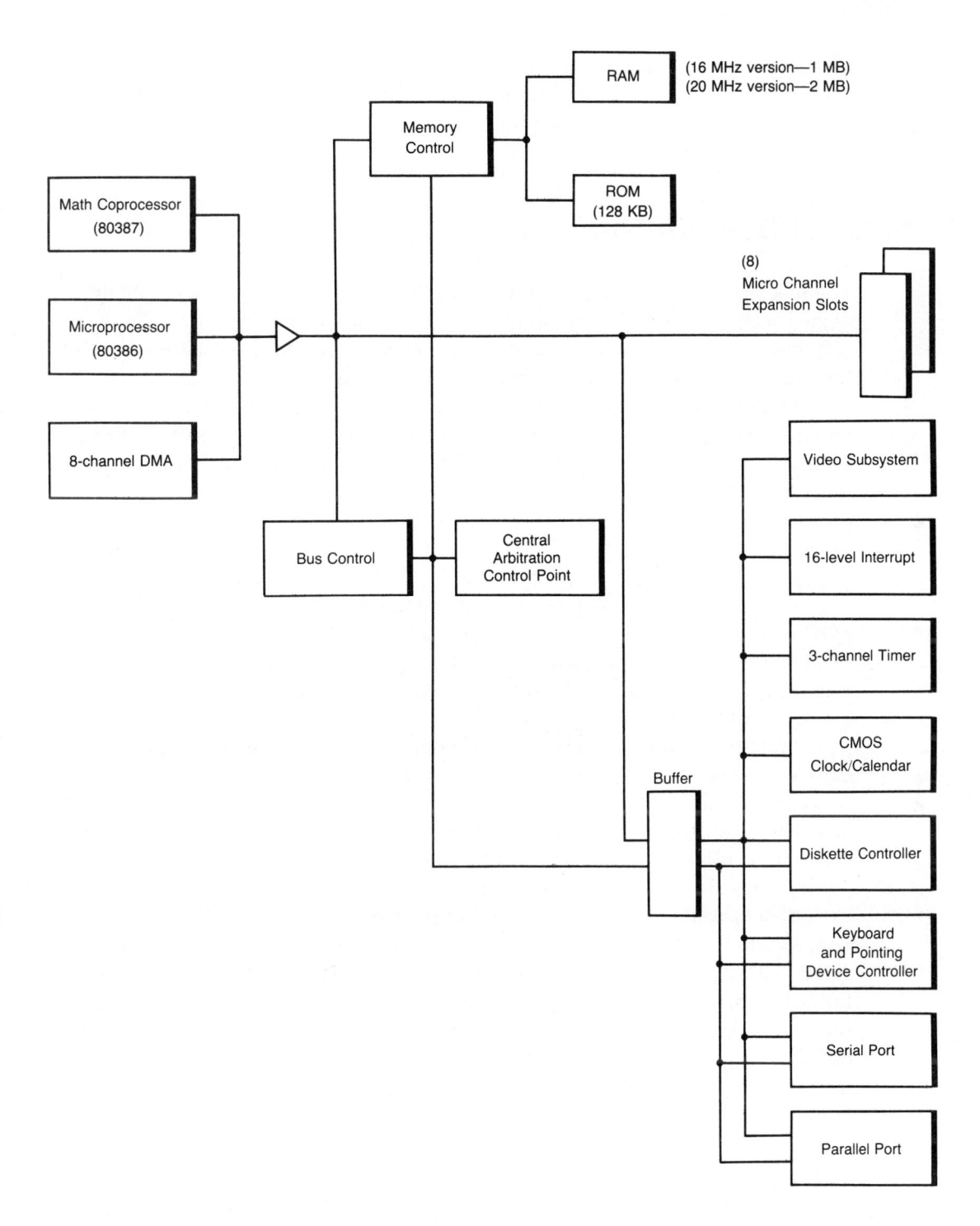

Figure 8-2 System Block Diagram. *The system board includes a central arbitration control point used for assigning control of the bus to competing devices.*

machine conform with the Micro Channel power specifications, the power supply should be able to handle a fully loaded machine. This was not the case for the Model 30 when IBM's laser printer card was installed.

Physical Description

The machine is a floor-standing tower, 23.5 inches high by 6.5 inches wide by 19 inches deep. It is fairly heavy, weighing in at over 45 pounds with only one hard disk installed, more if additional disks are present. A fold-down handle on the top, however, allows it to be slid across a floor fairly easily.

The simplicity of design used elsewhere in the Personal System/2 family is also evident in the Model 80. There are five sections on the front panel. The top section contains the red on/off power switch, along with two LEDs. The off position of the switch is indicated by a circle in the molding of the case, and the on position is a vertical straight line. The two LEDs are to the left of the switch; the top green light indicates that the power is on and the lower yellow light indicates that one of the hard disks is being accessed. The LEDs are labelled on the case by a light bulb outline for the power and a cylinder outline for the hard disk. The part number and serial number for the machine are also located in this section of the front panel. The ability to read the serial number without diving into a multitude of cables is a simple, yet highly desirable feature that more manufacturers should adopt.

Under the power switch there is room for two 3.5-inch, 1.44 MB diskette drives. The top drive is installed as standard; it has a yellow LED to show when it is being accessed and a blue eject button. The eject button is also labelled 1.44 MB to distinguish it from other 3.5-inch drives. Although 720 KB diskettes can be read from and written to, the magnetic media in the diskette itself is different from the media in the 1.44 MB diskettes. It is physically possible to format and use either diskette, but the reliability of the data is questionable if the diskettes are formatted at the wrong density. This situation is not the same as using single-sided, 5.25-inch diskettes as double-sided diskettes. Here the media is the same for both situations, and the operating system will mark out any bad blocks so they are not used. IBM recommends that any diskette formatted at the wrong density be discarded and not reformatted for use. If the diskettes are passed over a good quality bulk eraser, there is no reason why they cannot be reformatted at the correct density; without the bulk erasing, however, they may prove to be unreliable. If in doubt, remember that it is always the most important diskette that fails to work; not the tens of other backup diskettes that were used for experimentation. If the data is worth keeping, it is worth protecting properly.

If a second 3.5-inch disk drive is not installed, there is a facia plate in its prospective position. This can be punched out through the front after removing the front panel. The position below the disk drives appears to be the right shape for a 5.25-inch, 1.2 MB diskette drive, but IBM is not offering this as an option and there is no diskette controller to support it in the machine. The second hard disk occupies the internal space at this location if one is installed, or the IBM 3363 Internal Optical Disk Drive can be used instead.

The right side of the unit has ventilation grills at the top that draw in air for cooling. Also on this side is a recessed hand hold that can be used to pull the unit out from under a desk. The top of the unit has a strong fold-down handle that makes carrying the unit possible and when not in use stores flush with the top of the system.

The base of the unit has two swivel legs that can be set flush with the base, if the unit is to be packed in a box, or turned 90 degrees to make feet that protrude 3 inches from either side of the unit. One of the few regrettable features of this machine is that, according to the installation instructions, the cover must be removed to turn these legs. As all switches have been removed from inside the unit, it is a pity that something as simple as swiveling a mounting leg requires the cover to be removed. In fact, it is possible to turn the legs without removing the cover, but it requires more brute force.

Rear Panel Connectors

The rear of the unit, shown in Figure 8-3, contains all of the connectors that are necessary for the system. The top connector, alongside another ventilation grill, is for the power cord. The grill takes in air that is then blown through the lower half of the system unit by the interior cooling fan. Beneath the power cord connector are the system board connectors. The keyboard connector is at the top and is labelled on the case with a small keyboard icon. Immediately below the keyboard connector is an identical auxiliary device connector, which can be used to connect a mouse or other pointing device. These connectors are not the same as their equivalents on the PC family of machines. They are both 6-pin miniature DIN connectors, and the keyboard connector that is inserted has a flat side to aid in its installation. (The flat side faces the removable cover on the unit.)

The parallel port connector is under the pointing device connector and is a female 25-pin D-type connector typically used to connect a printer. The serial port connector beneath the parallel port is a male 25-pin D-type connector typically used to connect a modem or, less commonly, a serial printer. On the PC family of machines, a mouse was also connected via a full RS232C port such as this; on the Personal System/2 machines, the dedicated pointing device port provides a preferable alternative. The final connector on the rear panel is a female 15-pin D-type connector with three rows of pins, which is used to connect the monitor. Each of the D-type connectors has two threaded standoffs that can be used to screw down the cable connectors to the rear panel.

Below the connectors on the rear panel are the rear panels of each of the

Chapter 8

Figure 8-3 Rear Panel of the Model 80. *The keyboard and auxiliary pointing device connectors, although identical, cannot be used interchangeably.* Photo by Bill Schilling.

expansion boards. There are eight expansion slots in the machine; each expansion board has a narrow rear panel that is exposed through the rear of the system unit. Although the arrangement resembles the rear panel of the PC family of machines, the physical dimensions and grounding constraints are different for the Personal System/2 machines. The physical requirements are discussed in detail in Chapter 4. If no expansion board is present in a slot in the machine, its position in the rear panel is occupied by a blank end piece similar to that used on the PC. The rear panels of the adapter boards are held in the system unit by two means: their physical design causes them to fit tightly into their allocated space;

they are then held in position by a knurled screw with a slotted end, which can be loosened for removal.

Inside the System Unit

The left side of the unit is the removable cover. There are two ventilation grill areas, one at the top and one at the base of this side. In addition, there is a recessed hand hold similar to the one on the right side of the unit. Two large retaining screws are located on each side of the cover about half way up the unit. The keylock is also on this side, at the center of the top; this is used to prevent unauthorized opening of the unit. Although the lock is similar to the keylock on the AT, it does not lock the keyboard. The cover can be removed from the unit by unlocking the keylock and unscrewing the two captive mounting screws with a screwdriver or coin. The cover then can be tilted away from the top of the unit and slid from the bottom retaining lip.

Once the cover has been removed from the system unit, the machine's components are exposed. (See Figure 8-4.) The power supply extends across the full width of the top of the unit and occupies the top fifth of the unit. One end of the power supply extends through the front panel and has the power on/off switch attached. The other reaches the rear panel where the power cord is connected. The base of the power supply contains a fan that draws air from the top and the base of the unit to cool the power supply, system board, and expansion boards.

Two connectors and a cable on the power supply are internal to the unit. Connector P1 provides the power for the hard disk. The second hard disk is powered by a similar cable, which is plugged into the other vacant connector on the power supply and then into the hard disk itself. This cable is supplied with the second hard disk. The system board is linked to the power supply via a sturdy power cable that has a plug at the system board end.

Beneath the power supply on the left side is the system board, which occupies about two-thirds of the width of the unit. This lies flat along the base of the case, which is the right side of the unit in the normal operating position. With just the cover removed, the system memory boards can be seen immediately below the left side of the power supply. Unlike its siblings, the Model 80 has plug-in boards with a connector for the system board memory. The other Personal System/2 machines use single-in-line package (SIP) memory modules that have gold plated board edge connectors for mounting. The Micro Channel expansion slots are located on the left side of the bottom half of the system board.

Just below the power supply, on the right, is the enclosure for the two diskette drives. The top position is filled with a standard 1.44 MB, 3.5-inch diskette drive. Below it, a second 3.5-inch drive can be installed. The power and data for these diskette drives is obtained from a ribbon cable. One end of the cable is attached to the system board and connects to the two drive bays in a daisy-chain fashion. When the diskette drive is inserted into the bay, the edge connector

Chapter 8

Figure 8-4 Inside the System Unit. *A second hard disk can be added to the system by placing the disk in the central frame, closer to the front of the unit than the first disk.* Photo by Bill Schilling.

mates with the connector at the rear of the bay and is held in place by a plastic mounting frame. The installation of a second 3.5-inch diskette drive consists of popping off the front panel, which is held in place by plastic extrusions from the front panel into the chassis, and sliding in the drive.

Beneath the diskette drive enclosure is the position for the hard disks. A frame extends from the front of the unit to the back. With one disk installed, the rear of the enclosure is occupied. A second disk can be installed in front of the first, oriented 180 degrees from the first. The hard disks are held in place with a metal clamping frame, each with two large blue plastic knobs that are used to tighten down the clamping frame. The clamping frame also contains the screw holes for the captive screws that are in the cover. The data and address ribbon cables for the hard drives extend from the disk controller expansion board to connectors on the hard disks, located near the center of the machine.

The rest of the system board extends below the level of the hard disks to the base of the unit. Like the Model 60, the Model 80 features eight Micro Channel connectors. The lowest expansion connector is number 1 and the highest, closest to the hard disk is slot 8. Slot 8 contains the hard disk controller (either ST-506 or ESDI) provided standard with the system. The remaining area in the system unit, in the bottom right-hand corner, contains the lithium battery that maintains the real-time clock, the system configuration RAM, and the speaker. A plastic scoop arrangement on the cover of the unit channels air that is sucked from the top of the unit to the base. The air is forced to pass over the expansion boards and either exits through the side panel or the front of the unit. The inside of the cover also contains various strips of foam, which offer extra support for the two system board memory cards and the expansion boards. The remaining strips run along the base of the power supply and touch the disk drives. Presumably, they reduce any rattling that might occur and reduce the overall system unit noise slightly. It seems unlikely that they would be able to eliminate any damage resulting from a physical shock to the system.

The mounting screws for the legs can be seen at the bottom of the unit. The front leg has a metal lock pin that should be raised before the front leg is turned, according to the installation instructions; however, forcing the leg is possible, so the cover does not need to be removed to prepare the unit for shipping. The cover is reinstalled by placing the two protrusions at its base over the bottom of the case, sliding the base of the cover under the outer lip of the main unit and closing the unit. The two mounting screws should be tightened and the keylock locked, to secure the cover in three places.

Removing the System Board

The system board in the Model 80 can be removed for repair. To do so, all expansion boards must first be removed. The screw that holds the rear panel of each expansion board to the rear panel of the unit should be loosened. The two blue handles that are on the expansion board can then be used to pull the board out of its slot. If a second hard disk is installed in the system, it should be removed next. The front panel should be popped off to allow the disk to slide out of its housing. The two blue knobs that clamp the hard disk in position should be loosened and the drive slid forward. The whole mounting frame, including the second drive, can then be removed. The four screws that are in the front half of the mounting frame should be removed, then the frame slid forward slightly and lifted out. The rear of the frame has two small metal protrusions that hold it in place. Sliding the frame forward gives these protrusions clearance for removal of the frame.

Once the hard disks are removed, the entire system board is visible. All installed system board memory cards can be removed from their sockets. The cable that supplies power to the system board can be removed by squeezing the sides of the housing to clear the plastic flanges in the socket and pulling upwards. The cables attached to the rear panel, such as the monitor and keyboard cables, should be disconnected, as should the ribbon cable to the 3.5-inch diskette drives in the top right-hand corner of the system board, and the 4-pin connector attached to the battery and speaker at the bottom right of the system board.

Nine screws hold the system board in place. Eight of these are in the board itself. One attaches the bracket that supports the external connectors, such as the keyboard and video connectors, to the rear panel. Once these screws are removed, the board can be lifted slightly to clear some plastic locating pins and then slid towards the front of the case, to clear the external connectors through the rear panel.

The System Board

Most of the devices on the system board incorporate surface-mount technology. In addition, several of the components are VLSI devices, providing considerable functionality in a single IC package. The board is densely populated, yet its overall size of 11 inches wide by 16 inches high is not excessive for the machine's capabilities. Figure 8-5 shows the system board.

The connectors for the keyboard, auxiliary device, parallel port, serial port, and video are in the top left corner of the board, on the left edge. When the board is installed in the machine, they protrude through the rear panel for connection to their respective peripheral devices. A metal frame around the connectors ensures that the space around them is sealed when installed, to reduce RF emissions outside the case. This frame is part of the shielding skin of the chassis, which is discussed in detail in Chapter 4. The large 15-pin connector for the DC power supply output is at the top edge of the board, just left of the center. Immediately below are the two 96-pin connectors used to install system board memory. On all versions of the machine, the connector that is closest to the power supply, at the top of the board, is used for the first memory board; the second memory board is installed in the lower socket. On the 16 MHz machines, each card holds 1 MB of memory; on the 20 MHz machines, each board holds 2 MB. Figure 8-6 shows a memory board from a 16 MHz Model 80.

The controller for the 3.5-inch diskette drives is an integral part of the system board. The output from this controller is connected to the diskette drives through the connector in the top right corner of the board. The other connector in addition to the expansion slots on the board is for the speaker and battery. It is located at the front edge of the board in the lower half.

The layout of the system board is different for the 20 MHz Model 80 than

The Model 80

Figure 8-5 Model 80 System Board. *In layout the system board for the 16 MHz machine (pictured) is similar to that of the 20 MHz machine, except for a few differences, such as the processor's position.* Photo by Bill Schilling.

for the 16 MHz version; however, the connectors on each board are in similar locations. On the 16 MHz machine, the 80386 is located in the lower right portion of the board, with the 80387 socket alongside and closer to the edge of the board. On the 20 MHz machines, the 80386 and its math coprocessor are in the

Chapter 8

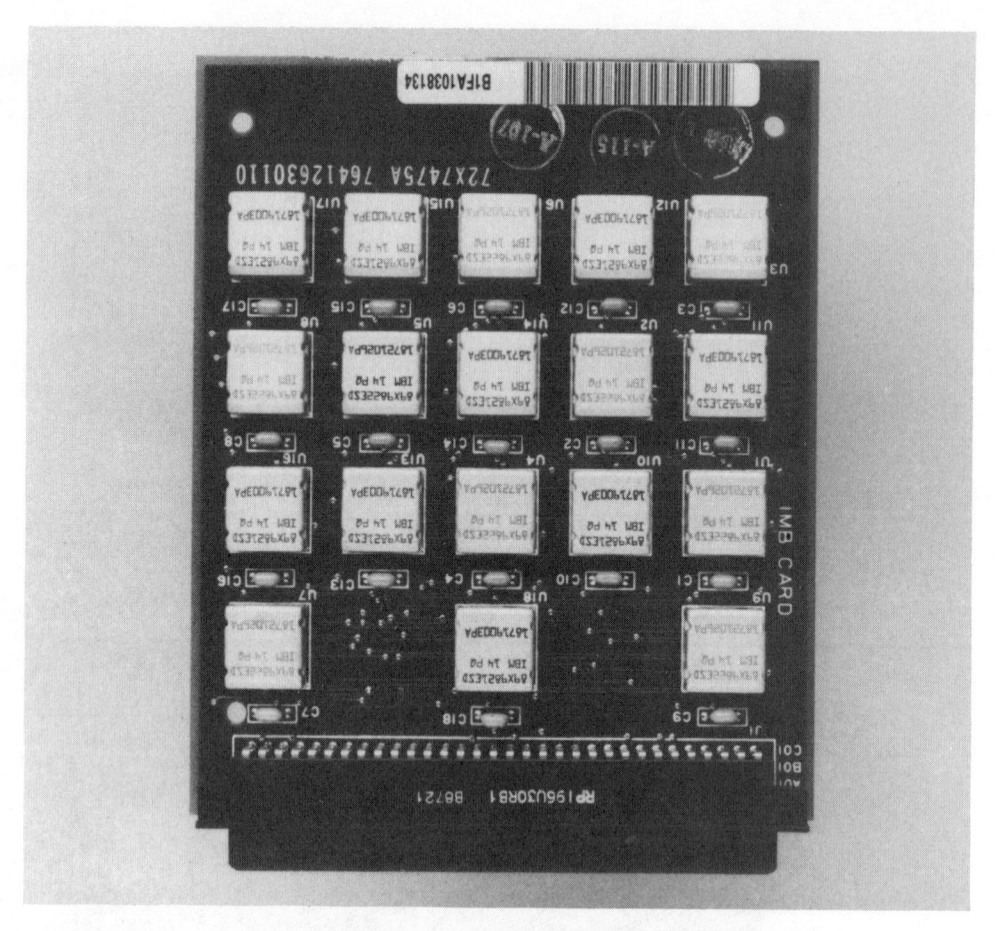

Figure 8-6 16 MHz Model 80 Memory Board. *This board provides memory to the system board on the local bus. Up to two boards, each containing 1 MB of memory, can be added to this machine.*
Photo by Bill Schilling.

top right quadrant of the board, below the diskette drive connector. The 80386 is to the left of the socket for the 80387.

The Micro Channel expansion slots are in the lower left quarter of the board. There are eight slots in all, numbered from the bottom of the board upward. Slot number 8 is always occupied in the standard configuration by the hard disk controller. There are three 32-bit slots: slots 1, 2, and 4. These are longer at both ends than the 16-bit slots. The extension to the right is for the additional address and data signals and a few of the 32-bit control signals. The smaller extension to the left is for boards that can support Matched Memory cycles. This allows memory boards that can transfer data faster than the standard Micro Channel cycle to signal to the microprocessor controlling the bus and accelerate the cycle. This feature is described in detail in Chapter 4 and is only supported on the 16 MHz versions.

The 20 MHz machines do not drive the appropriate control signals on the bus, but the connector is long enough to physically accommodate the 32-bit boards. The 80386 32-bit memory board can be used; the Matched Memory cycle that it supports is not used on the 20 MHz machines. IBM does not provide the required signals for the faster cycle.

Slots 3, 5, 6, 7, and 8 are all 16-bit slots that function in the same way as the slots in Models 50 and 60. Slot 6 has a different extension to the left of the connector, which is called the video extension. This is provided for any video boards that are added to the system. The signals present allow the VGA on the system board to be enabled or disabled. They also provide the means for a video expansion board to use the digital-to-analog converter (DAC) on the system board and to output video to the monitor through the system board video connector. Video boards and 32-bit boards need to be inserted in the appropriate slots; all slots can accept 16-bit boards.

The ICs present on the system board include several custom VLSI devices. Most are IBM proprietary, but some are commercially available. Although the major components on the 16 MHz version of the Model 80 are similar to those on the 20 MHz machine, the system board layout is different for each machine.

On the 16 MHz Model 80, the VGA IC itself is in a large square silver case just above slot 8 of the expansion slots. The DAC, which is used to generate the analog video signal, is an INMOS[†] device at the left edge of the board, below the video connector. The rest of the video circuitry, including the video crystal clock, is in the same area of the board as the VGA and DAC. The BIOS resides in four ROMs on the center right edge of the board. The diskette controller is immediately above the ROMs and the 8042 keyboard controller is on the top edge of the board, just right of center. The custom VLSI microprocessor support device is just above the 80387.

On the 20 MHz machine, the VGA is located just to the right of expansion slots 7 and 8. The DAC is still at the edge of the board, close to the video connector. The BIOS is located in two ROMs that are positioned centrally at the right side of the board. The diskette controller is directly below the keyboard controller, which is in the same position as on the 16 MHz machines. The custom VLSI device that supports the microprocessor is beneath the 80386 and directly to the left of the ROM.

Functional Description

The PC-based roots of the Model 80 can be seen in its architecture. The 80386 microprocessor is driven by a 16 MHz or 20 MHz clock, and is in parallel with the 80387 math coprocessor, the eight-channel DMA, and 128 KB of ROM. The mem-

---

[†]INMOS is a division of Thorn EMI.

ory controller is directly connected and buffered to the microprocessor and controls the RAM that is on the system board. On the 16 MHz machines, this is either 1 MB or 2 MB; on the 20 MHz machines, this can be 2 or 4 MB. The Micro Channel connectors are buffered from the microprocessor and are directly connected to the memory controller. The eight other I/O devices are buffered to the microprocessor as well as to the Micro Channel. The I/O devices are the diskette controller, the keyboard and auxiliary device port controller, the RS232 serial port controller, the parallel port controller, the three-channel timer, VGA, the real-time clock and associated RAM, and the sixteen levels of interrupt.

On the 16 MHz Model 20, the ROM BIOS is located in the top segments of the 1 MB of 8086 memory, as seen on previous machines. On the 20 MHz machine, however, the ROM BIOS is remapped to RAM (at the same memory location) during POST and the ROM is then disabled. This results in a higher performance for BIOS activities, as the RAM is much faster than ROM. This remapping has been seen in other machines, such as the Compaq 386, but is the first occurrence on an IBM personal computer. The top 128 KB of the first 1 MB of RAM is used for this purpose. If desired, this feature can be disabled through POS configuration. It is also possible to split the first 1 MB of memory so that 512 KB is available as contiguous DOS memory, instead of 640 KB. Regardless of the address used as the split address for the RAM, the top 128 KB in the first 1 MB is used for ROM-to-RAM mapping, unless the ROM is enabled specifically.

Additional 32-bit memory can be added to the system via the 80386 memory expansion options. Each of these boards can hold up to 6 MB of memory. However, the DMA controller is only capable of addressing memory with 24 bits, so the maximum amount of memory that can be addressed in the system is 16 MB, which may prove to be a limitation in the future, as more memory is required by applications. Up to 4 MB of this memory may be on the system board, depending on the version of the Model 80 and the system configuration.

Improved Microprocessor

The Model 80 is IBM's first use of the 80386 microprocessor in a personal computer. If current applications and DOS are used, its function is simply that of a faster AT; however, the 80386 has far more potential, which can be used both now with existing applications and in the future with new operating systems and applications. The 32-bit-wide data paths speed up the memory transfers and the faster clock speed improves the overall performance of the machine.

The 80386 has several modes of operation, which allow improvements in performance and functionality that were not possible on 80286- and 8088-based machines. The real-address mode of the 80386 is the mode used with DOS; it provides a contiguous 1 MB of memory that can be addressed. The protected virtual-address mode extends the addressing space that is available with the 80286. In addition, the virtual 8086 mode allows several existing applications to be run at one time, a feature not seen before on the 8086 family of microprocessors. In this mode, the 80386 gives each application a 1 MB environment of its own that behaves exactly like an 8086 environment.

The 80386 itself has six main functioning areas that can work independently of each other. The interconnection between the elements is 32 bits wide. This independence allows the execution of various stages of an instruction to be overlapped, as well as allowing several instructions to be manipulated at one time. While one instruction is being executed, the next can be decoded and yet another can be fetched from memory. When the instruction sequence is linear, as it is for the majority of instructions in an application, this pipelining is beneficial. It is only with instructions that are not sequential, such as indirect jump instructions, that they can only be dealt with one at a time.

Certain areas of the microprocessor have dedicated functions that speed up instruction processing. A barrel shifter can shift up to 64 bits in one clock cycle, and a dedicated multiply-and-divide unit can do 32-bit multiplications in as few as nine clock cycles. These features can be accessed using the 80386 instruction set. The memory management unit necessary for an effective multitasking system is incorporated into the microprocessor itself, allowing memory paging to be implemented by the operating system, if desired. The paging mechanism is accessed through the protected virtual-address mode of the processor.

There are eight 32-bit registers within the 80386. These are EAX, EBX, ECX, EDX, ESP, EBP, ESI, and EDI. In order to accommodate code that has been written for the 80286 and 8086, the registers can be accessed 16 bits at a time, producing the eight 16-bit registers in the low order ends of the 32-bit registers. Code to use the upper and lower half of the 16-bit registers is supported through the usual 8086 conventions. Any operations that are requested behave as expected for the type of register used, so an 8-bit operation produces results that are 8 bits wide and includes the expected supporting data, such as the carry flag.

If the 80386 is operating in real-address mode, up to 1 MB of contiguous memory can be accessed. This memory is divided into segments that are each up to 64 KB in size (minimum size is 16 bytes). The memory is addressed by using 20-bit addresses. The 20-bit addresses are formed by adding the 16-bit offset to the segment register values, which have been shifted left 4 bits. This results in segment boundaries only on multiples of 16 bytes.

For the protected mode of the 80386, a much larger address space is available, as the segmentation is performed differently. The memory address, as seen by the application, is referred to as the virtual address. This virtual address is given to the segmentation unit within the 80386, which translates it into a linear address that references the desired position in memory within the current segment. This is the physical memory address, or, if the paging mechanism is enabled, it is translated into the physical memory address.

The virtual memory address is composed of a 16-bit selector and either a 16-bit offset (as on the 80286) or a 32-bit offset. The 16-bit selector points to a segment descriptor table. The appropriate entry from the segment table is then added to the 32-bit offset, resulting in a 32-bit linear address. This provides a linear address space of 4 GB (2^{32} bytes). Because there can be 2^{14} different numbers in the 14 most significant bits of the segment descriptor, up to 64 TB (2^{46} bytes) of virtual address space is available. The other 2 bits in the segment descriptor can be used to restrict access to the different segments.

There are two segment descriptor tables: the global descriptor table and the local descriptor table. The local descriptor table contains the addresses being used by the current task that is being executed. The global descriptor table contains the addresses of more general segments that may be used by any task. The operating system controls which addresses are available in the descriptor tables and, if necessary, changes the local descriptor table as it changes the task being executed. This allows different programs to be in the physical memory, but access for a particular task can be limited to the areas of memory that are appropriate for that task.

In addition to this segmentation capability, the 80386 has a paging mechanism available within the protected mode. This is an optional mode that can be enabled or disabled as required. It allows the linear address that is generated to be different from the physical address that is actually accessed. This is useful, for example, in virtual 8086 mode. In this mode, each 8086 task has its own 1 MB of address space. Due to the paging mechanism, the physical location of these areas of memory can be different for each task, allowing these tasks to be independent of each other. The paging mechanism can translate the linear address that is requested into a physical memory address where the required data is stored. The 32-bit linear address that is obtained from the segmentation unit is split into three parts. The 10 most significant bits (MSBs) of the linear address are used to dictate the offset into the directory table. The directory table consists of 4 KB and allows 1024 32-bit page directory table entries. The starting address for the page directory is stored in the CR3 control register. This is combined with the 10 MSBs from the linear address, to point to one of the directory table entries. The 32-bit value in the page directory gives the base address for the page table and statistical information about the page table. The page table is also 4 KB and allows 1024 32-bit entries. This base address is combined with the next 10 MSBs of the linear address (bits 21 through 12) to create the offset into the page table. The 32-bit output from the page table contains the page frame address and statistical information about the page frame. The upper 20 bits of the page table entry are combined with the remaining 12 bits of the linear address, to obtain the physical address of the data required.

If this was the whole of the paging mechanism, severe performance reductions would be incurred by enabling paging, with no apparent gain. However, the process itself does allow segments of data to be protected. The additional data stored in the page directory table and page frame table can restrict access, allowing multiple tasks to protect their data. Also, the design of the paging mechanism allows a faster access to frequently used pages than it does to less frequently used pages. This is accomplished through a *translation lookaside buffer* (TLB). Addresses of the thirty-two most recently accessed pages are stored in the TLB. Any linear address that is requested is checked against the buffer. If, as is likely, it is one of the most recently used pages, the physical memory translation is obtained immediately. Only if the address requested does not lie in one of the thirty-two most recently accessed pages, does the two-level page translation occur. According to Intel, statistically a miss occurs approximately 2 percent of the time.

The other mode of the 80386, which is exciting for current applications that run under DOS, is the virtual 8086 mode. This allows several 8086-based applications to be run at the same time. Because each of these applications is able to run with its own area of memory, compatibility for multiple tasks in the same system is guaranteed. Addressing in the virtual 8086 mode is similar to real-mode addressing. The 20-bit address consists of two parts: the segment (contained in the segment register) and the offset. The 20-bit physical address is generated by the CPU, by shifting the segment left 4 bits and then adding the 16-bit offset to it. Only 1 MB of memory can be accessed by each task; however, the linear address space of the 80386 is 4 GB, and through paging each task can have a linear address that starts at zero. The contents of the page directory base register can be different for different tasks and can provide any necessary isolation for each task being run.

Memory Architecture

The different versions of the Model 80 all have the 80386 as their CPU, and all of the different modes are available to the operating system. However, there is a difference in the physical memory architecture between the 16 MHz and 20 MHz machines. On these machines, the first megabyte of system board memory is attached to the system microprocessor in a linear fashion, and the physical address that is sent to the dynamic RAM is used to access the required data. On the 16 MHz machines, it does not make any difference which byte of data is being accessed. The time taken to obtain the data is the same. On the 20 MHz machines, the memory is arranged in 2 KB pages. If the address that is requested is in the same 2 KB page as the last address accessed, the data can be obtained very quickly; but if the address is outside this 2 KB page, the data is obtained more slowly, because the current page needs to be changed.

In order to arrange the memory in this paged fashion, additional control circuitry is needed to support the RAM. The principle is that the memory addresses can be considered as consisting of two parts—say, a row and a column. If a particular address is accessed, one row and one column are activated. If the column is kept active until the next memory access, the DRAM has the data available as long as the column does not change. Only the required row need be activated.

On the 16 MHz machines, system board memory accesses require one wait

Chapter 8

state (three 62.5 ns clock cycles, totalling 187.5 ns). On the 20 MHz machines, accesses to system board memory take zero wait states for a read request within the same page and two wait states outside a page (two and four 50 ns clock cycles, totalling 100 and 200 ns respectively). On the 20 MHz Model 80, writes to the same page of system board memory require one wait state, and writes to system board memory outside the current page require two wait states (three and four 50 ns clock cycles, totalling 150 and 200 ns, respectively). The paged memory used with the 20 MHz Model 80 is effectively faster than the 16 MHz Model 80's linear memory arrangement, because, although more time is required to access memory outside the page, the majority of accesses are within the page, and the resulting faster access more than compensates.

Additional System Components

The fused power supply on the Model 80 has automatic ranging to accommodate both European and U.S. voltage and frequency requirements. The low range accepts a voltage of 90 V_{ac} to 137 V_{ac}, and the high range accepts 180 V_{ac} to 265 V_{ac}. The acceptable frequency is 50 to 60 Hz with a 5 percent tolerance. The power supply powers the system board, the Micro Channel adapters, the disk and diskette drives, the mouse, and the keyboard. The design of the power supply is such that it can recover if ac voltage is lost for a time and then restored. There are three DC power outputs from the power supply: + 5 V, + 12 V, and − 12 V. All are supplied to the system board, and the + 12 V line is also on the cable to the hard disk.

The keyboard is attached to the system unit via a cable that is 10 feet long. Although the keyboard itself is the same as the Enhanced Keyboard on the AT, the cable used to attach it to the system unit is different. The connector at the keyboard end is similar in that it is a flat 6-pin self-locking connector. The connector on the rear of the system unit is much smaller, however; it is a 6-pin miniature DIN with a convenient flat edge on its molding to help with orientation as it is plugged in. Because the keyboard itself is the same as the Enhanced Keyboard, it would be easy to exchange keyboards in order to get one with, for example, a less tactile feel to the keys. It is surprising how emotional the issue of keyboard layout and feel can be. Although IBM has not been particularly consistent with the feel and layout of the keyboards for the PC, the Personal System/2 machines are somewhat consistent, and third parties will be willing to supply any other variations that are desirable.

The keyboard supplied with the unit is the AT Enhanced Keyboard with 101 keys. For a variety of different countries in the European market, IBM offers a 102-key keyboard. The keyboard is linked to the system unit via a bi-directional serial interface. This keyboard contains a 16-byte FIFO buffer, which is used to store keystrokes until they are serviced by the system. The keys are all make/ break and typematic keys, with the exception of the Pause key, so the keyboard detects which keys are depressed and sends an appropriate scan code to the system. The typematic feature refers to the repeated sending of the scan code to the system at a rate of 10.9 characters per second, after a delay of one-half second. When the keys are not depressed, the scan code that indicates that the keys have been released is sent to the system. The Pause key is different; if depressed, it causes the system to pause until another key is pressed. The typematic delay and rate can be changed on the keyboard by the system.

The scan codes for the Enhanced Keyboard are an extension of the codes on the PC standard keyboard. There are three different sets of scan codes that can be set. The system defaults to set number 2, which assigns a unique 8-bit scan code to each key. When the key is released, the scan code is re-sent to the system, but this time with an 8-bit prefix (F0H) indicating the break condition. If the key is held down long enough for the typematic effect to occur, the same scan code is sent to the system as when the key was first depressed. The remaining keys that are not in the QWERTY part of the keyboard have scan codes that are sensitive to the state of the Ctrl, Shift or Alt keys as well as the state of the Num Lock key. These scan codes have an additional 8-bit prefix of E0H.

Scan code set 1 sends scan codes to the system for the majority of the keys that are not dependent on the state of the Shift key. The scan code is an 8-bit code and the break code, which is sent on release of the key, is 80H higher than the scan code. As with scan code set 2, the remaining keys that are not in the QWERTY part of the keyboard have scan codes that are sensitive to the state of the Ctrl, Shift, or Alt keys, as well as the state of the Num Lock key. Again, these scan codes have an additional 8-bit prefix of E0H.

Scan code set 3 sends scan codes that are different for each key on the keyboard. Each key is assigned an 8-bit scan code when pressed. This scan code is prefixed by F0H when the key is released. The state of any key does not affect the scan code that is sent for any other key. In addition, the scan codes that are assigned to each key can be changed.

The scan codes for the keyboard are sent to the system as an 11-bit serial data string. The first bit is the start bit (0), then the 8-bit scan code with the least significant bit first, then the parity and stop bit. The ROM BIOS within the system converts the scan codes into the extended ASCII character sets. In addition, translation of certain key combinations—such as Ctrl-Alt-Del for reset—are translated into different functions. The other keys that are treated exceptionally are the Pause key discussed previously, the Print Screen key (which causes the print screen interrupt to be set and invoked), and the System Request key (which is used to access INT 15H in the BIOS).

The keyboard controller on the system board is an 8042 device. This IC not only controls the keyboard port but also controls the auxiliary device port. Password protection is available for the keyboard, which compensates to a certain extent for the keylock on the cover not disabling the keyboard. Although IBM supplies a program to set the password for the system, it cannot be used to disable or enable the keyboard while an application is running.

The connector that is used for the auxiliary device port is the same as the

Chapter 8

connector for the keyboard. Although these two cables can be interchanged on the Model 30, they cannot on the Model 80. Interchanging the two connectors results in a 301 error when the system is powered on. The auxiliary device port, although commonly used for a mouse, can connect a variety of input devices, such as a touchpad, trackball, or keyboard.

The 3.5-inch diskette drive that is installed as standard in the machine is similar to the drive in Models 50 and 60. It has a maximum capacity of 1.44 MB, but can read and format diskettes that have a formatted capacity of 720 KB. This density is required for data to be read by a Model 25 or 30 or by a 3.5-inch drive that has been added to a PC or AT. The bay that is available to accommodate the drive has a single connector that links the drive to the system board. This connector contains the power and data lines that are needed for the drive to operate.

The hard disk that is used in the 16 MHz Model 80, part number 8580-041, is a 44 MB capacity with an interface similar to that of the hard disk drives in the XT and AT. The larger disk drives, 70 MB and over, have an Enhanced Small Device Interface (ESDI). This improves the performance of the disk drive and requires a different hard disk controller than the smaller drives. (See Figure 8-7.) The ESDI is described in more detail in Chapter 7. Care should be taken when interpreting the results of benchmarks that provide track-to-track and average seek times, as these will not perform in the same way on the different disk drive types. In order to function correctly under DOS, the information that is given to the configuration RAM by the disk drive is not the correct physical numbers. This allows the disk drive controller to perform more of the address translation for moving the drive head than is done with the ST-506 interface (the interface used for smaller drives). Benchmarks that use DOS to time the track-to-track time are not actually moving the head over one physical track; they are moving the head over what DOS *thinks* is a single track. Any average track-to-track time test will not use the same random

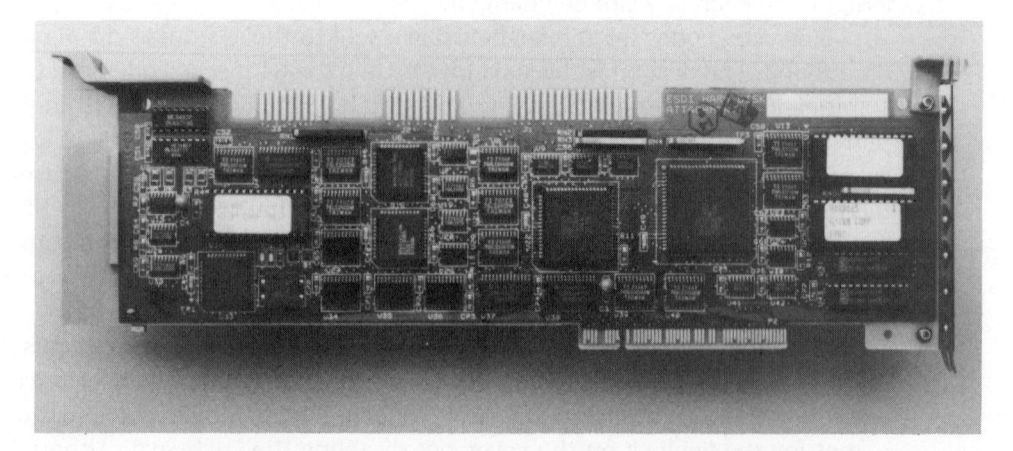

Figure 8-7 ESDI Hard Disk Controller. *All IBM disks that are 70 MB or over in size use an ESDI interface in the Personal System/2 machines.*
Photo by Bill Schilling.

number selection for the head movement, because one track to DOS does not equal one track on the disk itself. The validity of these low level benchmarks needs to be examined carefully so that valid comparisons are made.

Installation and Configuration

The Model 80 machines are similar to their siblings in that they are easy to configure and set up. The machine needs to be unpacked from its shipping cartons and the legs swivelled on the system unit. Any internal options should be installed according to the installation instructions; for many situations there will not be any additional options because the system board is well equipped. The monitor, keyboard, mouse or other pointing device, printer, and modem should be plugged into the rear of the unit. Finally, the power cord should be installed. The system is configured for use with the supplied Model 80 Reference Diskette. This is similar to the diskette supplied for other Personal System/2 machines, and should be used to initialize the system from drive A.

There are six sections to the program. The "learn about the computer" section is a pictorial tutorial that describes hardware, software, communications, troubleshooting, and miscellaneous information. This is elementary for experienced users, who can quickly examine the features that are on the system; but it will prove invaluable for the novice. So often, the basic introduction to a personal computer is missed by novices and they suffer in the long term when they want to use the computer as more than a word processing device.

Other functions offered on the Reference Diskette include producing a backup diskette; copying an option diskette, supplied for each option installed in the machine; setting the date and time, passwords, and keyboard speed; and finally, setting the configuration for the system. The *Set Configuration* option is used to automatically set up all of the expansion boards, system memory, port assignments, and options (such as a math coprocessor) that have been installed in the system.

The present configuration of the system can be viewed or backed up onto a copy of the Reference Diskette. The configuration option allows the setting of port assignments—for example, setting the parallel port as LPT1, LPT2, or LPT3. The math coprocessor is detected through this option, and any system memory is identified. These automatic or change configuration procedures are substitutes for the switches that expansion boards for the PC and AT used to sprout. Each expansion board, if it follows IBM's suggested design, incorporates software-settable configuration registers. These registers can identify the card that is present in a particular slot, identify and correct any addressing conflicts between expansion boards, and (if desired) shut down a particular expansion board. This latter feature is particularly desirable for critical items. In the event of a communications board failure, for example, the second communications

Chapter 8

board, which was not being used in the system, can be enabled by the user through this configuration facility and the defective board can become the inactive board. This board can be removed at a later, less critical time. The machine does not have to be opened to restore operation.

The *Restore Configuration* option is used in the event that the battery is replaced or, for some reason, removed. The previous configuration is maintained on the backup Reference Diskette as well as in the battery-backed configuration RAM in the machine.

Once configured, the date and time can be set on the system and the machine is ready for the hard disk to be formatted and the operating system installed. Further assurance of the functionality of the machine can be gained via the *Test the computer* option on the main menu of the Reference Diskette or through the use of advanced diagnostics. The advanced diagnostics are similar in structure and functionality to the advanced diagnostics for the AT, and are accessed by pressing Ctrl-A from the main menu. See Chapter 7 for additional information.

Security

On the AT, the keylock on the front panel can be used to prevent input from the keyboard as desired. On the Personal System/2 Model 80, the keylock is only used to hold the cover on the machine and has no interaction with the system itself. Password protection is available; however, this only operates from the DOS prompt, so the ability to lock the keyboard with an application running is lost.

Three levels of password security are available. These are set via the Reference Diskette. The first level of security is a power-on password, which requires a user to type a password up to seven characters in length before the operating system prompt can be obtained. The second password is called the network server mode. This locks the keyboard of the system, but allows other users on a network to access the hard disk on the machine. Typing the server password at the keyboard returns the machine to normal operation. The third level of password protection can be entered from the DOS prompt. The keyboard password program needs to be copied from the Reference Diskette to the hard disk with the DOS operating system. Executing the KP program then allows a keyboard password to be set for the session. Reinitializing the computer causes this password to be lost; however, the system cannot be re-initialized by entering Ctrl-Alt-Del when the keyboard password is enabled. The other passwords—power on and network server—can be deleted through the use of the Reference Diskette, if they can be remembered, or by removing the battery and reconfiguring the system.

Also on the Reference Diskette is the IBMCACHE program. This is a hidden file, for some mysterious reason, which can be used to cache the hard disk. There are several settings available, and the cache can be set up in DOS memory or extended memory. As the minimum configuration of the 16 MHz Model 80 has

1 MB of RAM, there is 384 KB of RAM that can be used for the cache without sacrificing precious DOS RAM. See Chapter 6 for additional information. Note that on the 20 MHz machine, 128 KB of this memory is taken up as the ROM is mapped to RAM during POST.

Documentation

The Model 80 is supplied with a Quick Reference guide that is comparable to the guide supplied with its siblings. A Technical Reference guide, which gives detailed information on the 16 MHz machine, is available at extra cost. This is a well-organized manual that is similar in structure to the Model 50/60 Technical Reference manual. Where applicable, additional information that pertains specifically to the Model 80 is available. For example, Matched Memory cycles and other 32-bit Micro Channel signals are discussed; these are not in the Model 50/60 version. This manual is also generally applicable to the 20 MHz machine. A supplement for the 8580-111 and 8580-311 machines (part number 68X2285) describes the difference between the various Model 80 versions. This document is not self sufficient and requires the Technical Reference guide for the Model 80, but together, they provide a complete reference.

Performance and Compatibility

The Model 80 offers predictably high performance. The system clock runs faster than the other Personal System/2 machines, the 80386 microprocessor performs better than its earlier siblings, and the 32-bit architecture also gives improved performance. The limitations of the machine are currently due to the surrounding I/O devices, such as the hard disk and diskette drives, but the increase in performance is still significant.

The 16 and 20 MHz versions of the Model 80 were tested with typical applications—a word processor, a database, a spreadsheet, and a graphics program. The application programs that were used were WordPerfect from WordPerfect Corporation, dBase III PLUS from Ashton-Tate, Lotus 1-2-3 from Lotus Development Corporation, PC-KEY-DRAW from OEDWARE, and Microsoft Macro Assembler from Microsoft Corporation. Details of the specific tests and results for all of the compared machines are given in Appendix A. For comparison purposes, the results discussed in this chapter are for the Model 80s and their closest competitors.

Communications packages were tested but not timed, because the operation speed of this type of application is not related to the performance of the machine for most situations. Asynchronous communications at 2400 bits per

Chapter 8

second—for example, a bulletin board—is not significantly different on a PC or an 80386 class machine.

The typical application results show that the hard disk was faster than the diskette. More interestingly, the performance of the diskette drive was improved in most situations by the use of 3.5-inch diskettes that were 1.44 MB (18 sectors per track) in capacity, rather than the lower density, 720 KB (9 sectors per track) diskettes. In some cases, this produced a 25 percent improvement.

Note, however, that with machines other than Models 50, 60, and 80 the interchange of data requires some planning. In order to use a media that is compatible with all of the machines, the highest density for Models 50, 60, and 80 cannot be used; but better performance is obtained by using the higher density. It is necessary to evaluate each specific situation to determine the best use of diskette densities. As most machines now have hard disks, the performance of the removable media diskettes is less critical, because they are only needed for backup and transfer of data.

Table 8-1 shows the non-disk-dependent results for the machines. The word processor and the spreadsheet could perform some of their operations solely in RAM, without accessing the disk. The results show the difference between the system board architecture of the various machines.

Table 8-1 Non-Disk-Dependent Results. *The Model 80, although fast, was not as fast as the Compaq Deskpro 386/20.*

Test Machine	Equipment	WordPerfect Move Top to Bottom	Move Bottom to Top	Search and Replace	Lotus 1-2-3 Recalc— Add	Recalc— Multiply	Recalc— Mix
AT 339	No math	28	21	32	5	8	45
	Math	28	21	32	5	5	7
Compaq 286	No math	28	21	34	5	7	45
	Math	28	21	34	5	5	7
Model 50	No math	22	16	27	4	6	36
	Math	22	16	27	4	4	5
Model 60	No math	22	16	27	4	6	36
	Math	22	16	27	4	4	5
AST 10 MHz	No math	16	12	20	3	4	28
	Math	16	12	20	3	3	4
16 MHz Model 80	No math	12	9	14	2	3	20
	Math	12	9	14	2	2	3
20 MHz Model 80	No math	11	8	13	2	3	16
	Math	11	8	13	2	2	2
Compaq 386/20	No math	9	6	10	2	2	13
	Math	9	6	10	2	1	2

All results are in seconds.

The word processor non-disk-dependent results involved manipulation of a large document. Once the file was loaded into the word processor, the time to move from the top to the bottom and from the bottom to the top was measured. This did not require access to the disk drives. The 20 MHz machine was slightly faster than the 16 MHz Model 80, and both were twice as fast as Models 50 and 60, which ran at 10 MHz. The AT 339 and Deskpro 286 were 2.5 times slower than the 20 MHz Model 80 for the same tests, and the AST Premium/286 was 1.47 times slower. The Compaq Deskpro 386/20 was 1.27 times faster than the 20 MHz Model 80, with its high speed SRAM memory cache.

The final word processing test was a large search and replace operation, where every occurrence of the letter "j" was replaced with the letter "k". Again, this was performed in RAM and the relative results were similar to those obtained from the other word processing tests. The Compaq Deskpro 386/20, which also has a microprocessor clock speed of 20 MHz, was faster than the 20 MHz Model 80. The AT 339, Compaq 286, Model 50, and Model 60 were all slower than the Model 80s.

Lotus 1-2-3 was used to perform recalculations on spreadsheets that fit into the available RAM. Recalculation times were assessed for three different spreadsheets: one containing only addition, one containing multiplication and division using the constant Pi, and one with a mixture of functions in a repetitive formula. These results were affected by the presence or absence of a math coprocessor. The Model 80s recalculated the supplied spreadsheets an average of 3.5 times faster with the 80387 installed.

When no math coprocessor was installed, the 16 MHz Model 80 was 2.32 times faster than the AT 339 and Deskpro 286, 1.84 times faster than Models 50 and 60, but only 1.4 times faster than the AST Premium/286. When the math coprocessor was installed, the 16 MHz Model 80 was 2.43 times faster than the AT 339 and Deskpro 286, 1.86 times faster than Models 50 and 60, but only 1.43 times faster than the AST Premium/286.

The other benchmarks involved disk access. The results were obtained from running each application from each of the media. A RAM disk was used in extended memory on each machine (80286-based or above) by using the IBM supplied VDISK. This provided another method of assessing performance independent of the magnetic media used. Because VDISK operates by putting the microprocessor in and out of protected mode, the results included some overhead due to this operation, but this was minimal compared with disk access time. Table 8-2 shows the RAM disk results.

The tests performed using the RAM disk included loading a large document file into WordPerfect and the addition spreadsheet into Lotus 1-2-3. The results show that the 16 MHz Model 80 was 2.36 times faster than the AT 339 and the Deskpro 286, and 1.86 times faster than Models 50 and 60 and the AST Premium/286. The 20 MHz Model 80 performed this test at a speed similar to the 16 MHz Model 80. Both were 0.71 times the speed of the Deskpro 386/20.

A typical application for a database in an office environment involves

Table 8-2 RAM Disk Results. *VDISK supplied with DOS was used to generate a RAM disk in extended memory.*

Test Machine	WordPerfect Load File	Lotus 1-2-3 Load File	dBASE III PLUS Sort on Title	dBASE III PLUS Sort on Disk ID	PC-KEY-DRAW Run Sample Macro	MASM Assemble File
AT 339	11	22	9	10	520	8
Compaq 286	9	22	10	11	473	8
Model 50	9	17	7	8	406	6
Model 60	9	17	8	8	406	6
AST 10 MHz	10	16	7	8	384	5
16 MHz Model 80	5	9	4	5	277	3
20 MHz Model 80	5	8	4	4	249	3
Compaq 386/20	4	6	3	3	189	2

All results are in seconds.

sorting a list for a mailing. This was simulated using dBase III PLUS. A database with over 1200 records was sorted in two different ways and the sort time was measured. The sorted database was written to the disk, so the timings obtained were dependent on the speed of the mass storage devices. When run from a RAM disk, the test only took 4 seconds on the Model 80, so the increase in time for the other mass storage devices is due to the performance of the different disks. The RAM disk results for sorting a large database file show that the Compaq 386/20 was slightly faster than the Model 80s, and all of the other machines were slower by a factor of between one and two.

The graphics program was used to assess the video performance of the machines. The sample macro supplied with the program was run from each of the mass storage devices on the machine. The time taken to run the macro from the RAM disk was within 2 percent of the time taken to run the macro from the hard disk without any cache installed. This shows that although the macro is mass storage device dependent, it is not dependent enough to obscure the relative video performances of the different machines. The 16 MHz Model 80 performed the test 1.87 times faster than the AT 339 and the Deskpro 286, 1.47 times faster than Models 50 and 60, 1.39 times faster than the AST Premium/286, but 0.9 times the speed of the 20 MHz Model 80 and 0.68 times the speed of the Deskpro 386/20.

The final test in RAM disk was the time taken to assemble a program using Microsoft's Macro Assembler. The Model 80s ran over twice as fast as the AT and the Deskpro 286, twice as fast as Models 50 and 60, but slightly slower than the Deskpro 386/20.

The same disk-dependent tests were run from the hard disk on each machine. The performance of the Model 80 was improved by the use of cache on the hard disk test. During a sort operation, the database is only rearranged not changed, so it is feasible that data may still be present in the cache when the data is used more than once in the sort process. The cache improved the performance in this circumstance by an impressive 20 percent. Table 8-3 gives the hard disk test results.

Table 8-3 Hard Disk Results. *The AT, Premium/286, and Deskpro 286 were not supplied with a disk cache program as standard, so the tests do not show these results.*

Test Machine	Equipment	WordPerfect Load File	Lotus 1-2-3 Load File	dBASE III PLUS Sort on Title	Sort on Disk ID	PC-KEY-DRAW Run Sample Macro	MASM Assemble File
AT 339	No cache	12	29	25	26	526	11
Compaq 286	No cache	11	28	21	22	477	9
Model 50	No cache	10	27	29	29	409	7
	Cache	10	20	20	21	405	7
Model 60	No cache	10	26	24	21	406	7
	Cache	10	19	22	21	406	7
AST 10 MHz	No cache	8	17	25	24	388	9
16 MHz Model 80	No cache	7	13	26	26	283	4
	Cache	7	11	22	21	277	4
20 MHz Model 80	No cache	7	13	19	18	249	3
	Cache	7	10	12	12	249	4
Compaq 386/20	No cache	5	7	12	12	189	2
	Cache	4	6	10	11	185	3

All results are in seconds.

The time taken to load a file into an application with no cache present showed the Model 80s to be twice as fast as the AT and Compaq, 1.8 times faster than Models 50 and 60, 1.25 times faster than the AST Premium/286, and 0.6 times as fast as the Compaq Deskpro 386/20. When cache was installed, the Model 80s were 1.67 times faster than Models 50 and 60, and 0.56 times as fast as the Deskpro 386/20.

The database sorting exercise shows the difference between the performance of the Model 50 and 60 hard disks. With cache installed, the 16 MHz Model 80 performed at a speed comparable to Models 50 and 60, and was slower than the 20 MHz Model 80. The 20 MHz Model 80 was comparable in speed to the Deskpro 386/20. Without the cache installed, the relative performance of the Model 50's hard disk dropped, and the AT and Deskpro 286 were comparable to the 16 MHz Model 80.

The graphics intensive results again show that the Deskpro 386/20 was faster than the Model 80s, and the other machines were slower. The effect of disk cache was minimal, showing the low disk dependency of this test. The 16 MHz Model 80 was 1.9 times faster than the AT and Deskpro 286, and 1.47 times faster than Models 50 and 60. The 20 MHz Model 80 was 1.1 times faster than the 16 MHz version, but was 1.35 times slower than the Deskpro 386/20.

The final test was with Microsoft's Macro Assembler. The VDISK program supplied with DOS 3.2 was assembled. The timing for this test included the time to load the macro assembler and to perform the assembly. A listing file and cross reference file were not generated. The time mark program supplied with the Norton Utilities from Peter Norton Computing was used in a batch file to time the operation. The compile only took 3 seconds when run from a RAM disk and 20 seconds from a low density 3.5-inch diskette, showing the high component of I/O transfers that are necessary for this test.

The time to assemble VDISK showed the Model 80s to be 2.75 times faster than the AT and Deskpro 286, 2.25 times faster than the Premium/286, 1.75 times faster than Models 50 and 60, and marginally slower than the Deskpro 386/20.

The same disk-dependent tests were run from the diskette drive, using a high density diskette. Note that the Personal System/2 machines have 3.5-inch diskette drives and the other machines have 5.25-inch diskette drives. This affected the relative results that are seen in Table 8-4. The Model 80 was almost as slow in loading files as Models 50 and 60, and was 8 percent slower than the AT loading the same file.

Table 8-4 High Density Diskette Results. *The Personal System/2 machines all have 3.5-inch diskette drives, whereas the other machines have 5.25-inch diskette drives.*

Test Machine	WordPerfect Load File	Lotus 1-2-3 Load File	dBASE III PLUS Sort on Title	dBASE III PLUS Sort on Disk ID	PC-KEY-DRAW Run Sample Macro	MASM Assemble File
AT 339	29	112	204	223	583	19
Compaq 286	33	112	171	169	543	21
Model 50	29	128	204	207	469	20
Model 60	29	128	204	204	468	20
AST 10 MHz	20	108	199	223	465	16
16 MHz Model 80	28	125	196	194	339	18
20 MHz Model 80	28	125	194	193	309	16
Compaq 386/20	44	107	171	172	334	20

All results are in seconds.

The poor performance of the 3.5-inch diskette drives is also seen in the database results. The 16 MHz Model 80 was only able to sort the database 1.1

times faster than the AT, which has a microprocessor running at half its speed. The lack of difference in the results between the various machines shows how dependent these computers are on the performance of their I/O devices, such as the disk drives. These results are reflected throughout the high density diskette test results. The machine is only as fast as its disk drive for disk-intensive purposes.

Table 8-5 shows the results of the same tests run with a low density diskette. Again, the performance of the different machines was much closer than their architecture alone would suggest. The degradation in performance between the use of a low density 3.5-inch diskette and the higher density 3.5-inch diskette, although tangible, was only a couple of percent. The effect was much more pronounced in the loading of the word processing file. It is interesting to note that in most situations, the absolute time taken to perform these tests was slower with the low density diskettes than with the high density diskettes.

Table 8-5 Low Density Diskette Results. *Although using the highest density diskette supported gives the best performance, this density may not be suitable for exchanging data between different machine models, due to the variety of diskette densities supported on the machines.*

Test Machine	WordPerfect Load File	Lotus 1-2-3 Load File	dBASE III PLUS Sort on Title	dBASE III PLUS Sort on Disk ID	PC-KEY-DRAW Run Sample Macro	MASM Assemble File
AT 339	43	115	193	184	572	30
Compaq 286	43	115	181	182	546	27
Model 50	45	134	209	211	480	22
Model 60	45	134	209	211	465	23
AST 10 MHz	40	112	176	180	468	26
16 MHz Model 80	39	125	209	202	337	20
20 MHz Model 80	45	125	195	197	317	25
Compaq 386/20	45	111	184	185	337	28

All results are in seconds.

The timings for the machines with less powerful and slower CPUs show that the effect of the disk drives was less critical for these machines. As the processing capability of the machine has improved, the performance of the I/O devices, such as the mass storage devices, has not increased in quite the same proportion. Although this is clear for the removable mass storage devices, the hard disk, which can store much more data than earlier hard disks could, has maintained a very respectable performance ratio with the processing capabilities of the machines.

Summary

The Model 80 machines are very fast and incredibly powerful. The personal computer industry has changed dramatically since the introduction of the first PC. The power that is available in the Model 80 has only been seen before in mini and mainframe computers. But this machine is not stuck in a remote, temperature-controlled room only accessible to the privileged few MIS people. This computer can sit at the desk and is accessible to anyone with enough money to buy one. Through LANs or micro-to-mainframe links, no functionality is lost; machines can be linked and data exchanged. The larger machines can be used far more productively for larger database manipulation and they are no longer burdened with the trivial tasks of supporting word processors for office memos. The Compaq Deskpro 386/20 outperformed the IBM Model 80, and for situations that do not require the Micro Channel architecture or the IBM label, it is the machine of choice. But to many, IBM is the preferred company, and with Micro Channel adapters becoming more available, the Model 80 is still worth considering.

The personal computer revolution is here to stay. With machines as versatile as the Model 80s, and with an architecture like Micro Channel that can be expanded for future needs, office automation will become even more prevalent. Compatibility with the past and growth potential for the future are incorporated in the hardware. The software still needs to grow, to make use of the features available in the machines, but this is beginning to occur. The 80386 microprocessor is a more desirable platform for software development than the 80286; in the next few years there will be plenty of opportunity for the growth of applications.

9

Personal System/2 Options

The Personal System/2 family is a highly integrated system that does not require many additional items to produce a fully functional unit. The system comes standard with a keyboard; the only additional equipment needed is a monitor. IBM offers four analog monitors for use with the system, each of which meets different requirements.

There are, however, a variety of add-on options that increase the functionality of the system. A display adapter that can be used in the Model 30 and the PC family of machines provides the VGA capabilities that are standard on the Micro Channel equipped models in the line. The graphics capabilities of Models 50, 60, and 80 can be enhanced by an 8514/A Display Adapter that, with the correct hardware configuration and software drivers, increases the maximum resolution of the displayed image to 768 pixels vertically by 1024 horizontally.

Each machine includes a socket for the installation of an appropriate math coprocessor, which improves the mathematical calculation performance of the machines for programs designed to use these capabilities. Apart from Models 30 and 25, the Personal System/2 machines are capable of addressing more than the 1 MB of memory that DOS uses. OS/2 and other protected mode operating systems require significant amounts of memory over 1 MB. OS/2 requires at least 1.5 MB of memory to run and 2 MB of memory when a DOS compatibility environment is required. Even more memory is required to run multiple tasks efficiently, so the memory expansion capability of the machines is significant. Expanded memory specification boards, needed to handle the memory requirements of multi-tasking in the 3270 Workstation Program, are available for the Model 30 and PCs as well as the Micro Channel machines.

There are several expansion boards for the Micro Channel that expand the communications capabilities of the machines. The IBM PC Network and Token-Ring network are supported. There is a board for the baseband network standard, the broadband network standard, and the Token Ring. In addition, communications boards that allow 5250 and 3270 terminal emulation are available.

In order to accommodate the media incompatibility problems of the Personal System/2 family, an external 5.25-inch diskette drive is offered for use with the Personal System/2 computers.

The first third-party hardware for the Micro Channel was quick to appear on the market. The RamQuest 50/60 board from Orchid Technology was available in June of 1987, less than three months after the Personal System/2 announcement. Since that time, more add-in boards are being offered by third-party vendors. There has not, however, been a flood of boards, due mainly to the highly integrated nature of the machines. The ubiquitous multi-function board that can be found for the PC is not the typical expansion requirement for Personal System/2 machines. The display capabilities of the Personal System/2 family have reduced the need for display board options for a typical business user. Lack of detailed information early on about OS/2 and the hesitancy of vendors to predict its popularity have stifled the growth of the add-in board market that IBM is welcoming for the Personal System/2 family.

The whole issue of Personal System/2 clones is uncertain at this time. One of the major market opportunities for the PC was in the compatible machine market. IBM has made it clear that many aspects of the new family of machines are patented, including the Micro Channel. Although from a technological viewpoint it is possible to copy a Personal System/2, there are several negating factors. One is a possibility of features on the Personal System/2 machines that are not yet "activated," which could be used at a later date to make any compatible machine that may be on the market incompatible. Then there is the question of the legality of producing a clone. The current press analysis seems to be that the viability of Personal System/2 clone manufacturers will be determined in a courtroom and not in the marketplace.

Monitors

IBM offers four analog monitors for the Personal System/2 machines: the 8503, 8512, 8513, and 8514. Each will run on any of the Personal System/2 models except the Model 25, which has a monitor integrated into the system unit. The monitors range in price and performance from $250 for the 12-inch Monochrome Display to $1,550 for the 16-inch Color Display. (See Figure 9-1.)

The Model 30 has the MCGA as its standard video system. This can be expanded to offer VGA capabilities through the addition of the Personal System/2 Display Adapter or a similar expansion board. This board can be used in the PC family of machines. Models 50, 60, and 80 provide VGA as the standard display system. The CGA and EGA boards that are typical on a PC-class machine present a

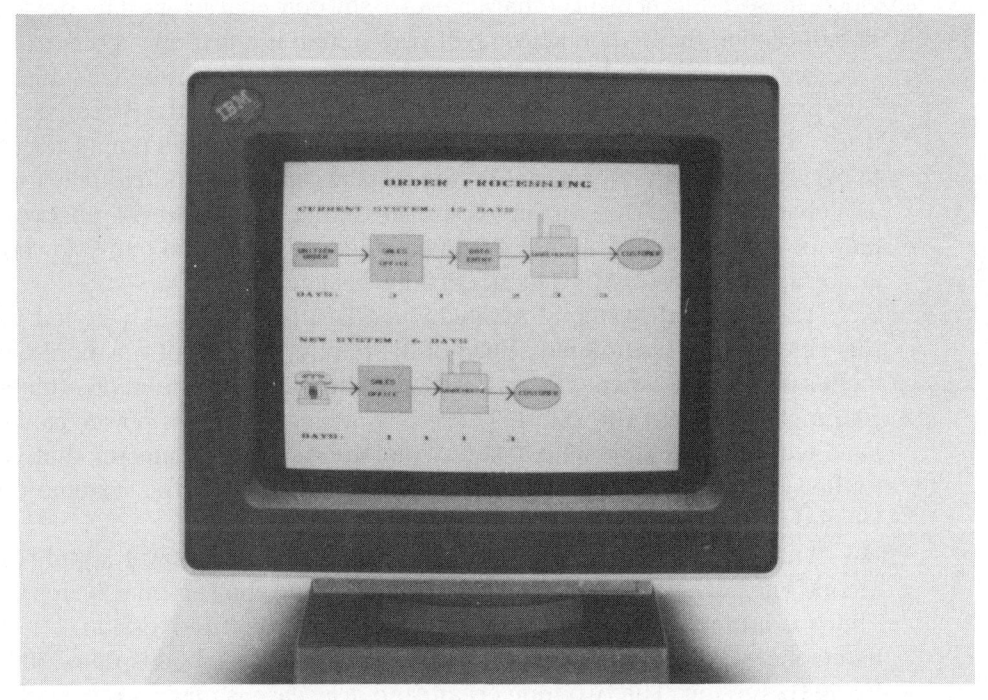

Figure 9-1 Personal System/2 Monitor. *Each of the analog monitors offered by IBM has similar styling.* Photo courtesy of International Business Machines Corporation.

digital output to the monitor. This digital signal is converted to an analog drive signal within the monitor itself in order to drive the CRT. The Professional Graphics Controller is the only IBM display card available for the PC family that provides an analog output. It is important to note that the output from the PGC and the digital output from the CGA and EGA cannot drive the Personal System/2 monitors nor can the Personal System/2 monitors work with the PGC. The output from the Monochrome Display and Printer Adapter (MDA) is also incompatible with the new monitors.

8503 Monochrome Display

The 8503 is a 12-inch diagonal monochrome monitor that displays white on black or black on white. It can also produce 64 shades of gray for graphics applications. The monitor has a tilt-and-swivel stand as standard and, at $250, is the lowest price monitor in the family. This monitor is best suited to applications that are mostly text, where the graphics features of the display are only used occasionally.

The Monochrome Display/Printer Adapter that is used with the PC family cannot produce graphics. It is only capable of producing text characters on the screen. Although characters can be underlined, highlighted, or even displayed in reverse video, these different display characters are still only characters. This has led to the misconception that a monochrome display system is a text-only system.

Hercules offers a board that is often referred to as a monographics board. This produces an image that is closer to the image seen on the Personal System/2 displays. It not only displays characters but is capable of all points addressable (APA) graphics, where each dot on the display can be set individually instead of as a character cell. This concept is similar to that of the 8503 Monochrome Display for the Personal System/2. EGA boards can also be used to display both text and graphics on an IBM PC Monochrome Display.

The Personal System/2 Monochrome Display is a sleekly styled monitor that reflects the streamlined design of the family. The size of the monitor is 12.6 inches wide by 12.2 inches deep; it is 12.5 inches high with its stand attached and 10.9 inches without the stand. There is a detachable 3-pin power cable and a hard-wired signal cable with a 15-pin miniature D-type connector that attaches to the system unit. The signal cable is 6 feet long to allow the system unit to be located on the floor under a desk.

Three versions of the 8503 are used, with different power supply requirements. The power supply is not self switching as on the Personal System/2 units. There is a northern hemisphere 100 to 125 V_{ac} 50/60 Hz version, a northern hemisphere 200 to 240 V_{ac} 50 Hz version, and a southern hemisphere 200 to 240 V_{ac} 50 Hz version. The two-tone cream and gray case is made from rigid plastic, with the darker gray around the CRT face, which is slightly recessed from the front panel. This, combined with the etched dark faceplate, provides reasonably good protection from glare.

The standard tilt-and-swivel stand can be removed without tools if desired, but the monitor tends to be low even with its stand attached. Unless it is to be placed on a shelf above desk level, it is more likely to be used with the stand installed. The stand allows the monitor to be tipped from a low angle of 5 degrees below the horizontal to a high position of 20 degrees above the horizontal, and rotated 180 degrees—90 in each direction.

There are only three controls on the unit. The red on/off switch is located on the top right-hand side of the monitor toward the front, and is a reasonably large rocker switch. The on and off marking is below the switch and is the standard open circle for off and vertical straight line for on. The power-on indicator is a green light located in the lower right-hand corner of the front of the display. The other two controls are the brightness and contrast controls, which are at the top of the left-hand side of the monitor, again toward the front. Both controls are blue, easy-to-use ridged wheels. The contrast control is marked with a circular symbol that is half filled in. The brightness control is marked with a symbol that resembles a light bulb. The default brightness setting is at the central notched position.

The nominal size of the CRT is 12 inches; the active area where the image is displayed (not including the border) is 8.15 inches horizontally by 6.10 inches vertically. The monitor has a horizontal dot addressability of 320, 640, and 720 dots and a vertical dot addressability of 350, 400, and 480 lines, which is the limit for the VGA. The 200-line modes of the VGA are double scanned to give 400 lines on the screen. The horizontal deflection rate is 31.5 KHz and the bandwidth is 30 MHz. The vertical deflection rate varies with the display mode being used: for the 350- and 400-line modes the frame rate is 70 Hz; for the 480-line mode it is 60 Hz.

The Monochrome Display, with its paper white medium persistence phosphor coating on the CRT, provides clear, crisp image quality. Up to 64 shades of gray can be displayed on the monitor in the graphics mode, with the gray scale conversion performed by the system BIOS. The monitor type is detected by the system unit and the presence of the 8503 causes all graphics images to be automatically converted to a gray scale. The 64 shades of gray provide an excellent shading feature for graphics images. The human eye cannot distinguish between each of the shades and so smooth images are possible.

The 8503 provides a highly acceptable, low cost monitor for the Personal System/2 family. For applications that are mainly text oriented, the white on black or black on white image is quite good. In fact, it may be more desirable than the least expensive color monitor that is available.

8512 Color Display

The 8512 is a 14-inch diagonal color monitor with a stripe pitch of 0.41 mm. This produces pixels that are made up of three vertical stripes instead of the triad of three dots used on traditional color monitors. It produces color images at all of the resolutions available on the VGA, resulting in simultaneous display of up to 256 colors in the appropriate mode. This monitor is best suited for applications that are mostly text, where the graphics feature is used only occasionally.

The Personal System/2 8512 Color Display has a width of 13.97 inches, a depth of 15.51 inches, and a height without its stand of 11.97 inches. The optional tilt-and-swivel stand can be snapped onto the base of the monitor, increasing the height of the unit to 14.57 inches. The stand is slightly different from that of the 8503, but the tilt-and-swivel capabilities are the same. The stand allows the monitor to be tipped to 5 degrees below and 20 degrees above the horizontal plane and rotated 90 degrees in either direction. The price of the monitor is $595 and the stand is $35. Again, the ergonomics of the monitor are such that the stand is really a necessity for comfortable use of the monitor.

The monitor has the same cream and gray rigid plastic case as the 8503. The area around the slightly recessed screen is gray and the remainder of the case is cream. There are three controls on the monitor: the on/off power switch, the contrast control, and the brightness control. The red on/off switch is located on the right side of the monitor, at the top and toward the front. This rocker switch is labelled with an open circle to represent the off position and a vertical line for the on position. The power-on light indicator is located in the bottom right-hand corner of the display. The contrast control is a blue wheel located on the left side of the unit, at the top and toward the front. The brightness control is directly behind the contrast control; it has a similar type of adjustment, except that there is a positive feel to the center position of the switch. The brightness switch is marked with a light bulb outline, the contrast switch with a half-filled circle.

As with the 8503, there are three versions of the monitor for different power requirements: a northern hemisphere 100 to 125 V_{ac} 50/60 Hz model, a northern hemisphere 200 to 240 V_{ac} 50 Hz model, and a southern hemisphere 200 to 240 V_{ac} 50 Hz model. The power cable is separate from the unit and is 6 feet long, whereas the 6-foot signal connector is hard wired into the monitor. The connector is the same 15-pin miniature D-type connector, but the allocations for pins 11 and 12 are reversed. On the 8512, pin 11 is monitor sense ground and pin 12 is monitor sense open. This allows the system unit to differentiate between the monochrome and color monitors.

The 8512 has a 14-inch CRT with an effective viewing area of 9.45 inches horizontally by 7.09 inches vertically, including the border, in all modes of operation. The horizontal dot addressability is 320, 640, and 720 dots; the vertical dot addressability is 350, 400, and 480 lines. The 200-line modes of the VGA are double scanned on the monitor, giving 400 lines of dots that make up the image. The bandwidth of the monitor is 28 MHz and the horizontal line rate is 31.47 KHz. The frame rate is 70 Hz for the 350 and 400-line modes and 60 Hz for the 480-line mode.

Each of the pixels displayed on the screen has the same vertical and horizontal separation when the monitor is used in 640-pixel by 480-pixel graphics mode. This means that if a square image is required on the screen, the program specifies the same number of pixels in the horizontal direction as in the vertical direction.

The stripe format of this monitor produces an image that is less crisp than its 8513 counterpart. It uses the same P22 phosphor, which provides scrolling without smearing the image, but the image itself is not as distinct. For this reason the graphics images that are produced are not as clear as on the 8513, but for text-only purposes the 8512 may be the monitor of choice. Personally the authors preferred the 8513 for both text and graphics, but they also prefer the NEC Multisync monitors in preference to the Sony MultiScan for similar reasons.

The 8512 seems to be designed to provide the lowest cost possible in a color monitor, but with the 8503 and the 8513 available, it is questionable how useful a role it really plays in the Personal System/2 family.

8513 Color Display

The 8513 is a 12-inch diagonal color monitor with a dot pitch of 0.28 mm. This produces a crisper image than that of the 8512, but the dots and therefore the displayed characters are smaller. The tilt-and-swivel stand is standard on the monitor, which costs $685. This monitor produces crisp graphics output and is desirable if graphics are a significant requirement. Up to 256 colors can be displayed on this monitor when the system unit display subsystem is in a 256-color mode.

The 8513 case is cream and gray, as on the other models in the display family. The gray portion of the case is around the recessed screen and helps to reduce glare. The three controls are the on/off power switch and the contrast and brightness controls. The power switch is at the top front of the right side of the unit; the off position is marked with a circle and the power-on position with a vertical line. The brightness and contrast controls are at the top of the left side of the monitor, with the contrast control in front of the brightness control. Both controls are blue with a ridged wheel type of adjustment. The central position of the brightness control can be detected by the increased tactile feel of the control at that point. The brightness control is identified by a light bulb beneath the dial; the contrast control has a half-filled circle as its identification.

There are three models of the 8513, each with different power requirements. The specifications are the same as for the 8503 and 8512. The detachable power cable is 6 feet long, as is the signal cable. The connector on the signal cable is a 15-pin miniature D-type connector that has the same pin configuration as the 8512. The monitor is 12.6 inches wide by 14.5 inches deep by 12.5 inches high with its standard stand attached. The stand can be removed without tools; when removed, the monitor is 10.9 inches high. The weight of the monitor is 23 pounds, compared to 33 pounds for the 8512 (with stand).

The CRT of the 8513 is a 12-inch diagonal display with an active viewing area that varies slightly with the mode of operation. The horizontal dot addressability is 320, 640, and 720 dots. The vertical dot addressability is 350, 400, and 480 lines. The 200-line modes of the VGA are double scanned on the monitor giving 400 lines of dots that make up the image. When the monitor is in the 350- and 480-line modes, the viewing area is 8.15 inches horizontally and 6.10 inches vertically. In the 400-line mode, this becomes 8.15 inches horizontally and 5.91 inches vertically. The bandwidth of the monitor is 30 MHz and the horizontal deflection rate is 31.47 KHz. The frame rate for the 350- and 400-line modes is 70 Hz, and 50 Hz when the monitor is operating in 480-line mode.

The CRT has a type P22 phosphor coating, as does the 8512, to optimize the brightness of the image and prevent smearing. The 640-pixel by 480-pixel, 60 Hz graphics mode has a unity aspect ratio; a square image is produced by programming the same number of pixels in the horizontal direction as the vertical direction.

This monitor has excellent clarity for both text and graphics. Some people complain that the text character size is too small, but we feel that small and clear is preferable to fuzzy. This monitor also comes with its stand as standard. Since we have started to use a monitor with a swivel stand, it now seems to be a prerequisite. IBM has stated that the height of its monitors was decided after ergonomic tests in Europe, but they are simply too low without their stands. Overall, this is a very nice monitor that is suitable for all applications except those that demand the higher resolution, larger size, CAD/CAM workstation-type functionality that is provided by the 8514.

8514 Color Display

The 8514 is a 16-inch diagonal color monitor. It can work with the VGA in all of its modes and it has the additional capability of a maximum addressability of 1024 by 768. This can be obtained by using the 8514/A Micro Channel Display Adapter to increase the performance of the VGA. The maximum number of colors that can be displayed at once is 256; this is available at a maximum resolution of 1024 by 768 when the monitor is used in conjunction with the 8514/A and its associated memory expansion kit. This monitor also has a standard tilt-and-swivel base and a price of $1,550. It is suitable for high resolution graphics applications such as CAD.

The case of the 8514 is cream and gray, as on the other monitors; the area around the screen is the darker color. The screen is slightly recessed into the monitor to reduce the glare from ambient light. The controls include a power on/off switch, which is located at the top of the right side of the unit, and the contrast and brightness controls at the top of the left side of the unit. The blue contrast control is in front of the brightness control and both have detent at their central position.

The monitor is 15.75 inches wide, 16.34 inches deep, and 12.60 inches high without its stand. The tilt-and-swivel stand is standard with the unit and, when attached, increases the height of the monitor to 14.17 inches. This is not a petite display, weighing in at 44 pounds. The tilt-and-swivel stand offers the same rotation and tilt as the other models—90 degrees rotation in either horizontal direction and a tilt of 5 degrees below and 20 degrees above the horizontal.

The CRT of the 8514 is nominally 16 inches, with an active display area of 11.1 inches by 8.3 inches. This area is used to display the image regardless of the mode of the display adapter. This results in a large character size in low resolution character modes. However, although the full active area of the display is used in all modes, the aspect ratios of the pixels remain the same as on the smaller monitors. The monitor has a horizontal dot addressability of 320, 640, 720, and 1024 dots and a vertical dot addressability of 350, 400, 480, and 768 lines. The 1024-pixel by 768-pixel mode is only available when the monitor is used in association with the 8514/A Display Adapter. The horizontal deflection rate is 31.47 KHz for all modes except the 1024 by 768, where it is 35.52 KHz; the maximum video clock frequency is 44.91 MHz. The frame rate for the 350- and 400-line modes is 70 Hz. For the 480-line mode, the frame rate drops to 60 Hz; for the 1024 by 768 mode, the frame rate is 43.5 Hz.

The larger size of this monitor makes it ergonomically pleasant to use. Its stand is necessary, but even without it the display does not give the impression of being too low, as do the smaller models. The CRT is etched to reduce the glare from ambient light and medium persistence phosphors are used to reduce the smearing effect. The video connector is a 15-pin miniature D-type, as on the other monitors, but the pin-out configuration is slightly different to allow the system unit to identify the 8514.

In general, this monitor is acceptable for its specifications. It provides a clear color image for CAD/CAM workstation use and is acceptable for all of the video modes of the VGA. This is the top-of-the-line unit for the Personal System/2 family, but it is not the high resolution monitor that many people have bought for dedicated workstation use. It can be argued quite correctly that if you are going to use a high resolution screen for CAD/CAM applications, it should be able to display at least 1024 pixels by 1024 pixels. This resolution is not currently offered by IBM for the Personal System/2 family of machines. The VGA is capable of producing a 1024 by 1024 bit pattern and the part of the bit pattern that is above the 768 line can be used to display overlaid windows as desired. Until an alternative option is available, the 8514 is an expensive but acceptable solution to high resolution requirements.

Model 30 Display Adapter Requirements

The video system in the Model 30 is a subset of the full featured VGA in the Model 50 and above. It can only emulate the CGA and offer two additional graphics modes: 320 pixels by 200 pixels with 256 colors and 640 pixels by 480 pixels with 2 colors. It does not have a bit-plane architecture as does the EGA and so cannot provide EGA emulation. This, for many environments, is a serious deficiency. EGA resolution has become a strong standard in PC-based systems, not only for its improved graphics resolution but for its text-based use as well. A user of a word processor or other text-based system quickly gets used to the well-formed characters and choice of colors of the EGA. There will be great reluctance to return to CGA standards.

To accommodate this type of user, IBM is offering an expansion board that can be installed in an expansion slot in the Model 30. This provides full VGA emulation and consequently produces text modes that are EGA emulations, as well as modes that improve on this standard. When coupled with a Personal System/2 monitor and appropriate software, this expansion board also works in the PC family of machines. This gives the PC family 256 color modes, which were not previously available.

The expansion board is called the IBM Personal System/2 Display Adapter (see Figure 9-2) and should not be confused with the IBM Personal System/2 Display Adapter 8514/A, which is a Micro Channel board that enhances the VGA. The cost of the Personal System/2 Display Adapter is $595, and it must be used with one of the analog monitors in the Personal System/2 series. The monitor is plugged into the display connector on the rear connector of the expansion board.

Chapter 9

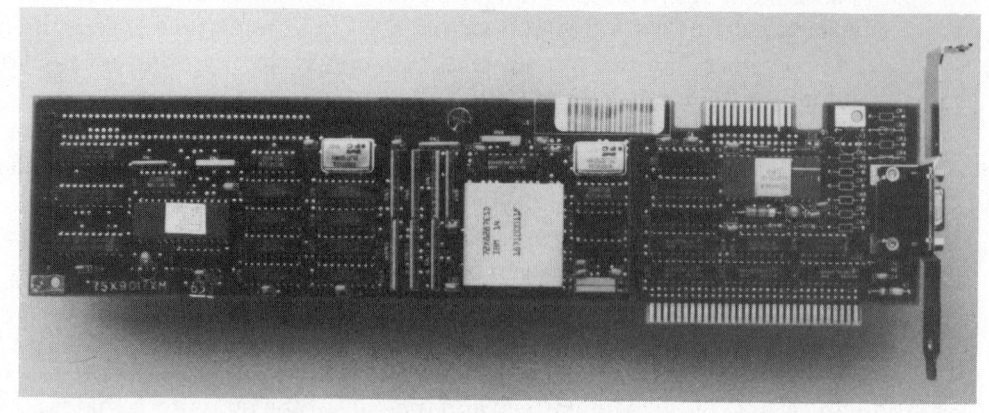

Figure 9-2 Personal System/2 Display Adapter. *This adapter can be used to bring VGA compatibility to the Model 30 or the PC family of machines.* Photo by Bill Schilling.

The Personal System/2 Display Adapter is able to function in all VGA modes, but there are a few small differences in the architecture of the board, because it is used in a machine with PC bus and system architecture rather than Personal System/2 Micro Channel and system architecture. The adapter board itself contains the VGA BIOS. On Models 50, 60, and 80, this BIOS is part of the main system BIOS. On the Micro Channel machines, one Micro Channel connector has a video extension. This extension provides the signals used by the VGA when an additional video card is used in the machine (e.g., an 8514/A coexisting with the VGA). On the Personal System/2 Display Adapter, this video extension cannot be part of the connector on the card that plugs into the PC bus, because the video signals required are not available on the PC bus. These signals are provided instead through a connector located on the top edge of the Personal System/2 Display Adapter board. This allows a PC-style expansion board that uses these video extension signals to be designed for the PC bus. This board would plug into the PC bus in the same way as the Personal System/2 Display Adapter, and a ribbon cable would be used to link the two boards.

The Personal System/2 Display Adapter has different design constraints than a board that is designed for the Micro Channel. One of the major reasons for choosing a new bus architecture for the Personal System/2 machines is to enable third-party vendors and IBM to design add-in boards that do not have to deal with the problems found on the PC. The most common of these problems are due to interrupt conflicts. To ease this problem, the Personal System/2 Display Adapter does not support the vertical sync interrupt on interrupt level 2. This allows boards such as the 3270 Communications Adapter to coexist with the VGA board. Applications that would normally use the vertical sync interrupt to read the current state of the display can use a register that is on the Personal System/2 Display Adapter.

There are three differences between a VGA on a Personal System/2 Model 50, 60, and 80, the MCGA on the Model 30, and the Display Adapter card in a PC or Model 30. The first is transparent to an application and is simply that the video enable/disable register is provided through an intermediate register. This is fully supported by the BIOS and only affects applications that unwisely do not use the BIOS.

It cannot be emphasized enough that avoiding the use of BIOS is simply not acceptable for machines that are to be used in a multi-tasking environment. Devices must be in a predictable state when each application receives control. This is a factor for DOS applications, as well as applications designed for native multi-tasking systems such as OS/2. DOS applications are also being multi-tasked, using multi-tasking operating environments and environments such as the OS/2 DOS compatibility environment.

The second discrepancy between the VGA, the MCGA, and the Personal System/2 Display Adapter relates to the BIOS equipment flag support. On the VGA and MCGA this flag is set automatically and indicates the current mode setting of color or monochrome. This is not the situation on the PC, and the display adapter has been designed to provide compatibility with other PC-based applications. This means that applications need to set this flag when changing modes.

Finally, the Personal System/2 Display Adapter card displays monochrome mode reverse video intensified characters as white on white instead of the expected black on white of the VGA and MCGA.

In addition to the small differences between the VGA and the Personal System/2 Display Adapter, there is one extra difference between it and the MCGA on the Model 30. When a monochrome monitor is attached to the display adapter, the default video mode is mode 7.

It is possible to use this board in a dual-display configuration when it is installed in a member of the PC family. It can coexist with either a CGA or a Monochrome Display and Printer Adapter, but there are certain limitations. The Personal System/2 Display Adapter is always the primary display, but the Monochrome Adapter always displays the monochrome modes and the Personal System/2 adapter the color modes, irrespective of the attached monitor. An installed CGA always displays the color modes, and in this configuration the Personal System/2 adapter supports the monochrome modes, displaying them on whichever display is attached to the board.

Not all of the video BIOS calls are fully supported in dual-screen configurations.

There appear to be many advantages to using the Personal System/2 Display Adapter as an upgrade route for existing PCs. If an IBM solution is necessary, then this board is an acceptable alternative to an IBM EGA board. However, most text-based applications currently support the EGA standard, clone EGA boards are very cheap, and third-party vendors are moving to add VGA compatibility to their boards' repertoires. Display drivers for applications that are im-

Chapter 9

proved by a higher resolution image are often bundled with the board, so third-party vendors are still a viable alternative. Just as with the EGA, however, IBM is the keeper of the standard. The clone manufacturers need to prove that they can provide full compatibility, and this requires applications that use all the additional features of the VGA.

Display Adapter 8514/A

The capabilities of the VGA on Personal System/2 machines with the Micro Channel architecture can be expanded by the IBM Personal System/2 Display Adapter 8514/A. (See Figure 9-3.) There are two elements to this board: the board itself and a memory expansion kit. The board works with any of the Personal System/2 displays, but it is designed to be part of a high resolution workstation where the normal display would be the 8514. This monitor is the only one that is capable of supporting the 1024-pixel by 768-pixel mode. Using the other monitors—8503, 8512, and 8513—with this expansion board increases their video capabilities to include improved text and graphics functions.

Figure 9-3 8514/A. *This Micro Channel board can display up to 1024 by 768 pixels in some display modes.* Photo by Bill Schilling.

The 8514/A is installed in the expansion slot on the Micro Channel bus that includes the video extension. On the Model 50, this is the second slot from the power supply end, the first slot being occupied by the hard disk controller. On Models 60 and 80, the board is installed in the third slot down from the top, slot number 6, with the top slot containing the hard disk controller. The monitor is then plugged into the rear connector of the adapter board, instead of into the VGA video connector on the rear panel of the Personal System/2 machines.

The memory expansion kit for the 8514/A is optional. However, it is similar to the memory expansion kit for the EGA in that, although all functions are available without it, it is really a necessity to gain full advantage from the board. The memory expansion kit increases the number of colors that can be displayed simultaneously from 16 to 256. It costs $270.

The features that are offered with the 8514/A adapter board include VGA compatibility, bit-block transfer (BIT-BLT), additional text fonts, and an increase in the numbers and sizes of the fonts. The graphics features of the display system are also increased. There is a 1024-pixel by 768-pixel video mode as well as a 640-pixel by 480-pixel mode. Hardware assistance for line drawing and area-fill and a rectangular scissor function are provided.

VGA Capability

The VGA capabilities are supplied by the VGA on the system board. When the video system is in VGA mode, the output from the VGA is routed through the 8514/A adapter board to the rear connector on the adapter. The bit-planes for the VGA are independent of the bit-planes for the 8514/A. This results in a smooth and rapid mode switching ability between the VGA and 8514/A. The functions that are available on the 8514/A are called advanced function mode operations.

Advanced Function Modes

When in advanced function mode, the addressability of the adapter is either 640 pixels by 480 pixels or 1024 pixels by 768 pixels. The 8514 Color Display is necessary for the 1024-pixel by 768-pixel mode. There are two types of character modes available with the advanced functions: alphanumeric and advanced text. The alphanumeric mode displays fixed character cell sizes at one time, the advanced text mode allows varying sizes on the screen at once. When additional memory is installed, the number of colors that is supported increases from 16 to 256 out of a palette of 256K. The video memory arrangement for this board is in bit-planes that are 1024 by 1024. The lower 1024 by 768, or 640 by 480, is displayable on the screen. The remaining area can be programmed directly, although

IBM does not support this feature. In general, this area is used by the adapter as a scratch data area, as a cache for programmable character sets, or as storage for area fill information. With the additional memory, there are eight bit-planes for each of the display modes (1024 by 768 or 640 by 480); without this memory, there are four bit-planes for the 1024 by 768 mode and two separate sets of four bit-planes for the 640 by 768 mode.

Alphanumeric Mode

The alphanumeric mode offers several different cell sizes. In the 640-pixel by 480-pixel mode, the character cell is 8 dots wide by 14 dots high, resulting in a display 80 characters wide and 34 rows high. This can be compared to the EGA cell size of 8 dots by 14 dots with a display of 80 characters in width but only 25 lines in height. The higher resolution mode of 1024 by 768 offers two character cell sizes: 12 dots wide by 20 dots high or 7 dots wide by 15 dots high. The former produces 85 characters and 35 rows on a screen, whereas the latter gives 146 characters on a line and 51 lines of text on the screen. If you consider that the CGA has a character cell of 8 dots wide by 8 dots high, it is obvious that even the densest image produced on the 8514 monitor using the 8514/A adapter has very clear characters.

Three character sets on the adapter—one for each of the modes described and one additional user-generated font—can be addressed at a particular time. The user-generated font can either be described in bit-map form or in a short vector format, which is fully described in the development toolkit supplied with the board. Note, however, that although character attributes such as foreground and background color and intensity are supported, blinking characters are not.

The resident fonts on the adapter support different languages. There are five different code pages, one for each of the following languages: 437 multilingual, 850 multilingual, 860 Portuguese, 863 Canadian/French, and 865 Nordic.

Advanced Text Mode

The advanced text mode of the 8514/A allows characters of varying sizes to be displayed and placed as desired on the screen. The bit-maps for the characters, which are stored in the main system memory, are defined in a special short vector format and can be up to 255 pixels square. Although this mode is treated as a text mode, the APA graphics capabilities of the board are used to create this mode.

It is possible to generate proportionally spaced fonts for display in this mode. The width of the character for a proportional font can be defined in a table. The character cell width is then changed depending on the actual character to be displayed. If proportional fonts are not being used, this width definition table is not necessary.

All Points Addressable Graphics

In addition to the advanced function modes, the 8514/A provides all points addressable graphics modes that extend the functions available with the VGA. Some of the functions that are supported in the hardware include: BIT-BLT, line drawing, area fill, patterns, color mixing, and rectangular scissoring.

BIT-BLT

The growth of PCs has caused an ever increasing demand for more information to be visible on displays at any one time. This can be achieved using old techniques in which bits of data are manipulated individually and screen updates are only performed during a screen refresh cycle; however, as the amount of data increases, the time required to change the image increases. This results in unacceptably slow video times. The advancement of technology and the increase in the level of integration that is possible in ICs has allowed progress in video adapters. Video adapters are available for the PC that offer advanced display techniques not possible five years ago.

Along with the new advances in display adapters comes a new vocabulary of terms. Many of these have been used before in larger workstation environments, but nearly all are new to PC systems. The term *bit-block transfer* (BIT-BLT) is one of these terms. A BIT-BLT instruction can be used to move a rectangular array of data in memory. This data can be located in system memory and transferred to the display adapter memory, or the data can be moved around within a bit-plane of the display memory or from one bit-plane to another. The implementation of these higher level commands is possible because of the increase in the sophistication of the hardware.

The 8514/A has various BIT-BLT instructions that can be used to manipulate graphics data. The area of memory that is to be moved must be rectangular, but the data can be moved to or from the system memory, to or from a bit-plane, between bit-planes, or simply around the same bit-plane. It is possible to overlap the area that is to be moved with the new position for the data. In addition, the data that is moved can be mixed with the data that is already present in the bit-plane. These mix functions include: logical AND, OR, and XOR; overpaint and underpaint; and arithmetic add, subtract, mean, maximum, and minimum.

Line Drawing and Area Fill

A feature that has been available on the more sophisticated boards in the market and is now available on the 8514/A is line drawing. The adapter can draw lines or polylines (several lines linked together), and their width and line type, such as dotted or dashed, can be specified. The area fill function allows areas to be filled with a solid color or a pattern. The bit-map that is to be used for the pattern can be specified by the programmer. The logical operations that can be used with BIT-BLT can also be used with the area fill functions.

Scissoring

The rectangular scissor function, also available in the hardware, works with all of the drawing functions. It is used to cut a particular image. The image outside the rectangle is not displayed, but the area inside is. This function also allows the display of multiple screen windows, where the displayed image is only part of the whole stored image.

Math Coprocessors

Each of the Personal System/2 machines can be equipped with a math coprocessor designed for use with the main CPU in the unit. Models 30 and 25 support an 8087 math coprocessor, Models 50 and 60 have a socket for an 80287, and the Model 80 can accommodate an 80387. A math coprocessor can perform arithmetic functions up to 100 times faster than the main processor. This increased performance can be obtained only with applications designed to use a math coprocessor. Applications such as CAD systems that are based on floating point arithmetic benefit greatly from the addition of a math coprocessor.

If a math coprocessor is to be used on a Personal System/2 machine, it is necessary to purchase not only the version of the IC that is appropriate for the machine architecture, but also an IC that is rated for use at the required clock speed. The 8087 used in a 4.77 MHz PC is not rated to work at the 8 MHz of the Model 30 or 25. In addition, a 10 MHz version of the 80287 is needed to run in a Model 50 or 60; the 8 MHz version that can be used in an IBM AT Model 339 should not be used. This is not an option that lets you try the lower rated part and see if it will work; the IC may work perfectly for most operations, but randomly drop a bit. On an 80-bit floating point multiplication, 1 bit being dropped may go undetected until it becomes the crucial digit in a calculation. If you need the numerical accuracy offered by these coprocessors, do not skimp by buying underrated parts. It is simply not worth it.

The 8087 used in all Model 30 and 25 configurations is a coprocessor that works in tandem with the 8086. Any instructions that are supported by the math coprocessor are executed by the 8087 instead of by the 8086. The IC is installed in the supplied socket in the system board and does not occupy an expansion slot. Floating point arithmetic functions, extended integer functions, and BCD arithmetic can be executed on this IC, increasing the performance of the overall system. Applications that can make use of the enhanced performance of the 8087 need to check for the presence of the math coprocessor; then, over forty additional instructions can be used.

Models 50 and 60 use an 80287 math coprocessor in the supplied socket on the system board. This works in tandem with the 80286 that is the main CPU in these machines. Like the 8087, the 80287 performs floating point arithmetic, extended integer arithmetic, and BCD arithmetic. As configured with the 80286 in the Personal System/2 machines, it fully conforms with the IEEE floating point arithmetic standard. Over fifty instructions are added to the 80286 instruction set by its presence, and it is compatible at the object code level with any application that is written for the 8086/8087 machines. This means that applications that make use of the 8087 in Models 30 and 25 can make use of the coprocessor in Models 50 and 60.

The Model 80 is available in 16 MHz and 20 MHz versions; each requires a different version of the 80387. The Model 80 part numbers 8580-041 and 8580-071 are 16 MHz machines and require a 16 MHz part; the 20 MHz versions, part numbers 8580-111 and 8580-311, require an 80387 that is rated to operate at 20 MHz. This IC can perform floating point arithmetic, extended integer functions, and BCD functions. In addition, the 80386 instruction set is expanded to enable it to directly perform trigonometric, logarithmic, exponential, and arithmetic functions for all data types. It is important to note, however, that the architecture of the 80386 makes it imperative that the presence of the math coprocessor be detected before any of the extended instruction set is used. The 80386 goes into an infinite loop if it is asked to perform some of the additional math coprocessor functions without a math coprocessor present in the system. The IC is upwardly compatible and can execute code written for the 8087 and 80287 systems.

Each of the math coprocessors for the Personal System/2 can be end user installed into the unit by removing the cover and inserting the part into the vacant socket on the system board. Instructions are given in the installation guide. The setup disk needs to be run after installation and before use so that configuration memory can be updated. These coprocessors are a valuable asset for the processing of mathematical applications, including most graphics programs. Floating point arithmetic is inherently faster and can be more accurate than integer arithmetic (which is limited to the bus width of the main CPU) in the specific implementation of Personal System/2 machines. IBM warranty agreements require the math coprocessor to be supplied by them, but an alternative source for units not under warranty or service with IBM would be Intel Corporation, the original IC manufacturers. Because of the difference in frequency ratings that are available for these ICs and the difficulty in assessing whether a particular IC is working fully to specification, care should be exercised in considering any source other than IBM or Intel for the math coprocessor.

LAN Boards

IBM has enhanced its offering of LAN boards with products that round out the existing product line for the PC family and, consequently, Models 30 and 25, as well as products that operate in the Micro Channel based machines. The LANs that these products operate on cover the full existing product line. There are products for the baseband IBM PC Network, the broadband IBM PC Network, and the IBM Token-Ring network.

IBM PC Network—Baseband

There are two additional boards and a Baseband Extender for the baseband IBM PC Network. This baseband network is a carrier sense, multiple access LAN with collision detect system (CSMA/CD). It uses IEEE 802.2 logical link control (LLU) protocols.

The Baseband Adapter, part number 1221, is an expansion board designed for use in the Personal System/2 Models 30 and 25; it can also be used in the existing PC family of machines. It is a 2 megabit per second board that supports IEEE 802.2/LLC protocols when used with the IBM Local Area Network Support Program. The protocol and NETBIOS support is provided by software device drivers loaded in system RAM and executed by the main CPU. The board does contain ROM, but it is used for POST purposes as a method of failure detection. This board can be used in a daisy-chain or star topology. Up to eight workstations can be chained on a LAN, which can have a total length of 200 feet of IBM type 3, four-wire twisted-pair cable.

The equivalent board for the Personal System/2 family of machines with Micro Channel architecture is the IBM PC Network Baseband Adapter/A. It performs functions identical to those performed by the Baseband Adapter for the PC expansion bus, but it is designed to work on the Micro Channel. Up to eight boards can be linked together with a total of 200 feet of IBM type 3, four-wire, twisted-pair cable. Both baseband adapters provide the remote initial program load (RIPL) feature. This feature allows a new system without a hard disk to perform its initial program load from a server on the LAN rather than from its own diskette drive.

The IBM 5173 PC Network Baseband Extender extends the functionality and connectivity of the two baseband adapters. This self-contained unit provides a method of linking up to ten daisy chains of PC or Personal System/2 machines linked with the baseband adapters. The maximum length of any given daisy chain is also increased to 400 feet by the use of the Network Baseband Extender. This results in a LAN that can have up to 80 workstations attached.

IBM PC Network—Broadband

Two boards have been introduced for the broadband PC Network. These provide the same function as the baseband boards just described, but for the broadband network. The IBM PC Network Adapter II fits the PC family machines and Models 30 and 25. It performs the same functions as the earlier PC Network Adapter, but is designed to function with the higher microprocessor speeds of the new machines. It is a 2 megabit per second board with CSMA/CD protocol, and it can be linked to a PC Network LAN with the PC Network Protocol Driver. Alternatively, the IEEE 802.2/LLC protocols are supported via the IBM LAN Support Program.

The IBM PC Network Adapter II/A performs the same function as the Network Adapter II for the Micro Channel based machines. The new broadband adapters support the remote IPL of systems that do not have a hard disk. Not having to use the diskette drive on each of these machines for system initialization greatly eases maintenance of connected systems for a system administrator.

IBM Token-Ring Network

The IBM Token-Ring Network Adapter/A (see Figure 9-4) is a Micro Channel implementation of the IBM Token-Ring Network Adapter II available for the PC family of machines. It operates at up to 4 megabits per second, with protocols that conform with IEEE 802.5 and ECMA 89 standards. The board contains 16 KB of RAM; the RIPL feature is available as an extra cost option. The feature kit includes an 8 KB EPROM and a sample load program, along with installation, programming, and user documentation.

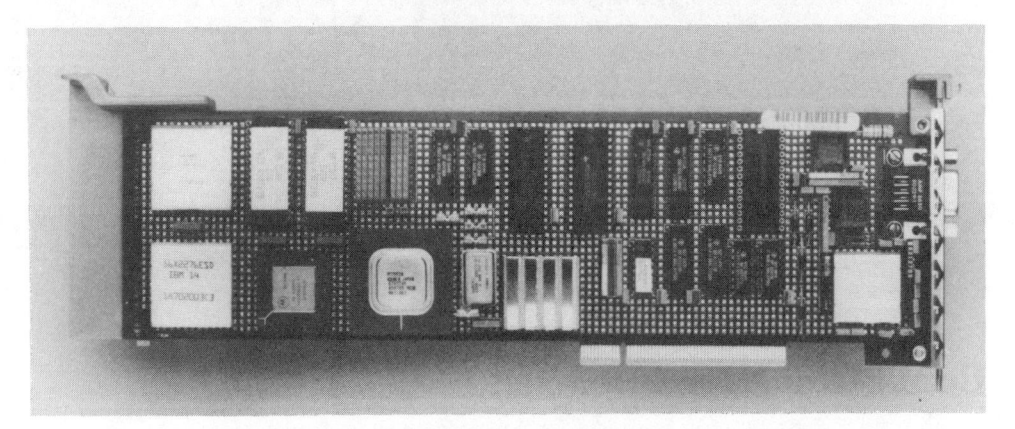

Figure 9-4 Token-Ring Adapter/A. *This is a Micro Channel implementation of the Token-Ring Network Adapter II, which is available for the PC family of machines.* Photo by Bill Schilling.

In order to accommodate the additional machines that are being supported by the Token-Ring, IBM has made changes to the Token-Ring Starter Kit and offers a Token-Ring Starter Kit/A for the Micro Channel based Personal System/2 machines. Each version of the starter kit supplies everything needed to set up a

Chapter 9

four station network on systems. The hardware consists of an access unit, adapters, cables, network software, installation aid, and instructions.

There have been many upgrades to the various network programs that can be run on the variety of network boards in both the PC family and Personal System/2 family.

Host Connection Boards

IBM has also announced Micro Channel versions of the boards used to connect PCs to System/3x and System/370 host computers for their Personal System/2 machines. Connections made to the System/3x are via twin-axial cable using the System 36/38 Workstation Emulation Adapter/A. Connection to the System/370 is made using the Micro Channel 3270 Connection Adapter (see Figure 9-5) to attach to a terminal controller via coaxial cable.

Figure 9-5 3270 Connection. *3270 display station emulation can be obtained via the 3270 Connection Adapter installed in Models 50, 60, or 80.* Photo by Bill Schilling.

Communications Boards

There are several communications boards for the Personal System/2 family of machines. The boards that are discussed in this book fall into two categories: asynchronous communications and multi-protocol boards. The IBM Personal System/2 Dual Async Adapter/A (see Figure 9-6) is a Micro Channel expansion board that has two independent RS232C serial ports. The hardware architecture of the systems is such that up to three of these boards can be added to any particular unit, although currently only four serial ports can be accommodated by PC DOS version 3.3 (COM 1, COM 2, COM 3, COM 4). The ports on this board supplement the serial port on the system board. Each board that is added to the system occupies an expansion slot.

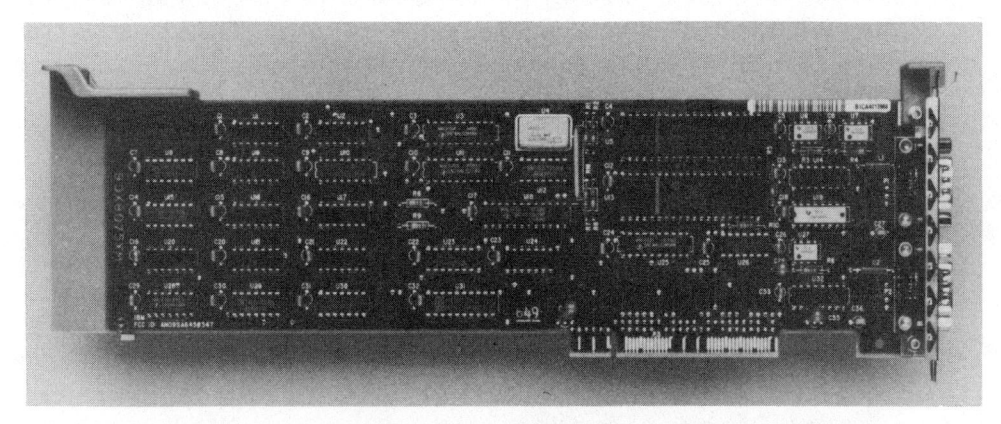

Figure 9-6 Dual Async Adapter/A. *Two independent RS232C serial ports are available on this board.* Photo by Bill Schilling.

A Hayes-compatible internal modem called the IBM Personal System/2 300/1200 Internal Modem/A (Figure 9-7) is available for the Micro Channel based Personal System/2 systems. This board occupies a single slot and supports 300 and 1200 bps using the Bell 212A communications standard and the Hayes AT command set. The modem comes with a 7-foot cable and a diagnostic diskette.

Figure 9-7 300/1200 Internal Modem/A. *The Hayes AT command set and the Bell 212A communications standard are supported.* Photo by Bill Schilling.

The Realtime Interface Coprocessor Multiport Adapter is a multiport communications adapter for use on PCs and the Personal System/2 Model 30. It supports up to eight ports operating asynchronously at up to 19,200 bps or 38,400 bps using HDLC/SDLC protocols. All eight ports can operate concurrently at up to 9,600 bps. Asynchronous or synchronous hardware support is provided on the remaining ports. The board features an Intel 80186 processor operating at 7.37 MHz and 128 KB or 512 KB of memory, with two-channel direct memory access between the processor and the first two ports. This adapter is used by the LAN Asynchronous Connection Server Program to provide multiple asynchronous port connections to a Token-Ring or PC network.

The Multi-Protocol Adapter/A (Figure 9-8) is a single slot board for the Micro Channel, which has a single-channel output that can be programmed to transmit in full-duplex or half-duplex. It supports asynchronous, bisynchronous, HDLC, and SDLC protocols. The protocol can be changed at startup via software and is programmable to support speeds of up to 19,200 bps.

Figure 9-8 Multi-Protocol Adapter/A. *The single channel output from this adapter supports asynchronous, bisynchronous, HDLC, and SDLC communications.* Photo by Bill Schilling.

Mouse

The pointing device port that is standard on the Personal System/2 system boards is not a standard RS232C port. It is designed to be used by a pointing device such as a mouse, digitizer, or trackball. To this end IBM is offering a two-button mechanical mouse, shown in Figure 9-9, for use with the Personal System/2 machines. The mouse comes with the necessary device driver on a diskette, which is executed as a DOS terminate-and-stay-resident file. The mouse has a shielded 9-foot cable that contains four conductors. The buttons on the mouse have a positive tactile feel and the size of the mouse allows it to fit comfortably in the hand. Personally, we have used better mice but this version is satisfactory and reasonably priced at $95.

Figure 9-9 Personal System/2 Mouse. *This mechanical mouse is attached to the system unit through the auxiliary pointing device port.*
Photo courtesy of International Business Machines Corporation.

Memory Boards

There are a variety of memory boards for the Personal System/2 machines. These include memory boards that are installed in the system board of the unit to expand the local memory capacity of the machines, as well as boards that fit into the Micro Channel. Models 50 and 60 come with 1 MB of memory as standard on the system board; no additional memory can be added to the system board. The Model 80 part number 8580-041 has 1 MB of memory as standard on the system board; an additional 1 MB of 80-nanosecond memory can be added to the system board using the IBM Personal System/2 System Board Memory Expansion Kit. The Model 80 part number 8580-071 has 2 MB of memory on the system board as supplied. The Model 80 part number 8050-111 has 2 MB of system board memory as standard and can be expanded to 4 MB of 80-nanosecond memory on the system board using the IBM Personal System/2 System Board 2 MB Memory Expansion Kit.

Although the system board memory expansion is limited to 1 MB on Models 50 and 60, additional memory can be added via expansion boards that fit into the Micro Channel. Models 50 and 60 can use the 80286 Memory Expansion Option, shown in Figure 9-10. This has 512 KB of memory as standard; up to three expansion kits, each with 512 KB of memory, can be added to this single board, making a total of 2 MB of memory. The Model 50 can accommodate up to three fully populated boards, and the Model 60 can accommodate up to seven of these boards.

The Model 80 can make use of the 32-bit memory option available. The 80386 Memory Expansion Option has 2 MB of 80-nanosecond memory as standard; up to two Memory Expansion Kits, each with 2 MB of additional memory, can be added.

Figure 9-10 80286 Memory Expansion Option. *Each single-in-line memory module contains 256 KB of memory. Up to 2 MB of memory can be installed on the board.* Photo by Bill Schilling.

(See Figure 9-11.) A maximum of 16 MB of system memory can be addressed by the system, and this can be obtained without adding the System Board Memory Expansion Option that is available for the 8580-041 and the 8580-111. Each board occupies one 32-bit slot in the Model 80 and each board can contain up to 6 MB. This memory can be enabled and disabled in 1 MB blocks by POST.

These memory expansion boards will become more significant as operating systems that can address more than the 1 MB addressable by DOS become available. For example, OS/2 can run with as little as 1.5 MB of memory, but 3 to 4 MB is required for its effective use, particularly if the extended edition is used.

The memory expansion boards that increase the amount of memory available to the main CPU for operating system use should not be confused with the IBM Personal System/2 80286 Expanded Memory Adapter/A. This board is for use in a Model 50 or 60 with the IBM 3270 Workstation Program. It allows PC DOS to use this memory as an area for up to six virtual disks, or the 3270 Workstation Program can use the memory for an expanded memory session and one virtual disk session.

External 5.25-Inch Diskette Drive

The Personal System/2 machines have only a 3.5-inch diskette drive and no 5.25-inch diskette drives, such as are found on the PC family of machines. In addition, there are two standards for the 3.5-inch diskette drives on the Personal System/2 machines; the Model 30 has a capacity of 720 KB, and Models 50, 60, and 80 have a 1.44 MB capacity. This inconsistency causes a variety of problems that are com-

Figure 9-11 80386 Memory Expansion Option. *Each piggyback board contains 2 MB of memory. Up to three can be installed on each adapter.* Photo by Bill Schilling.

pounded as the mixture of machines being used in a given installation becomes more varied. For most mixed environments, where machines are of both PC and Personal System/2 origin, the external 5.25-inch diskette drive is a necessity. It is important to realize that the use of an external 5.25-inch diskette drive precludes the use of a second internal 3.5-inch diskette drive.

There are two parts to the diskette drive: the adapter card and the drive itself. The IBM Personal System/2 5.25-inch External Diskette Drive Adapter is used in Models 30 and 25, and the Personal System/2 5.25-inch External Diskette Drive Adapter/A is used in Models 50, 60, and 80.

The Model 30 version is a half-length board that occupies one slot in the machine. The external drive can only be configured as drive B and eliminates the use of a second 3.5-inch drive in Models 30 and 25. The diskette drive is attached to the adapter card via a 37-pin D-type connector, which provides the same signals as the connector on the PC XT diskette controller board. The adapter card not only plugs into the expansion slot, but has an additional connector on top of the expansion board, which connects to the drive B cable that is in Models 30 and 25 for the second disk drive.

An expansion board with a Micro Channel style of interface is used in Models 50, 60, and 80. There are three parts to the assembly. The expansion board itself occupies a slot in the machine, and a ribbon cable assembly links the board to an adapter card that, in turn, plugs into the socket/connector that would be used by a second internal 3.5-inch diskette drive.

Chapter 9

The diskette drive itself is an oversized, overweight box that contains a half-height 360 KB diskette drive and power supply. (See Figure 9-12.)

Figure 9-12   5.25-Inch External Diskette Drive. *This drive, although it does much to reduce media interchange problems, is oversized and overweight for its purpose.* Photo by Bill Schilling.

IBM seems to have made some strange decisions with this option. It is understandable that, to ensure that the 3.5-inch diskette becomes the standard for the personal computer market, the 5.25-inch drives should not be a standard option. The 3.5-inch media is a superior product; it is far less prone to damage than the 5.25-inch diskettes. The fact still remains that there are millions of PCs in use that do not support 3.5-inch media. The discrepancy between the media densities on the various machines, however, is acceptable for the movement of materials to a more powerful machine; the Model 30 disks can be read on the Model 50 for example. Often though, the machines are purchased on a trickle-down system, where the power user gets the latest machine and the old machine is used to upgrade the next user down the chain. This necessitates the exchange of information in both directions from the more powerful machine to the less powerful machine and back. This is where IBM's solution is not very efficient.

IBM is offering a Data Migration Facility for transferring the data from a PC to a Personal System/2 machine. This is done with an adapter (Figure 9-13) that allows the Centronics connector end of a printer cable to be plugged into the parallel port on a Personal System/2. The other end of the printer cable is attached to the sending PC and the supplied software is used to make the transfer. Data migration is discussed in more detail in Chapter 14.

Other IBM Mass Storage Options

An external 3363 Optical Disk Drive, shown in Figure 9-14, can be attached to any of the Personal System/2 machines. Up to 200 MB of data can be stored on a 5.25-inch removable disk. This is write once, read many times data due to the

PS/2 Options

Figure 9-13 Data Migration Adapter. *This adapter is supplied with software to allow the transfer of data between PCs and Personal System/2 machines.* Photo by Bill Schilling.

Figure 9-14 3363 Optical Disk Drive. *This external disk drive requires an adapter board, which is installed in the system unit itself. One adapter board can drive up to two external Optical Disk Drives.* Photo courtesy of International Business Machines Corporation.

Chapter 9

technology involved. The disk drive requires installation of an adapter in an expansion slot of the machine. Two drives can be attached to the adapter card. If the drive is to be installed in a Model 30, the part number is IBM 3363 A01 for the first drive and adapter card. For the Micro Channel version, the part number is IBM 3363 A11 with the adapter card. The second drive has the same part number, IBM 3363 B01, for any machine.

On Models 60 and 80 an internal Optical Disk Drive, feature number 8700, is available. This requires an expansion slot for the adapter card and fits into the position that would be occupied by the second hard disk. The 5.25-inch media is then removed through the front panel opening below the 3.5-inch diskette drives.

A Micro Channel version of an adapter card to drive the IBM 6157 Streaming Tape Drive is available as feature number 4160. This board occupies a single slot in the system unit. The 6157 Streaming Tape Drive is shown in Figure 9-15.

Laser Printer Controller Card

The PC laser printer that was announced in April, at the same time as the Personal System/2 machines, does not connect to a standard serial or parallel port. The printer itself does not contain any of the intelligence necessary for page composition. Instead, page composition is done by an Adobe System's PostScript page-description language processor board that fits into the expansion bus. At the time of the printer's announcement, the PC version of the expansion board was available; the Micro Channel version was announced later in October. The board, the IBM 4216 Personal Pageprinter, a mouse, and Aldus' PageMaker publishing software together with the Windows operating environment are purchased as the Publishing SolutionPac Option/A for $5,888.

Other Printers

Three other printers are now being offered for use with the Personal System/2 machines. The 4201 Proprinter II Model 002 (Figure 9-16) provides near letter quality, dot matrix output at better speeds than the Model 001. The fast printing speeds include a draft quality mode that can print up to 240 characters per second (cps), an emphasized mode at 120 cps, and a near letter quality mode of 40 cps. The narrow carriage machine can accommodate continuous forms that are between 3 and 10 inches wide; single sheet paper can be up to 11 inches wide. The maximum printable line is 8 inches wide. The printer is attached to the system unit via a parallel port; an extra-cost serial interface card allows attachment via a serial port.

PS/2 Options

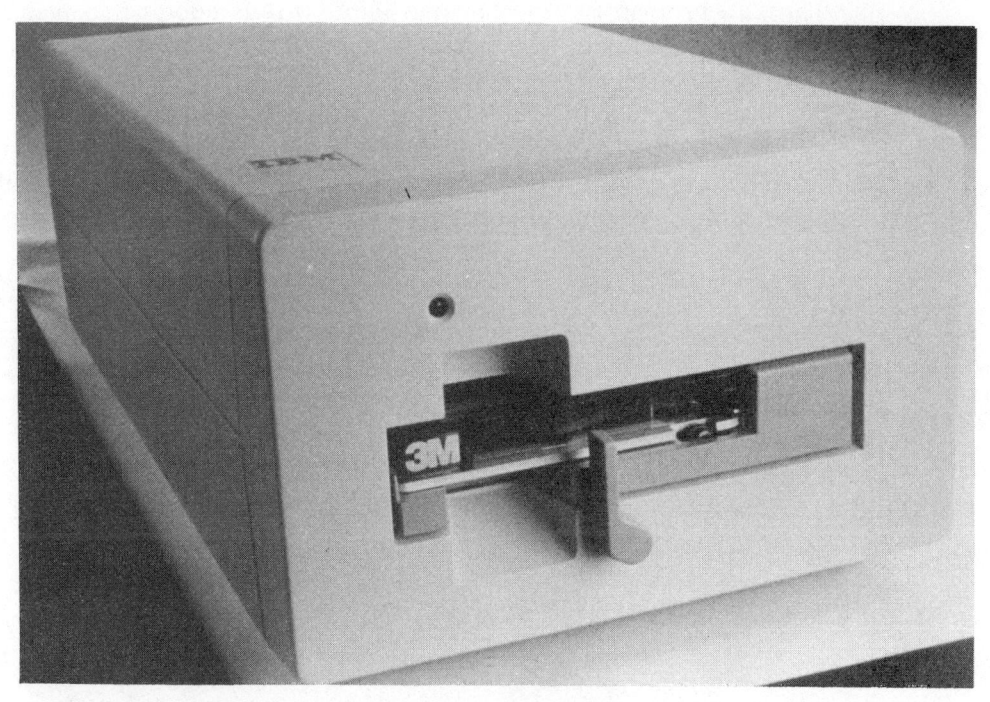

Figure 9-15 6157 Streaming Tape Drive. *This tape drive is used on the Personal System/2 via the 6157 Tape Drive Adapter.*
Photo courtesy of International Business Machines Corporation.

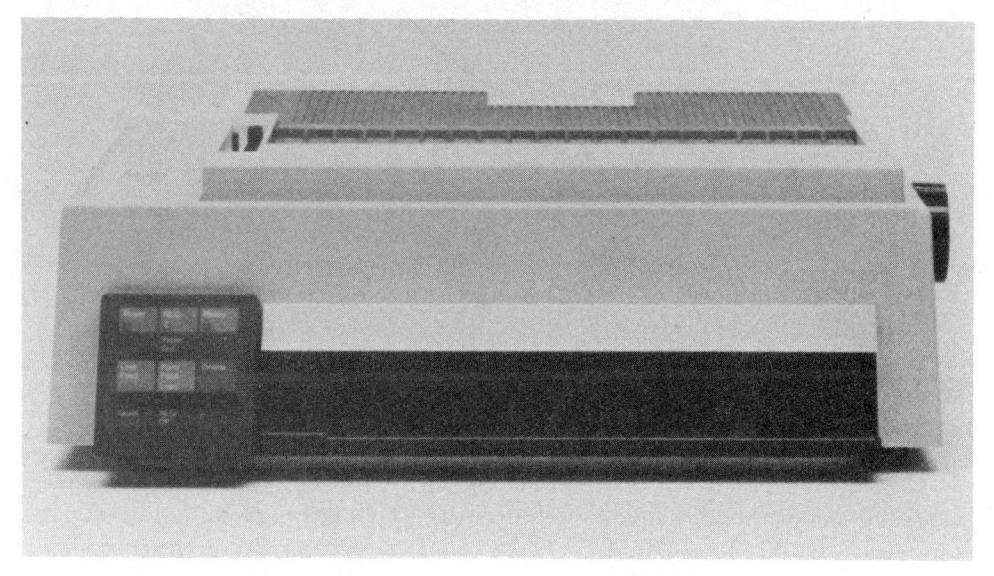

Figure 9-16 Proprinter II. *Speeds of up to 240 cps in draft quality mode are possible with this printer.*
Photo courtesy of International Business Machines Corporation.

Chapter 9

The 4207 Proprinter X24, shown in Figure 9-17, is a dot matrix printer that extends the functions of its sibling, the 4202 Proprinter XL. It can provide letter quality printing, because the print head is 24 wires high instead of 9 wires as in the Proprinter XL and II. In letter quality mode the printer can achieve speeds of 67 cps. The additional wires in the head also allow high resolution graphics, which are not possible on the other Proprinters. Note that the printer ribbons for this machine are not interchangeable with the ribbons for the other Proprinters.

Figure 9-17 Proprinter X24. *The 24-pin-high print head allows letter quality output from this dot matrix printer.*
Photo courtesy of International Business Machines Corporation.

The Quietwriter III is a resistive-ribbon, nonimpact printer that can produce letter quality and all points addressable graphics on textured papers. (See Figure 9-18.) This printer, although expensive, can print in draft mode at 160 cps and letter quality at 80 cps. There are four standard fonts on the machine: Courier 10, 12, and 17.1 pitch and a bold proportionally spaced font. Additional fonts are available. This is a wider carriage printer than the Proprinter, accommodating cut paper up to 16.54 inches, and continuous forms that are 14.5 inches between the pin feed holes. Lines up to 13.2 inches wide can be printed.

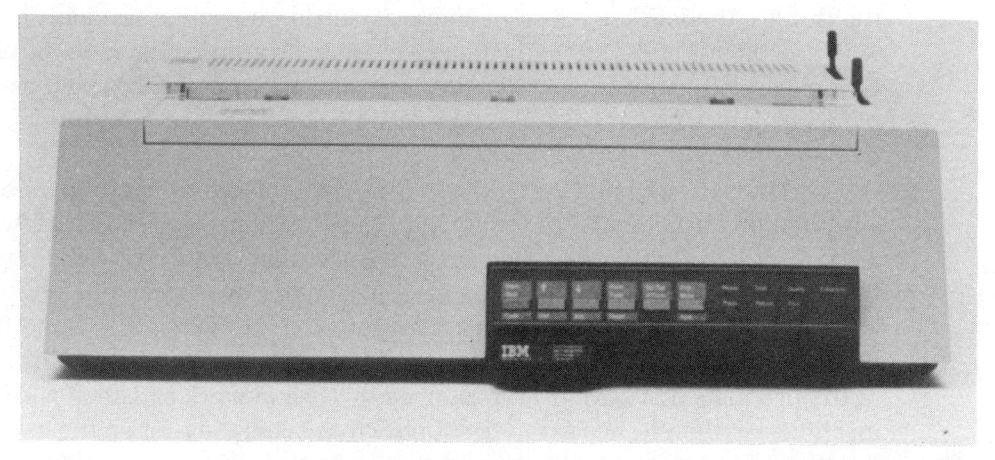

Figure 9-18 Quietwriter III. *Letter quality printing on textured paper is possible with this printer.*
Photo courtesy of International Business Machines Corporation.

Third-Party Hardware

For the first few months after the IBM announcement there was very little movement by third-party vendors to announce products that supported the Micro Channel. Statements to the press varied from those by companies that stated they were investigating the feasibility of Micro Channel products to those that said they were waiting to see which way the market was going. By November of 1987 even Compaq, which had vehemently stated that it was not working on a product incorporating the Micro Channel, announced that it was developing the capability *should the market demand it.*

The add-in market for the Personal System/2 machines has changed from the PC market, because the multi-function boards that were the best sellers on PCs are not required in the same form on the Personal System/2 machines. The real-time clock is now an integral part of the system board, as is the first parallel port, a serial port, and a port for a mouse. This does not eliminate the need for add-in boards, but it does require them to be more specialized. The major applications for add-in boards seem to be memory expansion, backup, or connectivity boards, judging from the boards being offered by the manufacturers. A sampling of the products that have been announced or are available does not contain many surprises.

Third-Party Memory Boards

The RamQuest 50/60 board from Orchid Technology in Fremont, California was the first on the market. It is a memory expansion board for Models 50 and 60 with a 16-bit bus and 2 MB of memory. The board is compatible with the original LIM expanded memory specification (version 3.2) and can be used as an extended memory board with OS/2, in a fashion similar to the XMA board offered by IBM.

Orchid is a company that has been in the PC industry for a long time. It has a reputation for being the first to bring a product to market, particularly in the area of accelerator boards. Due to its aggressive marketing strategy of trying to be first, the results have not always been perfect for every machine on the market. A particular accelerator board may not work in the generic clone PC that has been purchased. This is a typical problem with the PC market. However, all of the products that have been produced by Orchid work satisfactorily on IBM machines, so there is no reason to suppose that being the first on the market for the Personal System/2 family should be a limitation.

Orchid's conservative decision to build a board that contains only 2 MB of memory proved to be wise in a few months. IBM did not specify how to implement add-in memory boards that contain over 2 MB of memory until November 1987. As a result, some manufacturers were burned by designing products that did not meet the full specification. IBM stated that the omission from the documentation was simply an oversight on their part. The addition refers to the need for device drivers to provide extended memory to OS/2.

Other manufacturers have joined Orchid in offering Personal System/2 memory boards. Quadram produces a 16-bit memory board, the Quadmeg PS/Q, for Models 50 and 60. This board, shown in Figure 9-19, can contain up to 4 MB of memory. The company claims compatibility with the OS/2 extended memory specification, which allows the memory to be used by an 80286 microprocessor operating in protected mode. In addition, the LIM expanded memory specification Version 4.0 is supported via software drivers. This board has three ID numbers that can be used on the Micro Channel to reduce the chances of conflict with other boards. This ID selection is performed by a toggle switch on the adapter board's mounting bracket, which extends through the rear panel. This is contrary to the intent of the "switchless" system suggested by IBM, but is a far preferable solution to the internal jumpers to which IBM has had to resort on the 8514/A adapter. The toggle switch should not present a problem, provided the system unit is not located in a position where it can be bumped or altered by an end user.

Several other manufacturers are offering memory boards for Models 50 and 60. The company names are familiar and it appears that these boards are not difficult to develop and manufacture. The list of manufacturers includes STB Systems of Richardson, Texas, which has a memory board that allows up to 10 MB of extended or expanded memory, and Tecmar of Solon, Ohio, which offers a memory board (the MicroRAM board shown in Figure 9-20) that accepts

PS/2 Options

up to 8 MB of extended or expanded memory (using the LIM expanded memory specification 4.0 or the enhanced expanded memory specification from AST).

Figure 9-19 Quadram QuadMeg PS/Q. *1 MB (shown), 2 MB, or 4 MB of RAM can be installed in this board. Upgrades are possible.*
Photo courtesy of Quadram Corporation.

Figure 9-20 Tecmar MicroRAM. *Up to 8 MB of memory can be installed in the RAM board.* Photo courtesy of Tecmar, Inc.

Chapter 9

The best known of all add-in board manufacturers is AST Research, Inc. of Irvine, California. In the past couple of years they have expanded their product line to make top quality machines, but they have a well-deserved reputation for producing the most compatible add-in boards for the PC and the generic clones. (The AST Premium/286 is one of the comparison machines that was used to evaluate the Personal System/2 machines for this book.) If an AST board doesn't work in a particular machine, the machine is considered incompatible rather than the board. This company's introductions for the Personal System/2 market are maintaining the same level of quality.

AST Research has a 32-bit memory board that can be configured to contain 1 MB, 4 MB, or 8 MB. In addition, a daughterboard that contains a serial port and parallel port can be added. This board, shown in Figure 9-21, can use the matched memory cycle of the 32-bit bus and is in direct competition with the memory card offered by IBM.

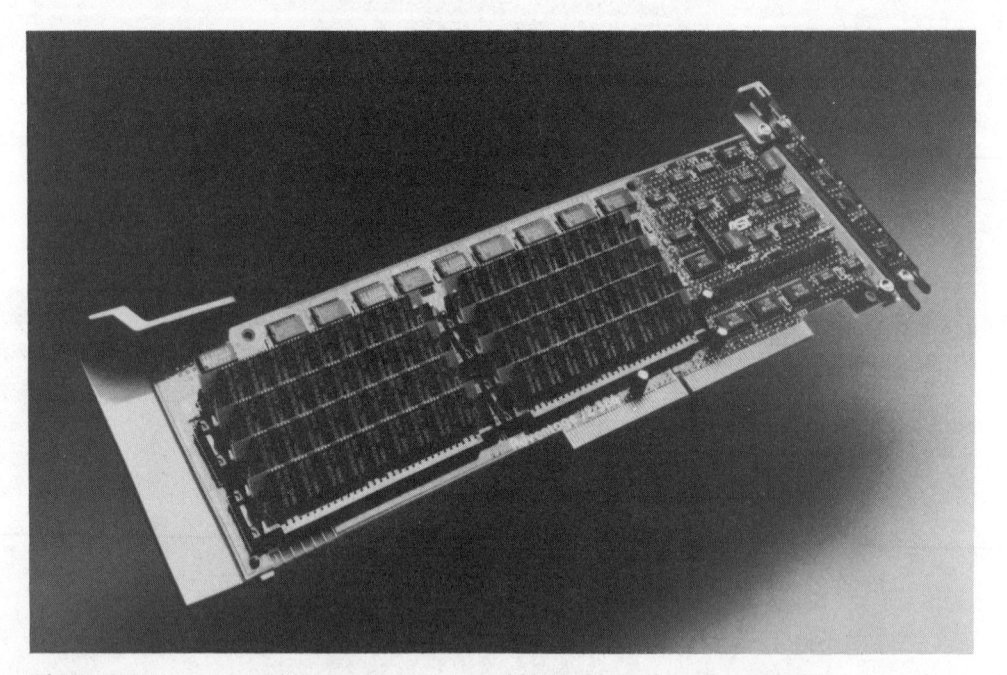

Figure 9-21   AST Advantage/2-386. *An additional board can be added to provide a serial and parallel port.*
Photo courtesy of AST Research, Inc.

Other manufacturers of 32-bit memory boards for the Model 80 include Quadram and Orchid. The true compatibility of the memory boards, however, will require time, because the LIM 4.0 applications and OS/2 applications needed for their use are not yet commonly available.

In some ways, this parallels the development of add-in boards for the PC family. In that case, the standardization was across many versions of the machine; for the Personal System/2 machines, the standard is across the many different applications that will run with different operating systems.

Additional Storage Options

The storage capability of IBM Personal System/2 machines can be expanded using third-party hardware. Products available include internal 3.5-inch diskette drives, internal hard disks, internal backup options, external 5.25-inch or 3.5-inch diskette drives, and external backup options.

Priam of San Jose, California offers external hard disks for the Model 50, which compensate for the low hard disk capacity that is offered by IBM. The products' drive sizes include 42 MB, 62 MB, and 133 MB; the necessary controller card and software are shipped with the unit. The company's offering for Models 60 and 80 is an internal disk drive with a capacity of 330 MB. The 330 MB drive uses the internal ESDI hard disk drive controller that is present in the machines as standard.

External backup requirements may include a tape backup unit to provide a fast method of protecting a system from irrevocable data loss. A single tape backup unit can be used across many machines to provide a backup. This is done by installing an adapter in each machine and moving the tape drive itself from machine to machine.

This system may appear cumbersome, but it is infinitely more reliable than trying to persuade end users to back up their hard disks onto diskettes at the end of the working week. Having a tape drive and making one person responsible for backing up machines means that backups are far more likely to get done.

With the IBM Personal System/2 machines, the situation does not need to change. An add-in board that functions in the Micro Channel is necessary; then the same drive can still be used across all of the machines. Various manufacturers, including Maynard Electronics of Casselberry, Florida, have announced adapter boards for the Micro Channel that can be used with the same tape drive that is used on the PC family of machines.

For the user who wishes to purchase a dedicated tape drive that will fit into the second 3.5-inch diskette drive position, such units are available from third-party vendors such as identica from Mountain View, California.

Communications Products

Although the system board on the IBM Personal System/2 machines includes a serial port and parallel port, there is still a great need for communications products for the Micro Channel. Personal computers are increasingly being used in environments where they are connected. This connection can be to a mainframe, minicomputer, or other personal computers via a LAN. The connectivity requirements have not changed from the PC family except for the use of the Micro Channel as the expansion bus.

Third-party vendors have produced boards that can meet the communications needs for the IBM Personal System/2 machines. A selection of these boards is described to illustrate the diversity of boards that are available and to show that the major vendors of add-in products for the PC have appended products for the IBM Personal System/2 machines to their line.

Advanced Transducer Devices of Sunnyvale, California offers an internal 2400 bps modem for the Micro Channel. This company has manufactured add-in products for the PC using the company president's name, Matt Zuckerman, in the product name.

Micro Channel based boards are required for linking Personal System/2 machines with existing PCs on a LAN. IBM has enhanced its product line with Micro Channel versions of its boards, and other manufacturers have also produced additional products. Tiara Systems, Inc. has a Lancard A-II for an Arcnet network that contains a boot PROM, thus allowing a machine to be configured as a diskless workstation.

The AST-3270/CoaxIIA is a 3270 emulation board that runs in the Micro Channel of Models 50, 60, and 80 and supports the IBM standard for 3270 emulation or the IRMA standard from DCA. These two standards are prevalent in the PC environment, so support for both is significant. In addition, the design of the board is such that it is possible to change the communication standard used by the board. This is done through custom processors that can load instruction sets from diskette.

The MainLink II from Quadram Corporation is a 3270 emulation board that uses the Chips and Technologies chip set and has a Micro Channel interface. (See Figure 9-22.) It is the first board to be released using the CHIPSlink IC. This VLSI component is the first single component protocol controller for 3270 emulation.

IDEAssociates is also offering a third-party alternative to the IBM 3270 Connection board. The IDEAcomm 3278/MC connects to a 3270 terminal controller via coaxial cable. (See Figure 9-23.) It can be installed in the Personal System/2 Models 50, 60, or 80 and provides extensive terminal emulation, including 3278 Model 5 with 132-column and 28-line applications.

The first third-party 5251 emulation board for the Micro Channel based machines was from IDEAssociates. The IDEAcomm 5251/MC (Figure 9-24) allows IBM Personal System/2 machines to be linked to IBM System/3x machines. The twin-axial connection is made via the adapter in the IBM Personal System/2 machine to a System/3x or a 5294 remote workstation controller. Several terminals can be emulated using this board including the 5251, 5291 and 5292 Model 2, and 3180.

PS/2 Options

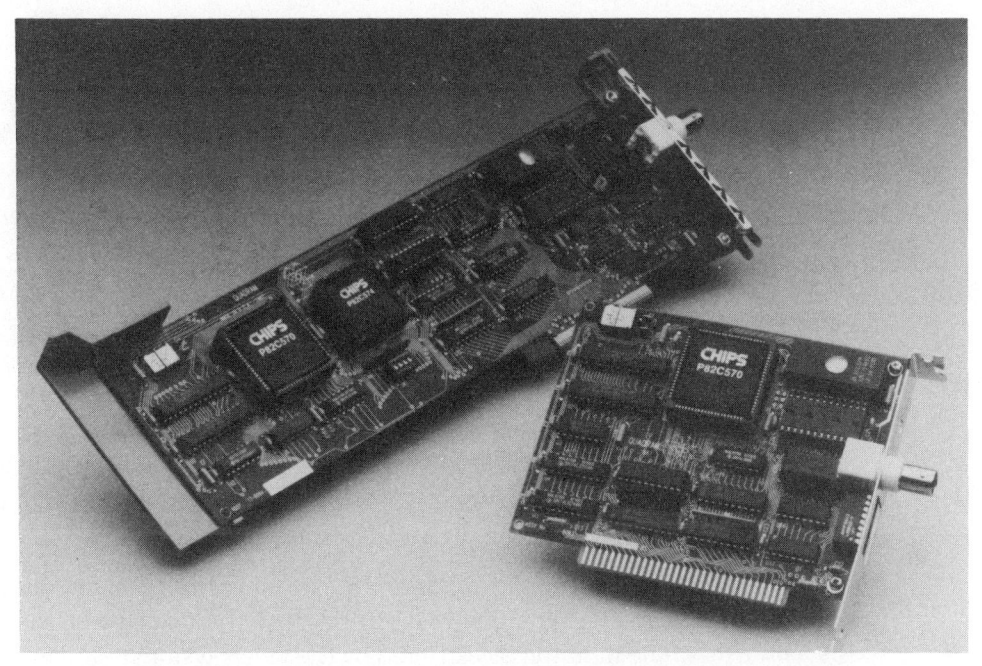

Figure 9-22 Quadram MainLink II. *3270 emulation is provided using the CHIPSlink 3270 protocol controller from Chips and Technologies.* Photo courtesy of Quadram Corporation.

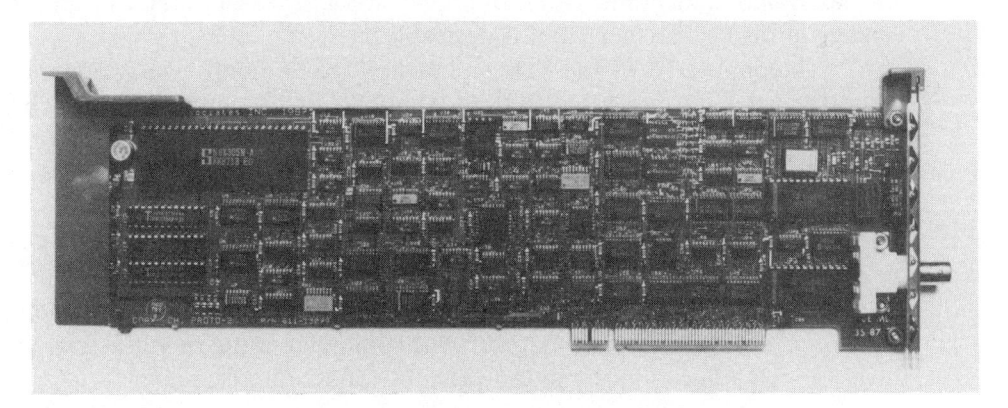

Figure 9-23 IDEAssociates IDEAcomm 3278/MC. *This third-party alternative to the IBM 3270 Connection board offers terminal emulation of 3278 terminals.* Photo courtesy of IDEAssociates, Inc.

Monitors

With the introduction of new monitors for the IBM Personal System/2 machines, the monitor manufacturers were quick to offer direct substitutes for the IBM

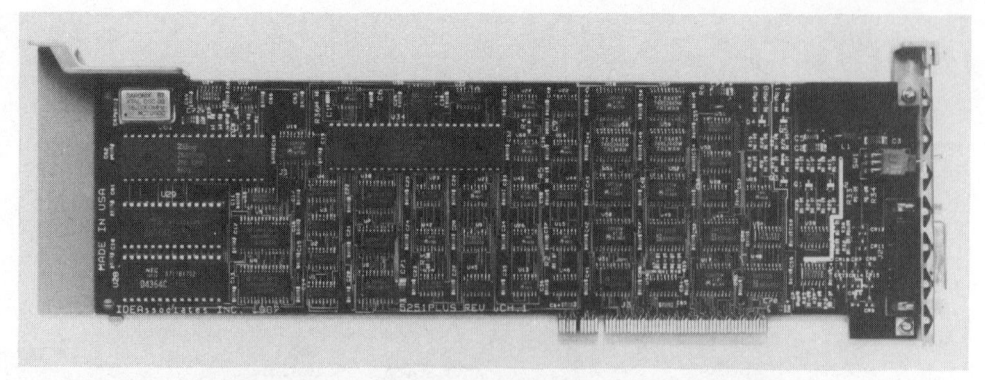

Figure 9-24 IDEAssociates IDEAcomm 5251/MC. *This third-party alternative to the IBM System 36/38 Workstation Emulation Adapter/A offers emulation of several terminals.*
Photo courtesy of IDEAssociates, Inc.

displays. Although the technology of monitors is improving, it has been hard to justify the purchase of a new monitor just because the display is clearer or it costs a little less. The monitor is restricted by the video system that drives it and in many respects is not an innovative product. New monitors are purchased only when the reliable, long lived old ones cannot be repaired. The need to use an analog monitor for the VGA standard was seen as a negative aspect of the Personal System/2 machines, even though it was a necessary step to gain the large range of displayed colors that is desirable.

The popularity of the NEC Multisync type of monitor in recent years has breathed new life into a staid market. These adaptable sync monitors, which can be plugged into several different types of video boards and can accommodate all of the different video modes, shook the market. They represented an investment in the future because the monitor would be able to cope with any new standard produced, assuming that any new standard would be an incremental step over existing standards. The addition of the VGA as a video standard did not cause the adaptable sync monitors to become obsolete. A new cable to attach to the monitor was necessary, so that the VGA could detect the monitor type, but this was quickly offered for a modest price by the manufacturers.

Various new models of monitors, both direct substitutes for the IBM offering and enhanced adaptable sync types, are now being offered by manufacturers such as NEC, Sony, Amdek, and Princeton Graphics.

Video Boards

The VGA standard for the Micro Channel based machines fills many of the needs of a typical end user; however, for applications such as CAD workstations, additional resolution is necessary. This is provided to a certain extent by the IBM 8514/A Display Adapter, which provides a maximum resolution of 1024 by 768 pixels. As in the PC market, this is not sufficient for some applications. Third-party vendors are making specialized higher resolution boards to support the specialized dedicated graphics workstation environment.

One of the most famous of the companies that fill this need for the PC environment is Matrox from Dorval, Quebec, Canada. The fact that they have introduced a high resolution board for the Micro Channel is not surprising. This board is a vector driven board that uses the Texas Instruments 34010 graphics processor along with additional custom gate arrays that increase the vectors per second that can be drawn by the board to 100,000. The board also incorporates three-dimensional hardware accelerators, which aid in the vector drawing speed of the board. With a list price of $6,995 this board is not inexpensive, but can meet the specialized needs of a dedicated workstation in a manner not possible using the IBM VGA or the 8514/A.

PC Add-In Boards

The introduction of the IBM Personal System/2 machines has affected the PC product line in a limited fashion. Most of the existing add-in boards for the PC work in Models 30 and 25, the biggest restriction being the size constraint (oversized boards will not fit) and the limited number of slots that are available for expansion. The Micro Channel architecture and the system board design has affected the add-in boards in one obvious way and has provided some more subtle third-party opportunities.

The video system on the IBM Personal System/2 machines is the VGA. Models 30 and 25 have a subset of the VGA called the MCGA, and IBM is offering a PC expansion bus add-in board that expands the machine's capabilities to a fully fledged VGA. This board also works in the PC family of machines, enabling the VGA to be a standard video system across all of the machines. This has produced a flood of VGA-compatible boards, mostly from manufacturers that already offer EGA-compatible boards at a fraction of the cost of the IBM version. The manufacturers include: AST with their AST-VGA board, Quadram with their Quad-VGA board, Orchid with their Orchid Designer VGA and Orchid Designer VGA-2, and Everex Systems, Inc. from Fremont, California with their EVGA.

These boards do not just provide the basic VGA capabilities. As with the EGA clone boards, they provide added functionality. An argument for not using the new VGA boards is that the application programs do not currently support the new video modes and an analog monitor is required for use with the board. With time this situation will change and the new applications or upgrades to existing applications will make use of the new modes. VGA boards will become more prevalent in the PC market.

Chapter 9

Some third-party vendors have made the upgrade to the new IBM video standard more gradual with their product offerings. The AST board and the Quadram board allow the adapter to be used as an EGA, with a digital monitor attached to a second connector on the rear bracket. Then when applications that require the VGA video modes are used, the requisite analog monitor can be added to the system instead of the digital monitor. The Quadram QuadVGA is shown in Figure 9-25.

Figure 9-25   Quadram QuadVGA. *An upgrade path is possible with this board. It can be used with an EGA digital monitor or an analog VGA-compatible monitor.* Photo courtesy of Quadram Corporation.

Other third-party vendors have chosen to enhance the VGA standard by adding additional video modes. This was done with the EGA standard and proved to be successful. At first sight, however, this approach appears strange, because special drivers are necessary for each application that makes use of the additional modes. The Orchid Designer VGA and the Orchid Designer VGA-2 offer desirable additional modes, such as 640 by 480 pixels with 256 colors and 800 by 600 pixels with 16 colors, as well as 1024 by 768 pixels with 16 colors on a suitable monitor. The Orchid VGA-2 has an additional mode of 800 by 600 pixels in 256 colors.

The Everex EVGA offers different additional modes. There is an 800 by 600 pixel mode with 16 colors, but the additional 256-color mode has a resolution of 640 by 400 pixels. The EVGA is shown in Figure 9-26.

Figure 9-26 Everex EVGA. *The EVGA offers additional modes with a higher resolution than the standard VGA.*
Photo courtesy of Everex Systems, Inc.

The market for video add-in boards for the PC is changing rapidly, as each manufacturer updates its products to ensure that its additional features are the best. However, the trend in the market is obvious. The VGA clone boards for the PC offer full VGA compatibility and either a migration path, by supporting digital monitors, or additional video modes that enhance the performance, or both. As the VGA becomes the more prevalent standard, the emphasis will move from the migration path for users purchasing new equipment to the availability of drivers for the additional modes of operation.

Bi-Directional Port

One obvious application for an expansion board for the PC environment is a bi-directional parallel port to allow transfer of data in both directions between a Personal System/2 machine and a PC. This would most logically come from a company that sells low cost serial port LANs. Advanced Transducer Devices has produced the Zuckerlink LAN, which includes an expansion board for the PC that provides this capability, as well as the software and hardware to link the various computers.

Personal System/2 Clones

Although many aspects of the Micro Channel are patented by IBM, IBM is willing to license some of these patents to other manufacturers. Manufacturers that have announced Micro Channel based products include Tandy Corporation and Dell Computer Corporation. There is, however, uncertainty as to whether clone manufacturers will be able to legally produce a machine, with or without a license. This legal restriction has not prevented several integrated circuit manufacturers from announcing that they are developing chip sets that emulate the Personal System/2 system functions. These chips would perform the same function for the Personal System/2 clones that the Chips and Technologies PC chip set did for the PC. The set's presence in the marketplace would drive the cost of developing a clone machine low, and the cost of producing a machine in mass quantities would also be reduced. These IC manufacturers include Chips and Technologies, Inc., Western Digital Corporation, Intel Corporation, and Zymos Corporation.

10

Disk Operating System (DOS), Version 3.30

The IBM Disk Operating System (DOS) Version 3.30 was introduced by IBM in April 1987 along with the IBM Personal System/2 family of computers. This new version of PC DOS, which replaces versions 3.1 and 3.2, can be used on the IBM PC family of computers as well as on IBM Personal System/2 computers. It is the only version of PC DOS offered for use with the Personal System/2 family. It contains several useful enhancements in addition to providing support for new features offered by the Personal System/2 computers. PC DOS 3.30 includes the BASIC interpreter, Version 3.30.

DOS is a single-user, single-task operating system used on PC and compatible computers. It now faces competition from multi-tasking operating systems such as Operating System/2, which are available for use with the Personal System/2 Models 50, 60, and 80, as well as with the AT and XT 286. Because these operating systems cannot run on PCs or the Personal System/2 Models 25 and 30, DOS will be used on millions of computers for years to come.

History

The Disk Operating System (PC DOS) and the IBM PC have been close companions since the introduction of the PC. Version 1.0 of PC DOS was introduced with the PC, and although IBM offered PC versions of CP/M-86 and the UCSD p-System, PC DOS quickly became the dominant operating system for the PC.

Version 1.0 of PC DOS was IBM's licensed version of Microsoft's MS DOS 1.10, which was itself a direct derivative of the SCP 86-DOS operating system that Microsoft had acquired from Seattle Computer Products. Versions 1.0 and 1.1 of

PC DOS were very similar to the CP/M (Control Program/Microprocessor) operating system, which was the dominant operating system for many of the 8-bit microcomputers that preceded the PC.

Then, as now, PC DOS operating system functions were provided by the interactive command and batch file processor COMMAND.COM, supported by disk file and general purpose I/O routines contained in the hidden files IBMDOS.COM and IBMBIO.COM. Simple commands such as for deleting, renaming, and copying files were performed directly by COMMAND.COM. More complex commands, such as for formatting and copying entire diskettes, were accomplished using one of the external programs provided with PC DOS. Version 1.1 added support for double-sided diskette drives and the EXE2BIN command, for a total of eighteen internal and external commands.

With Version 2.0, PC DOS became more than just a simple control program for the PC. Introduced along with the PC XT, PC DOS 2.0 provided forty-six internal and external commands, and its memory resident portion was twice as large as that of Version 1.1 (24 KB versus 12 KB). Several of the added commands were for the support of hard disks and for the support of hierarchical file directories on diskettes and hard disks. A new nine-sector-per-track diskette format was added, allowing double-sided diskettes to hold 360 KB, rather than the 320 KB provided by PC DOS 1.1's eight-sector-per-track format. Also added were commands to provide input/output redirection and other UNIX-like features, as well as background printing and enhanced batch processing commands.

Perhaps one of the most significant features provided in PC DOS 2.0 was that of installable device drivers. This feature allows device handlers to be installed at system initialization rather than requiring them to be built into the operating system itself. Device drivers are added to a system by specifying the directory and filename of the driver in the CONFIG.SYS file on the system disk. During system initialization, IBMBIO.COM loads the routines in these files into memory, where they remain resident to control their devices. This allows new types of devices to be added to a system without modifying the operating system, and saves memory since drivers for unused devices need not be loaded.

Version 2.1, like most new releases of PC DOS, was introduced to support new hardware—the IBM Portable PC and PCjr and the half-height 360 KB diskette drives featured on them. It provided no new commands and an insignificant increase in the amount of memory used.

Version 3.0 of PC DOS was introduced in April 1984 along with the PC AT, to support its 1.2 MB diskette drive and 20 MB hard disk. Although it was the only version of PC DOS offered for the AT, it did not take advantage of the 80286's protected virtual-address mode, and was also offered for other members of the IBM PC family. In addition to supporting the AT's new storage devices, it provided date, time, and currency formats for fifteen different countries and supported keyboard layouts for five countries other than the U.S. Also, for the first time, a RAM disk driver (VDISK.SYS) was provided. Because the memory resident portion of Version 3.0 was considerably larger than that of Version 2.1 (36

KB versus 24 KB), Version 2.1 remained available for use on IBM PCs other than the AT.

Version 3.1 was introduced in March 1985 to provide support for the IBM PC Network. Version 3.1 augmented the file sharing capabilities introduced in Version 3.0. Two new commands (JOIN and SUBST) were added. Respectively, these commands allowed files on more than one drive to be accessed using a single drive specifier, and a drive letter to be used in place of a directory path name.

More importantly, Version 3.1 allowed most DOS commands executed on a PC to operate on file directories, disks, and printers on other PCs on an IBM PC Network, as if they were on the local PC. Memory requirements for the resident portion of the operating system did not change, but the size of the portion of COMMAND.COM that could be overwritten if required when programs were loaded did increase slightly.

Version 3.20 of PC DOS was introduced in April 1986 along with the IBM PC Convertible. This version of PC DOS provided support for the Convertible and for the 720 KB, 3.5-inch diskette drives featured on it and offered for attachment to other IBM PCs. It also provided support for the IBM Token-Ring network. Two very useful commands—REPLACE and XCOPY—were added to PC DOS with Version 3.2. REPLACE provides for the replacement of all occurrences of a file on a disk; XCOPY selectively copies groups of files, including subdirectories.

The STACKS command was provided to override default values for the number and size of stack frames used by hardware interrupt service routines. DOS 3.2 allocates a stack frame for use by the interrupt service routine each time an interrupt occurs, and returns the frame to the pool once the interrupt has been processed.

More significant for users of systems with hard disks was the alteration of the FORMAT command to always require the specification of a target drive letter when executed. This change greatly lessened the likelihood of the default disk or diskette drive being accidentally reformatted. Because the resident portion of Version 3.2 was considerably larger than that of Version 3.1 (44 KB versus 36 KB), Version 3.1 remained available for use on IBM PCs not requiring Version 3.2's special features.

Features

As befits the most current version of PC DOS, Version 3.30 provides several new commands as well as refinements and enhancements to existing ones. Five new commands have been added: APPEND, CALL, CHCP, FASTOPEN, and NLSFUNC. The ATTRIB, BACKUP/RESTORE, COUNTRY, DATE/TIME, FDISK, FORMAT, GRAFTBL, GRAPHICS, MODE, STACKS, and SYS commands have been enhanced.

PC DOS 3.30 provides an input/output and system services interface to PC and Personal System/2 computers for users and applications programs. It provides a disk and diskette file management system, and the ability to initiate the execution of system and applications programs, as well as to communicate with external devices.

Version 3.30 supports the use of the 1.44 MB, 3.5-inch diskette drives featured on the Personal System/2 Models 50, 60, and 80; the 720 KB, 3.5-inch diskette drives used on the Personal System/2 Models 25 and 30 and PCs; and the 5.25-inch diskettes at all capacities supported by previous versions. Hard disks may now be partitioned into multiple logical drives using FDISK. The total size of the disk may be greater than 32 MB, but logical drive sizes may not be larger than 32 MB. See *Installation and Use*, later in this chapter, for additional information.

Several of the new commands have to do with data and batch file facilities. **APPEND** is a new command that allows data files in a directory other than the current directory to be accessed, in much the same way that program and batch files are accessed once a PATH command has been executed. This terminate-and-stay-resident utility is based on the APPEND command provided with the IBM PC Network Program and IBM PC LAN Program. APPEND searches for files in specified alternate directories if files are not found in the directory specified by the application.

FASTOPEN is another new terminate-and-stay-resident command that provides fast access to recently accessed files, by maintaining the locations of directories of recently opened files in memory. FASTOPEN also keeps directory entries for closed files. If the file is reopened soon after it is closed, FASTOPEN usually keeps its location in memory, enabling the file to be opened without disk I/O. FASTOPEN support is provided only for user-specified, nonremovable drives.

CALL is a new command that is used in one batch file to call another. When execution of the second batch file completes, control returns to the first at the statement following the CALL. Batch files may now use environment variables created using the SET command; a variable is referenced by enclosing its name in percent signs (%). The display of any line in a batch file, including the ECHO OFF batch command may now be suppressed by starting it with an ampersand (@). (A program whose name begins with an @ may be run by preceding its name with a drive letter—@C:@PGM.)

Several file management commands have been enhanced. **ATTRIB** has been expanded to be able to modify most file attributes for single files, selected files, or for all files at a given directory level. ATTRIB can also be used to find all occurrences of a particular filename; the command ATTRIB ABC.EXT /S finds and lists all occurrences of file ABC.EXT on the current drive.

The **BACKUP/RESTORE** commands have been enhanced to provide greater performance and convenience. A /F parameter has been added to BACKUP to support the formatting of previously unformatted diskettes. For better performance, BACKUP now writes only two files on each backup diskette (CONTROL.nnn and BACKUP.nnn, where nnn is the diskette number). The /L option generates a log file that contains a record for each file processed and identifies the diskette number of the backed-up file.

RESTORE can restore backup files generated by previous versions of BACKUP as well as those generated by BACKUP 3.30. RESTORE has been modified so that it no longer restores COMMAND.COM and the system files IBMBIO.COM and IBMDOS.COM. This allows RESTORE to be used to load files onto a hard disk without the often undesirable result of replacing the operating system files.

FORMAT has been modified to support the formatting of 1.44 MB, 3.5-inch diskettes. Lower capacity, 720 KB, 3.5-inch diskettes are formatted using the /N:9 and /T:80 (9 sectors per track and 80 tracks) parameters. It is the users responsibility to specify parameters and use the correct capacity diskettes; FORMAT indiscriminately attempts to format any diskette inserted in the drive at the highest capacity supported by the drive.

The **DATE** and **TIME** commands set DOS's date and time as in previous versions, but they now also update the date and time stored in the permanent clock memory of those systems that feature IBM standard clocks. This includes all Personal System/2 and AT models. The date and time can only be updated using these commands once it has been set initially, using the system Reference or Diagnostic Diskette.

GRAPHICS, the DOS graphics print screen utility, has been updated to provide support for the PC Convertible's thermal printer and LCD (liquid crystal display) display.

MODE now supports four asynchronous ports and allows communications rates of up to 19,200 bps to be set.

Version 3.30 supports eleven national languages—U.S. English, U.K. English, German, French, Italian, Spanish, Danish, Dutch, Norwegian, Portuguese, and Swedish. The **COUNTRY** command now indicates the location of a file (COUNTRY.SYS) used to provide extended national language support. If the location of the file is not specified, the file is assumed to be in the root directory of the system disk.

A new load keyboard command, **KEYB**, can be used to replace the BIOS resident keyboard program to support other than U.S. English keyboards. Tables for converting keyboard scan codes to ASCII codes are contained in the file KEYBOARD.SYS. KEYB assumes this file is located in the root directory of the system disk, unless otherwise indicated when the command is invoked. The KEYB command supports seventeen keyboards and replaces the five individual KEYBxx programs provided with earlier versions of DOS. AUTOEXEC.BAT files that use one of the KEYBxx programs must be changed to use KEYB instead.

NLSFUNC is a new terminate-and-stay-resident command that, when used in conjunction with the MODE and CHCP commands, allows code page selection and switching if hardware support is available. A code page is a set of characters and symbols appropriate to a given country. The **MODE** command is used to select a code page to be used with a particular device; **CHCP** is used to select the code page that DOS and as many devices as possible will use. Devices that support code page selection currently include the 4201 Proprinter, 5202 Quietwriter III, EGA/VGA devices, and the Convertible LCD display.

STACKS has been modified to allow the allocation of stack frames to hardware interrupt service routines by DOS to be disabled. Disabling this allocation causes hardware interrupt service routines to use the interrupted program's stack area, as was the practice with versions of DOS before 3.20.

SYS has been upgraded to transfer the DOS system files to a diskette containing a previous version of DOS only if there is sufficient space for a successful transfer. Previous versions of SYS overwrite existing system files before determining if sufficient space is available for the new system file, which can result in a diskette that can no longer be used to initialize the system.

DISPLAY.SYS and **PRINTER.SYS** are two new device drivers provided with DOS 3.30 to support code page switching. DISPLAY.SYS provides code page support for the CON system device; PRINTER.SYS provides code page switching support for the LPT1, LPT2, and LPT3 printer devices. Both support the download of up to twelve code page images, selection of one of the code page images, and query of the active code page and the list of possible code pages.

DOS 3.30 provides several new interrupt 21H DOS functions. These functions support the use of extended country information, code page switching, and enhanced reliability and performance. The **65H** (Get Extended Country Information) function is used to request extended country information (such as date, time, number, and currency format) for any country in any code page. NLSFUNC must be loaded in order to access information for any country/code page other than the one currently in use. The **66H** (Get/Set Global Code Page) function changes the code page for the current country. NLSFUNC must be installed when this function call is used.

DOS 3.20 applications were limited to twenty files per process; using the **67H** (Set Handle Count) function, an application can specify the maximum number of files that an application can open (up to 255). The **68H** (Commit File) function enables an application to write all modified file information to disk, including any information in data buffers, without closing the file. This provides the reliability of device storage of modified data without the performance degradation involved in repeatedly closing and reopening a file.

The resident portion of DOS 3.30 is larger than that of Version 3.20 (46 KB versus 44 KB). Additional memory is required if the DISPLAY.SYS, FASTOPEN, NLSFUNC, and PRINTER.SYS extensions are installed.

Packaging and Documentation

DOS Version 3.30 is provided in a standard IBM slip-case package. The package includes the software (contained on a single 720 KB, 3.5-inch diskette and on two 360 KB, 5.25-inch diskettes), a spiral-bound User's Guide, a User's Reference manual in a three-ring binder, and a Quick Reference Card. Also included is a pam-

DOS Version 3.30

phlet entitled "Read This First," which directs the user to different sections of the User's Guide or Reference manual for installation instructions, depending on the type of system on which DOS is to be installed. Documentation for BASIC 3.30, which is included with DOS 3.30, is provided in the IBM BASIC Reference Version 3.3. This manual is available at additional cost.

Also available at extra cost is the Disk Operating System Version 3.30 Technical Reference manual. This manual is a necessity for systems programmers and application developers. It provides technical information on the operation of DOS, including descriptions of DOS interrupts and function calls.

The Technical Reference is provided in an IBM standard slip-case package that matches the DOS package. The package includes the Technical Reference manual in a three-ring binder and a Technical Quick Reference Card. Also provided are a 720 KB, 3.5-inch diskette and a 360 KB, 5.25-inch diskette. Both diskettes contain the DOS utilities DEBUG.COM, EXE2BIN.EXE, LIB.EXE, and LINK.EXE, and VDISK.ASM, which is the assembly language source for the VDISK RAM disk device driver provided with DOS. DEBUG.COM is also included on the DOS diskettes, but the documentation for it is only included in the Technical Reference manual.

Updates

An update for DOS 3.30 dated August 31, 1987 is available from IBM and authorized dealers to remedy several problems. One part of the update is a replacement for the BACKUP utility, and should be applied to all installed DOS systems. Another part of the update concerns deficiencies in the ROM BIOS of the Personal System/2 Models 50, 60, and 80 when used with DOS 3.30, and should be applied only to those systems.

The BACKUP utility (dated 3-18-87) provided with DOS 3.30 will not properly back up a large number of subdirectories in a given directory. The update diskette provides a new version of BACKUP.COM (dated 7-24-87) and a batch file for installing it on the user's DOS diskette or hard disk. The new version of BACKUP.COM is installed by inserting the update diskette in drive A, making A the default drive, and entering the command INSTALL X:, where X is the drive letter of the DOS disk or diskette to be updated. The INSTALL batch file contained on the diskette uses the DOS REPLACE utility (also contained on the diskette) to replace all occurrences of BACKUP.COM on the hard disk. Users who have both DOS and OS/2 system files on their hard disks should manually copy this new version of BACKUP.COM to the DOS system file directory. If INSTALL is used, it replaces the OS/2 version of BACKUP.COM in the OS/2 system file directory as well as the DOS version in the DOS system file directory.

The second part of the update has to do with deficiencies in the ROM BIOS of the Personal System/2 Models 50, 60, and 80 when used with DOS 3.30. The update provides a DOS device driver to remedy problems that display the following symptoms:

- ▶ Intermittent Not Ready Error Reading Drive A message is displayed when performing DIR or COPY commands from drive A.
- ▶ Intermittent problem diskette drive Not Ready or hard disk General Failure messages appear.
- ▶ FORMAT fails with a Track 0 bad or invalid media message when a user replies Y to the prompt Format another (Y/N)?, in order to format a second 3.5-inch diskette. This problem occurs only if the system has been initialized from the hard disk.
- ▶ When power is interrupted or switched on and off quickly, error codes 301 and 8602 may be displayed.

The device driver provided to remedy these problems is DASDDRVR.SYS and is dated 8-21-87. It requires less than 1 KB of memory when loaded. It is installed by inserting the update diskette in drive A, making A the default drive, and entering the command FIXBIOS X:, where X is the drive letter of the DOS disk or diskette to be updated. The FIXBIOS batch file copies DASDDRVR.SYS to the root directory of the target disk; it then creates a CONFIG.SYS file containing the command DEVICE=\DASDDRVR.SYS, or adds this command as the first line of the existing CONFIG.SYS file and renames the existing CONFIG.SYS file CONFIG.OLD. Users wishing to keep the root directory of the hard disk uncluttered can copy DASDDRVR.SYS from the update diskette to the DOS system file directory, and insert the DEVICE= statement with the full path name of the DASDDRVR.SYS file in the CONFIG.SYS file, using a text editor.

A second version of DASDRVR.SYS (Version 1.20, dated 9-28-87) addresses the intermittent display of 162/163 errors during system initialization and loss of the correct time while the system is powered off. It also remedies the problems described above, and should be installed and used in place of the earlier version.

The symptoms listed above can be caused by factors other than the BIOS deficiencies that this update corrects. Real disk and diskette failures can occur, as well as spurious ones. If symptoms persist after the installation of the update, normal problem determination procedures should be followed.

Installation and Use

Installation and use of DOS 3.30 requires an IBM Personal Computer or Personal System/2 computer with an IBM or equivalent display; at least one diskette drive capable of reading 360 KB, 5.25-inch diskettes or 720 KB, 3.5-inch diskettes; and at least 128 KB of memory. Although not advertised or guaranteed by IBM, DOS

3.30 can also be successfully installed and used on most IBM PC-compatible computers. Compatible computers cannot, however, use IBM's BASIC because it operates using portions of ROM BASIC, which is available only on IBM PC and Personal System/2 computers.

Instructions for the installation of DOS 3.30 on a diskette-only system are provided in Chapter 4 of the User's Guide. Instructions for installing DOS 3.30 on a hard disk that currently has a version of DOS installed are provided in Chapter 3 of the Reference manual. Procedures for installing DOS on a hard disk that has not been used for DOS before are provided in Chapter 4 of the User's Guide.

Installation of DOS on a diskette-only system or on a hard disk not previously used for DOS is performed using the SELECT command. The installation process is initiated by inserting the DOS Startup/Operating Diskette in drive A and turning on the computer or pressing Ctrl-Alt-Del (if the computer is already on). The user should enter the current date and time when prompted if the displayed date and time are incorrect.

Diskette-Only System

On a diskette-only system, the command SELECT XXX YY should be entered after the system is initialized. XXX is one of the twenty-one supported country codes and YY is one of the seventeen supported keyboard codes. See Table 10-1 for a list of supported country and keyboard codes.

Table 10-1 DOS 3.3 Country and Keyboard Codes. *International support is increased in DOS 3.3 to include twenty-one different countries and seventeen keyboards.*

Country	Country Code	Keyboard Code
Arabic	785	US
Australia	061	US
Belgium	032	BE
Canada (English)	001	US
Canada (French)	002	CF
Denmark	045	DK
Finland	358	SU
France	033	FR
Germany	049	GR
Hebrew	972	US
Italy	039	IT
Latin America	003	LA
Netherlands	031	NL
Norway	047	NO
Portugal	351	PO
Spain	034	SP
Sweden	046	SV

Table 10-1 (cont.)

Country	Country Code	Keyboard Code
Switzerland (French)	041	SF
Switzerland (German)	041	SG
United Kingdom	044	UK
United States	001	US

The installation procedure then informs the user that the process will destroy any data on the target diskette to be inserted in drive B. Once the user acknowledges this by pressing Enter, the DOS FORMAT utility is activated; the user is directed to insert a diskette in drive B and press Enter. FORMAT then begins operation, displaying the head and cylinder number being processed as the format operation progresses. After a minute or so, a message indicates that the format operation is complete and that the system has been transferred, and asks if the user wishes to format another diskette. The user should reply N for no; SELECT then copies the contents of the diskette on which it resides to the diskette in drive B. It also generates CONFIG.SYS and AUTOEXEC.BAT files on the diskette in drive B.

The CONFIG.SYS file contains a COUNTRY statement indicating the country and keyboard codes specified when SELECT was invoked. If the country code and keyboard selected were 001 and US, respectively, this statement can be removed from the CONFIG.SYS file, because these are the system default values. The AUTOEXEC.BAT file contains statements to set the path to the root directory and request the DATE and TIME, and a KEYB command to load the keyboard definition specified in the SELECT command from the file KEYBOARD.SYS. The DATE and TIME commands are not required if the DOS diskette is to be used on an AT or Personal System/2 computer and may be removed; the KEYBD command is not required for systems employing the U.S. country and keyboard codes and may likewise be removed.

This completes the process for installing DOS from the 3.5-inch diskette. The procedures described here work on a single- or dual-diskette machine. The only difference is that, for a single-drive system, each time (and there will be several times) a message requests that a target diskette be placed in drive B, the user must remove the source diskette from drive A, place the target diskette in drive B, and press Enter . . . and vice versa when a message requests that the source diskette be placed in drive A.

This completes the installation process. The original DOS diskettes should be put away for safekeeping, and the DOS Startup or DOS Startup/Operating diskette created during the installation process should be used to initialize the system for day-to-day use. For a system with only a 360 KB drive or drives, the generated DOS Operating diskette will also be required whenever it is necessary to execute the DOS programs stored on it.

Hard Disk Systems

Installing DOS on a hard disk not previously used for DOS is a four-step process—initializing the system using the DOS 3.30 diskette, preparing the hard disk using FDISK, selecting the keyboard and country codes and transferring system files using the SELECT command, and transferring the remainder of the DOS files using a COPY command.

Installing DOS on a hard disk that currently has a version of DOS installed is mainly a matter of replacing existing system files. In either case, the first step is to insert the 3.5-inch DOS Startup/Operating Diskette in drive A and turn on the computer, or press Ctrl-Alt-Del if the computer is already on. The user should enter the current date and time when prompted, if the displayed date and time are incorrect.

The next step for a hard disk on which DOS has not been previously installed is to run FDISK. This program is started by entering the command FDISK after the DOS A> prompt is displayed. This command is used interactively to create a primary DOS partition on the disk and to make it the active partition (the hard disk partition from which the system is initialized.) Any attempt to install DOS on a hard disk that has not previously been prepared using FDISK will result in an Invalid drive specification message. Once FDISK is loaded into memory, the menu shown in Figure 10-1 is displayed.

```
IBM Personal Computer
Fixed Disk Setup Program Version 3.30
(C)Copyright IBM Corp. 1983,1987

FDISK Options

Current Fixed Disk Drive: 1

Choose one of the following:

     1. Create DOS partition
     2. Change Active Partition
     3. Delete DOS partition
     4. Display Partition Information




Enter choice: [1]




Press ESC to return to DOS
```

Figure 10-1   FDISK Main Menu. *The hard disk is partitioned using FDISK.*

Choosing the default selection (*Create DOS Partition*) produces the menu shown in Figure 10-2. From this menu the user should select the default option

(*Create Primary DOS Partition*). Selection of this option displays a message asking the user to indicate if the largest DOS partition possible on the hard disk (32 MB or the size of the disk, whichever is smaller) should be created and made active. (See Figure 10-3.) If DOS is being installed on a hard disk that is 32 MB or less in size, the default option (Y) should be selected, unless the user wishes to use the hard disk as multiple logical drives (see below). Once that selection is made, a message is displayed indicating that the system will be restarted after any key is pressed.

```
Create DOS Partition

Current Fixed Disk Drive: 1


    1. Create Primary DOS partition
    2. Create Extended DOS partition




Enter choice: [1]



Press ESC to return to FDISK Options
```

Figure 10-2   Create DOS Partition Menu. *This menu can be used to create a primary or extended DOS partition.*

Using DOS Version 3.30, a hard disk can be accessed as more than one logical drive by defining an extended DOS partition on the disk. The primary DOS partition is limited in size to 32 MB; an extended DOS partition must be defined in order for DOS to use more than 32 MB of space on a hard disk. An extended partition may be defined immediately after FDISK is used to define the primary partition, or it may be defined later, even after DOS has been completely installed on the primary DOS partition (provided there is room remaining on the hard disk).

An extended DOS partition can be defined immediately after the primary DOS partition is defined, if the primary partition is defined and made active manually. This is often desirable if the user wishes to have the hard disk divided into logical drives approximately equal in size, rather than permitting FDISK to define the size of the primary DOS partition.

If FDISK is allowed to allocate the space for the primary partition, it will allocate 32 MB (if the disk is large enough). This can result in the extended DOS partition and any logical drives defined in it being rather small (12 MB on a 44 MB

```
Create Primary DOS Partition

Current Fixed Disk Drive: 1

Do you wish to use the maximum size
for a DOS partition and make the DOS
partition active (Y/N).........? [Y]




















Press ESC to return to FDISK Options
```

Figure 10-3 Create Primary DOS Partition Screen. *Answering Y causes the largest DOS partition possible to be created on the hard disk.*

hard disk, for instance). Also, when the partitions are formatted using the DOS FORMAT command, a partition less than 16 MB in size is organized using 4 KB clusters, rather than the 2 KB clusters used with larger partition sizes. Because a cluster is the smallest increment of disk space that can be assigned to a file, this can result in a significant amount of wasted disk space if the partition contains a large number of small files.

The primary partition is defined manually by entering N when the *Create Primary DOS Partition* screen is displayed. This displays a screen that specifies the number of cylinders on the disk and the size in cylinders of the largest primary DOS partition that can be created on the disk. The size of the primary DOS partition is limited to 32 MB; thus, the maximum size will be the largest number of cylinders that is less than 32 MB (if the size of the disk is more than 32 MB) or the total number of cylinders on the disk (if the disk is smaller than 32 MB). (See Figure 10-4.) The user must press Enter to accept the default value displayed or indicate the size (in cylinders) of the primary DOS partition to be created. Most users will probably find it convenient to make the primary and extended DOS partitions approximately the same size.

Once the user enters the desired size and presses Enter, a message is displayed indicating that the primary DOS partition has been created. Pressing Esc returns control to the FDISK main menu, so that *Create DOS Partition* and then *Create Extended DOS Partition* can be selected. This results in the menu shown in Figure 10-5 (actual cylinder values, of course, depend on the size of the disk and the size of the primary partition created). The size of the extended partition is specified in cylinders, and may occupy the entire remainder of the hard disk. The

Chapter 10

extended partition can be more than 32 MB in size; however, there can be only one primary and one extended DOS partition on a hard disk. Entering a size in cylinders and/or pressing Enter causes the extended DOS partition to be created and a message screen to be displayed showing the location and size of each partition.

```
Create Primary DOS Partition

Current Fixed Disk Drive: 1




Total disk space is  731 cylinders.
Maximum space available for partition
is  550 cylinders.

Enter partition size...........: [ 550]


No partitions defined

Press ESC to return to FDISK Options
```

Figure 10-4 Specify Primary Partition Size. *The size of the primary DOS partition is limited to 32 MB.*

```
Create Extended DOS Partition

Current Fixed Disk Drive: 1

Partition Status    Type   Start  End  Size
    C: 1            PRI DOS    0   365   366




Total disk space is  731 cylinders.
Maximum space available for partition
is  365 cylinders.

Enter partition size...........: [ 365]




Press ESC to return to FDISK Options
```

Figure 10-5 Create Extended DOS Partition. *The system cannot be initialized from the DOS extended partition.*

DOS Version 3.30

Pressing Esc displays the *Create Logical DOS Drive(s)* menu, shown in Figure 10-6. Pressing Enter with less than the total number of disk cylinders in the extended DOS partition specified causes that number of cylinders to be assigned to the next available drive letter, and the user to be repeatedly prompted until all of the space in the extended partition is assigned to a logical drive or all available drive letters are used (Z is the last available). Logical DOS drives can be up to 32 MB in size. Multiple logical drives are necessary in order to utilize the full capacity of disk drives on the Personal System/2 Models 60 and 80 that are greater than 64 MB in size.

```
Create Logical DOS Drive(s)




No logical drives defined


Total partition size is   365 cylinders.

Maximum space available for logical
drive is   365 cylinders.

Enter logical drive size........: [ 365]


Press ESC to return to FDISK Options
```

Figure 10-6 Create Logical DOS Drive(s). *Logical drives can be up to 32 MB in size; they are necessary for DOS to utilize the entire capacity of large hard disks.*

Once the logical drives have been assigned, the user can press Esc to return control to the FDISK main menu. *Change Active Partition* should then be selected, and the primary DOS partition should be selected as the partition to be made active. The resultant display is shown in Figure 10-7. The primary partition must be made active in order for DOS to be loaded from it each time the system is initialized. It is made active automatically if FDISK is instructed to create the maximum size DOS partition possible in the *Create Primary DOS Partition* screen, but must be done manually if the primary DOS partition is defined manually. Omitting this step results in ROM BASIC being loaded whenever system initialization from the hard disk is attempted. Once the primary DOS partition has been selected as the partition to be made active, the user can press Esc to return to the FDISK main menu, then press Esc again followed by any key to restart the system.

```
Change Active Partition

Current Fixed Disk Drive: 1

Partition Status    Type   Start  End  Size
  C: 1      A      PRI DOS    0   365   366
     2             EXT DOS  366   730   365




Total disk space is  731 cylinders.







Partition 1 made active

Press ESC to return to FDISK Options
```

Figure 10-7 Change Active Partition Screen. *The active DOS partition is the partition from which the system is initialized.*

Once the system is restarted, the command SELECT C:\DOS XXX YY should be entered. DOS is the subdirectory on drive C into which the majority of the DOS files will be copied. Designation of such a subdirectory, though not mentioned in the User's Guide installation procedures, is highly desirable. It significantly reduces the number of files in the root directory, allowing the user to see subdirectory names, uncluttered by fifty DOS filenames, when a DIR is executed. XXX is one of the twenty-one supported country codes and YY is one of the seventeen supported keyboard codes listed in Table 10-1. SELECT informs the user that the process will destroy any data on the target disk (the primary partition of drive C). Once the user acknowledges this by pressing Enter and confirms the acknowledgment by pressing Y and Enter, the DOS FORMAT utility is activated.

FORMAT then begins operation, displaying the head and cylinder number being processed as the format operation progresses. After several minutes, a message indicates that the format operation is complete and the system has been transferred, and asks if the user wishes to enter a volume identification label for the DOS partition or press Enter for none. The volume label is not required but is recommended, because FORMAT will not format a hard disk with an existing volume label until the volume label is entered. This feature provides an additional check to assure that the user really wishes to format the drive specified on the FORMAT command. Once the volume label is entered and/or Enter is pressed, SELECT copies the contents of the diskette on which it resides to the root directory of drive C. It also generates CONFIG.SYS and AUTOEXEC.BAT files on drive C.

As noted above, if the country code and keyboard selected were 001 and

US, respectively, the COUNTRY statement placed in the CONFIG.SYS file by SELECT can be removed, because these are the system default values. Also, the DATE and TIME commands placed in the AUTOEXEC.BAT file are not required on an AT or Personal System/2 computer and may be removed; the KEYB command is not required for systems employing the U.S. country and keyboard codes and can be removed as well.

This completes the installation of DOS in the primary partition. Any logical drives created in the extended DOS partition must be formatted using the FORMAT command before they are used. Use of the /V option, so that a volume label can be assigned to the logical drive, is recommended for the reason mentioned above.

It is essential that all logical drives be formatted using the FORMAT command. If this step is omitted it may be possible to store files on the drive, but reads and writes to defective disk blocks can occur, because the defective blocks have not been identified by FORMAT. This usually results in a `General Failure error writing drive x` message, when data is written to a defective block. Even worse, data can sometimes be written to a marginally defective block, but then cannot be successfully read. One way of checking to make sure that the drive has been formatted is to use the CHKDSK command on the logical drive. If it has not been formatted, CHKDSK displays a `Probable non-DOS disk` message.

Users who wish to use code page switching should modify the CONFIG.SYS and AUTOEXEC.BAT files generated during the installation process, using EDLIN or the text editor of their choice. DISPLAY.SYS and PRINTER.SYS `DEVICE=` statements must be added to the CONFIG.SYS file, and the NLSFUNC and several MODE commands must be added to the AUTOEXEC.BAT file. Devices that currently support code page selection include the 4201 Proprinter, 5202 Quietwriter III, EGA/VGA devices, and the Convertible LCD display.

This completes the installation process except for any customization of the AUTOEXEC.BAT and CONFIG.SYS files desired by the user. Each time the system is turned on or Ctrl-Alt-Del is pressed, the system is initialized from the copy of DOS 3.30 installed on the hard disk (unless a system or diagnostic diskette is in drive A, in which case the system is initialized from the software contained on the diskette).

Enhanced User Interfaces

The user interface for DOS Version 3.30 is the familiar command line interface provided by earlier versions. No menu facilities are provided, and only the previous command line entered can be recalled and edited or reentered. This simple command interface, while adequate for PCs and their users in 1981, is cryptic and burdensome for many of today's users.

Fortunately, many commercially available and custom applications provide user-oriented interfaces that minimize the user's interaction with the DOS command interface. General purpose interfaces that extend or replace the DOS command interface are also commercially available. Products available include Bourbaki's 1dir+, Executive System's Xtree, and the WordPerfect SHELL program included in the WordPerfect Library. These products, called user interfaces or shells, allow the user to execute DOS commands, perform file operations, and execute programs, by selecting items from menus or lists rather than entering commands. For programmers and system developers intimately familiar with DOS commands, this multiple-choice approach may not be as efficient as DOS's fill-in-the-blank approach, but it is efficient and helpful for many DOS users.

Some interfaces, such as the WordPerfect SHELL, allow context switching. That is, they allow the user to suspend the execution of a program, return to the shell, initiate another program, then later return to the suspended program, and vice versa. Many word processors have provided the ability to suspend execution in order to execute another program, but the ability to leave multiple programs in a suspended state and then return to them is a fairly new and useful feature.

Users who want to run multiple applications concurrently and/or see the output of more than one application on the screen at the same time should consider more sophisticated products known as *operating environments*. Products available include Digital Research's GEM, IBM's TopView, Microsoft Windows, and Quarterdeck's DESQview.

These products allow well-behaved DOS applications (typically ones that request system services through DOS and BIOS, rather than accessing hardware devices directly) to share system resources, including the CPU, memory, and the display screen. The user can start the execution of applications from menus or lists. The output of these applications can be viewed as a full display screen or in multiple screen windows, if the application adheres to rules specific to each environment. Microsoft Windows, for instance, only supports the concurrent execution of applications developed using the Windows Application Programming Interface (API).

These environments all provide restricted multi-tasking of applications, because DOS and the 8088 and 8086 processors are not designed for multi-tasked use. General purpose multi-tasked execution of PC applications is available using a multi-tasking operating system and an 80286-based computer (such as the AT, XT 286, or Personal System/2 Models 50 and 60) or an 80386-based computer such as the Personal System/2 Model 80. Multi-tasking PC operating systems offered by IBM include Operating System/2 for IBM 80286-based computers, which is described in the next chapter, and AIX for the Personal System/2 Model 80, which is described in Chapter 12. These systems provide for the multi-tasked execution of programs written using their respective APIs.

The multi-tasked operation of existing DOS applications can be accomplished using the virtual machine capabilities of 80386-based computers and appropriate control software. Control software (both operating systems and operating environments) that provides this capability when used with the 80386-based Personal System/2 Model 80 is described in Chapter 13.

11

Operating System/2 (OS/2)

    Beginning with the announcement of the 80286-based AT in 1984, there were many rumors and much speculation about a soon-to-be-released, advanced version of DOS. No official announcements were made, however, until April 1987 when IBM introduced the Personal System/2 family of computers. A new version of DOS (3.30) was announced; however, although required for use on Personal System/2 computers, it did not significantly improve upon the features provided by DOS 3.2. This disappointing news was offset by IBM's concurrent announcement of a new operating system called Operating System/2 (OS/2).

    Though primarily developed by Microsoft, OS/2 has an IBM name, complete with a slash. In its most complete configuration it provides services similar to those provided by other IBM operating systems, and supports extensive sharing of data and resources among other IBM computer systems, as well as among PCs.

    OS/2 is a single-user, multi-tasking operating system for use on 80286- and 80386-based personal computers. It is offered by IBM for use on the 80286-based Personal System/2 Models 50 and 60 and the 80386-based Personal System/2 Model 80. OS/2 can be used on the PC AT and the XT 286, but not on other IBM PC models. OS/2 is more hardware specific than DOS. IBM's version of OS/2 is designed for use only on IBM computers; it cannot be used reliably on computers from other vendors. Several major vendors of AT-compatible computers, including AST Research and Compaq Computer Corporation, offer a version of OS/2 tailored to their computers. OS/2 cannot be used on the 8086-based Personal System/2 Models 25 and 30 or on other personal computers that do not feature an 80286 or 80386 central processor.

    IBM is offering OS/2 in two configurations—standard and enhanced. The standard edition provides large memory support, simultaneous execution of multiple applications, and graphics and windowing capabilities. The extended edition provides all the capabilities of the standard edition, plus a communications manager and a database manager. Both configurations of OS/2 can run multiple OS/2 applications simultaneously with one DOS 3.30 application.

The standard edition is being released in phases. Version 1.0, which provides all the features just mentioned except graphics and windowing capabilities, was released in December 1987. Version 1.1 is to be released in October 1988. This version includes the OS/2 Presentation Manager, which provides graphics and windowing capabilities. Hard disk support for files and partitions greater than 32 MB in size and a system editor are also provided.

The extended edition will also be released in phases. Version 1.0, available in July 1988, provides a relational database manager and a communications manager, as well as all the capabilities of OS/2 Standard Edition, Version 1.0. The OS/2 Database Manager is consistent with the IBM host family of relational database products: Database 2 (DB2), Structured Query Language/Data System (SQL/DS), and Query Management Facility (QMF). The Communications Manager allows personal computer systems to host network communications, using a variety of methods and protocols including synchronous data link control (SDLC), Distributed Function Terminal (DFT) mode to a 3x74 cluster controller, Token-Ring network and IBM PC Network, and asynchronous links.

OS/2 Extended Edition, Version 1.1, is to be available in November 1988. It provides support for the OS/2 LAN server program, provides concurrent communications services for Token-Ring network and PC Network, and provides programming interfaces for NETBIOS and LAN IEEE 802.2, in addition to services provided by Version 1.0 of the extended edition and Version 1.1 of the standard edition of OS/2.

Languages and development tools for use with OS/2 were announced at the same time as the operating system. Available languages include C/2, COBOL/2, FORTRAN/2, Macro Assembler/2, Pascal Compiler/2, and BASIC Compiler/2. An OS/2 Programmer Toolkit and an OS/2 Graphics Development Toolkit are available to aid in the development of OS/2 applications. Features provided by each edition and version of OS/2 are shown in Table 11-1.

Table 11-1 OS/2 Features. *OS/2 is a multi-tasking operating system that makes use of the protected mode of the 80286 processor.*

Feature	Standard Edition 1.0	Standard Edition 1.1	Extended Edition 1.0	Extended Edition 1.1
Database Manager			•	•
Communications Manager			•	•
Files larger than 32 MB		•		•
Presentation Manager		•		•
Online contextual help	•	•	•	•
DOS compatibility	•	•	•	•
Multi-tasking	•	•	•	•
Large memory support	•	•	•	•

System Architecture

OS/2 is an operating system designed to provide concurrent execution of large programs. It is designed to provide new services for the user and the application developer by exploiting the Intel 80286's protected mode operation features. Features provided by OS/2 include:

- Large memory support
- Concurrent execution of multiple applications
- National language support
- High-level programming interface
- DOS 3.30 compatibility
- Ease of use facilities
- Systems application architecture support
- Presentation manager

Additionally, the extended edition provides a communications manager and a database manager.

Memory Management

Unlike DOS, OS/2 exploits the 80286 processor's capability of supporting up to 16 MB of physical memory and up to 1 gigabyte of logical address space when running in protected mode. This means that the size of applications written for OS/2 is not limited to 640 KB, as DOS applications are, nor is it limited to the size of physical memory. An application can use up to 16 MB of memory, with OS/2 maintaining only the most active segments in physical memory. OS/2 maintains the contents of the remainder of the application's memory on hard disk and transfers segments into physical memory as required.

OS/2 applications can allocate a large number of data segments, each up to 64 KB in size; but OS/2, not the application, controls the location of the segments. OS/2 maintains a *local descriptor table* (LDT) for each application. This table contains the physical memory location and length of each of the application's code and data segments.

As instructions are executed, the actual physical address referenced is determined by using the value in the segment portion of the address to reference an entry in the LDT. The referenced entry is then combined with the offset portion of the address to calculate the actual physical address. A *global descriptor table* (GDT) is also maintained, which defines code and data segments that are available to all tasks.

Chapter 11

OS/2 assigns memory address space to applications by creating and maintaining an LDT for each application, as well as the GDT. All accesses to memory by an application must be made through its LDT or the GDT (neither of which can be modified by the application). Therefore, an application cannot access memory outside the address space defined by the descriptor tables assigned to it by OS/2.

OS/2 applications can execute anywhere in the 80286's physical memory; DOS applications can only execute in the lower 640 KB of physical memory. OS/2's use of physical memory is shown in Figure 11-1. If OS/2 is not configured to run a single DOS application also, all of the OS/2 system code is stored in the lower 640 KB of memory, rather than being split between that area and the space above 1 MB.

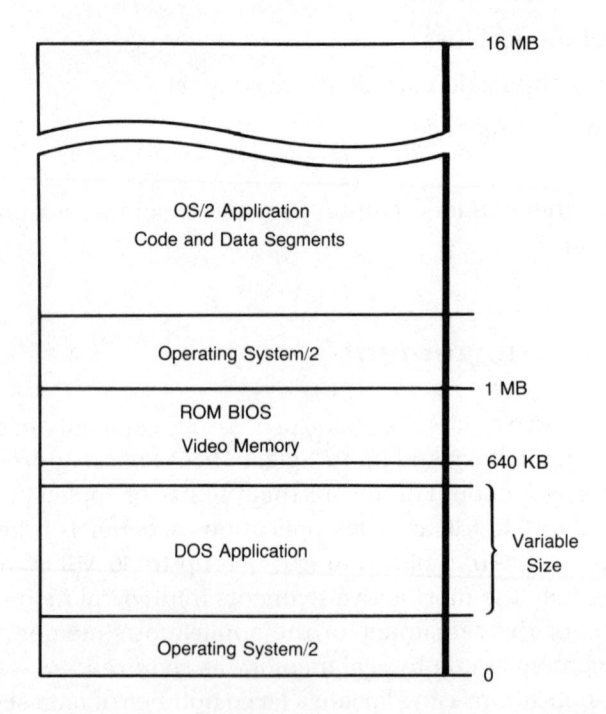

Figure 11-1 OS/2 Physical Memory Map. *An application can use up to 16 MB of memory; OS/2 maintains only the most active segments in physical memory.*

Multi-Tasking

The real power of OS/2 comes from its ability to support the concurrent execution of multiple applications, in addition to allowing each of them to use more than

640 KB of memory. In OS/2 Version 1.0, up to twelve OS/2 sessions can be active concurrently and each session can be an independent OS/2 application. For example, a user can start a text editor as one session, a spreadsheet as a second, and a compiler as a third. Applications are protected from each other by the multi-level design of OS/2 and by the 80286's segmented memory architecture.

OS/2 is organized in layers. At the lowest level are the device drivers, which interface with the system hardware. Above them comes the kernel of the operating system, which provides basic system services. Above basic system services are the various subsystems with which applications can interact via the OS/2 function call application program interface (API). The overall system structure of OS/2 is shown in Figure 11-2.

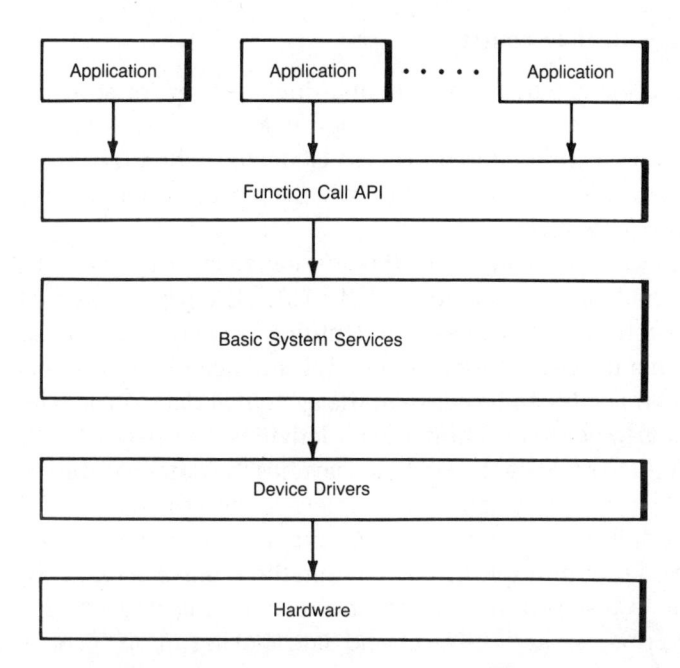

Figure 11-2   OS/2 System Structure. *OS/2 is organized with multiple system layers between the application and the hardware.*

Session Management

The highest level OS/2 multi-tasking element consists of one or more processes and is called a *session*. A session is sometimes referred to as a screen group (a group of processes that share the logical screen buffer and the keyboard). Each session requires its own screen group unless it is running under the control of the Presentation Manager. When a session is in the foreground, it writes directly to the screen and receives input from the keyboard; all sessions not in the foreground are said to be in the background. Background OS/2 sessions continue to execute until they become blocked for some reason, such as waiting for keyboard input. The single DOS compatibility environment session executes only when it is in the foreground.

Sessions are typically initiated and switched among, using the OS/2 Program Selector. An application can be initiated directly from the Program Selector if it is listed, or it can be initiated from the OS/2 command prompt once a command processor session has been initiated. Accessing a session using the Program Selector causes it to become the foreground session; the contents of its logical screen buffer are copied to the display and data entered from the physical keyboard is directed to the session's logical keyboard.

Processes and Threads

An application can take the form of a single process or several cooperating processes. A *process* is an executing program and all the system resources assigned to it. An application's initial process is started when its session is started; this process can start additional processes. Each process has a process ID and a local descriptor table (LDT).

Processes are made up of *threads*, which are the smallest OS/2 entities that can be scheduled for execution by the CPU. Every process contains at least one thread. Each thread has its own identifier, priority, and processor state, but all threads in a process share a single LDT and access to files, pipes, and memory. Up to 255 threads can be active in the system at once. Processes can start multiple threads to perform simultaneous activities. For instance, a data manager can be designed to create a thread that searches the database for requested information while another thread accepts queries from the user.

An application can perform simultaneous functions using either cooperating processes or multiple threads within the same process; however, the amount of overhead involved in creating and terminating threads is considerably less than for a process, and resource and data sharing among processes is more complicated than among threads.

OS/2 provides concurrent execution of threads (and thus applications) by allocating the CPU to each for a small amount of time. When the CPU is switched to a thread for the duration of its execution time slice (248 milliseconds by default), the local descriptor table (LDT) for the thread's process becomes the current LDT. This enables it to access the memory allocated to the process by OS/2 as specified in the LDT, and to access code and data segments available to all tasks with the appropriate privilege level as specified in the global description table.

All memory segments in an OS/2 system, whether system or application, have a privilege level associated with them. OS/2 and its device drivers run at privilege level 0; special purpose routines that require input/output privilege (IOPL) run at level 2; and applications run at privilege level 3. Level 1 is currently unused. (See Figure 11-3.)

Operating System/2

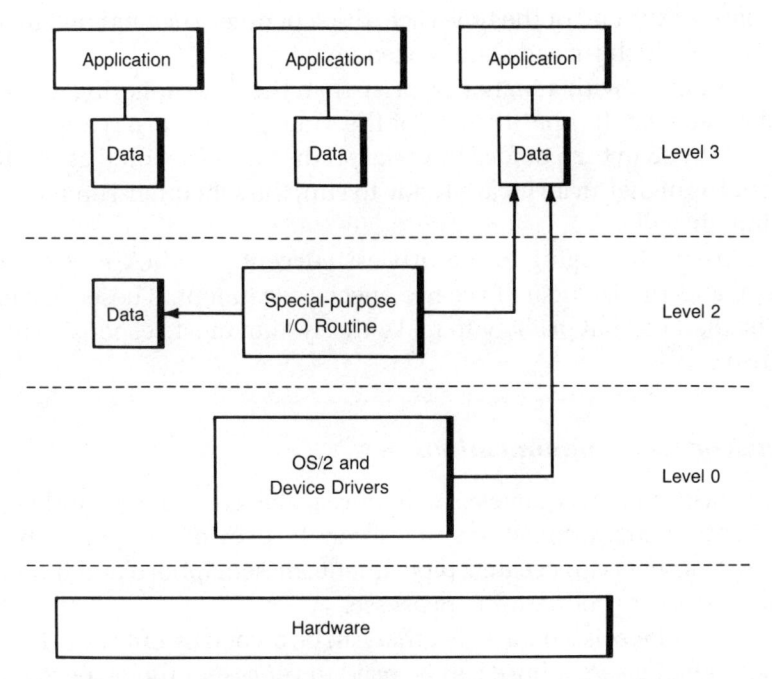

Figure 11-3 OS/2 Protection Hierarchy. *OS/2 and its device drivers run at level 0, special purpose I/O routines run at level 2, and applications run at level 3.*

Data at each privilege level can only be accessed by code at the same or a lower numbered (indicating higher privilege) level. By default, an application can only access its own data, which is also at level 3. For example, a special purpose I/O routine at level 2 can access a client application's data at level 3, as well as its own data at level 2. The 80286 assures that a task only accesses data with a privilege level number equal to or greater than its own, by comparing the task's privilege level with the requested segment's privilege level as stored in the LDT or GDT.

Application tasks are also precluded from executing certain privileged instructions reserved for operating system use when the 80286 processor is operating in protected mode. I/O instructions can only be executed by tasks with the proper privilege. This privilege level is determined by code executed at privilege level 0. Together, these features provide for the safe and efficient sharing of system services and devices among simultaneously executing OS/2 applications.

Scheduling

OS/2 schedules the execution of threads using a preemptive, multi-level priority scheduling algorithm. Each thread is assigned a priority level when it is created. The OS/2 scheduler allows a thread to execute for a short period of time called a *time slice*. At the end of the time slice, the scheduler regains control of the CPU to determine which thread should execute next.

If a thread with a higher priority than the one executing is ready to run, it is executed next. If other threads of the same priority are ready to run, they are each executed in turn before the original thread is executed again. If no threads of equal or higher priority are ready to run, the scheduler returns control to the original thread.

Threads belonging to the process currently in the foreground (using the display and the keyboard) receive special treatment. These threads are given higher than normal priority to make the system most responsive to the interactive user.

Interprocess Communications

OS/2 provides pipes, queues, semaphores, shared memory, and signals for the interprocess communications necessary to provide cooperation among processes. A process can create a pipe, queue, or semaphore and then pass its handle or address to cooperating processes.

A *semaphore* is a data object that can be owned by only one thread at a time. Ownership of a semaphore can be used to represent the ownership of a device that cannot be shared. If two threads request ownership of a semaphore, OS/2 grants ownership to the one requesting it first, and the other thread must wait. Two types of semaphores are provided, one for communication between threads within a process and one for communication between processes. The RAM semaphore is a high performance facility best suited for thread use; the system semaphore is a high function facility particularly suited for process use.

Pipes are file-like facilities used to communicate information between processes. A pipe acts as a fixed length first-in-first-out (FIFO) circular queue for transferring information to a process. More than one process can write to a pipe, but only the process that creates the pipe can read from it. If the pipe becomes full of data, then the next process attempting to write to the pipe must wait until the process reading from the pipe has removed enough information for the new data to be added. Although accessing a pipe is similar to accessing a file, no external device is involved; data is written to and read from a segment in memory. The size of a pipe is specified when it is defined, with a limit of 64 KB.

Queues provide a more sophisticated method of interprocess communication than pipes. Multiple processes can add data to a queue, but only the creator can remove data from it. When a queue is created, the method of removing elements from it is defined as FIFO, last-in-first-out (LIFO), or according to a priority associated with each element. The owner of the queue can examine elements in the queue, remove elements, purge all elements in the queue, or delete the queue entirely. Queues provide better performance than pipes in that data is passed as a shared memory segment, rather than being copied.

OS/2 provides facilities for threads in different processes to share the same data segment in memory. Once a process has allocated a shareable data segment, OS/2 provides threads in other processes with a selector with which to access the segment, upon being provided the name of the shared data segment. OS/2 maintains a record of each request for the shared segment, and does not free the segment until all threads have indicated that they no longer require access to the segment.

Signals notify processes of the occurrence of an external event. A process can define routines to handle specific signals or it can let OS/2 take whatever default action it deems appropriate. A typical signal is the one indicating that the user has pressed Ctrl-Break to terminate the foreground process. If the process has defined a handler for this signal, control is passed to it upon receipt of the signal; otherwise OS/2 takes appropriate action. A Ctrl-Break signal handler can be useful, for instance, for a process that needs to do certain cleanup processing, even when it has been terminated from the keyboard.

I/O Services

OS/2 provides a complete set of character and block I/O services. These services are accessed using the OS/2 function call application program interface. OS/2 functions can be called from high level language programs in the same way that a program calls its own routines. Some devices, such as the keyboard and video display, are accessed through calls specific to the device. Video, keyboard, and mouse services are three separate dynamic link packages, each of which can be replaced by the user. File management calls are used to access disks and character devices such as printers and serial ports. The following device names are reserved for devices supported by base OS/2 device drivers:

CLOCK$	Clock
CON	Console keyboard and screen
COM1–COM3	First through third serial ports
KBD$	Keyboard
LPT1 or PRN	First parallel printer
LPT2	Second parallel printer
LPT3	Third parallel printer
MOUSE$	Mouse
NUL	Nonexistent (dummy) device
POINTER$	Pointer draw device (mouse screen support).

With the exception of COM1–COM3, which are reserved device names only when the ASYNC (RS232C) Device Driver is loaded, these device names take precedence over filenames. A filename that matches the device name cannot be opened, because the device will be opened instead.

OS/2 can be customized by the user for use in twenty-five different countries, using seventeen different keyboard layouts. Primary and alternate code pages can also be specified, with run-time code page switching provided. A code page defines the time and date format, language collating sequence, and language case mapping, as well as the characters and symbols appropriate to a specific country. OS/2 code page management can accept keyboard input and generate display and printer output for concurrent applications that use different code pages. Devices that support code page selection currently include the 4201 Proprinter, 5202 Quietwriter III, and Enhanced Graphics Adapter (EGA) and Video Graphics Array (VGA) devices.

OS/2's video I/O services support the Color Graphics Adapter (CGA), EGA, and Personal System/2 Display Adapter (on a PC AT or XT 286), and the VGA and 8514/A (on the Personal System/2 Models 50, 60, and 80). Video I/O services write to the physical display when a session is in the foreground, and to the logical display buffer when the session is in the background.

When a session is switched to the foreground, OS/2 saves the contents of the video buffer, and other video parameters if the video adapter is operating in text mode. If the previous foreground application is an OS/2 application, it continues to run in background with display output going to the logical display buffer; if it is a DOS application, its execution is suspended while it is in the background. Video information is restored to the video adapter when the application is again switched to the foreground. If the adapter is operating in graphics mode, OS/2 directs the application to perform the save/restore operation.

For some DOS compatibility environment EGA applications, the screen contents will be incorrect when the DOS mode application is switched to the foreground after having been in the background. This is true for EGA graphics applications that use more than one display page or write directly to EGA adapter registers. This situation may be remedied by modifying the application so that it saves the EGA state upon receipt of an INT 2FH, AX = 4001FH, and restores them upon receipt of an INT 2FH, AX = 4002H. These are new multiplex interrupts issued by OS/2 to indicate, respectively, that the DOS mode application is to be moved to the background and has been moved to the foreground. Alternately, the application can be modified to use the EGA Register Interface described in Chapter 9 of the OS/2 Technical Reference. This problem does not occur when using OS/2 with a Personal System/2 computer or when using a PC AT or XT 286 with a Personal System/2 Display adapter, because video registers are readable as well as writable.

Since OS/2 is a multi-tasking system with multiple task sharing system devices, it does not allow individual applications to revector interrupts to their private interrupt handlers in order to monitor characters coming from or going to a device. This facility is provided instead through OS/2 device monitor services.

OS/2 character device monitors allow any registered process to remove, insert, or modify information passing through a device driver. Examples of device monitors are applications that monitor keystrokes for particular sequences, and the OS/2 print spooler, which intercepts characters being written to the printer and writes them to a disk file for later printing.

The file I/O services provided by OS/2 are similar to those provided by DOS. Although the file system supported is identical to that of DOS, OS/2 provides the ability to perform asynchronous I/O operations. Thus, a program can initiate an I/O operation and proceed with other processing while the I/O operation is performed by the I/O subsystem.

Device Drivers

Unlike most DOS device drivers, OS/2 device drivers are interrupt driven, so they can surrender the CPU (for use by applications) while waiting for I/O operations to complete. Because OS/2 supports the DOS compatibility environment, OS/2 device drivers must be able to support DOS mode as well as OS/2 applications.

Applications access a device driver through the file system interface, subsystem interface, or I/O control (IOCtl) interface. Device drivers are accessed via the file system interface using I/O function calls. OS/2 applications use dynamic link calls to request file I/O services; DOS mode applications use INT 21H file I/O function calls. I/O function calls are used mainly for block devices such as hard disks; however, some I/O function calls such as *DosOpen*, *DosClose*, and *DosRead* can also be used with character devices.

A subsystem interface allows an application to perform I/O to a logical device, and the interface manages the physical device I/O. Examples of subsystem interfaces are the file, video, keyboard, and mouse subsystems.

The IOCtl interface is a means for sending device-specific control commands to a device driver. This interface can be used by an application or a subsystem to send commands to block or character device drivers.

OS/2 device drivers can support multiple synchronous and asynchronous I/O requests. If the request is for synchronous I/O (the requesting thread cannot continue until the operation completes), the service is provided immediately if the device is not busy, or the request is added to the queue of requests to be performed and then performed when the device is not busy. In either case, execution of the requesting thread is blocked until the device drivers interrupt routine indicates that the request is complete. When asynchronous I/O is requested, the OS/2 device support creates an I/O thread to perform the operation, and the requesting thread continues executing. When the I/O operation completes, the I/O thread uses a RAM semaphore to indicate that the I/O operation is complete, and terminates.

A device driver manages I/O for its device, whether for OS/2 applications or the DOS mode application. DOS I/O requests using INT 21H calls are translated to the same request packets to which OS/2 I/O service requests are translated, and handled accordingly. The situation is somewhat more complicated, however, for DOS I/O using ROM BIOS or I/O requests sent directly to the device. The OS/2 device driver must intercept BIOS software interrupts, serialize access to the device, and protect critical sections of BIOS execution from suspension. Serializing access to the device can be accomplished using semaphores to indicate that the device is busy with a BIOS request and consequently cannot accept another request. The driver must protect critical sections of BIOS execution from suspension, because certain I/O operations performed in the BIOS cannot be suspended, which is what occurs when the DOS mode application is switched to the background.

The Personal System/2's PC compatibility BIOS (CBIOS) is of little use to OS/2 and its device drivers for anything other than supporting BIOS calls from DOS mode applications. The Personal System/2's advanced ROM BIOS (ABIOS), however, provides useful multi-tasking support features that would otherwise have to be provided in precious real-mode RAM. OS/2 device drivers can issue I/O function requests to ABIOS rather than directly accessing hardware devices.

Device drivers request ABIOS services using the advanced BIOS transfer convention and the operating system transfer convention. OS/2 internal device drivers use the operating system transfer convention; user-written, installable device drivers can use either convention. Device driver helps (DevHlps) are available for both conventions.

DevHlps or device driver helper services are system functions provided by OS/2 that help the driver manage device I/O. Device drivers access these services via the DevHlp interface. When a device driver is initialized at system initialization, it receives a pointer to the DevHlp interface routine. DevHlp services are obtained by loading a function code into the DL register and making a FAR CALL to the DevHlp interface routine. Currently, fifty-seven function codes are defined; six of which are reserved for future use. Services provided include process management, management of semaphores, request queues and character queues, interrupt management, physical and virtual memory management, and access to ABIOS services.

Device drivers that use the operating system transfer convention request ABIOS services via the *ABIOSCall* DevHlp service. The device driver passes the ABIOS request block pointer, the logical ID (LID) of the device driver, and the ABIOS function to be performed to ABIOSCall. This places the information on the stack and causes the requested ABIOS function to be called. The request block is a parameter block containing fields that specify the I/O device and details of the request, including memory areas involved.

Device drivers that use the ABIOS transfer convention request ABIOS services via the *ABIOSCommonEntry* DevHlp service. The device driver passes the mode specific pointer to the ABIOS request block, and the primary function requested to ABIOSCommonEntry. This places the information on the stack and causes the requested ABIOS function to be called. Primary functions are common start, common interrupt, and common timeout.

Application Program Interface

Application programs request services from OS/2 using a CALL-RETURN interface, rather than the software interrupts used by DOS and the system BIOS. Request parameters are passed on the application's stack using a convention compatible with that used by many high level language compilers. Parameters are pushed onto the stack before a call is issued, and copied from the requestor's stack to the called system routine's stack by the hardware.

Unlike DOS and BIOS software interrupts, which typically must be invoked using assembly language, OS/2 system functions are accessed by directly calling system routines from application programs written in a high level language. Thus the application requests system services using the same mechanism that it uses to call its own routines.

The OS/2 system routines called by the application are dynamically linked to the application and loaded when called. An application's own routines may also be linked and loaded dynamically. Dynamic linking and loading saves memory because infrequently loaded routines are only loaded when they are needed, and one copy of a routine can service more than one application; it saves disk space because copies of commonly used routines do not have to be stored in the executable file of each application; and it allows commonly used routines to be updated without having to update every application that uses them.

The use of the same CALL-RETURN interface to access dynamically linked system routines allows system functions to be extended or updated easily by adding new modules or replacing modules in dynamic link libraries. The CALL-RETURN API provides a single, consistent method of accessing system services and extensions; moreover, it is an API that can be easily expanded while maintaining compatibility with existing applications.

Programs that execute in either DOS or OS/2 mode can be developed using a subset of the OS/2 API called the DOS Family API. A program written to the specifications of the Family API can be linked and bound, to produce a file that can be executed using either DOS or OS/2. When used with DOS, code bound to the application translates the Family API calls to software interrupt requests; when used with OS/2, this code is not loaded into memory and the Family API calls are processed directly by OS/2.

DOS 3.30 Compatibility

OS/2 can run a single DOS 3.30 application concurrently with multiple OS/2 applications. This feature is provided to support the use of existing DOS applications until OS/2 versions of them are available, and to support DOS applications whose function or use does not merit their conversion to full OS/2 operation.

Because the DOS application is not aware that OS/2 applications are executing (and using the keyboard and display), it cannot be allowed to execute when it is not the foreground application. Thus, when the user switches control of the display and keyboard to an OS/2 application, the DOS application is suspended until it is once again the foreground application.

Many DOS applications can run unchanged in the OS/2 DOS compatibility environment. OS/2 services all DOS function calls and BIOS software interrupts. This feature permits the sharing of system resources and services, such as the file system, among the DOS application and OS/2 applications.

Applications that may not run in the compatibility environment include applications that are time dependent, or require device driver or communications support not currently available for OS/2. Applications that require a full 640 KB system will also have problems, because a portion of OS/2 as well as DOS is loaded into the lower 640 KB of memory, thus reducing the amount of memory available for DOS applications. OS/2 Standard Edition installed on a Model 60 with default parameters provides 510 KB of free space for a DOS application, whereas 585 KB of memory is available if the system is initialized under the exclusive control of DOS 3.30.

Although OS/2 is a new operating system, it provides the same familiar commands and user interface as DOS. The user interface of its command processor (CMD.EXE) is nearly identical to that of COMMAND.COM (the prompt can even be made the same if desired), and it supports almost all of the DOS commands. So even though applications must be recompiled or rewritten to run using OS/2, the basic operation of the system remains the same.

A program can be executed by entering its name and pressing Enter at the OS/2 command prompt; files can be deleted and renamed using DEL and REN; the contents of a directory can be displayed using DIR; and so forth. The OS/2 batch command language is the same of that of DOS, only with extensions. OS/2 batch files are named with the extension CMD, to distinguish them from DOS batch files. Most system utilities are written using the Family API, allowing them to be run in OS/2 mode or in the DOS compatibility environment of OS/2.

The OS/2 Programmer Toolkit supports the creation of applications, using the DOS Family API, that can run both under DOS and OS/2. Such applications can be made portable between DOS and OS/2 by using a specific subset of the capabilities of OS/2.

OS/2 uses the same file system formats as DOS. Files are completely interchangeable between DOS 3.30 and OS/2, and can exist in the same file system on the same hard disk. OS/2 can be installed on a hard disk that contains DOS, without reformatting the disk and without destroying any files except for the DOS hidden system files IBMBIO.COM and IBMDOS.COM, which are replaced by their OS/2 versions. Files and directories created by OS/2 can be accessed and used by DOS as if they had been created under the control of DOS and vice versa.

On a system with OS/2 installed on the hard disk, the system can be initialized from a DOS system diskette and run under the exclusive control of DOS, with access to all system files, as if OS/2 were not even installed on the system. This feature is particularly useful when a system has been converted to OS/2 use, but needs to be run using DOS for particular applications that will not run in the DOS compatibility environment. Using this procedure is preferable to running OS/2 from a diskette; if the OS/2 system is initialized from a diskette, the diskette must remain in drive A so that dynamically loaded system files are available.

Ease of Use Facilities

OS/2 provides a menu as well as a command line interface. Upon system initialization, the user is presented with a Program Selector menu, from which applications can be executed. A comprehensive help facility, accessible from the Program Selector and its submenus, provides information about the particular item that is highlighted. Help information about system error messages is available from the command prompt, and an optional help line at the top of the command prompt session reminds the user how to return to the Program Selector or access help. An interactive tutorial on OS/2 is automatically installed on the hard disk along with OS/2 and is accessible from the Program Selector menu.

The facilities described above are available on all OS/2 implementations, including Standard Edition Version 1.0. Presentation Manager, which is part of Version 1.1 (October 1988 release), provides a sophisticated graphic user interface for applications developed to use it. Presentation Manager users can perform input functions using a point-and-select approach rather than a command entry approach, and view the output from several applications (or the same application) in multiple overlapped windows on the display screen. Application windows can be sized, moved, and even hidden from view dynamically. Most data entry is done through pull-down menus and dialog boxes (a sophisticated form of pop-up menus).

Systems Application Architecture Support

OS/2 is included in IBM's Systems Application Architecture (SAA). SAA is a collection of software interfaces, conventions, and protocols designed to act as a framework for the development of consistent applications across the IBM System/370, System/3x, and IBM PC computing environment. SAA consists of four elements—common user access, common programming interface, common communications support, and IBM-developed common applications.

At present, OS/2 participates in the first three elements of SAA. Elements of

Chapter 11

OS/2 that support SAA are the Presentation Manager, Communications Manager, Database Manager, Query Manager, and the OS/2 languages C/2, COBOL/2, and FORTRAN/2.

Presentation Manager

OS/2 Version 1.0 provides multi-tasking support, but the user can only see the display output of one application at a time. Presentation Manager, which is incorporated in Version 1.1 of OS/2, allows the output of multiple Presentation Manager applications to be visible on the screen at the same time in overlapped windows. Additionally, each of these applications can have multiple windows, as shown in the model in Figure 11-4.

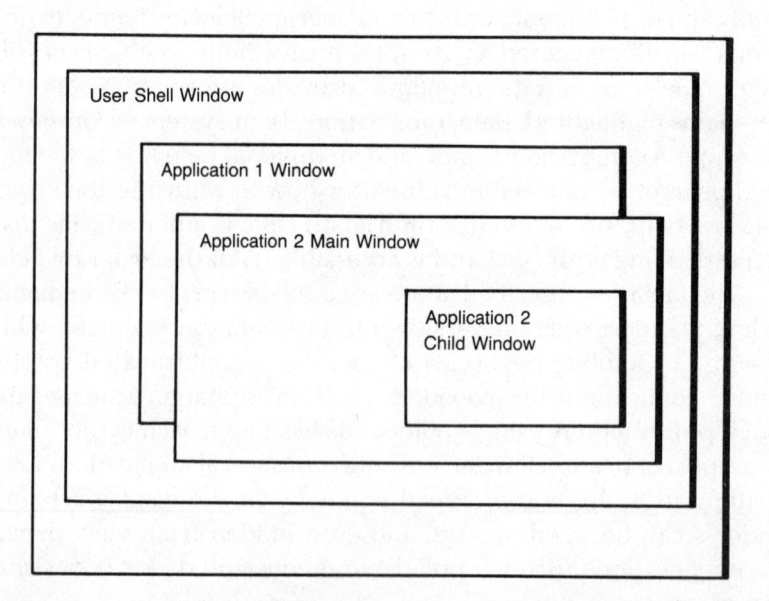

Figure 11-4 Presentation Manager Screen Model. *Several applications can be displayed on the screen at once; individual applications can have multiple screens.*

   Presentation Manager provides a graphical user interface similar to that provided in Microsoft Windows 2.0. This interface is radically different from the DOS command line user interface, and more flexible and versatile than the OS/2 Program Selector. Presentation Manager is a multiple-choice rather than a fill-in-the-blank user interface. Programs to be executed, files to be operated upon, and most control and data information is input by selecting items displayed on the screen using a pointing device or the keyboard, rather than keying in text that specifies commands and options.

Presentation Manager provides support for all-points-addressable (APA) graphics. Graphics and text can be loaded, saved, and printed on a deferred basis. Utilities are available for the display of picture files, and for conversion of data as necessary for exchange with other systems.

An APA graphics adapter with an appropriate display is required for use with the Presentation Manager. Video adapters and modes supported include: CGA (640 by 200 pixels, 2 colors), EGA (640 by 350 pixels, 16 colors), VGA (640 by 480 pixels, 16 gray scales or 16 colors), and 8514/A (640 by 480 pixels, 16 gray scales or 16 colors; 640 by 480 pixels, 256 color; 1024 by 768 pixels, 16 colors; 1024 by 768 pixels, 256 colors).

Presentation Manager provides recommendations for an application's use of the display, keyboard, and mouse. Standard key assignments for help, accessing menus, and exiting an application are provided. This provides consistent interface for the user across applications and, through Presentation Manager's consistency with the SAA Common User Interface, across IBM computer product lines as well. Tools and services are provided that enable software developers to provide these features through the Presentation Manager application program interface (API) calls.

The Presentation Manager API provides a high level CALL interface for use in the development of Presentation Manager applications. Applications developed using this interface support character and graphics input/output, screen windows, mice, text with typographic fonts, bit-maps, and standard menus and dialogs. The OS/2 Standard Edition Version 1.1 Programmer Toolkit is available at extra cost for use in developing Presentation Manager applications.

In order to support the shared use of the screen and other resources, Presentation Manager applications must adhere to a number of structural and operational rules. Adherence to many of these rules is accomplished by performing I/O using Presentation Manager input and output functions. These functions support output to printers, plotters, and data files, as well as output to the display and input from the keyboard and mouse.

Although the use of these functions requires additional programming initially, less programming effort should be required in the long run, because Presentation Manager and OS/2 device drivers (not the application) provide the hardware interface. Therefore, as hardware devices are switched or upgraded, the application continues to work using its standard Presentation Manager I/O calls, and any software changes required are made at the device driver level, independent of the application software developer.

Command and Control

Presentation Manager and Presentation Manager applications are controlled through the Presentation Manager User Shell. This program, which appears in the initial screen window after Presentation Manager is loaded, includes the capabilities provided by the MS DOS Executive included in Microsoft Windows plus additional capabilities and ease of use facilities. The execution of applications and the manipulation of files and windows is controlled using the keyboard and, optionally, a mouse.

The Presentation Manager is a complex message switching system. The user communicates with the Presentation Manager and applications running under its control via the mouse and the keyboard. Input from the mouse and the keyboard is processed in the form of messages that are usually passed along to the application. Presentation Manager reserves some key combinations for receiving requests from the user, and thus does not pass these keystrokes along to applications. These include Alt-Esc, which switches control to the next application window, Ctrl-Esc, which makes the task manager the active window, and Alt-Space, which causes the system menu to drop down.

Part of the consistent user interface provided by Presentation Manager and its applications is the consistent use of the mouse. Use of the mouse within Presentation Manager is intuitive and consistent. Applications developed for use with Presentation Manager should observe the recommended guidelines for mouse usage whenever possible. Guidelines provide for the selection of items by clicking (quickly pressing and releasing button 1 of the mouse when the screen pointer is on the item), moving objects by dragging (holding down button 1 when the screen pointer is on the item, and then moving the mouse), and the selection of an item and the performance of its default action by double clicking when the screen pointer is on the item. The use of button 2 on the mouse can be defined by the application. Clicking mouse button 3 transfers control to the Presentation Manager Task Manager, and double clicking button 3 transfers control to the next task.

Window Appearance

Presentation Manager windows are designed to appear and operate in conformance with the Common User Access definition of SAA. All windows are defined by a set of parameters such as position, size, and visibility. *Position* is simply the position of the window on the screen or relative to the parent window if a child window. *Size* is the horizontal and vertical dimensions of the window; however, every window has three size definitions, one each for its normal, maximum, and minimum sizes. Windows also have a *border*; the border can be thin, normal, or wide. The area inside the window border, where the application's output is displayed, is called the *client area*. Windows can be invisible as well as visible. The window definition also specifies if a window remains visible when overlapped by another, and the shape the screen pointer is to take when it is within the window.

Windows can have a number of optional features in addition to the properties common to all windows just described. Optional features include a title bar, action bar, scroll bars, dynamic operations, and a window procedure. Some of these features have to do only with the appearance of the window, but others determine how the window operates and/or how it can be operated upon. Optional window features are shown in Figure 11-5.

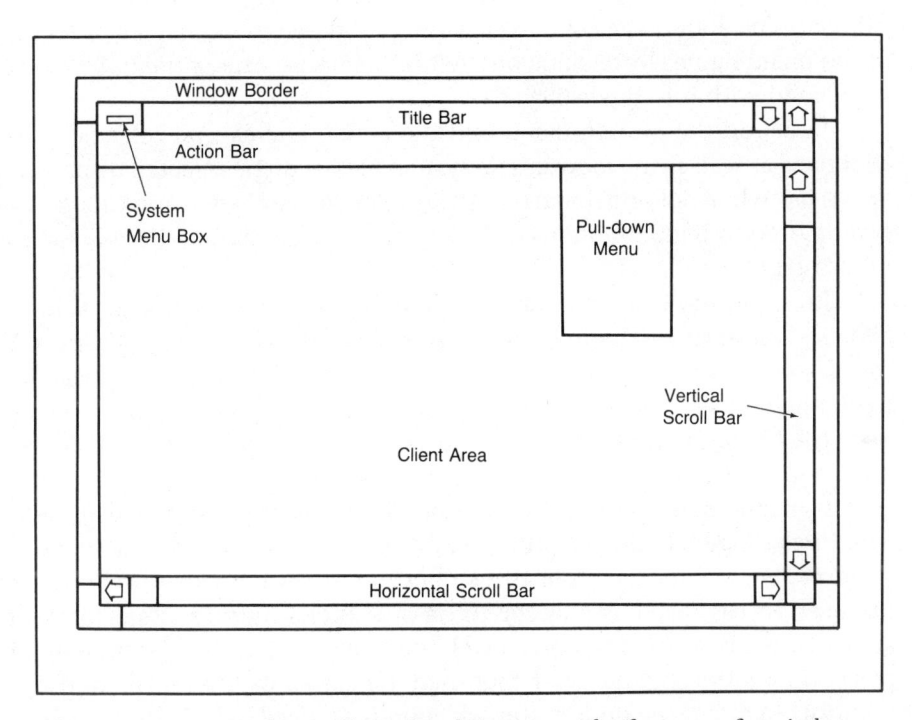

Figure 11-5 Window with Optional Features. *The features of a window conform to the Common User Access definition of SAA.*

The title bar is a bar across the top of the window that contains the name of the window, the system menu box, and the maximize/minimize/restore boxes. The window name is simply a character string used to identify the window. The system menu box is the area at the left end of the title bar in which the screen pointer is placed in order for the system menu to appear when mouse button 1 is clicked. The system menu also appears when the Space Bar is pressed while holding down the Alt key.

The system menu contains a list of system functions such as move, size, and close window. The maximize, minimize, and restore boxes are areas in which the screen pointer is placed in order for the window to be maximized (to full screen size), minimized (to icon size), or restored to normal size (from full screen size), when mouse button 1 is clicked.

The maximize and restore symbols are mutually exclusive symbols that appear in the same box. When a window is its normal size, the maximize symbol appears in the box if the window can be maximized. Placing the screen pointer on the maximize symbol and pressing mouse button 1 causes the window to expand to fill the entire screen and the restore symbol to appear in the box. Subsequent clicking on the restore symbol causes the window to return to its normal size.

The action bar appears at the top of the window, immediately below the title bar. The action bar contains a list of pull-down menu names. Selecting one of the menu names by clicking on it with the mouse causes the pull-down menu associated with it to be displayed.

Items on the system menu and the action bar can be selected using the keyboard as well as the mouse. The system menu can be selected by pressing the Space Bar while holding down the Alt key. The Cursor Down and Cursor Up keys can be used to highlight items on the menu; the highlighted item is selected by pressing Enter.

Items on the system menu include minimize, maximize, and restore, thus allowing these functions to be performed from the keyboard if desired. When the system menu is displayed, the Cursor Right and Cursor Left keys can be used to highlight desired action bar items. Pressing Enter causes the menu associated with the highlighted item to be displayed. Menu items are selected from these menus in the same way as from the system menu.

Dynamic operations are operations that the user can perform on a window, including changing its position and size. The application that controls the window determines which of these operations can be performed and enables the user access facilities only for those operations that it will perform. For instance, if a window cannot be moved, the *Move* item in the system menu is displayed in a different character font than items that can be selected. If the user attempts to select the option, the selection is ignored.

A window procedure is a parcel of program code associated with a window. There is no direct correspondence between it and the window's appearance; however, it affects the way a window operates internally. If a window has a window procedure associated with it, the window can be treated as a programming object that handles messages sent to it in a given way. Applications can be made up of multiple window procedures, each of which controls a particular part of the application's user interface or screen display.

Data and control information can be entered for Presentation Manager applications using dialog boxes. Dialog boxes are windows that support the free-format entry of data and control information. Dialog boxes can contain buttons that can be selected (usually to perform actions), check boxes that indicate the selection of items, radio buttons that indicate mutually exclusive selection of items (after the fashion of station selection buttons on an automobile radio), list boxes from which items can be selected, input fields for entering text, and scroll bars.

Using Presentation Manager

Presentation Manager's Filing System provides a graphic display of system files and directories. The Filing System can be used to display the contents of files and directories (using filing windows) and to start the execution of applications (using the STARTUP window).

A particular application is chosen to receive input from the keyboard by bringing its window to the top of the stack of windows (and thus making it completely visible), using either the Alt-Esc or Alt-Tab key combination to move through the list of windows, in much the same way that one moves through regular OS/2 applications using the Program Selector.

Pressing Ctrl-Esc, as one would to invoke the OS/2 Program Selector, invokes the Presentation Manager Task Manager. The Task Manager shows the entire list of applications active in the OS/2 system. From the Task Manager, the application to next receive keyboard and mouse input can be selected using the keyboard or the mouse. If the application selected is a Presentation Manager application, its window is moved to the top of the stack of windows on the display. If it is a non-Presentation Manager application, the task's screen group is switched to the physical display. If Ctrl-Esc is later pressed from the non-Presentation Manager application, the Presentation Manager screen group is displayed with the Task Manager as the top window.

Presentation Manager utilities are provided for cutting and pasting text and graphics information between applications, and for the spooled printing of graphics information. Utilities are also provided for customizing Presentation Manager's appearance and operation.

OS/2 Extended Edition

The extended edition of OS/2 is designed to provide an environment for the execution of stand-alone applications, as well as applications that are distributed across a network. OS/2 Extended Edition provides comprehensive communications and database management support in addition to all the capabilities provided by the standard edition of OS/2.

Communications Manager

The Communications Manager is provided only with OS/2 Extended Edition. It is a comprehensive communications system that provides capabilities for communicating with other IBM PCs and Personal System/2s, System/3x computers, and System/370 mainframe computers. It uses OS/2's multi-tasking capabilities to support multiple, concurrent communications sessions.

Connectivity with other systems is provided using asynchronous links, Distributed Function Terminal (DFT) mode connection to an IBM 3x74 terminal control unit, Token-Ring and PC Network local area networks (LANs), and synchronous data link control (SDLC). Various communications protocols are supported, including asynchronous communications, 3270 data stream (LU 2), and LU 6.2.

The Communications Manager provides terminal emulation and file transfer services for communications and data interchange with other systems. Emulation for both asynchronous and synchronous terminals is provided. Asynchronous terminals emulated include the IBM 3101 (Model 20) and DEC VT100. Keyboard remapping allows the keyboard interface to be arranged according to host system demands and user preference. Communications with the host computer may be performed using switched or nonswitched lines, or CCITT V24/28 (RS232) direct connection.

Synchronous terminals emulated include the IBM 3278 (Models 2-5), IBM 3279 (Models S2A and S2B), and IBM 3178, Model 2. The emulation supports all data stream functions including extended attributes, extended data stream, and interactive screen. Seven colors and extended highlights are supported, as well as keyboard remapping and file transfer.

Text and binary files can be transferred among personal computers and IBM host computers, using the Communications Manager. XMODEM and pacing protocols are provided for asynchronous communications transfers. The 3270 PC File Transfer Program works with both asynchronous communications and 3270 communications.

With the Communications Manager, an OS/2 system can communicate with multiple computer systems of different types using different communications protocols, at the same time. Communications may be direct, using DFT or Token-Ring network connection, or remote, using asynchronous or SDLC communications. Communications supported on a particular PC or Personal System/2 system are, of course, subject to the availability of the appropriate communications adapter and communications link.

The Communications Manager supports communications across LANs, using the IBM NETBIOS and the IEEE 802.2 data link control interfaces. Applications using these interfaces can be developed with IBM Macro Assembler/2, Pascal/2, and C/2. Existing DOS applications that use these interfaces need to be modified in order to work with the Communications Manager.

Applications can also be developed using the APPC (Advanced Program-to-Program Communications) and SRPI (Server-Requester Programming Interface) programming interfaces. APPC implements the LU 6.2 architecture; it supports data-stream-independent as well as data-stream-dependent communications. It can be used to develop applications that communicate with System/370s running MVS/CICS and VSE/CICS, and with System/3x, System/88, PC RT, and Series/1 systems, as well as PC and Personal System/2 systems.

The SRPI interface is part of the Enhanced Connectivity Facilities (ECF) support provided by the Communications Manager. It supports requests from a PC or Personal System/2-based requester program to a System/370 host server program. Host server support is available on MVS/TSO and VM/CMS systems. Communications is supported using the LU 2 protocol. These same functions are provided by the family of DOS IBM PC 3270 Emulation programs (see Chapter 14). Any DOS application written to use the DOS APPC interface needs to be modified; DOS SRPI applications need to be recompiled or revised. APPC and SRPI applications can be developed using IBM Macro Assembler/2, Pascal/2, and C/2.

Communications Manager includes an asynchronous communications device interface (ACDI), in addition to the previously described synchronous communications interfaces. The ACDI interface is designed to allow the hardware-independent exchange of data over asynchronous links. Required device-specific programming modules for each supported device type are included with the Communications Manager. Applications for this interface can likewise be developed using IBM Macro Assembler/2, Pascal/2, and C/2.

The Communications Manager provides extensive communications network management facilities. It can transmit problem determination data over the network, and it can send communications and systems management (C&SM) alerts to the host computer when SDLC, asynchronous data link, PC Network or Token-Ring network errors are detected. Applications written to the Communications Manager's API can make use of this ability to send alerts to the host system.

The user can display extensive information about the SNA communications activities used and controlled by the Communications Manager. Upon request, data can be displayed about which programs are using which communications sessions and data links, as well as detailed information about the sessions and other active resources. Control facilities are provided for the activation and deactivation of sessions and data links.

Capabilities for logging, dumping, and displaying active communications and error logs are provided. Accesses to programming interfaces, data units, and system events can be detected, logged, displayed, and printed.

Database Manager

The OS/2 Database Manager, which is provided only with the OS/2 Extended Edition, is based on the relational data model. Data tables are defined and manipulated using the Structured Query Language (SQL) data manipulation language supported by IBM's System/370 mainframe database managers DB2 and SQL/DS. The Database Manager provides a common interface on OS/2 and System/370 systems, facilitating application portability and ease of use, training, and maintenance.

The maximum table size is limited only by the amount of hard disk storage available; however, a database must reside on a single diskette or hard disk partition (currently 32 MB maximum size). Data types supported by the Database Manager include character (fixed and variable length), integer, floating point, packed decimal, date, time, and timestamp.

Access to databases is supported using record and table locking. Password protection is provided for database security. Database integrity is maintained using the COMMIT and ROLLBACK features. Backup utilities are provided to offer protection against data loss or invalidation.

Applications can be created using menus, screens, and procedures. Menus can be used to support the execution of predefined queries, screens, or additional menus. A screen generator allows the development of customized display screens for data entry, search, and database update. Procedures allow the execution of a sequence of data management statements to be initiated with a single command. An SQL application program interface (API) allows SQL statements to be embedded in C/2 source code. A precompiler is provided that converts the SQL statements into code that can be compiled and executed. This API is a subset of the Systems Application Architecture's common programming interface.

A data entry facility allows data to be inserted and updated in data tables. Data definition facilities that support the creation and deletion of tables, views, and indexes are supported.

Utilities are provided to perform the following functions:

- Backup/restore database or changes
- Convert OS/2 files in other formats to or from data table format
- Restore/save a single data table
- Reorganize a data table for more efficient access
- Update table or index statistics

A query manager is provided to allow the user to retrieve data from databases. Queries are performed using SQL generated by the query facility, based on user input. Interactive and batch query processing is supported. A report generator is also provided.

Planned Enhancements

IBM has announced several planned enhancements for the Communications Manager and Database Manager. Many of the planned Communications Manager enhancements merely provide multi-user versions of single-user communications services currently available for use with DOS. Database Manager enhancements include functions that are currently not offered by IBM DOS-based database manager products.

The Communications Manager will be enhanced to provide support for 3270 gateways from LANs to SNA networks, additional ECF features, 5250 workstation emulation, X.25 support, and a high level language 3270 API (HLLAPI). Programs written using the Entry Emulator High Level Language application program interface (EEHLLAPI) of the IBM PC 3270 Emulation Program, Entry Level (see Chapter 14) will work on the Communications Manager HLLAPI interface upon recompilation.

A precompiler will be provided to support the processing of SQL statements embedded in IBM Pascal/2 and IBM COBOL/2 source programs. A data import facility to support the entry of nondelimited ASCII files into a database will be provided. Token-Ring and PC Network support will be added to the Database Manager to support the access of a common database by multiple workstations.

Languages

A new operating system is of little use without programming languages with which to develop applications to use it. At the same time that it announced DOS 3.30 and OS/2, IBM also announced new compilers that can be used with both DOS 3.30 and OS/2. These products, which became available in November 1987, include C/2, COBOL/2, FORTRAN/2, Macro Assembler/2, Pascal Compiler/2, and BASIC Compiler/2.

These compilers and assembler generate code for use on OS/2 or DOS 3.30. On 80286- and 80386-based machines they can be executed using OS/2 or DOS 3.30. On 8088- and 8086-based machines they must be executed using DOS 3.30. The generated code is linked with the libraries for the operating system on which it is executed; or if written to the specifications of the Family API, it can be linked and bound so that it will run with both OS/2 and DOS.

The compilers and assembler, and the code they generate, can be run concurrently with other applications in the OS/2 multi-tasking environment. Each contains a set of tools including a linker, library manager, and a debugger.

When using DOS, at least 384 KB of memory must be available to use the assembler and these compilers. The minimum memory requirements for OS/2 are adequate for their use with OS/2. A hard disk is recommended for use with these products (required if they are used with OS/2). When used with DOS, a dual diskette system can be used but is impractical, particularly in the case of the C/2 and COBOL/2 compilers.

Each product package contains the software on both 5.25-inch, 360 KB diskettes and 3.5-inch, 720 KB diskettes. Each package contains a program license agreement, a Fundamentals manual, a Language Reference manual, and a Compile, Link, and Run manual.

IBM C/2, COBOL/2, and FORTRAN/2 can be used to write applications that conform to the Systems Application Architecture Common Programming Interface (CPI).

BASIC Compiler/2

The IBM BASIC Compiler/2 is a BASIC language compiler for use with OS/2 or DOS 3.30. The BASIC Compiler/2 and the code it generates can be run concurrently with

Chapter 11

other applications in the OS/2 multi-tasking environment. The compiler provides compatibility with the BASIC 2.0 compiler, while providing enhancements such as user definable record data structures, control structures (including block IF/THEN/ELSE, CASE, and DO loop), and functions with variables that are PUBLIC to other modules. DECLARE, a new statement for the identification of external procedures or forward references to functions, is also provided.

Index Sequential Access Method (ISAM) support is enhanced to allow simultaneous file access by multiple OS/2 tasks. Full record and file locking support for remote ISAM files is provided when using DOS 3.30.

BASIC/2 features a common language interface that allows routines written in C/2, Pascal/2, or Macro Assembler/2 to be called from a compiled BASIC/2 program. An interactive debugger that supports the isolation of source code errors while the compiled BASIC application is executing is provided. The following can be performed while executing a BASIC/2 program under control of the debugger:

- Display and change the value of variables
- Set conditional or unconditional breakpoints
- Single-step execution of source or generated assembly code
- Symbolic references to local and global variables
- View memory, registers, and the application output screen
- View source or assembly code during application execution

Documentation provided for BASIC/2 includes a Fundamentals manual, a Language Reference manual, and a Compile, Link, and Run manual. The IBM BASIC Reference Version 3.3, which describes the operation of the BASICA interpreter provided with DOS 3.30 and OS/2, is also available from IBM at extra cost. This manual, which is not required for use of the BASIC/2 compiler, is a revised version of the IBM BASIC Reference Version 3.2.

C/2 Compiler

C/2 is a C language compiler for use with OS/2 or DOS 3.30. The C/2 compiler language is in partial compliance with the proposed ANSI standard for C. The C/2 compiler is compatible with IBM PC C Compiler Version 1.0, while providing additional features such as additional compiler options, multiple memory models, and new library functions. The C/2 compiler and the code it generates can be run concurrently with other applications in the OS/2 multi-tasking environment.

C/2 provides Intel math coprocessor support; math coprocessor emulation can be used if a math coprocessor is not available. Utilities provided with the compiler include an enhanced MAKE facility and EXEMOD, which allows the modification of the file header of DOS .EXE files. An interactive debugger (with the same features as the BASIC/2, Macro Assembler/2, and Pascal/2 interactive debugger) is also provided.

COBOL/2 Compiler

COBOL/2 is a COBOL language compiler for use with OS/2 or DOS 3.30. COBOL/2 is designed according to the specifications of the ANSI X3.23-1985 COBOL standard, intermediate level. IBM intends to submit COBOL/2 for certification of compliance with the standard. The COBOL/2 compiler provides an environment for the development and maintenance of applications for use on the IBM PC and Personal System/2. The COBOL/2 compiler and the code it generates can be run concurrently with other applications in the OS/2 multi-tasking environment. Applications can also be developed for compilation and execution on System/370 MVS and VM/CMS operating system environments.

COBOL/2 supports large COBOL programs, restricted in size only by available real memory, with both data and procedure divisions greater than 64 KB. Small, medium compact, large, and huge memory models are supported. It is upward compatible with IBM PC COBOL Version 1.00 and 2.00 source code, and also provides extensions developed by MicroFocus, Ltd. COBOL 2.00 is an early version of COBOL/2, released in April 1987 for use with the DOS 3.2 and 3.30 operating systems only. Registered licensees of COBOL 2.00 can receive a copy of COBOL/2 at no extra charge. The COBOL/2 compiler recognizes and supports subsets of the COBOL language variants supported by System/370 IBM VS COBOL II Release 1.0 and OS/VS COBOL Release 2.4 compilers.

A fast SORT/MERGE module and Index Sequential Access Method (ISAM) support is provided. Split keys and alternate keys with duplicates are supported. One primary key and sixty-three alternate keys are supported. A utility program to help recreate damaged files is included. DOS file handles and fully qualified filenames, including path specifications, are supported.

COBOL/2 includes an interactive source level debugger named Animator. The following can be performed while executing a COBOL/2 program under the control of Animator:

- ► Change the flow of program execution
- ► Debug called programs to any level of nesting
- ► Read, write, and monitor the value of data items
- ► Set conditional or unconditional breakpoints
- ► Single-step execution of source statements
- ► Skip execution of subroutines or called programs
- ► Start and stop program execution

Chapter 11

- ► Use split screen to view two different source code areas
- ► View source code, memory, registers, and variables
- ► View the application screen

COBOL/2 supports use of the IBM PC Network in the DOS environment. File and record locking for shared data, as well as communication between programs via the network, are supported. An installation procedure for PC Local Area Network 1.2 is provided with the product.

FORTRAN/2 Compiler

FORTRAN/2 is a FORTRAN language compiler for use with OS/2 or DOS 3.30. FORTRAN/2 is designed according to the specifications of the ANSI X3.9-1978 FORTRAN standard. IBM intends to submit the FORTRAN/2 compiler for certification of compliance with the standard. FORTRAN/2 provides additional features while maintaining compatibility with IBM PC Professional FORTRAN Version 1.30.

FORTRAN/2 and the code that it generates will run concurrently with other tasks in the OS/2 multi-tasking environment. Math coprocessor support and math coprocessor emulation are provided, and several enhanced data types are supported. An interactive debugger is provided.

Macro Assembler/2

Macro Assembler/2 is an assembler language processor for use with OS/2 or DOS 3.30. It is functionally compatible with previous versions, and produces object files that can be linked with object files produced by previous versions. The Macro Assembler/2 supports the full 80286/80287 instruction set, as well as that of the 8088/8087. It provides pseudo-ops for the assembly of 80286 protected mode instructions, and conditional pseudo-ops for improved runtime error checking. Its use of a greater than 64 KB internal work area allows the assembly of larger source files and more extensive use of macros. The Macro Assembler/2 and the code it generates can be run concurrently with other applications in the OS/2 multi-tasking environment.

Utilities included with the Macro Assembler/2 include a MAKE utility, the CREF cross reference utility, and the EXEMOD utility, which allows the modification of the file header of DOS .EXE files. The existing Structured Assembly Language Utility (SALUT) includes a new statement—$SALUT. This statement allows formatting specifications to be embedded in the assembler source file, and also allows each procedure to have its own label prefix. An interactive debugger (with the same features as the BASIC/2, C/2, and Pascal/2 interactive debugger) is also provided.

Pascal Compiler/2

The Pascal Compiler/2 is a Pascal language compiler for use with OS/2 or DOS 3.30. The Pascal Compiler/2 and the code it generates can be run concurrently with other applications in the OS/2 multi-tasking environment. The compiler maintains a high degree of compatibility with the IBM PC Pascal Compiler Version 2.00 and 2.02. In addition, Pascal/2 provides medium memory model support and compatibility with the C/2 large memory model, math coprocessor support, math coprocessor emulation if no math coprocessor is available, and spawning of C/2 language programs. Decimal math support allows the conversion of earlier Pascal compiler floating point operations to IEEE floating point format. Network support is provided, including file sharing and record locking using Local Area Network Program, Version 1.2.

Language interfaces are provided that allow a Pascal/2 program to call programs written in languages such as Macro Assembler/2 and C/2, and allow a Pascal/2 program or subroutine to be called from programs written in Macro Assembler/2, compiled BASIC/2, or C/2. An interactive debugger (with the same features as the BASIC/2, C/2, and Macro Assembler/2 interactive debugger) is also provided.

Language compilers for use with OS/2 are available from Microsoft, Lattice, and other software vendors. These compilers work with OS/2 and support features provided in previous offerings. Several of the Microsoft products are very similar but not identical to those offered by IBM.

Packaging and Documentation

OS/2 Standard Edition, Version 1.0 is provided in a standard IBM slip-case package. The package includes the software, contained on 1.44 MB, 3.5-inch diskettes or 1.2 MB, 5.25-inch diskettes, a User's Guide, and a User's Reference manual. Also included is a pamphlet, entitled "Read This First," which directs the reader to the User's Guide for information about OS/2 and the degree of DOS compatibility it provides. The pamphlet also describes how to prepare a diskette that can be used to initiate system operation under the exclusive control of DOS after OS/2 is installed on a system's hard disk. (See *Installation* section, below.) Documentation for BASIC 3.30, which is included with OS/2, is provided in the IBM BASIC Reference Version 3.30, which is available at extra cost.

The OS/2 Standard Edition Programmer Toolkit provides tools for the creation of OS/2 and family applications. The Toolkit, which is available at extra cost, contains Presentation Manager tools used to create application display windows, tools for creating family applications, Macro Assembler/2 and C/2 sample programs, macro libraries, and message preparation utilities. The Toolkit includes the OS/2 Programmer Guide, which discusses the use of the various tools and sample code provided to convert existing applications or create new OS/2 applications. Like OS/2, Standard Edition, the Toolkit is being released in two phases. Version 1.0, which became available in December 1987, contains all the tools described except the Presentation Manager tools. Version 1.1, which includes the Presentation Manager tools, is scheduled to be available in October 1988. Version 1.0 Toolkit licensees will receive a copy of the Version 1.1 Toolkit at no additional cost.

The Version 1.1 Toolkit provides tools necessary for the development of Presentation Manager applications, in addition to the tools provided in the Version 1.0 Toolkit. The Version 1.1 Toolkit includes sample programs that illustrate the use of various aspects of the Presentation Manager application program interface. Provided along with these sample programs is a set of "include" files that provide definitions for functions and data structures necessary for writing Presentation Manager applications.

Three development tools (themselves Presentation Manager applications) are provided. They are the dialog editor, the bit-map/cursor editor, and the font editor. These tools are similar to comparable tools provided with Microsoft Windows. The dialog editor allows the interactive creation and editing of dialog boxes, exactly as they will appear in the running application. Once created, a dialog box definition is saved in a disk file for later inclusion in an application. The bit-map/cursor editor is used to create and edit bit-maps, cursor/pointers, and icons. The font editor supports the creation and editing of the characters in a font. Individual characters can be modified, and entire character fonts can be created. Once defined, bit-maps, cursor/pointers, icons, and fonts are stored in disk files that are made into resource files to be linked with the application program. These resource files are produced by a Toolkit utility called the resource compiler.

The OS/2 Standard Edition Technical Reference provides detailed, technical information about OS/2. Topics covered in this two-volume publication include application program interface, device driver architecture, function and return call definitions, input/output subsystems, system architecture, and use of the OS/2 linker. While OS/2 Programmer Toolkit publications refer to the Technical Reference, it is not included in the Toolkit but is available at extra cost. It also is being issued in two phases—1.0 and 1.1—corresponding to those releases of OS/2. Version 1.0 became available in December 1987; Version 1.1 is scheduled to be available in October 1988. Version 1.0 Technical Reference licensees will receive a copy of Version 1.1 of the publication at no additional cost.

Installation

An IBM Personal Computer AT, XT 286, or Personal System/2 Model 50, 60, or 80 with a hard disk with at least 5 MB of free space, and at least 1.5 MB of memory is required for the installation and use of OS/2 Standard Edition Version 1.0. At least 2 MB of memory is required if both DOS 3.30 and OS/2 applications are to be run.

Minimum recommended memory for running the extended edition configured to run only OS/2 applications is 3 MB. Additional memory required depends on the application mix. IBM advises potential users of the extended edition who anticipate the use of large data bases, large numbers of programs or files, or the execution of several concurrent applications to be sure that their computers' mass storage can be expanded beyond 20 MB. Users with such needs will likely not be able to use the Model 50 with its 20 MB hard disk capacity limitation. Models 60 and 80, however, have larger mass storage devices.

Users who have DOS installed on the system hard disk should create a diskette that can be used to run DOS on the system once OS/2 is installed. Such operation may be required in order to run applications that are time dependent, or require device driver or communications support not currently available for OS/2. The procedure for creating such a diskette is provided in the "Read This First" pamphlet included in the OS/2 package.

The procedure consists of segregating the files necessary for the operation of DOS into a subdirectory (with a suggested name of DOS) on the hard disk, and then copying the contents of the subdirectory onto a DOS system diskette (created using the DOS FORMAT /S command). This isolation of DOS system files in a subdirectory is required because the OS/2 installation process places OS/2's CONFIG.SYS file in the root directory of the system hard disk, and replaces DOS files (such as COMMAND.COM and AUTOEXEC.BAT) in the root directory with their OS/2 DOS environment equivalents.

The DOS system's CONFIG.SYS and AUTOEXEC.BAT files are copied to the DOS Starter Diskette, for use when the system is initialized from diskette. These files are modified as necessary to indicate the full drive and path locations of any referenced device drivers or executable files. Additionally, the command SHELL=C:\DOS\COMMAND.COM /P should be added to the CONFIG.SYS file. The command SET COMSPEC=C:\DOS\COMMAND.COM should be added to the AUTOEXEC.BAT file, and C:\DOS should be included in the path statement. The last command in the AUTOEXEC.BAT file should be C:, to change the default drive from A to C.

The DOS Starter Diskette should be tested to be sure that it can be successfully used to initialize the system, before proceeding with the installation of OS/2. Once OS/2 is installed, the DOS Starter Diskette can be used to run the system under the exclusive control of DOS, as necessary. Version 3.30 of DOS can be started by turning on the computer or pressing Ctrl-Alt-Del with the Starter Diskette in drive A. Earlier versions of DOS can only be successfully started by inserting the DOS Starter Diskette in drive A while OS/2 is running, and pressing Ctrl-Alt-Del.

The installation of the OS/2 software is described briefly in the OS/2 User's Guide. The guide simply instructs the user to insert the Installation Diskette in drive A, turn on the computer (or press Ctrl-Alt-Del if the computer is already on), and follow the instructions displayed on the screen. Also, an illustration of

Chapter 11

the Program Selector screen that should appear after the installation is successfully completed is provided.

Initializing the system from the Installation Diskette causes the display of an IBM logo screen bearing the version number of the software being installed. Pressing Enter displays a screen that asks the user to press Enter to continue or Esc to abort the installation process. The screen also advises users with other than a U.S. English keyboard to refer to the U.S. English keyboard template in the OS/2 User's Reference manual. Pressing Enter yields another explanatory screen, detailing key usage conventions to be used during the installation process—Cursor Up and Cursor Down keys to select items, Enter to proceed to the next screen, Esc to return to the previous screen, and F1 for help.

The contents of the next screen displayed depends on whether DOS or OS/2 has been installed on the system's hard disk. If neither have been installed on the hard disk (or if it has been freshly formatted using the advanced diagnostics low level format routine described in Chapter 7), the screen shown in Figure 11-6 is displayed. This screen prompts the user to accept a predefined primary partition size (32 MB or the size of the disk, whichever is smaller) or to select item 2 in order to specify sizes of the primary and extended OS/2 partitions.

```
                        Preparing the Fixed Disk

     Preparing the fixed disk is a two-step process.  The first step,
     partitioning, separates the fixed disk into areas called partitions.
     The second step, formatting, prepares the first partition to accept
     data.

     If the fixed disk is larger than 32MB, it can be separated into
     multiple partitions, but the maximum size of any partition is 32MB.



     Select an option:

       ▸ 1. Accept predefined partition

       • 2. Specify your own partitions



     Enter   Esc=Cancel   F1=Help
```

Figure 11-6   Preparing the Fixed Disk Menu. *This screen only appears if the hard disk does not contain an operating system.*

Selecting *Specify your own partitions* (item 2) causes the display of the OS/2 hard disk setup program (SYSDISK.COM) main menu, shown in Figure 11-7. (A later part of the installation process stores a copy of this program in the C:\OS2

Operating System/2

directory under the filename FDISK.COM.) The setup program has the same user interface and functions the same as the DOS FDISK program. Primary partitions, extended partitions, and logical drives in the extended partition are created exactly the same as with the DOS FDISK program. See Chapter 10 for additional information.

```
IBM Personal Computer
Fixed Disk Setup Program Version 1.00


FDISK options



Choose one of the following:

    1.  Create an IBM Operating System/2 partition
        or a logical drive
    2.  Change the active partition
    3.  Delete an IBM Operating System/2 partition
        or a logical drive
    4.  Display the partition data


Enter choice: [1]




Press Enter to continue or
Esc to return to IBM Operating System/2
```

Figure 11-7 Fixed Disk Setup Program Screen. *This screen is used to create or change the partitions on the hard disk.*

Contrary to the information contained in the screen shown in Figure 11-6, the maximum size of any partition on the disk is not 32 MB. The primary partition is limited to 32 MB and the logical disk drives defined in the extended partition are limited to 32 MB each, but the extended partition can be as large as the amount of space remaining on the disk once the primary partition has been defined. Multiple logical drives are necessary in order to utilize the full capacity of the Personal System/2 Model 60 and 80 disk drives that are greater than 64 MB in size.

Once the partitions have been defined, the user presses Esc to return to the OS/2 installation process. (SYSDISK.COM automatically makes the primary partition the active partition during the definition process.) Upon doing so, a message directs the user to press Ctrl-Alt-Del (with the OS/2 Installation Diskette still in drive A) to continue the installation process.

Once the system is reinitialized from the Installation Diskette, the installation procedure displays a message telling the user to wait while the hard disk is formatted, and begins formatting the primary disk partition defined in the previous step. If a DOS or OS/2 partition (there is no difference) was already defined on the hard disk before the OS/2 installation process, the process just described is skipped entirely and the menu screen shown in Figure 11-8 is displayed instead.

```
                      Formatting the Fixed Disk


    A fixed disk must be formatted before information can be
    placed on it.  The fixed disk you are using has a partition.
    This partition may need to be formatted before it can be used.

    Note:  If the fixed disk partition is already formatted and
    contains information, reformatting destroys your information.




    Select an option:

       ▸ 1. Do not format the partition

       • 2. Format the partition


  Enter   Esc=Cancel   F1=Help
```

Figure 11-8   Formatting the Fixed Disk. *This menu appears first if the hard disk has been previously partitioned.*

This menu asks the user to indicate whether the hard disk partition on which OS/2 is to be installed should be formatted. The default option is *not* to perform a formatting operation. As the screen and its help screen indicate, formatting the partition destroys any information currently stored in it. Formatting the partition should *only* be necessary on a disk on which neither DOS nor OS/2 has previously been installed, or in the unusual event that the user wishes to remove all directories and files from the disk before installing OS/2.

If the user elects to format the hard disk, another clear warning message is displayed, indicating that the process will destroy the data on the hard disk. The user is given the choice of pressing Enter to initiate the format process or Esc to cancel it.

Once initiated, the format process displays a message indicating that the hard disk is being prepared for OS/2, and proceeds to format it. A running display of the cylinder and surface number being formatted is not displayed, as it is when a format is performed using FORMAT.COM from the command line. The formatting process is rapid, requiring just under two minutes for the Model 50's 20 MB disk and proportionally longer times for larger disks.

The format process formats only the primary DOS partition on the hard disk. If the system has not previously been used for OS/2 or DOS, each logical drive defined in the extended DOS partition must be formatted using the OS/2 or DOS Environment FORMAT command. It is essential that all logical drives be formatted. If this step is omitted it may be possible to store files on the drive, but reads and writes to defective disk blocks can occur because the defective blocks have not been identified by FORMAT.

An attempt to write to a defective block results in a somewhat cryptic SYS1187: ... cannot access target file error message (when using the OS/2 command processor) or a File creation error (when using the DOS environment command processor). One way to find out if the drive has been formatted is to use the CHKDSK command on the logical drive. If it has not been formatted, CHKDSK displays a blizzard of The first cluster number of ... is incorrect error messages.

If the format process is bypassed, a screen is displayed telling the user that the CONFIG.SYS and AUTOEXEC.BAT files in the root directory of the hard disk will be renamed to CONFIG.BAK and AUTOEXEC.BAK, unless the user specifies other new names for them. This step is omitted if the format process is initiated, because any existing CONFIG.SYS and AUTOEXEC.BAT files are eliminated by the format process.

The next step is copying the OS/2 system files from the Installation Diskette to the hard disk. This process begins upon completion of the format process or after the user enters any desired filenames for the existing CONFIG.SYS and AUTOEXEC.BAT files and presses Enter (if the format process was bypassed). A message indicating that the OS/2 files are being copied to the hard disk is displayed. After about two minutes of file copying, a message indicates that OS/2 system files have been installed on the hard disk, and directs the user to remove the Installation Diskette from drive A and press Ctrl-Alt-Del to continue the installation process.

Initializing the system as directed causes the display of a screen that instructs the user to insert the OS/2 Diskette 1 into drive A and press Enter. The installation process then copies files necessary to start the operating system and the Program Selector, from the diskette to the hard disk. When that operation is complete, Diskettes 2 and 3 are likewise requested and the OS/2 external commands, device drivers, and support files contained thereon are copied to the hard disk.

When the file copying operation is complete, the user is instructed to press Enter. Doing so displays the screen from which the country and keyboard to be used for the system is selected. The user may either select the predefined country and keyboard values (United States for both) or select from a list of twenty-five countries and seventeen keyboards. Primary and alternate code pages may also be specified during this process.

The next choice is what printer will be used on the system (one of several IBM models or *Other*), and what mouse is to be used. (Choices are IBM Personal System/2 mouse, Microsoft serial mouse, Mouse Systems mouse, Visi-On mouse, or none.)

Chapter 11

The next screen asks the user to indicate if serial device support needs to be provided. Users who plan to use a serial port for anything other than a serial mouse should select this option; for example, if a modem or serial printer is to be connected to a serial port. Once this selection is made, a screen display asks the user to either accept OS/2's predefined configuration, or view and/or change it.

The screen shown in Figure 11-9 is displayed when *View configuration* or *Change configuration* is selected from the menu. The screen shows the default values for each OS/2 mode configuration parameter and the allowed range or choice of values. These parameters, along with descriptions, are listed in Table 11-2. The configuration process stores these parameters and their values in the CONFIG.SYS file in the root directory of the system disk. After the initial installation is complete, the system can be reconfigured by changing the parameters in the CONFIG.SYS file and reinitializing the system.

```
                        OS/2 Mode Configuration



            BUFFERS. . . . . [ 30]            (1 - 100)
            DISKCACHE. . . . [  64]           (64 - 7200)
            MAXWAIT. . . . . [   3]           (1 - 255)
            MEMMAN MOVE. . . [MOVE   ]        (MOVE/NOMOVE)
            MEMMAN SWAP. . . [SWAP   ]        (SWAP/NOSWAP)
            PRIORITY . . . . [DYNAMIC ]       (DYNAMIC/ABSOLUTE)
            PROTECTONLY. . . [NO ]            (YES/NO)
            SWAPPATH . . . . [C:\                               ]
            THREADS. . . . . [  64]           (32 - 255)
            TRACE. . . . . . [OFF]            (ON/OFF)
            TRACEBUF . . . . [  4]            (1 - 63)




   Enter   Esc=Cancel   F1=Help
```

Figure 11-9 OS/2 Mode Configuration Menu. *The range or choice of values and the default values for each parameter are shown.*

Names of some device drivers stored in the CONFIG.SYS file are model group dependent. The three model groups on which OS/2 runs are the PC AT model group, the XT 286 model group, and the Personal System/2 model group (not including Models 25 and 30). If the user manually inserts DEVICE= statements in the CONFIG.SYS file, the user must ensure that the indicated device driver is correct for the model group of the system on which OS/2 is to be started.

Table 11-2 OS/2 Configuration Commands. *The system can be reconfigured by changing parameters in the CONFIG.SYS file.*

Command	Purpose
BREAK	Sets Ctrl-Break checking for DOS mode
BUFFERS	Sets the number of disk buffers
CODEPAGE	Selects system code pages
COUNTRY	Identifies country for country-dependent information
DEVICE	Specifies device driver to be installed
DEVINFO	Prepares a device for system code page switching
FCBS	Sets number of file control blocks for DOS mode
IOPL	Allows I/O privilege to be granted to processes
LIBPATH	Specifies locations of dynamic link libraries
MAXWAIT	Maximum idle time before a process receives higher priority
MEMMAN	Selects storage allocation options for OS/2 mode
PRIORITY	Sets priority calculation options
PROTECTONLY	Controls creation of the DOS compatibility environment
PROTSHELL	Specifies OS/2 mode command processor
REM	Allows remarks in CONFIG.SYS file
RMSIZE	DOS mode size
RUN	Loads and starts a program at system initialization
SHELL	Specifies DOS mode command processor
SWAPPATH	Location of file to contain swapped out memory
THREADS	Maximum number of threads
TIMESLICE	Sets length of processor time slice
TRACE	Selects system trace
TRACEBUF	Sets the size of the circular trace buffer

Online help is available for each item displayed. A detailed description for the parameter selected by the cursor can be displayed by pressing F1 (see Figure 11-10). Each parameter is also described in Chapter 4 of the OS/2 Reference manual.

Changes are made in the configuration by selecting the field to be changed using the Tab or cursor keys, and entering the new value. Pressing Enter causes the changes to be accepted. The DOS Mode Configuration screen is then displayed, as shown in Figure 11-11. If the user disables creation of the DOS compatibility environment by setting **PROTECTONLY** to YES, this screen is skipped. Control can be returned to the previous menu at any time by pressing Esc.

After Enter has been pressed to indicate the user has completed the configuration screen(s), the Spool Support screen is displayed. This screen asks for the name of the device whose output is to be spooled, and the name of the device to which the output is to be sent. The default for both is LPT1.

Pressing Enter to end data entry for the Spool Support screen displays a screen that asks the user to press Enter to complete the OS/2 installation process. When that is done, the configuration parameters are written to the hard disk and the user is instructed to remove the OS/2 diskette from drive A and to press Ctrl-Alt-Del to start OS/2.

```
                         OS/2 Mode Configuration



            BUFFERS. . . . . [ 30]             (1 - 100)
            DISKCACHE. . . . [  64]            (64 - 7200)
        ┌─────────────────────────────────────────────────────┐
        │                    BUFFERS Help                     │
        │                                                     │
        │  This number, times 512, specifies the number of bytes of storage
        │  reserved for temporarily storing data to be written to the fixed
        │  disk.  The number can be from 1 to 100; it is preset at 30,
        │  which is sufficient for most programs.
        │
        │
        │
        ├─────────────────────────────────────────────────────┤
        │ Esc=Cancel    F1=Help    F5=Index    F9=Keys        │
        └─────────────────────────────────────────────────────┘
```

Figure 11-10   Online Help for Configuration Parameters. *Online help information is available during the installation process.*

```
                         DOS Mode Configuration



          BREAK. . . . . . . .  [OFF]    (ON/OFF)
          OPEN FCBS. . . . . .  [ 16]    (1 - 255)
          PROTECTED FCBS . . .  [  8]    (0 - 255)    Must be less than or
                                                      equal to OPEN FCBS
          RMSIZE . . . . . . .  [640]    (256 - 640)




        Enter   Esc=Cancel    F1=Help
```

Figure 11-11   DOS Mode Configuration Menu. *The DOS compatibility environment can be configured, as well as the OS/2 mode.*

Upon initialization, OS/2 automatically loads the correct base device drivers for the system hardware's model group. The OS/2 kernel automatically adapts itself to hardware-dependent features that are not device driver related. It then attempts to read the CONFIG.SYS file in the root directory for system configuration information. If CONFIG.SYS is not present, default system configuration information is used. Next, it reads the batch file STARTUP.CMD in the root directory of the system disk and executes commands contained therein. When this is done, the OS/2 Program Selector is loaded.

Using OS/2

Once the system is initialized and OS/2 is loaded into memory, the OS/2 Program Selector menu is displayed. (See Figure 11-12.) The Program Selector is the primary user interface for Version 1.0 of OS/2. Programs that can be run from this menu initially include the OS/2 and DOS command processors (*OS/2 Command Prompt* and *DOS Command Prompt*), and a program called *Introducing OS/2*, which is installed on the hard disk when OS/2 is installed.

```
 Update                                                        | F1=Help

                           Program Selector
         To select a program, press →, ↑, or ↓.  Then, press Enter.
         To select Update, press F10.  Then, press Enter.

  ┌──────────────────────────────────┐  ┌──────────────────────────────────┐
  │ Start a Program                  │  │ Switch to a Running Program      │
  │ ─────────────                    │  │ ───────────────────────────      │
  │                                  │  │                                  │
  │ • Introducing OS/2               │  │ • DOS Command Prompt             │
  │ • OS/2 Command Prompt            │  │                                  │
  │                                  │  │                                  │
  │                                  │  │                                  │
  │                                  │  │                                  │
  │                                  │  │                                  │
  └──────────────────────────────────┘  └──────────────────────────────────┘
```

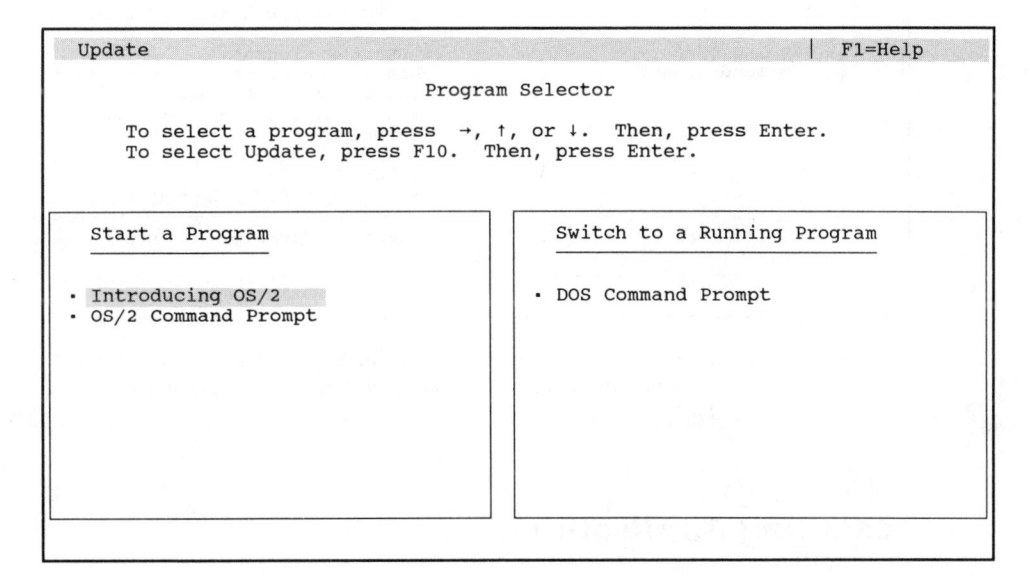

Figure 11-12   Program Selector Main Menu. *OS/2 contains an introductory tutorial guide to its operation, which can be accessed using the Program Selector.*

The *Introducing OS/2* program can be started by moving the highlight to it using the Tab or cursor keys and pressing Enter, or by pressing mouse button 1 twice in quick succession while the screen pointer is on its entry in the *Start a Program* menu. (The screen pointer changes shape whenever it is on a selectable item.) *Introducing OS/2* is a menu-driven program that provides information about getting started using OS/2, running programs, and managing information on an OS/2 system.

Help information is available to the user anytime F1=Help is displayed. The OS/2 help facility can be accessed by pressing F1 or mouse button 1 when the screen pointer is anywhere on F1=Help. Help information about a specific item is accessed by requesting help when that item is highlighted. Figure 11-13 shows the help display for the *Start a Program* menu. The More: ↓ (Cursor Down) symbol indicates that the help display contains additional information that may be viewed by pressing the Cursor Down or PgDn key.

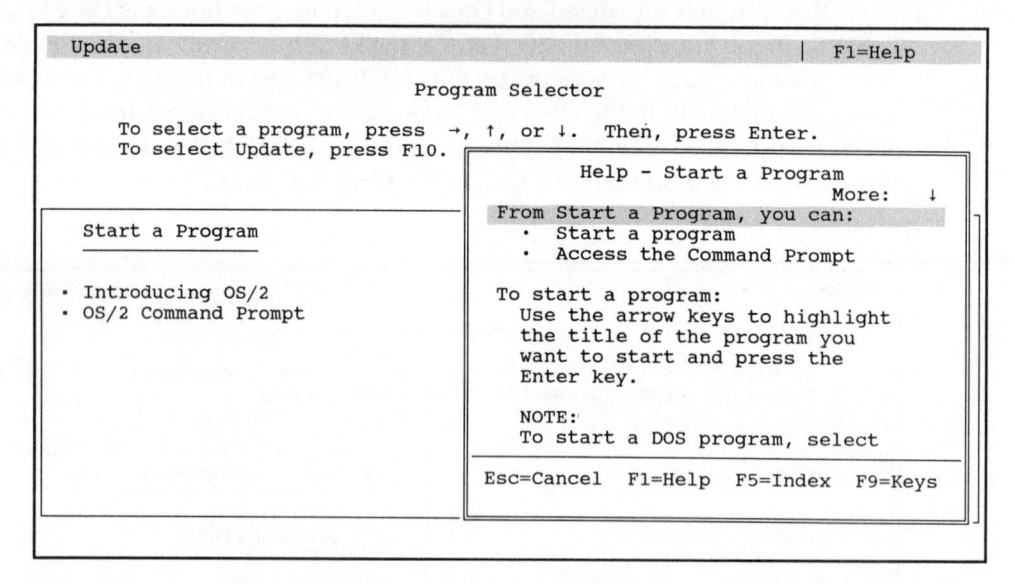

```
Update                                                        | F1=Help
                         Program Selector
    To select a program, press  →, ↑, or ↓.  Then, press Enter.
    To select Update, press F10.
                                     ┌─────────────────────────────────┐
                                     │   Help - Start a Program        │
                                     │                    More:  ↓     │
                                     │ From Start a Program, you can:  │
    Start a Program                  │  • Start a program              │
    ───────────────                  │  • Access the Command Prompt    │
                                     │                                 │
    • Introducing OS/2               │ To start a program:             │
    • OS/2 Command Prompt            │   Use the arrow keys to highlight│
                                     │   the title of the program you  │
                                     │   want to start and press the   │
                                     │   Enter key.                    │
                                     │                                 │
                                     │ NOTE:                           │
                                     │ To start a DOS program, select  │
                                     ├─────────────────────────────────┤
                                     │ Esc=Cancel  F1=Help  F5=Index  F9=Keys │
                                     └─────────────────────────────────┘
```

Figure 11-13   Start a Program Help Screen. *Additional help information can be obtained by pressing the PgDn key or the Cursor Down key.*

Installing Applications

Additional OS/2 programs installed on the system can be added to the menu using the Program Selector Update function. This function should be used only to add OS/2 programs to the menu. If a DOS program is added, an error message indicating that the program cannot be run in OS/2 mode will be generated when its title is selected from the *Start a Program* menu. The Update function is selected by pressing F10 or mouse button 1 while the screen pointer is on Update in the Program Selector action bar. (See Figure 11-14.)

Selecting *Add a Program Title* . . . from this menu displays the Add a Program Title screen (Figure 11-15). The user must then enter information about the

Operating System/2

program to be added to the list of programs that can be started from the Program Selector. Help information about a field may be obtained by pressing F1 when the field is selected.

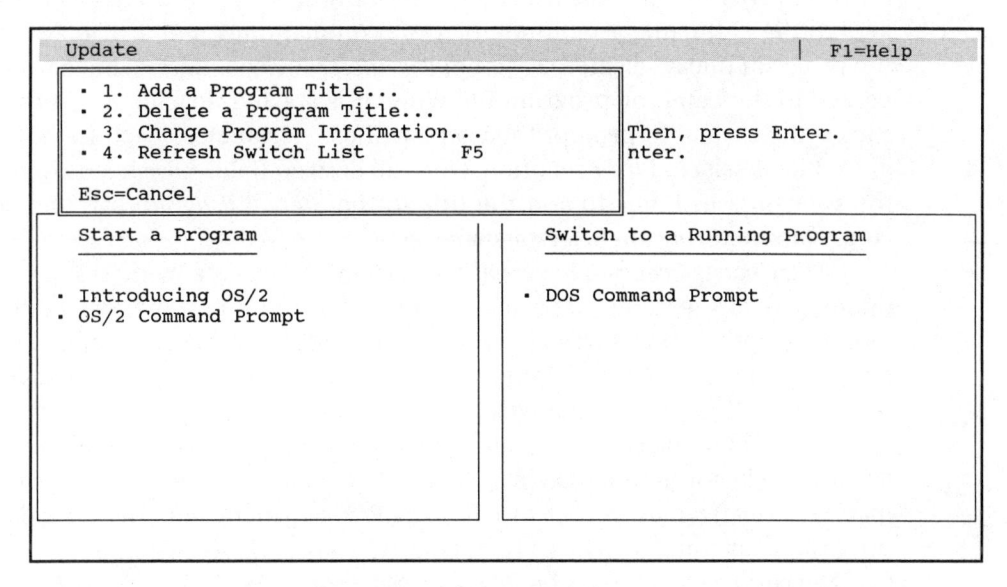

```
┌─────────────────────────────────────────────────────────────────────────┐
│ Update                                                      │ F1=Help   │
│   ┌──────────────────────────────────────────┐                          │
│   │ • 1. Add a Program Title...              │                          │
│   │ • 2. Delete a Program Title...           │                          │
│   │ • 3. Change Program Information...       │ Then, press Enter.       │
│   │ • 4. Refresh Switch List        F5       │ nter.                    │
│   │                                          │                          │
│   │ Esc=Cancel                               │                          │
│   └──────────────────────────────────────────┘                          │
│   ┌──────────────────────────────┐  ┌──────────────────────────────┐    │
│   │  Start a Program             │  │  Switch to a Running Program │    │
│   │                              │  │                              │    │
│   │  • Introducing OS/2          │  │  • DOS Command Prompt        │    │
│   │  • OS/2 Command Prompt       │  │                              │    │
│   │                              │  │                              │    │
│   │                              │  │                              │    │
│   │                              │  │                              │    │
│   │                              │  │                              │    │
│   └──────────────────────────────┘  └──────────────────────────────┘    │
└─────────────────────────────────────────────────────────────────────────┘
```

Figure 11-14 Program Selector Update Menu. *Additional OS/2 programs can be added to the main menu.*

```
┌─────────────────────────────────────────────────────────────────────────┐
│ Update                                                      │ F1=Help   │
│                        Add a Program Title                              │
│           ┌───────────────────────────────────────────────────┐         │
│           │  Type the information below.                      │         │
│           │  Press the Enter key to save.                     │         │
│           │                                                   │         │
│           │  Program Title . . . . [                       ]  │         │
│       Sta │                                                   │         │
│           │  Program Pathname. . . [                      >]  │         │
│           │                                                   │         │
│           │  Program Parameters. . [                      >]  │         │
│    • Int  │  (Optional)                                       │         │
│    • OS/  │                                                   │         │
│           │  Enter   Esc=Cancel   F1=Help                     │         │
│           └───────────────────────────────────────────────────┘         │
│                                                                         │
└─────────────────────────────────────────────────────────────────────────┘
```

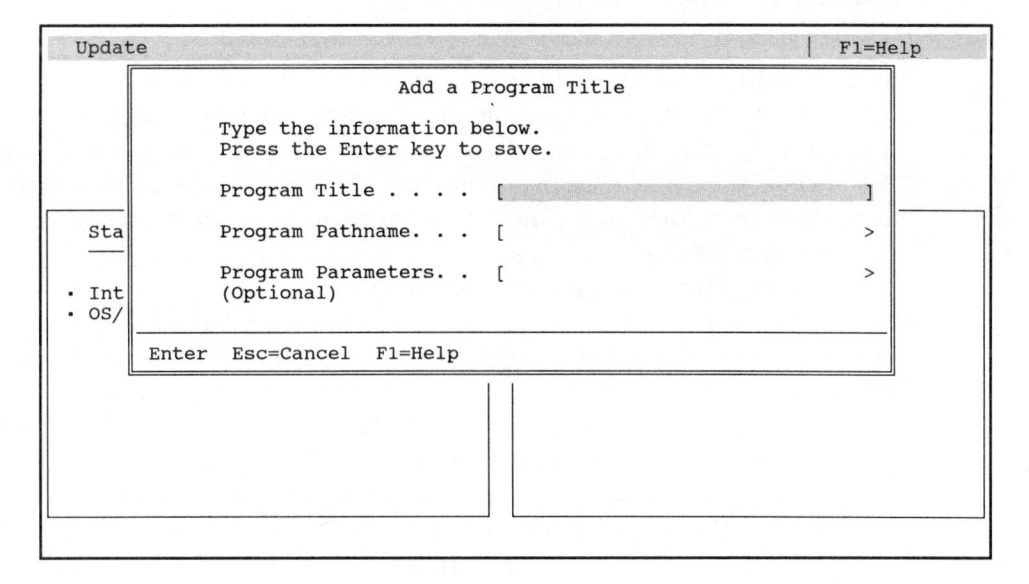

Figure 11-15 Add a Program Title Screen. *The program title is the name displayed on the Program Selector menu.*

375

Program Title is the name under which the program will be listed in the *Start a Program* list; it can be up to 30 characters long. Once the title is entered, the next field can be selected using the Tab key or Cursor Down key. *Program Pathname* must specify the full path and filename (including extension) of the batch or program file to be executed. The pathname can be up to 64 characters long. If the pathname is omitted, the OS/2 command prompt is displayed when the program title is selected. *Program Parameters* are command line parameters passed to the batch or program file when it is started. Placing a ? in this field causes the system to prompt the user for these parameters each time the program title is selected for execution. Once all desired fields have been completed, the user presses Enter to add the title to the *Start a Program* list, and return control to the Program Selector main menu.

The *Change Program Information* entry on the Update menu can be used to update any of the items just discussed for any program on the *Start a Program* menu, except the *OS/2 Command Prompt* item. Selecting *OS/2 Command Prompt* from the *Change Program Information* master list causes an error message to be displayed indicating that changes are not allowed.

DOS programs can be installed on the system but cannot be added to the Program Selector menu. DOS programs are installed by selecting the *DOS Command Prompt* from the *Switch to a Running Program* menu and then installing the program as if using a DOS system. Once the installation is complete, pressing Ctrl-Esc returns control to the Program Selector.

Running Applications

OS/2 applications are started from the Program Selector menu either by highlighting the title with the cursor and pressing Enter or by pointing to it with the mouse and pressing button 1 twice. Applications can be executable program files or OS/2 batch files. OS/2 batch files have CMD as name extensions, to distinguish them from DOS batch files. Titles of applications that can be executed from the *Start a Program* menu are listed in alphabetical order.

Selecting *OS/2 Command Prompt* from the *Start a Program* menu causes an OS/2 command processor (CMD.EXE) session to be started and brought to the foreground. Its screen output replaces the Program Selector display, and it receives input from the keyboard. At startup the OS/2 command processor executes the batch commands found in the file STARTUP.CMD, which is stored in the root directory of the system diskette by the OS/2 installation process. Commands in this batch file set the program and data paths for the command processor session and turn on the help line at the top of the screen.

CMD.EXE's user interface is similar to that of DOS's COMMAND.COM. (The prompt can even be the same if desired.) The user enters commands to be executed at the `C:\` prompt, just as in DOS. The usual DOS command buffer

Operating System/2

editing keys are available when using the OS/2 command prompt. See Table 11-3 for key definitions used by OS/2. As shown in Figure 11-16, OS/2 displays numbered error messages reminiscent of those provided by System/370 operating systems. Fortunately, online explanations can be obtained using the HELP command, and the explanations do not contain the infamous phrase "probable user error."

Table 11-3 Program Selector/Command Interface Key/Mouse Functions.
The usual DOS command buffer editing keys are available when using the OS/2 command interface.

Function	Keyboard	Mouse
Select item from a list	Move cursor to item, Enter	Button 1 double click on item
Select a nonlist item	Move cursor to item, Enter	Button 1 click on item
Enter selected information	Enter	Button 1 click on Enter
Return to previous menu/cancel command displayed	Esc	Button 1 click on Esc
Delete previous character	Backspace	
Insert a character	Ins	
Delete a character	Del	
Display help when available	F1	Button 1 click on F1=Help
Display previous command, one character at a time	F1	
Display characters in command, up to character entered after pressing F2	F2	
Display previous command	F3	
Delete all characters in a command before the one entered after pressing F4	F4	
Update the Switch to a Running Program list	F5	Button 1 click on F5
Display help index, when available	F5	
Accept the edited command as the current command	F5	
Display key assignments	F9	Button 1 click on F9=Keys
Move the cursor to Update	F10	
Switch to the next program	Alt-Esc	
Display the Program Selector	Ctrl-Esc, or Enter EXIT at OS/2 command prompt	
Return cursor to field for which help was requested	Alt-F6	
Stop display from scrolling	Ctrl-NumLock or Pause	
Cancel an OS/2 command	Ctrl-Break	
Print screen information	Shift-PrtSc or Print Screen	
Restart OS/2	Ctrl-Atl-Del	
Send end-of-file to print spooler	Ctrl-Alt-PrtSc	

```
 OS/2      Ctrl+Esc = Program Selector                    Type HELP = help
[C:\]PATH C:\;C:\OS2;C:\OS2\INSTALL;

[C:\]DPATH C:\;C:\OS2;C:\OS2\INSTALL;

[C:\]CALL HELP ON

[C:\]del myfile
SYS0002: The system cannot find the file specified.

[C:\]help 2

SYS0002: The system cannot find the file specified.

EXPLANATION: The file named in the command
does not exist in the current directory or search path
specified. Or, the filename was entered incorrectly.
ACTION: Retry the command using the correct filename.


[C:\]
```

Figure 11-16 OS/2 Command Prompt Screen. *The OS/2 command interface is similar to that of DOS.*

The only readily apparent new feature is the Help line at the top of the display, which indicates that the OS/2 command processor is in use and that the user can request help by pressing F1 or return to the Program Selector by pressing Esc while holding down Ctrl. The Help line can be eliminated by entering the command HELP OFF.

The ability to start programs from the Program Selector menu is not a powerful feature, in and of itself. Several third-party DOS shells allow commands to be selected from menus. The real power of OS/2 becomes evident, however, when the user presses Ctrl-Esc to return to the Program Selector. (See Figure 11-17.) The *Switch to a Running Program* menu now lists *OS/2 Command Prompt*. The title of the menu means just what it says: even though the user is interacting with the Program Selector, previously started programs continue to run.

Any time the Program Selector is displayed, the user can start additional programs (up to a maximum of twelve), switch to a running program by selecting it from the switch menu, or perform any other functions available from the menu. Multiple OS/2 command processor sessions can be active at once, as can multiple copies of the same application, as long as data and resource sharing rules are observed. Multiple copies of the same program (including CMD.EXE) are identified on the *Switch to a Running Program* menu by a 1 added to the title of the first instance, 2 to the second, and so forth, after the second copy of the program is started. (See Figure 11-18.)

Operating System/2

```
┌─────────────────────────────────────────────────────────────────┐
│ Update                                          │ F1=Help       │
│                    Program Selector                             │
│                                                                 │
│    To select a program, press →, ↑, or ↓.  Then, press Enter.   │
│    To select Update, press F10.  Then, press Enter.             │
│                                                                 │
│   ┌───────────────────────────┐   ┌───────────────────────────┐ │
│   │ Start a Program           │   │ Switch to a Running Program│ │
│   │                           │   │                           │ │
│   │ • Introducing OS/2        │   │ • DOS Command Prompt      │ │
│   │ • OS/2 Command Prompt     │   │ • OS/2 Command Prompt     │ │
│   │                           │   │                           │ │
│   │                           │   │                           │ │
│   │                           │   │                           │ │
│   │                           │   │                           │ │
│   └───────────────────────────┘   └───────────────────────────┘ │
└─────────────────────────────────────────────────────────────────┘
```

Figure 11-17 Program Selector with OS/2 Command Prompt Running. *Even though the user is interacting with the Program Selector, previously started programs continue to run.*

```
┌─────────────────────────────────────────────────────────────────┐
│ Update                                          │ F1=Help       │
│                    Program Selector                             │
│                                                                 │
│    To select a program, press →, ↑, or ↓.  Then, press Enter.   │
│    To select Update, press F10.  Then, press Enter.             │
│                                                                 │
│   ┌───────────────────────────┐   ┌───────────────────────────┐ │
│   │ Start a Program           │   │ Switch to a Running Program│ │
│   │                           │   │                           │ │
│   │ • Data Entry              │   │ • DOS Command Prompt      │ │
│   │ • Generate Report         │   │ • Data Entry 1            │ │
│   │ • Introducing OS/2        │   │ • Sort Data               │ │
│   │ • OS/2 Command Prompt     │   │ • Generate Report         │ │
│   │ • Sort Data               │   │ • Data Entry 2            │ │
│   │                           │   │                           │ │
│   │                           │   │                           │ │
│   └───────────────────────────┘   └───────────────────────────┘ │
└─────────────────────────────────────────────────────────────────┘
```

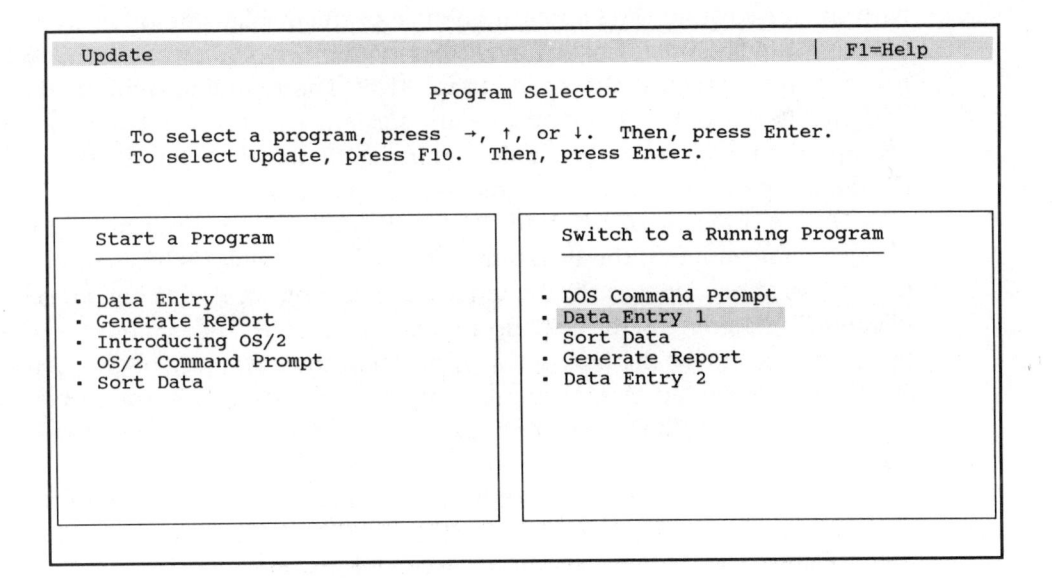

Figure 11-18 Multiple Sessions of the Same Application. *Multiple copies of the same program can be run provided data and resource sharing rules are observed.*

When a program has been brought to the foreground, either by starting it or by selecting it from the Switch menu, the user can switch another running application to the foreground without going back to the Program Selector. Pressing Esc while holding down Alt brings the next program on the Switch list to the foreground. Switching a different OS/2 program to the foreground does not suspend execution of the program switched out of the foreground (although it executes at a lower priority than the foreground program). The program running in the DOS compatibility environment is suspended when it is switched to the background.

The *DOS Command Prompt* can be brought to the foreground by selecting it from the Switch list using the Program Selector or using Alt-Esc from a running OS/2 program to move through the Switch list until the *DOS Command Prompt* entry is reached. It is always on the Switch list because it is always loaded in memory (in the region below 640 KB). The first time the DOS command prompt is selected from the Program Selector, it executes the AUTOEXEC.BAT file, which is stored in the root directory by the OS/2 installation process. Commands in this file set the program path for the DOS compatibility environment and turn on the Help line at the top of the screen.

DOS is not designed for multi-tasking; when it is executing it asserts full control over system devices, including the display. Therefore, it and the single application it runs are allowed to execute only in the foreground. Interference with the foreground OS/2 application's use of the display and other system resources would result if background DOS execution were allowed. The *DOS Command Prompt* screen is shown in Figure 11-19. The Help line reminds the user that this is the DOS command prompt screen. As with the OS/2 command prompt, the Help line can be eliminated using the HELP OFF command. Note that the default paths for the two command processors are different.

Once a program or batch file initiated from the DOS prompt is complete, the user can return to the Program Selector by pressing Ctrl-Esc, or go to the next active application on the Switch list by pressing Alt-Esc. When an OS/2 application initiated from the OS/2 prompt is complete, the user can return to the Program Selector or the next active application the same way, or the CMD.EXE session can be ended by entering EXIT when the session is in the foreground. Entering EXIT from the DOS prompt does not return the Program Selector to the display.

Most commands provided with DOS 3.30 are also provided with OS/2. A few commands (APPEND, ASSIGN, BREAK, GRAFTABL, JOIN, and SUBST) are not available in OS/2; however, these commands can be run from the DOS command prompt. Attempting to run one of the commands from the OS/2 command processor results in a SYS0191 message stating that the command cannot be run in OS/2 mode.

Several commands that run only in OS/2 mode are provided. These include ANSI, DETACH, DPATH, and PATCH. The ANSI command enables or disables ANSI extended display and keyboard support in OS/2 mode; ANSI support is

```
    DOS        Ctrl+Esc = Program Selector                    Type HELP = help
C>PATH C:\;C:\OS2;

C>CALL HELP ON
C:\>
C:\>
```

Figure 11-19 DOS Command Prompt Screen. *Note that the default program path for the DOS command processor is different from that for the OS/2 command processor.*

obtained in DOS mode by specifying `DEVICE=\OS2\ANSI.SYS` in the CONFIG.SYS file. DETACH starts and detaches an OS/2 process from its command processor; the detached process should not issue any I/O calls to the keyboard, mouse, or display. DPATH provides a search path to data files; it provides the same function as the DOS mode APPEND command. PATCH is a utility for applying binary updates to the operating system or other disk files. Patches can be entered from the keyboard or from a file containing patch commands and data.

Three new batch file commands—ENDLOCAL, EXTPROC, and SETLOCAL—are provided for OS/2 mode use. SETLOCAL saves the current drive, directory, and environment variables, so that the batch file can define these items temporarily. The temporary definitions remain in effect until an ENDLOCAL command is executed or the batch file ends. EXTPROC specifies the filename of the batch processor to run the batch file containing the EXTPROC statement; the EXTPROC statement must be the first line in the batch file.

The OS/2 SPOOL and START commands reflect OS/2's multi-tasking features. START initiates an OS/2 program in another session. This command can be entered from the OS/2 command prompt, but it is usually placed in the STARTUP.CMD file (the OS/2 equivalent of the DOS AUTOEXEC.BAT file), to start OS/2 programs automatically at system startup.

SPOOL manages the OS/2 print queue. It intercepts print output data, stores it in temporary files in a subdirectory (\SPOOL by default), and prints them on a designated printer. The OS/2 installation process places a `RUN` statement in the CONFIG.SYS file to initiate the SPOOL at system startup.

The SPOOL program copies output to a temporary file until it is instructed to close the file and print it. Most DOS programs are not configured to send the necessary instructions to the SPOOL programs, so their temporary files are not closed and printed until the program ends. The user can initiate printing of a spooled file while the DOS application is still running by pressing Ctrl-Alt-PrtSc when the DOS application is in the foreground. Care must be taken to assure that the application has actually completed printing to the spooler, or the hardcopy output will be printed in two separate print files.

Error Recording Facilities

Developing, upgrading, and maintaining a multi-tasking operating system is an ongoing error isolation and problem determination process. Realizing this, IBM has included several commands in OS/2 for gathering operational data to be supplied to IBM when reporting problems with the operating system. These commands are the TRACE and TRACEFMT commands, the TRACE and TRACEBUF CONFIG.SYS statements, and the OS/2 stand-alone memory dump utility.

TRACEBUF=n, where n represents the size of the circular trace buffer in kilobytes, can be included in the CONFIG.SYS file to create a buffer for the storage of variable length data. Such a buffer provides a historical trace of system events, such as tasking events and hardware interrupts. The statement TRACE=ON (or OFF) in the CONFIG.SYS file, optionally followed by a list of event codes separated by commas, specifies that trace data should (or should not) be written to the buffer each time one of the indicated event numbers occurs. If no event numbers are specified, then all are traced if ON is specified and none are traced if OFF is specified. Tracing can also be turned ON and OFF using the OS/2 TRACE command, which has the same syntax as the CONFIG.SYS TRACE statement (except that a space is substituted for the = sign).

The OS/2 TRACEFMT utility is provided to display the contents of the trace buffer. The utility reads the information in the trace buffer and displays the trace information in reverse time-stamp order. Time-stamps are displayed in the form *ss.hh*, where *ss* is seconds and *hh* is hundredths of seconds. The title, process ID, type (kernel or dynlink), and mode (OS/2 or DOS) of each event is printed. Tracing does not have to be active at the time this utility is used, but a trace buffer has to have been defined in order for events to be displayed.

Tracing system events can be useful, but it is sometimes necessary for systems support personnel to examine the contents of system memory at the time when a problem occurred, in order to diagnose problems. IBM has provided the stand-alone dump facility to allow the contents of memory to be saved to diskette for later analysis. This facility dumps the entire contents of physical memory to diskette, including the 640 KB to 1 MB address space. This facility is invoked by pressing Ctrl-Alt-Num Lock twice. A message displayed on the screen then directs the user to insert a diskette, previously prepared using the OS/2 create dump disk

(CREATEDD) utility, in drive A and press any key to begin the dump. When the dump completes, a message indicates the amount of memory dumped; the user is then instructed to remove the dump diskette and restart the system (by pressing Ctrl-Alt-Del). If more than one diskette is required to dump the contents of memory, additional diskettes that have been formatted using the OS/2 FORMAT utility should be used. The speed of this operation is limited by the speed of the diskette drive. Dumping a 3 MB Model 60 system takes about eight minutes.

Creating a dump diskette using CREATEDD is straightforward, but it must be done before a problem requiring a memory dump is encountered. Entering the OS/2 command CREATEDD x: causes the diskette inserted in drive x: to be prepared for stand-alone dump use. CREATEDD first formats the diskette using the FORMAT command and then creates two files on it. The first (RASDMP.COM) is a stand-alone memory dump program, which is started by OS/2 when Ctrl-Alt-Num Lock is pressed twice, as described above. The second is a file called DUMPDATA.001; this file is sized to fill the diskette so that other data cannot be written to the diskette by accident.

Summary

OS/2 is a product whose time has finally come. It has been awaited since the introduction of the AT in 1984, by those who want to do multi-tasking on PCs. For this class of users, OS/2 has much to offer. It is an operating system designed for multi-tasking, not a reworked version of DOS with multi-tasking added. Nonetheless, it is compatible with the DOS file system, and can run a single well-behaved DOS application along with OS/2 applications, both of which will ease the transition process as OS/2 applications become available.

With the features provided by the extended edition, OS/2 will be an important player in IBM's effort to provide consistency across its various product lines via the interfaces and protocols defined in the Systems Application Architecture. With the power and flexibility of OS/2 and the common user interface it offers, OS/2-equipped systems will become an integral part of the corporate information processing environment.

That is not to say that everyone is part of a large enterprise or that every personal computer in a corporation can or will run OS/2. In fact, millions of installed 8088-based PCs, and the hundreds of thousands of 8086-based Personal System/2 Model 30s sold to date cannot run OS/2. For these systems, DOS (possibly with a shell) will be the operating system platform for some time to come.

OS/2 cannot be all things to all people. Both it and DOS will be widely used for the foreseeable future. DOS will continue to be predominate on personal productivity workstations, particularly in smaller enterprises; OS/2 will be the operating system of choice for the development and operation of data processing systems.

OS/2's connectivity features will allow it to communicate with DOS equipped personal computers, as well as mini and mainframe computers. With follow-on products, OS/2-equipped systems will be able to provide data, file, and communications services for DOS systems, thus lessening the need for smaller workstations to be converted to OS/2 use.

Although its use is not yet widespread, OS/2 seems to be a solid, reliable, software product. With planned enhancements, such as the ability to use 32-bit mode and run multiple DOS applications on 80386-based computers, computer users will have an operating system that exploits the power of the 80286 and 80386, but still offers a degree of DOS compatibility.

In the interim, owners of 80386-based computers such as the Model 80 may wish to consider one of the several available DOS-compatible operating systems and operating environments described in Chapter 13. These products support the simultaneous execution of multiple DOS applications, using the 80386's virtual memory management capabilities. These systems, which provide for multi-tasking of current applications without the expense of conversion, can be very useful in a personal productivity environment.

Because OS/2 is a single-user operating system, not a multi-user one, users who require multi-user capabilities may wish to consider AIX (Advanced Interactive Executive), IBM's UNIX System V offering for the Model 80, or other available UNIX implementations. AIX, which was originally introduced for the RT PC in 1986, is discussed in the next chapter.

12

AIX Personal System/2 Operating System

The Advanced Interactive Executive (AIX) is a multi-user, multi-tasking, virtual memory operating system, which was introduced by IBM as the AIX/RT operating system in January 1986 along with the IBM RT PC. AIX Personal System/2 is a 32-bit implementation of the AIX operating system for the Personal System/2 Model 80. Announced in April 1987 along with the Model 80, AIX Personal System/2 is scheduled for release in September 1988.

The RT PC is a workstation designed for use by technical professionals. It uses an IBM developed microprocessor that is based on a 32-bit reduced instruction set computer (RISC) architecture. Compatibility with the IBM PC family is provided through an optional AT coprocessor board and control software; compatibility with the Personal System/2 family is provided through the AIX operating system and applications development languages common to the RT PC and the Personal System/2 Model 80.

Features

AIX is a UNIX-based operating system developed by IBM and INTERACTIVE Systems Corporation. AIX is compatible with UNIX System V Release 2. It is based on INTERACTIVE's INix product, which is based on UNIX System V, as licensed by AT&T Bell Laboratories. AIX Personal System/2 is based on the AIX/RT operating system; it supports up to sixteen users and includes selected Berkeley Software Distribution 4.3 extensions. It requires an IBM Personal System/2 Model 80 with at least 2 MB of memory for operation. The system can be accessed from the system console, asynchronous terminals, and LAN connected workstations. The languages provided for use with AIX Personal System/2—VS FORTRAN, VS

Pascal, and C—are source code compatible with the RT PC VS FORTRAN, RT PC VS Pascal, and C languages, respectively. Execution of DOS 3.30 applications is supported using the optional DOS Merge feature.

Memory and I/O Support

AIX Personal System/2 uses the flat demand-paged memory mode of the Model 80's 80386 to provide virtual memory support. This feature allows applications to have a virtual memory size of up to 4 gigabytes each, and allows the size of the operating system and applications to exceed the size of actual physical memory. Applications and data are accessed using a virtual memory scheme similar to that employed on IBM mainframes. Virtual memory space exists on an area of the disk reserved for paging, and pages are read into physical memory as required.

Device-independent input/output and virtual terminal support are provided, as well as virtual memory. Multiple virtual terminal support, which allows several full-screen virtual displays to share the same physical display, is provided via the High Function Terminal (HFT) driver. This feature allows users to alternate using a keyboard sequence (hot-key) between virtual AIX Personal System/2 and/or DOS 3.30 sessions. A mouse (pointing device) and sound output are also supported.

User Interfaces

Several user interfaces are available for use with AIX Personal System/2. The *Bourne shell* and the *C shell* are provided standard with the system. The Bourne shell is a widely-used command interpreter that receives commands from users, and then calls and executes the program or procedure named in the command. The C shell is a user interface used in Berkeley Software Distribution systems.

AIX Personal System/2 Usability Services is a separately priced product that provides a menu-driven interface to AIX Personal System/2 operating system functions. A subset of AIX commands can be accessed through a command bar and pop-up menus, using either the keyboard and/or the mouse. New windows can be created and a list of existing windows and their status can be displayed. Usability Services features are available at the system console and from any attached terminals.

X-Windows

AIX Personal System/2 X-Windows is another separately priced product (also available for the RT PC), which provides support for multiple, active windows on the system console. This product provides windowing functions as described in *X Version 11 Protocol*, distributed by the Massachusetts Institute of Technology. It allows the console screen to be divided into windows that act as independent terminals. Windows can be created, deleted, rearranged, and resized on the screen. Text entered from the keyboard is placed in the window in which the cursor is located; the cursor can be easily moved among windows. An IBM Personal System/2 mouse or compatible pointing device is required.

X-Windows allows users to simultaneously view and access applications and data from the local computer and other LAN-based computers that are using a compatible windowing product. AIX Personal System/2 X-Windows is compatible with AIX/RT X-Windows 2.1, and offers a mode compatible with AIX/RT X-Windows, Version 1.1.

DOS Merge

DOS 3.30 and applications can be executed using AIX Personal System/2 DOS Merge. DOS Merge is a separately priced product based on MERGE/386, which was developed by Locus Computing Corporation. It allows multiple users to execute DOS 3.30 and DOS applications, while maintaining the system access and file security provided by AIX.

DOS Merge allows DOS and AIX programs to be executed from remote terminals as well as the system console. Single or multiple DOS sessions may be executed along with single or multiple AIX sessions. AIX Personal System/2 files can be accessed from DOS applications and DOS files can be accessed from AIX Personal System/2. DOS files can be stored under the control of the AIX Personal System/2 file system, or they can be stored on a DOS diskette or in a DOS disk partition. ASCII files can be converted between DOS and AIX Personal System/2 file formats with no loss of information; data can be transferred between DOS and AIX Personal System/2 applications using UNIX pipes. DOS graphics applications can be run within the AIX Personal System/2 X-Windows environment.

Text Formatting System

The AIX Personal System/2 Text Formatting System is a separately priced product that provides text processing support for AIX Personal System/2 users. It provides commands and utilities intended for experienced UNIX text processing users. Included are the *nroff* and *troff* printing and typesetting formatting utilities, and commands for formatting mathematical text and tables, as well as for displaying and spell-checking documents.

Operating System Extensions

Several operating system extensions are available as a separately priced product called AIX Personal System/2 Operating System Extensions. These additional commands and utilities are designed for use in large application development projects and in the academic community. Features supported include the UNIX-to-UNIX copy program (to move files between UNIX systems) and the ability to execute tasks on remote AIX systems.

Terminal Support

AIX Personal System/2 supports IBM 3151, 3161, and 3163 ASCII terminals, DEC VT100 and VT220 compatible terminals, and personal computers using IBM 3101 terminal emulation. A system interface is also provided to implement support for additional ASCII terminals. A maximum of three IBM Personal System/2 Dual Async Adapters can be attached to the Model 80, thus providing a total of seven serial ports (including the one on the system board) to which terminals can be connected. These serial ports combined with the local keyboard and console provide support for eight system users. Additional users can access the system via TCP/IP (Transmission Control Protocol/Internet Protocol) on a local area network, up to a system maximum of sixteen users total.

Languages

Three separately priced, high level languages are available for use with AIX Personal System/2—AIX Personal System/2 VS FORTRAN, VS Pascal, and C. All three are compatible with their RT PC counterparts.

VS Fortran

AIX Personal System/2 VS FORTRAN is source code compatible with IBM System/370 VS FORTRAN Version 2 Release 2, but does not support the following features:

- Asynchronous input/output
- Data-in-virtual
- Dynamic file allocation
- Extended error processing
- FORTRAN-77 language mode
- Indexed file support
- Partition data sets

AIX Operating System

Quadruple precision floating point arithmetic

Systems Application Architecture (SAA) error flagging

6600 character length, 99 continuation lines

Most System/370 VS FORTRAN programs may be compiled by AIX Personal System/2 VS FORTRAN without modification. All System/370 VS FORTRAN directives are accepted, even if they perform no action in AIX Personal System/2 FORTRAN.

AIX Personal System/2 VS FORTRAN is source code compatible with Digital Equipment Corporation (DEC) VAX FORTRAN, but does not support the following features:

Alternate PARAMETER SYNTAX

Argument list built-in functions

DEFINE FILE statement

ENCODE/DECODE statements

ERRSNS subroutine

Expressions in FORMAT statements

FIND statement

Indexed file support

%LOC function /NOF77 interpretation of EXTERNAL statement

Non-ANSI keywords in input/output statements

Quadruple precision floating point arithmetic

Runtime range checking

RADIX-50 constants and character sets

Text libraries

AIX Personal System/2 FORTRAN is designed according to the American National Standard Programming Language FORTRAN, ANSI X3.9-1978 (FORTRAN 77) standard and International Organization for Standardization (ISO) 1539-1980 Programming Languages-FORTRAN standard. It provides bit-string manipulation functions as described in ANSI/ISO-S61.1.

VS Pascal

AIX Personal System/2 VS Pascal is source code compatible with IBM System/370 VS Pascal Version 1 Release 1 except:

Chapter 12

> Functions are not allowed in CONST declaration statements.
>
> Function result can only be a scalar, pointer, or string.
>
> Global labels in segment units are not indicated as errors.
>
> Invalid values for value assignments are not indicated.
>
> Pack and Unpack are recognized but not supported.
>
> Range checking for integer-to-character conversion is not supported.

Most VS Pascal statements, data types, and directives are supported. All System/370 VS Pascal directives are accepted, even if they perform no action in AIX Personal System/2 Pascal. AIX Personal System/2 VS Pascal conforms to the ANSI-83 definition of Pascal.

AIX C

The AIX Personal System/2 C language is source code compatible with RT PC C language. All C statements, data types, and directives are supported. IBM XENIX 2.0 applications, which are written in C (as defined in *The C Programming Language*, Brian W. Kernighan and Dennis M. Ritchie, Prentice-Hall © Bell Telephone Laboratories, Inc., 1978) and which use UNIX System V system calls, are source code compatible with and can be recompiled for execution on AIX Personal System/2. XENIX 2.0 applications written in FORTRAN that conform to the FORTRAN 77 language standard are likewise source code compatible with and can be recompiled for execution on AIX Personal System/2.

General Characteristics

The AIX Personal System/2 language compilers offer several advanced features. The compilers produce optimized object code by eliminating unused code, common subexpressions, and redundant load/store sequences; by removing invariant code from loops; and through algebraic simplification, subscript optimization, and efficient register usage. Optimization may be selected or deselected by the user at compilation time. All of the languages provide a high level interface to AIX system calls.

AIX Personal System/2 supports the storage of common routines in a shared object module library. Routines from such libraries are loaded at runtime rather than when programs are linked. This reduces the size of executable modules, and also assures that applications use the latest versions of shared routines.

Programs executed on the Model 80 using these languages may produce different results than when executed on the System/370, due to differences in system architecture and software implementation. Such factors include the differences between the IEEE floating point format used by the Model 80's Intel 80387 math coprocessor and that used by the System/370, internal representation of character data (ASCII on the Personal System/2 Model 80, EBCDIC on the System/370), and details of specific compiler implementations, such as the treatment of uninitialized variables.

The separately priced AIX Personal System/2 Application Development Toolkit provides tools commonly required for applications development. The Toolkit consists of an assembler, a symbolic debugger, and a source code control system (SCSS). The assembler is compatible with UNIX System V assembly language for the Intel 80386. It supports macro assembly, repeat block assembly, and conditional assembly directives. The symbolic debugger provides high level debugging support for AIX Personal System/2 FORTRAN, Pascal, and C. It allows breakpoints to be set on subroutines, lines, variables, and addresses. Both execution tracing and breakpoints are supported. Argument passing and standard I/O redirection are standard features.

The SCCS is a collection of utilities used to manage the creation and update of source and text files. These utilities support the storage of files and updates to them, so that common code is stored only once. They also support file security, automatic insertion of identifying information into source and object files, and restoration of files to previous versions.

Connectivity

IBM provides three separately priced program products to allow connectivity between the Personal System/2 Model 80 using AIX and other systems. They are the AIX Personal System/2 Workstation Host Interface Program, AIX Personal System/2 Transmission Control Protocol/Internet Protocol (TCP/IP), and AIX Personal System/2 INmail/INed/INnet/FTP. These products support communications and file transfer using SNA, Ethernet, and IBM Token-Ring network protocols. They require a Personal System/2 Model 80 with at least 3 MB of memory and an appropriate network adapter.

Workstation Host Interface Program

The AIX Personal System/2 Workstation Host Interface Program provides 3270 terminal emulation, and file transfer between an AIX Personal System/2 system and an IBM System/370 host. Hardware connection is provided via the IBM 3270 Connection adapter, which is cable connected to an IBM 3174/3274 terminal control unit port configured for Distributed Function Terminal (DFT) operation. For

Chapter 12

file transfer operations, the IBM System/370 IND$FILE program must be installed on the System/370 host.

The Workstation Program emulates the IBM 3278 Model 2 terminal or 3279 Model 2A/B or S2A/B color terminals. Up to five 3278/9 terminal sessions per adapter may be active at once. Because the Workstation Program can operate as an AIX Personal System/2 virtual terminal, the user may switch between host 3278/9 sessions and other AIX Personal System/2 virtual terminal sessions. The following 3270 terminal hardware-specific features are not supported:

Alternate cursor

APL/text character set

Attachment to port 0 of an IBM 3274 control unit

Cursor blink

Cursor select

Magnetic reader control and accessories

Monocase switch

Security keylock

Selector light pen

Kanji

Katakana

Numeric keylock

Programmed symbols

Response time monitor for an IBM 3274 control unit

Security keylock

Video output

3270 diagnostic dump

Utilities are provided to change keyboard mappings and color definitions. Support is provided for the thirteen national language keyboards supported by AIX Personal System/2. The IBM 3278/9 enhanced data stream feature is supported. 3270 print sessions may be set up to route files directly to a printer attached to the Model 80. Host screens may be copied to a printer attached to the Model 80 or to an AIX Personal System/2 file.

The FXFER file transfer utility transfers a file between the Model 80 and a host VM/CMS or MVS/TSO system. Files may be uploaded or downloaded; their contents may be text or binary data. FXFER optionally converts ASCII text files to EBCDIC during an upload operation, and EBCDIC text files to ASCII on a download.

AIX Operating System

FXFER may be initiated by the user from the Bourne shell or C shell, or from a workstation terminal session. It may also be initiated through its application programming interface, which is a library routine that is linked with a user application program. The application programming interface may be used by programs written in VS FORTRAN, VS Pascal, and C. It provides message level host support (for MVS/TSO and VM/CMS), access to emulator presentation space, optional translation between ASCII and EBCDIC, and a high function, application-to-application interface.

TCP/IP

The AIX Personal System/2 TCP/IP supports peer-to-peer or workstation-to-host data exchange between workstations using the popular TCP/IP protocols. With the appropriate network adapter, it supports direct attachment to Ethernet and IBM Token-Ring networks. It is functionally compatible with the IBM RT PC TCP/IP program. AIX Personal System/2 TCP/IP supports several remote host features including remote logon. Features provided include:

File Transfer Protocol (FTP)

Network maintenance services commands

Simple Mail Transfer Protocol (SMTP)

Telenet Protocol

Trivial File Transfer Protocol (TFTP)

An Ungermann Bass NET/ONE NICps/2 adapter or equivalent is required to attach the Model 80 to an Ethernet LAN; an IBM Token-Ring Adapter/A is required for attachment to an IBM Token-Ring network.

INmail/INed/INnet/FTP

The IBM AIX Personal System/2 INmail/INed/INnet/FTP program allows users to interactively enter commands to be executed at remote systems, and provides for the queued transfer of files and electronic messages between systems. This program uses asynchronous communications to transfer information between two or more users on the same AIX Personal System/2 system, or on other systems with INmail or compatible software installed. Use of a port on an IBM Personal System/2 Dual Async Adapter/A or the Model 80 serial port is required. AIX Personal System/2 INmail/INed/Inet/FTP is functionally compatible with the product of the same name offered for the IBM RT PC.

INmail provides an electronic mail system for the AIX Personal System/2 system. Each system user has a private mail box. Users can send reminders to other users at specified times or start a program at specified times. Users can use the INed editor, the UNIX *ed* editor, or no editor at all for composing messages.

The INed editor is a full-screen editor that can be used to create and update text files. Multi-window and multi-file support is provided. Word processing functions such as word wrap, cut and paste, move/copy, and vertical and horizontal scrolling are provided.

INnet performs communications between two or more systems with INmail/INed/Inet/FTP or compatible software installed. Users on connected systems can print on remote printers. A system can also relay output received from another system to a third system, thus eliminating the need for direct connection among all systems on an asynchronous network.

The File Transfer Program (FTP) supports the interactive transfer of files between systems, and also allows a user to interactively enter commands to be executed on remote systems. Once the communications link with a remote system has been established, a user can send, receive, rename, and delete files on the remote system.

Installation

IBM provides a well-structured, menu-driven set of installation procedures for AIX Personal System/2. That is not to say that installing the system and related products is as quick and easy as installing DOS, for instance. The fact that the AIX Personal System/2 system—complete with the system extensions, DOS Merge, Usability Services, X-Windows, and Text Formatting System—occupies twenty 1.44 MB diskettes provides a hint that installation will take longer than the average lunch hour.

Installation is quicker and smoother than with most UNIX systems, however, because IBM has provided an installation system that meaningfully assists the user in setting system parameters, by providing suggested default values based on the hardware configuration of the Model 80 on which the system is being installed. The installation tools provided with AIX Personal System/2 also provide a uniform method of installing and updating AIX Personal System/2 applications programs, as well as AIX Personal System/2 standard and optional programs.

Market Position

AIX Personal System/2 brings the power and flexibility of an excellent UNIX System V implementation to the IBM Personal System/2 family. IBM invested a large amount of time and manpower in developing this operating system for the IBM RT PC; however, that machine, though powerful (and expensive) in its own right, is not particularly PC compatible, even with the optional add-in AT coprocessor. Consequently, it is not widely used.

The combination of the Personal System/2 Model 80, the AIX Personal System/2, and the optional DOS Merge feature provides a powerful platform that allows multiple users to run and develop UNIX and System/370-based technical applications in harmony with the use of popular DOS applications. This is an attractive combination for technical users who can afford a Model 80, particularly in comparison to OS/2, which offers only limited compatibility with UNIX and System/370 technical applications and provides no multi-user capabilities.

13

Control Software for the Model 80

The 80386-based Personal System/2, Model 80 is in much the same situation as the 80286-based PC AT was, from its introduction in 1984 until the introduction of OS/2 in 1987—advanced hardware with little software to properly exploit it. The OS/2 operating system can be used on the Model 80, but it functions the same as it does on a Model 50 or 60. It uses the 80386 as if it were only a high speed 80286. OS/2 supports multi-tasking, but only of applications developed specifically to run under OS/2. It takes no advantage of the 80386's larger maximum memory segment size (4 GB versus 64 KB on the 80286), memory paging, I/O protection, or virtual 8086 mode. It allows only one PC DOS application to be run along with multiple OS/2 applications. Lastly, it is a single-user system, offering no multi-user capabilities.

AIX and UNIX operating systems make use of the Model 80's advanced capabilities to provide multi-tasked execution of UNIX applications for multiple users. Such features are more useful for current users of UNIX systems than for traditional PC users, who have invested large amounts of time and money in developing and using DOS applications. Many such users would like the capability to execute more than one DOS application at a time. The Model 80's 80386 processor provides features (not available on the 80286) that make that capability not only feasible, but practical as well.

Virtual 8086 mode is a form of protected mode available on the 80386 (but not on the 80286), which allows real-mode applications to execute within protected mode. In virtual 8086 mode, each real-mode application executes as if it were executing using a separate 8086 with 1 MB of conventional address space. In actuality, tasks can reside anywhere in the 80386's 4 GB physical address space, and be mapped to the first megabyte using the 80386's paging hardware. (See Figure 13-1.)

Chapter 13

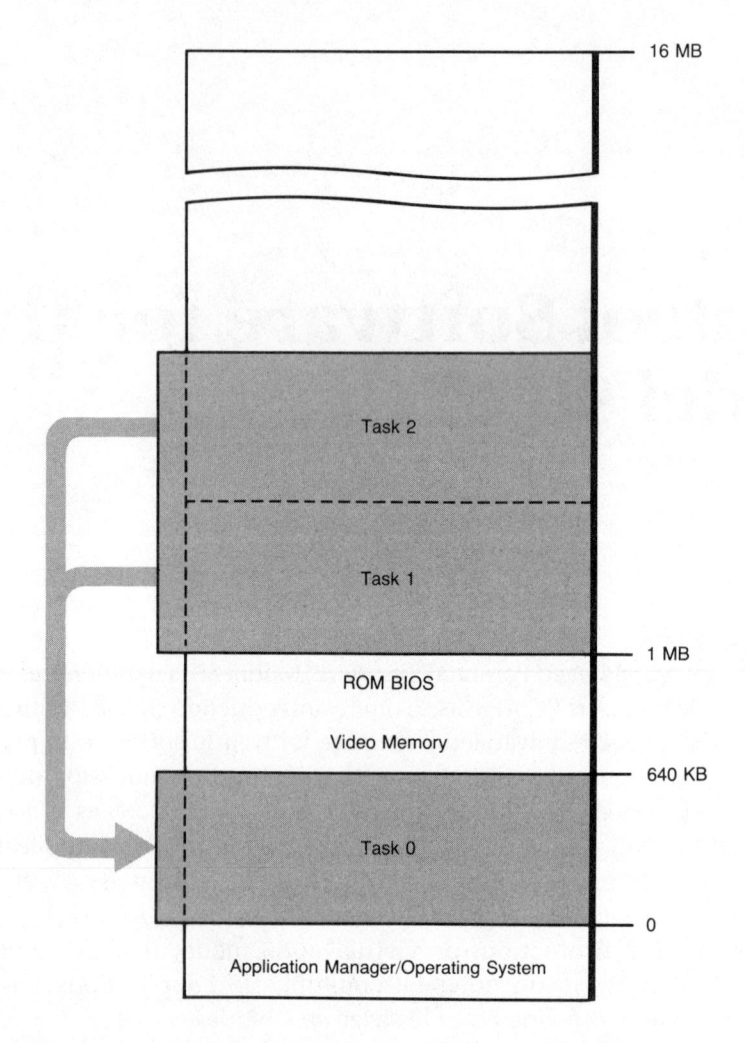

Figure 13-1 Virtual 8086 Memory Mapping. *The virtual 8086 mode of the 80386 offers the opportunity to run several DOS-based applications at the same time.*

When operating in virtual 8086 mode, all interrupts are vectored through the protected mode interrupt descriptor table, allowing interrupts to be controlled by a protected-mode interrupt handler outside the virtual 8086 application. This, combined with the 80386's ability to trap privileged instructions and maintain control over I/O ports, allows the construction of system software that can control the time-shared execution of multiple 8086 real-mode tasks on the 80386. With such control software, many users could use the power of the Model 80's 80386 to solve one of their pressing problems—the need to run more than one of their existing DOS applications at the same time.

As noted above, OS/2 currently does not provide this capability, nor is it expected to in the near future. Fortunately, software products are available that provide this capability. Such system software can take the form of an operating system that runs directly on the Model 80, independent of DOS, or an application manager that works with and adds functionality to DOS.

Developing a DOS-compatible, multi-tasking operating system is no small task. Several vendors have developed such products with varying emphases and degrees of DOS compatibility. In an effort to make their operating systems more attractive for the development of applications, some vendors have added multi-user capabilities to their products as well.

Available DOS-compatible, multi-tasking operating systems include Concurrent DOS 386 from Digital Research, Inc. and PC-MOS/386 from The Software Link, Inc. Both of these operating systems are designed for use on the IBM Personal System/2 Model 80, Compaq Deskpro 386, and compatible systems. PC-MOS/386 also operates on 8088/8086-based and 80286-based computers, but total memory for all tasks is limited to 640 KB unless special add-in memory management hardware is used. Multi-tasking application managers such as Quarterdeck's DESQView/386 and Microsoft Windows/386 that work with, rather than replace, DOS are also available. These products are discussed later in the chapter.

Concurrent DOS 386

Concurrent DOS 386 is a multi-user, multi-tasking operating system from Digital Research, Inc. It runs PC/MS-DOS 3.3, CP/M-86, and Concurrent DOS XM applications on the IBM Personal System/2 Model 80, Compaq Deskpro 386, and compatible Intel 80386-based computers. It requires a minimum of 512 KB of memory for a single-user system; 1 MB is recommended for each three additional users. A hard disk is not required, but is recommended to increase performance and to support multiple users. The Model 80 is supplied with a hard disk as standard equipment.

Concurrent DOS 386 Release 2.0 is the latest in a long line of operating systems from Digital Research, Inc., the company that developed CP/M, CP/M-86, and Concurrent DOS. It provides compatibility with DOS 3.3, with added capabilities that are useful for the development and operation of multi-user systems. It supports ten users performing up to a total of 255 tasks, and runs over 1,000 multi-user Concurrent DOS business applications. Using the Concurrent DOS System Builder's Kit, more than ten users can be supported, limited by available memory and disk space. The features provided by Concurrent DOS 386 are listed below.

Concurrency

Ten or more concurrent users

Up to 255 concurrent tasks

User Interface
- Four sessions on system console
- Two sessions on each remote terminal
- Window management
- Configurable menu system
- Online help facility
- Keyboard macros
- Command line editing and recall

Display Support
- VGA, EGA, CGA, MDA, Hercules
- Wyse 60 remote terminals

Memory Management
- Free memory per application >512 KB
- Code/data sharing
- Expanded memory support

File Support
- DOS and CP/M file formats
- File sharing

Other
- Print Spooling
- System Developer's Toolkit

Installation

Concurrent DOS 386 can support DOS and CP/M partitions on the hard disk. DOS partitions may be up to 32 MB in size; CP/M partitions can be up to 512 MB in size. Installation is straightforward; the user only needs to initialize the computer system using the first Concurrent DOS 386 diskette and press F10 as directed, to begin the installation process. If the disk has not been partitioned, it must be partitioned with either the Concurrent DOS 386 FDISK or the DOS FDISK command before the installation process can proceed.

The Concurrent DOS 386 installation process creates a directory CDOS on the disk that the user selects from a menu, and copies the system files into it. If the disk is a DOS system disk, the installation process creates or modifies the AUTOEXEC.BAT file to query the user each time the system is initialized, as to whether Concurrent DOS 386 is to be loaded. If the user enters Y, Concurrent DOS 386 is loaded; otherwise DOS is.

When loaded, Concurrent DOS 386 starts four sessions at the main console. The screen display tells the user to press Esc to enter the command line interface or F10 to see the next menu. The bottom line of the system display indicates which session is in the foreground, the primary task that is running in each session, and the time of day. A session is switched to the foreground by pressing its number (1, 2, 3, or 4) on the numeric keypad, while holding down Ctrl.

Concurrent DOS Commands

Pressing Esc after loading Concurrent DOS invokes a command line interface similar to that provided by DOS's COMMAND.COM. Concurrent DOS 386's faster video routines make its command processor more responsive than that of DOS, and its multi-line command buffer allows the user to scroll backward and forward through previous commands using the cursor keys. Concurrent DOS 386 provides many commands in addition to those provided by DOS. Commands for allocating resources for multi-tasking, creating windows and menus, and CP/M media management services are provided, as listed in Table 13-1.

Table 13-1 Concurrent DOS 386 Specific Commands. *In addition to DOS commands, Concurrent DOS 386 provides commands for multi-tasking, window, and menu support.*

Command	Purpose
8087	Indicates that a program uses the coprocessor
AUX	Selects auxiliary port 0 or 1
BANK	Controls program mapping to extended memory
CARDFILE	Stores and retrieves names and addresses
CHSET	Changes the command header of .CMD files
COPYMENU	Copies menus from one file to another
CPM	Provides access to CP/M files
DELQ (ERAQ)	Deletes files, upon confirmation
DREDIX	Creates and edits text files
DSKMAINT	Formats, copies, and verifies diskettes
EDITMENU	Creates, modifies, and deletes menus
FM	Allows commands to be selected from menus
FSET	Sets file and drive attributes and protections
FUNCTION	Assigns function and window switching keys
HELP	Explains Concurrent commands
INITDIR	Formats CP/M directories
LIMSIZE	Sets expanded memory limit for an application
LOAD386	Starts Concurrent from DOS
MEMSIZE	Limits conventional memory for an application
ORDER	Changes the command file search order
PASSWORD	Sets password protection on files or paths
PIFED	Defines system parameters for an application
PRINTER	Changes the current printer number
PRINTMGR	Controls the printing of files
REBOOT	Resets the system

Table 13-1 (cont.)

Command	Purpose
RUNMENU	Runs a menu
SDIR	Shows director and file status information
SETPORT	Configures the serial ports
SETUP	Modifies system characteristics
SHOW	Displays information about disk drives
STOP	Terminates a program
SUSPEND	Suspends a background program
SYSDISK	Shows/sets current system disk
TOUCH	Sets the time and date on groups of files
USER	Changes user number on CP/M media
VSET	Prevents an application's use of specific interrupts
WINDOW	Shows and modifies window characteristics
WMENU	Provides interactive manipulation of windows
XDEL	Selectively deletes groups of files
XDIR	Provides extended file directory display

Startup Menu

Pressing F10 at system initialization causes the display of the Concurrent DOS 386 Startup menu, shown in Figure 13-2. Function keys are used to initiate execution of the Help utility, File Manager, Subdirectory Tree Display, DR EDIX Editor, Cardfile utility, or Printer Manager. Pressing Esc terminates the menu and returns control to the previous menu.

The Help function, activated by pressing F1, can also be invoked from the command line interface by entering the command HELP. It displays a list of topics about which it can display information. Information about a topic is displayed by entering the HELP command with the name of the topic as its operand.

The File Manager utility, initiated by pressing F2, is a convenient interface for using Concurrent DOS 386. It may also be invoked from the command line interface by entering the command FM. It displays the names of the files in the current directory and a list of functions that can be performed on the files. (See Figure 13-3.) The function to be performed is located using the cursor keys and selected by pressing Enter, or by typing the initial letters of the command. Filename operands, when required, are selected in the same manner. Function keys can be used to select common commands; F1 provides help as described previously and F10 allows the user to enter a command directly without leaving the File Manager. Pressing Esc returns control to the command line interface.

Pressing F3 on the Startup menu causes the disk's directory hierarchy, or Subdirectory Tree Display, to be displayed. The size (in bytes) of each directory is displayed, along with the number of files in the directory. The total number of files found and bytes of disk space in use is also displayed. This display may also be activated by entering the command TREE from the command line interface.

Model 80 Control Software

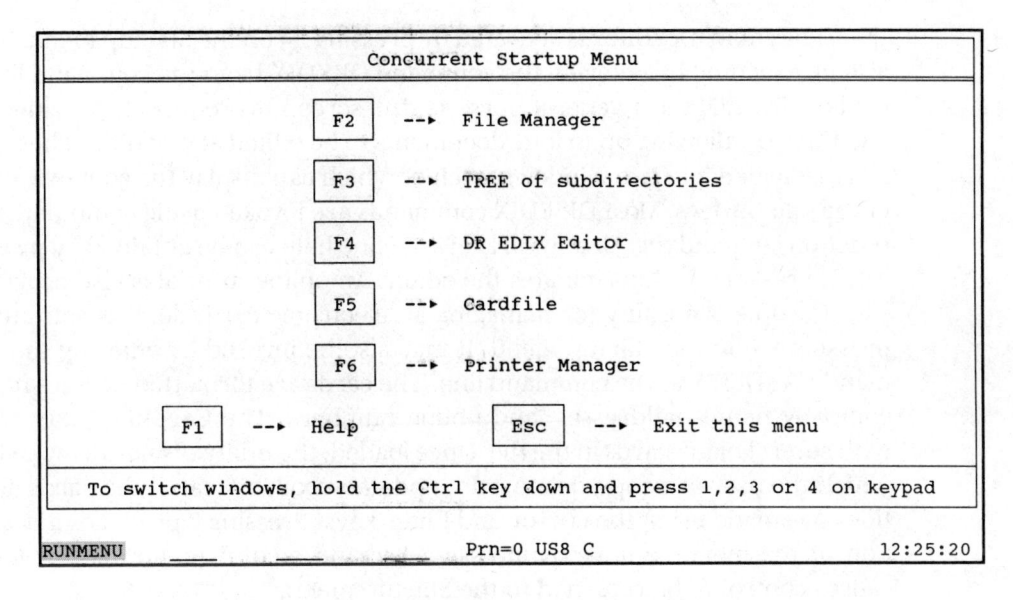

```
┌─────────────────────────────────────────────────────────────────┐
│                    Concurrent Startup Menu                       │
│                                                                  │
│                    ┌────┐                                        │
│                    │ F2 │  --▶  File Manager                    │
│                    └────┘                                        │
│                    ┌────┐                                        │
│                    │ F3 │  --▶  TREE of subdirectories          │
│                    └────┘                                        │
│                    ┌────┐                                        │
│                    │ F4 │  --▶  DR EDIX Editor                  │
│                    └────┘                                        │
│                    ┌────┐                                        │
│                    │ F5 │  --▶  Cardfile                        │
│                    └────┘                                        │
│                    ┌────┐                                        │
│                    │ F6 │  --▶  Printer Manager                 │
│                    └────┘                                        │
│                                                                  │
│         ┌────┐                    ┌─────┐                        │
│         │ F1 │ --▶ Help           │ Esc │ --▶ Exit this menu     │
│         └────┘                    └─────┘                        │
│                                                                  │
│  To switch windows, hold the Ctrl key down and press 1,2,3 or 4 on keypad │
├─────────────────────────────────────────────────────────────────┤
│ RUNMENU                      Prn=0 US8 C              12:25:20  │
└─────────────────────────────────────────────────────────────────┘
```

Figure 13-2 Concurrent DOS 386 Startup Menu. *Commands can be accessed through this menu; pressing Esc provides access to the command line interface.*

```
┌──────────────────────────────────────────────────────┬──────────────────────┐
│ DOS Media │ You are here ▶ C:\                       │▶Help                 │
│   2396k   │                                          │ File    Directory    │
│           │                                          │ Subset of Files      │
│           │                                          │ Drive Selection      │
│                                                      │ Type    File(s)      │
│ AUTO.BAT      AUTOEXEC.BAT   AUTOEXEC.SAV  BATCH.CMD │ Print   File(s)      │
│ BUSS.IDX      CARDFILE.DAT   CCDAUTO.BAT   CCPM.SYS  │ Copy    File(s)      │
│ CDOS.COM      COMMAND.COM    CONFIG.386    CONFIG.38E│ Rename  File(s)      │
│ CONFIG.38L    CONFIG.38M     CONFIG.480    CONFIG.C38│ Delete  File(s)      │
│ CONFIG.DV     CONFIG.FCM     CONFIG.HAR    CONFIG.MMU│ Backup/Restore       │
│ CONFIG.MOU    CONFIG.SYS     CONFIG.VP     CONFIG.W38│ Edit a File          │
│ DV.BAT        FIRST.IDX      HDMENU.DAT    IBMBIO.COM│ Run   a Program      │
│ IBMDOS.COM    JAS.BAT        JUNK          LOAD386.COM│ Copy    Diskette    │
│ LOADSYS.COM   MENU           MOSAUTO.BAT   MOUSE.SYS │ Format Diskette      │
│ N.CMD         NAME.IDX       PHONE.IDX     START002.BAT│ Free  Memory       │
│ WINDOW1.TXT   WINDOW2.TXT    WINDOW3.TXT   WINDOW4.TXT│ Size/Date ON       │
│ WSETUP.BAT                                           │ Set Up System        │
│                                                      │                      │
├──────────────────────────────────────────────────────┴──────────────────────┤
│ Command: Help, Concurrent DOS Help System                                    │
│ Esc=      F1=HELP       F3=Repeat      F5=Run      F7=Dir1    F9 =Directory │
│ EXIT      F2=Type       F4=Cancel      F6=Edit     F8=Dir2    F10=Command   │
│ FM                                     Prn=0 US8 C              12:25:22    │
└─────────────────────────────────────────────────────────────────────────────┘
```

Figure 13-3 Concurrent DOS 386 File Manager Menu. *The available functions are shown on the right and along the bottom of the screen.*

403

The DR EDIX Editor is activated by pressing F4 on the Startup menu. It may also be executed by entering the command DREDIX from the command line interface. DR EDIX is a general purpose, full-screen text editor. It provides four edit buffers, allowing up to four documents to be edited at one time. The screen can be divided into two windows, each of which can display the contents of one of the edit buffers. Most DR EDIX commands are invoked using combinations of function keys and the Alt and Ctrl keys. Online help can be obtained by pressing Alt-H. Pressing Alt-X terminates the editor. An online tutorial is also provided.

Cardfile is a utility for managing an electronic card file. It is activated by pressing F5 on the Startup menu. It may also be invoked by entering the command CARDFILE at the command line. The cards are formatted to store names, company names, addresses, and phone numbers. The Cardfile utility comes with several blank cards in the file. Once loaded, the utility displays a menu from which the user can copy, delete, add, update, scroll, or search the cards. Selections are made using the cursor and Enter keys. Pressing ? produces a description of the menu commands; pressing Esc and a confirmation key sequence causes control to be returned to the Startup menu.

The Printer Manager may be initiated either by pressing F6 from the Startup menu or by entering the command PRINTMGR from the command line interface. This utility can print multiple files, running one or more printers while another program is running in the same window. Once activated, the Printer Manager displays a menu from which function keys are used to start or stop the Printer Manager, add or delete files from the list to be printed, and display the status of the print queue. The Printer Manager manages up to five printers and a print queue of up to 254 files. Pressing Esc while in any Printer Manager program or menu returns control to the previous menu.

When Concurrent DOS 386 is initialized, it searches for and executes commands from files STARTnnn.BAT in the root directory of the system disk, where *nnn* is the number of each session. These files can contain commands to customize the operating environment for each session. If the STARTnnn.BAT file is not found for a session, the Concurrent DOS 386 initialization portion of the AUTOEXEC.BAT file is run instead. The AUTOEXEC.BAT file, as created or modified by the Concurrent DOS 386 installation process, displays the version number and current path in each session, and initiates the Concurrent DOS 386 Startup menu in session 1. The Startup menu may also be initiated in other sessions by entering the command `RUNMENU HDMENU.DAT` from the command line interface, or by placing it in the session STARTnnn.BAT file.

Defining Windows and Menus

By default, each Concurrent DOS 386 session begins with a full-screen window. The size of the window and its location on the main console display can be changed using the WMENU command. Once installed, WMENU is invoked (from any session) by pressing the numeric pad + key while holding down Ctrl. This causes the display of a one-line menu at the bottom of the screen from which window position, size, color, and other attributes may be changed.

Once the item to be changed is selected, adjustments are made using the cursor keys, and the results are displayed on the screen. When each window is set up as desired, the WMENU WRITE command can be used to create a batch file called WSETUP.BAT. This file contains the commands required to generate the windows as displayed. Entering the command WSETUP from the command line interface (or in the AUTOEXEC.BAT file) causes the saved windows' layout to be restored.

Concurrent DOS 386 provides a mechanism for creating, editing, deleting, and running menus. The EDITMENU command can be used to create, modify, and copy menus from a file, which may contain more than one menu. The COPYMENU command copies menus from one file to another, and the RUNMENU command runs a menu.

Running Applications

Once windows and menus have been defined, running applications under Concurrent DOS 386 is straightforward. Applications requiring up to 512 KB of free memory can be run in each session. Application requests for expanded memory are handled automatically by Concurrent DOS 386's built-in expanded memory manager. This manager uses the 80386's paging hardware and the system's extended memory to provide expanded memory services.

Switching between sessions using Ctrl and the session number (selected from the numeric keypad) is natural, and the display is updated quickly. Both text and graphics applications may be run full screen or in windows on the system console. Only one graphics session may be active at a time. If more than one window is displayed on the screen, the foreground window is surrounded by a two-line border, as shown in Figure 13-4. Each Concurrent DOS 386 remote terminal can display two full-screen windows.

Summary

DRI provides a User's Guide and Reference Guide with Concurrent DOS 386. A system builder's kit is available for manufacturers and resellers who wish to tailor Concurrent DOS 386 to their products; it includes DRI's DR NET networking utility, a System Guide, and a Programmer's Utilities Guide. A programmer's toolkit is available for software vendors developing Concurrent DOS applications; a System Guide, Programmer's Guide, and Programmer's Utilities Guide are included in the toolkit.

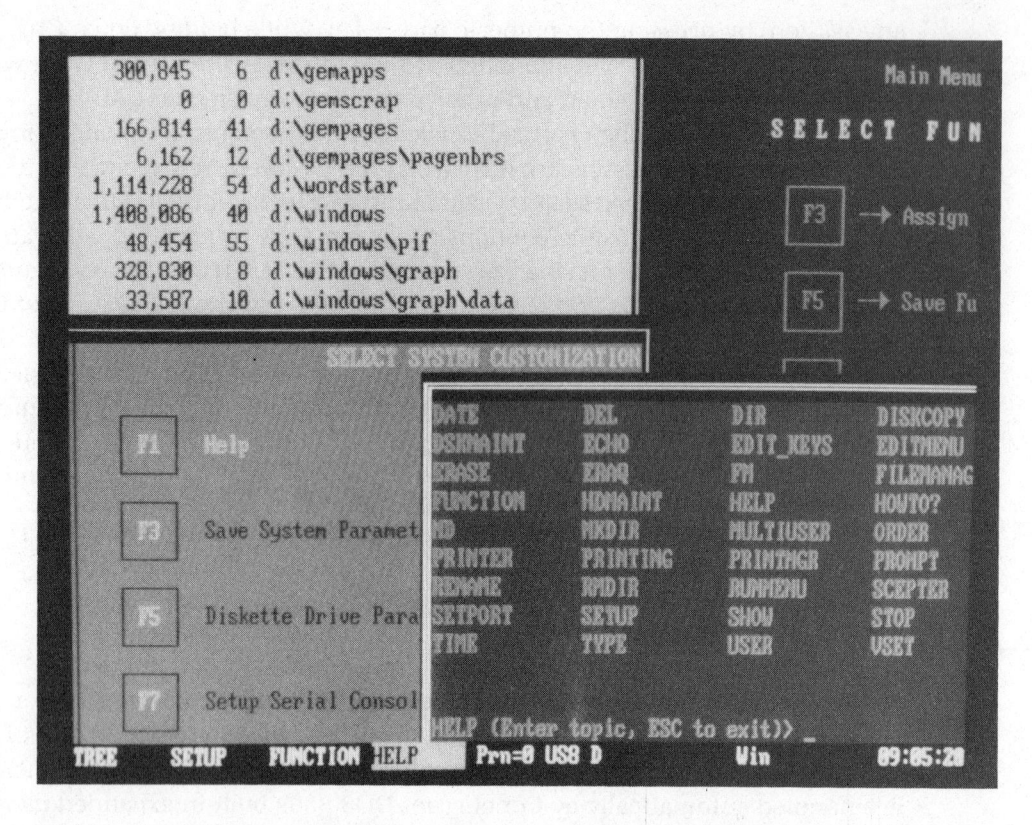

Figure 13-4 Concurrent DOS 386 Display Windows. *Several DOS applications can be run at once; the foreground window has a prominent border.* Photo courtesy of Digital Research, Inc.

Concurrent DOS 386 is an operating system that provides the fast context switching, file sharing, remote terminal support, intertask communication, and process synchronization features required for the development of multi-user, multi-tasking systems. These features, combined with its ability to run multiple DOS applications, make it a popular vehicle for the development of many accounting, point of sale, office automation, database, real-time process control, and communications systems.

PC-MOS/386

PC-MOS/386 is a DOS-compatible, multi-user, multi-tasking operating system from The Software Link, Inc. It supports up to 25 users and 99 concurrent tasks. PC-MOS/386 runs on the IBM Personal System/2 Model 80, Compaq Deskpro 386, and compatible Intel 80386-based computers. It requires a minimum of

512 KB of memory for a single-user system. PC-MOS/386 itself requires about 192 KB of memory. About 20 to 40 KB additional memory is required to manage each task, in addition to the task's own memory requirements. Extended memory is used as required for task storage and task management. A hard disk is not required, but is recommended to increase performance and support multiple users.

PC-MOS/386's features include:

Concurrency
- Up to 25 concurrent users
- Up to 99 concurrent tasks

User Interface
- No limitation on sessions per terminal
- Online help facility
- Keyboard macros
- Command line editing and recall

Display Support
- EGA, CGA, MDA
- Thirteen or more types of terminals

Memory Management
- Free memory per application >512 KB
- Code/data sharing
- Expanded memory support

File Support
- Disk partitions up to 256 MB
- File sharing
- File security and encryption

Other
- Print spooling
- NETBIOS emulation
- System Developer's Toolkit

PC-MOS/386 was developed for use on 80386 machines, but can also run on 80286-, 8086-, and 8088-based computers including the Personal System/2 Models 25, 30, 50, and 60, and the IBM PC. When run on these machines, it is limited to using conventional memory unless add-in memory management hardware is employed. It also provides no virtual device support when used on these computers because their processors do not have the 80386's virtual device capabilities.

Installation

PC-MOS/386 is available in single-user, 5-user, and 25-user versions. Its installation process is straightforward and well documented. If DOS is already installed on the computer's disk, PC-MOS/386 may be installed by copying its files to the disk and replacing the disk's DOS boot record using the MSYS command. If the computer's disk has not been set up for DOS, it must be partitioned and formatted using the PC-MOS/386 HDSETUP and FORMAT commands, or the DOS FDISK and FORMAT commands. PC-MOS/386 supports disk partitions up to 256 MB in size; however, some DOS applications may not work correctly in a disk partition larger than 32 MB.

PC-MOS/386 Commands

A number of commands can be placed in PC-MOS/386's CONFIG.SYS file to tailor the system. Commands are available to set the size of the pool of memory used for task management and device drivers, the number of time slices that each task receives in turn, and address areas between 640 KB and 1 MB that PC-MOS/386 may use to relocate parts of itself. Device drivers for a disk cache (supplied by Software Link), a memory manager that permits the system's extended memory to be used as expanded memory, inter-task communications, a RAM disk, and a terminal controller may be specified in CONFIG.SYS. Unlike DOS, PC-MOS/386 allows device drivers to be added and removed from memory while the system is running. If a mouse or other serial device uses a port and interrupt that the system's serial device driver can use, the mouse driver's DEVICE= statement must be placed after the DEVICE=$SERIAL.SYS statement in the CONFIG.SYS file.

    PC-MOS/386's command line interface is similar in appearance to that of DOS's COMMAND.COM. Its default prompt is the current drive and directory, enclosed in square brackets [ ]. Its faster video routines make it more responsive than DOS, and its multi-line command buffer allows the user to scroll backward and forward through previous commands, using the cursor keys. PC-MOS/386 provides many commands in addition to those provided by DOS. Commands for configuring the system for multi-tasking, resource allocation, and system security are provided. Useful commands such as EXCEPT and ONLY, which limit the action of other commands, are also provided. A list of commands is given in Table 13-2.

Table 13-2. PC-MOS/386 Specific Commands. *The action of other commands can be limited using the EXCEPT and ONLY commands.*

Command	Purpose
ABORT	Stops processing within a batch file
ADDDEV	Dynamically adds a device driver
ADDTASK	Dynamically creates a memory partition for a task
ALIAS	Substitutes a drive letter for a directory name
AUTOCD	Restores drive and directory previously redirected
BATECHO	Controls the initial state of ECHO
CLASS	Assigns or changes a directory's security class
COMPFILE	Compares the contents of two files
DESNOW	Drives color display without snow pattern
DIRMAP	Displays disk directory map
DISKID	Assigns a volume identifier to a disk or diskette
ED	Creates and modifies text files
ENVSIZE	Specifies minimum environment space size
EXCEPT	Allows files to be excluded from a command
EXPORT	Creates a compressed backup copy of files
FILEMODE	Changes read-only or archive attributes of a file
FLUSH	Clears command recall buffer
FREEMEM	Defines regions above B0000H that MOS may use
HDSETUP	Sets up and maintains hard disks
HELP	Displays information about MOS commands
IMPORT	Restores compressed backup files
INSERT	Specifies insert mode for command line editing
KEY	Prompts for a keystroke
MOS	Controls memory, display, and I/O configuration
MOSADM	Controls system scheduling and disk caching
MSORT	Sorts records in a file
MSYS	Writes a boot sector to a disk
ONLY	Limits the action of a command to specific files
REL	Displays the release of MOS in use
REMDEV	Dynamically removes a device driver
REMTASK	Dynamically removes a memory partition
SEARCH	Searches one or more files for a character string
SIGNOFF	Exits a secured mode of MOS
SIGNON	Allows access to secured items
SLICE	Sets number of time slices for each partition
SMPSIZE	Sets size of system memory pool
SPOOL	Specifies where print files are to be sent
STOP	Causes an immediate exit from a batch file
TEXT	Displays a video screen from a batch file
USERFILE	Specifies the location of the system security file
VERIFY	Detects and fixes file allocation table errors
WVER	Specifies that disk writes are to be verified

PC-MOS/386 internal system commands can be prefaced by a period (.) to indicate that the system is to execute the internal command rather than a file by the same name. Except for the occasions when a program must have the same name as a system command, the period can be omitted entirely.

Startup

When the system is initialized, PC-MOS/386 starts up one session only (task 0). Additional sessions (tasks) are initiated using the ADDTASK command. The system automatically assigns an identification number to each task; or the user can assign the task a specific number to facilitate its identification by other users. Memory size (the amount of free memory required by the task) must be specified when the task is created. Memory size must be at least 32 KB, up to a system-dependent maximum that is usually around 550 KB. The terminal device driver and serial port to which the task is to be assigned and a startup batch file may also be specified on the ADDTASK command.

Each PC-MOS/386 session has a full-screen virtual display associated with it. Users view that display (subject to security limitations) by entering the task's identification number from the numeric keypad while holding down the Alt key. This special use of the Alt key and numeric keypad can be disabled (or re-enabled by pressing Alt-999).

PC-MOS/386 provides an online help facility that is invoked by entering the HELP command. This utility displays a list of topics about which it can display information. (See Figure 13-5.) Information about a topic is displayed by selecting the topic using the cursor keys and then pressing Enter.

```
                    PC-MOS/386 HELP FACILILTY

        .ABORT           .DATE          .FLUSH         .MOSADM        .SET
        .ADDDEV          .DEBUG         .FOR IN DO     .MSORT         SHELL
        .ADDTASK         DESNOW         .FORMAT        .MSYS          .SIGNOFF
        .ALIAS           DEVICE         FREEMEM        .NEXT          .SIGNON
        .AUTOCD          .DIR           .GOTO          .ONLY          SLICE
        .BATECHO         .DIRMAP        .HDSETUP       .PATH          SMPSIZE
        .BREAK           .DISKCOPY      .HELP          .PAUSE         .SPOOL
        BUFFERS          .DISKID        .IF (NOT)      .PRINT         .STOP
        .CALL/.RETURN    .DOT           .IMPORT        .PROMPT        .TEXT/.ENDTEXT
        .CD              .ECHO          .INSERT        .RD            .TIME
        .CLASS           .ED            .KEY           .REL           .TYPE
        .CLS             .ENVSIZE       .KEYMAP        .REM           USERFILE
        .COMMAND         .ERASE         .MD            .REMDEV        .VERIFY
        .COMPFILE        .EXCEPT        MEMDEV         .REMTASK       VTYPE
        .COPY            .EXPORT        .MORE          .RENAME        .WVER
        COUNTRY          .FILEMODE      .MOS           .SEARCH


              Use "↑ ↓ →  to select choice and then press "<─┘" (enter)
                         Use "Esc" to return to MOS
```

Figure 13-5 PC-MOS/386 Help Facility. *The Help facility displays a list of commands and topics about which it can display information.*

Features

PC-MOS/386 provides several utilities to manage the multi-user, multi-tasking environment. The MOS command provides general purpose utilities that can be made available to all system users. (See Table 13-3.) These utilities display system information, and allow customization of partitions and the sharing of system resources. The PC-MOS/386 system administrator's command (MOSADM) provides additional utilities that should generally be reserved for the system administrator. This program can be protected using the PC-MOS/386 file security system. These utilities can change the number of $1/18$-second time slices a task receives in turn, and the processing priority of a task relative to others. A utility is also provided for turning the disk cache on and off.

Table 13-3 PC-MOS/386 MOS Utility Commands. *Additional utilities are available to the system administrator.*

Command	Function
MOS MAP	Displays partition map
MOS DIS	Disables task in keyboard loop
MOS NODIS	No disable task in keyboard loop
MOS USEIRQ	Reserves IRQ for application
MOS FREEIRQ	Frees IRQ reserved by application
MOS IRQ	Lists IRQs reserved on system
MOS WAIT	Waits for event before continuing
MOS VMODE	Sets video mode
MOS SERINIT	Initializes a serial port
MOS ROUTE	Redirects printer I/O
MOS RESIZE	Adjusts partition size
MOS INFO	Displays memory allocation information

PC-MOS/386 provides a DOS-compatible batch file processor, a print spooler, and a debugger. The batch file processor provides several enhanced commands, including commands that display predefined screens while the batch file is executing. The print spooler is versatile, supporting the use of multiple printers and print classes. Once executed in a partition, the SPOOL command intercepts any output to a printer. SPOOL command options determine the disposition (print and delete, hold, print and retain, print directly) of intercepted output. The PRINT processor is executed in a separate partition. It automatically prints files generated by SPOOL.

ED, the PC-MOS/386 editor, is provided for the creation and editing of text files. Files are displayed with line numbers, and can either be edited by line number or in full-screen mode. From within the editor, graphic characters can be assigned to function keys to allow their easy insertion into text files.

DEBUG allows programs in memory to be examined and debugged. DEBUG provides thirty-five commands, including commands to set and clear breakpoints. A command is provided to toggle the system console between output from DEBUG and the application being debugged. On systems with terminals, DEBUG's output can be viewed at the system console, and the application's output can be directed to a terminal.

PC-MOS/386 provides facilities for securing files, directories, and partitions. System security is based on twenty-six security classes that correspond to the letters A through Z. A security file can be set up that defines the type of access (none, execute only, read and execute only, unrestricted) to items with a particular security classification. To access such items a user must enter a SIGNON with his user name and password, as defined in the security file. A file's security class may be defined when it is created or later, using the COPY command, if the file already exists. The CLASS command assigns a security class to a directory or task partition. Files with a nonblank security class are stored on disk in an encrypted format. To provide greater security, PC-MOS/386 can be configured to require that the file decryption key be entered each time the system is initialized.

PC-MOS/386 provides several methods of intertask communication. One way is through *pipes*. Pipes are created by specifying the $PIPE device driver in the CONFIG.SYS file along with an eight character identifier and, optionally, a buffer size. A task may send input to the pipe, as to any other device; that input is retained in the buffer until retrieved by another task.

A more elaborate method of intertask communication is provided by PC-MOS/386's NETBIOS emulation. The Network Basic Input/Output System (NETBIOS) protocol is designed to allow computer-to-computer communication over networks. By specifying PC-MOS/386's $NETBIOS device driver in the CONFIG.SYS file, up to twenty-five tasks may communicate with each other using NETBIOS features, as if they were tasks running on independent computers. This allows the transparent execution of applications designed to communicate with other applications operating on a NETBIOS network.

Summary

A User Guide and Quick Reference manual are provided with PC-MOS/386. The User Guide contains explanatory information about multi-tasking and multi-user information, as well as reference material. The information provided is complete and well organized. Both the table of contents and the index are useful and comprehensive. An online help facility that provides descriptions of all PC-MOS/386 commands is also provided. A Technical Reference manual is also available. This manual describes PC-MOS/386's memory structure, system calls, and device driver functions.

In summary, PC-MOS/386 is a dependable, easy-to-use operating system for those who wish to use Model 80 or other 80386-based computers to execute multiple DOS applications, either for a single user or for multiple users operating remote terminals. It provides the performance, system security, and system administration features required for a successful multi-user operation.

DESQview/386

Unlike operating system products that replace PC DOS and thus must perform all its functions, application managers are used with DOS to provide additional services while assuring full DOS compatibility. Application managers provide user interfaces that are more flexible and powerful than DOS's command line interface, as well as some degree of multi-tasking support.

DESQview/386 from Quarterdeck Office Systems is a multi-tasking application manager that works with PC/MS DOS Version 2.0 or later on the Model 80, Compaq Deskpro 386, and other compatible Intel 80386-based computers. DESQview/386 supports the execution of most DOS applications without modification or the use of additional software. It supports programs written to the IBM TopView 1.1, Digital Research GEM 1.1, and Microsoft Windows 1.04 application program interfaces. (A commercial or runtime version of GEM or Windows is required to run GEM or Windows applications.) A list of the features provided by DESQview/386 includes:

Concurrency
>   Up to 250 active programs

User Interface
>   Multiple text and graphics windows
>   Online Help facility Keyboard macros
>   Mouse support

Display Suppprt
>   VGA, EGA, CGA, MDA, Hercules

Memory Management
>   Free memory per application $>$512 KB
>   Code/data sharing
>   Expanded memory support

File Support
>   File sharing

Other
    PIFs
    Cut-and-paste facility
    System Developer's Toolkit
    IBM TopView API compatibility
    Companion applications

DESQview/386 is the newest and most capable version of the DESQview application manager introduced in 1985. It is a single-user system that supports up to 250 graphics and text windows. It runs on the IBM Personal System/2 Model 80, Compaq Deskpro 386, and compatible computers. VGA, EGA, CGA, MDA, and Hercules video adapters are supported. Quarterdeck recommends 640 KB of conventional memory for use with DESQview/386; DESQview/386 itself requires 85 KB or less of conventional memory on 80386-based computers. Simulated expanded memory is used as required for storing programs and for task management. The maximum memory size for each task depends on the system configuration, but it is generally greater than 512 KB. In addition to the memory required for task storage, DESQview/386 uses expanded memory to store each task's virtual display image; this can require up to 256 KB per task in VGA high resolution video modes. Thus 1.5 to 2 MB of memory are required for efficient multi-tasking of other than very small applications.

Expanded memory is provided for use by DESQview/386 and for applications, by the separately priced QEMM 386 expanded memory manager, also sold by Quarterdeck. QEMM 386, Version 4.0 is an expanded memory driver that simulates expanded memory, using the system's extended memory. The driver is compatible with version 4.0 of the Lotus Intel Microsoft Expanded Memory Specification.

DESQview/386 is designed specifically for use on 80386-based computers. DESQview 2.0 is available for use on 8088, 8086, and 80286-based computers, including the IBM PC and Personal System/2 Models 25, 30, 50, and 60. This version executes applications stored in conventional or expanded memory. Since these computers do not provide the 80386's virtual device capabilities, applications that write directly to the display can only be run in a window or in background with the aid of application-specific loaders provided by Quarterdeck.

Installation

DESQview/386 is installed onto a diskette or hard disk using a menu-driven process. The installation process is automated and requires only a few minutes time. During the installation process, DESQview/386 searches the target disk for any of the nearly 100 applications about which it has resource requirements information. It then builds a .DVP program information file for each and adds its name to the DESQview/386 *Open Window* menu. Once the installation is complete, this process can be repeated for programs resident on other drives, by running AUTOINST with the drive name as an operand.

Near the end of the installation process, the user is asked to specify system options, either by answering a few simple questions—such as whether DESQview/386 is to use colors and a mouse—or by using an advanced options menu that permits the user to specify information about the system display adapter, size, color, and location of windows, and processor scheduling.

The processor scheduling options specify the number of $1/18$-second time ticks a window is given for execution when it receives its turn to execute, in round-robin fashion. At the end of that number of ticks, execution of the current task is interrupted, and the next task is allowed to execute. The default specifications are 9 ticks for the foreground task and 3 ticks for each background task. Users who do more processor-intensive work in the background than in the foreground may wish to increase the number of ticks that each background task receives. The number of ticks each window receives should not be reduced to less than 2; otherwise, the time slice is so small that very little gets accomplished.

Main Menu Options

Once installed, DESQview/386 is invoked by executing the DV.BAT file installed in the root directory of the disk/diskette. This causes the display of the DESQview/386 main menu, from which the user can open or close a window, ask for help, or quit DESQview. This menu is also used to zoom a window (to/from full screen size), switch between or rearrange windows, and mark and transfer data between windows. Items may be selected from this menu by highlighting them with the cursor keys or mouse, or by keying in the single character assigned to each function.

Selecting the *Open Window* menu displays a list of available applications. These include all programs identified by the installation process and any added by the user through the *Add a Program* option from this menu. Options for deleting and changing a program are also included. Small programs not on the menu may be executed using the predefined 128 KB DOS command window.

A DOS services program is included; this allows common DOS services to be selected with the cursor or mouse, or by entering a single character. The DESQview/386 memory status program included on this menu provides information on the amount of common, conventional, and expanded memory in use and available on the system. Common memory is the memory (usually about 12 KB) that DESQview uses to store information needed to manage windows, and mark and transfer operations.

Once a program is selected for execution from the *Open Window* menu,

DESQview/386 consults its .DVP file and allocates the system resources necessary for the program's execution. Unless the .DVP files specify otherwise, programs are started in a small window, using one of the sizes and screen locations defined in DESQview/386. The DESQview/386 menu may be displayed while a program is executing by pressing the DESQ key. (See Figure 13-6.) By default this is either of the Alt keys, but it can be changed to another key or key combination when the system is installed, or by running DVSETUP from DESQview/386.

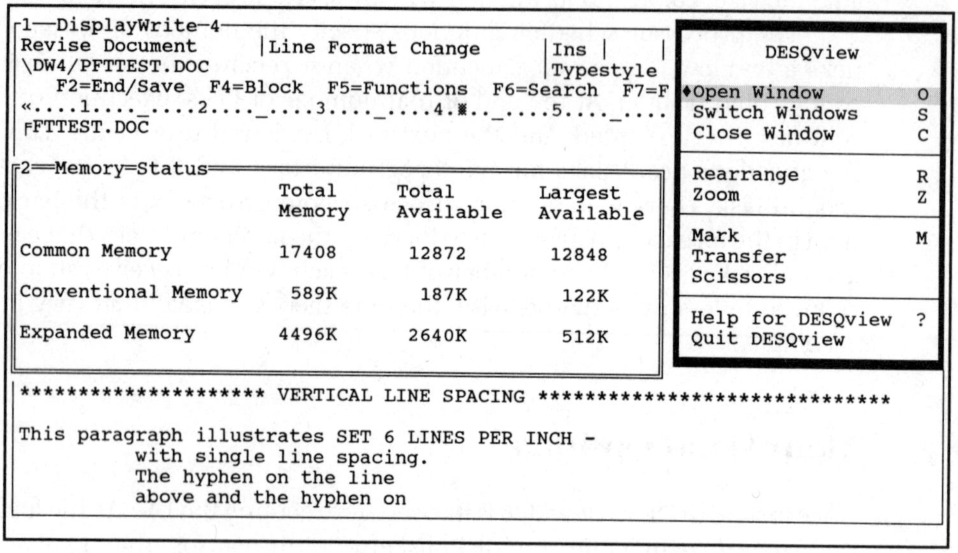

```
┌1─DisplayWrite─4─────────────────────────────────────┐┌─────────────────────┐
│Revise Document        |Line Format Change  |Ins |   ││     DESQview        │
│\DW4/PFTTEST.DOC                            |Typestyle│                     │
│   F2=End/Save   F4=Block  F5=Functions  F6=Search  F7=F│♦Open Window      O│
│«....1...._....2...._....3...._....4.*.._....5...._....│ Switch Windows    S│
│FFTTEST.DOC                                          ││ Close Window      C│
│┌2══Memory═Status════════════════════════════════════┐│                     │
││                                                    ││ Rearrange         R│
││                 Total     Total       Largest      ││ Zoom              Z│
││                 Memory    Available   Available    ││                     │
││                                                    ││ Mark              M│
││Common Memory    17408     12872       12848        ││ Transfer            │
││                                                    ││ Scissors            │
││Conventional Memory  589K   187K        122K        ││                     │
││                                                    ││ Help for DESQview ?│
││Expanded Memory   4496K    2640K        512K        ││ Quit DESQview       │
│└────────────────────────────────────────────────────┘└─────────────────────┘
│******************** VERTICAL LINE SPACING *****************************
│This paragraph illustrates SET 6 LINES PER INCH -
│         with single line spacing.
│         The hyphen on the line
│         above and the hyphen on
└──────────────────────────────────────────────────────────────────────────┘
```

Figure 13-6 DESQview/386 Display Windows. *The Alt key can be used to activate the DESQview/386 menu; however, this function can be assigned to a different key.*

Features

DESQview/386's built-in application program interface (API) is compatible with the IBM TopView Version 1.1 API. Programs written using this API can manipulate windows and subwindows, spawn subtasks, and communicate with other programs and subtasks. With appropriate runtime modules, DESQview/386 also runs GEM 1.1 and Windows 1.04 applications.

DESQview/386 can run 32-bit protected mode applications, developed with Phar Lap Software's 386/DOS-Extender, simultaneously with DOS applications. This allows applications such as Borland's Paradox 386 to take advantage of the full power of the Model 80's 80386.

A separately priced product—DESQview Companions 1—provides a calculator, datebook, notepad, and communications utilities. Both the calculator and notepad utilities can operate on information obtained from windows using DESQview's Mark and Transfer feature. The DESQview Link communications facility provides background host communications services as well as access procedures for electronic mail services.

Summary

The DESQview/386 manual is an excellent example of well-organized and useful documentation. It provides extensive reference as well as tutorial information. The manual is well illustrated and indexed; it gives the user extensive advice not only on how to do things, but on what to do when something goes amiss. The sections on video and memory mapping are particularly detailed and useful.

DESQview/386 is a useful tool for anyone who needs to execute more than one DOS application at a time, but has no need for multi-user capabilities. It allows multiple DOS applications to be conveniently executed in a windowed environment, isolated and protected from each other. Its user interface is easy to learn and use, and works well with or without a mouse.

As with any windowing environment, setting up the required program information files is a bit of work; however, Quarterdeck has already done that work for most popular applications. With its ease of operation and clarity of documentation, it can be set up, run, and maintained by end users much more easily than a whole new operating system.

Windows/386

Microsoft Windows/386 provides the convenience of the familiar Windows graphic user interface without the complexity and inconvenience of developing applications using the Windows application program interface. (See Figure 13-7.) Windows/386 is a multi-tasking application manager that works with DOS, rather than replacing it.

Windows/386 is a single-user system that manages tasks executing in up to thirty-two different virtual machines. Windows 2.0 runs in one of these virtual machines; the others are available to run unmodified DOS applications. Each virtual machine can be up to 640 KB in size, with over 512 KB of memory available for the DOS application running in the machine. Most DOS applications can be run in a window along with Windows applications, or full screen. Both text and graphics windows are displayed in graphics mode in Windows 2.0 overlapping windows. These windows can be moved, sized, or zoomed to a full screen using the keyboard or a mouse. The list of features provided by Windows/386 includes:

Concurrency
    Up to 32 active programs

User Interface
    Microsoft Windows graphic interface
    Mouse support
    Expanded keyboard buffer

Display Support
    VGA, EGA, CGA, Hercules

Memory Management
    Free memory per application >512 KB
    Code/data sharing
    Expanded memory support

File Support
    File sharing

Other
    Program information files
    Cut-and-paste facility
    Print spooler
    Windows software development kit
    Companion applications

Windows/386 runs on the IBM Personal System/2 Model 80, Compaq Deskpro 386, and compatible computers using DOS Version 3.1 or higher. It requires a graphics adapter and display. VGA, EGA, CGA, Hercules, and several other popular graphics adapters are supported. A mouse is recommended but not required for its operation. Windows/386 requires a minimum of 1 MB of memory; 2 MB or more is recommended. Extended memory is used for task storage and virtual display support using a built-in memory manager.

This built-in memory manager uses the 80386's paging hardware to provide expanded memory for use by Windows 2.0 and by DOS and Windows applications, as well as virtual memory for task management and storage. The expanded memory driver uses the 80386's paging hardware to simulate expanded memory using the system's extended memory. The driver is compatible with version 4.0 of the Lotus Intel Microsoft Expanded Memory Specification.

Windows 2.0 is available as a separate product for use on 8088-, 8086-, and 80286-based computers, including the IBM PC and Personal System/2 Models 25,

30, 50, and 60. Windows 2.0 executes Windows applications and well-behaved DOS applications using conventional and expanded memory and non-preemptive scheduling. Since these computers do not provide the 80386's virtual device capabilities, applications that write directly to the display can only be run in a full-screen window.

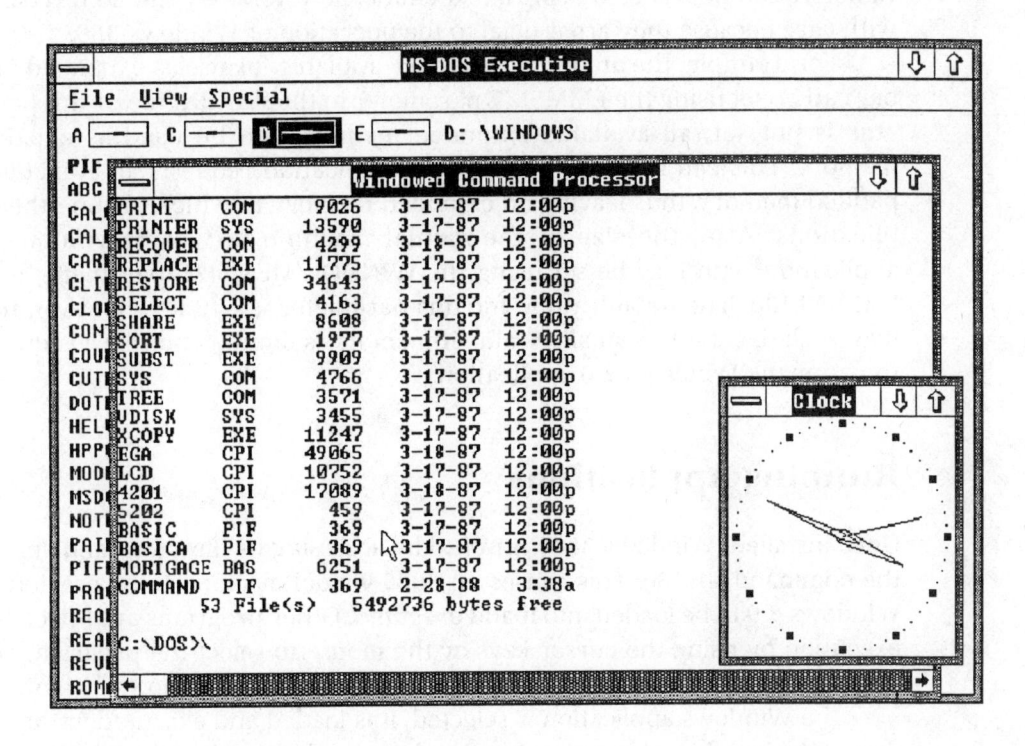

Figure 13-7 Windows/386 Display Windows. *Windows/386 uses the Windows 2.0 graphic user interface.*

Installation

Windows/386 is installed using the menu-driven process used by Windows 2.0. During the process the user must specify information about the system, such as the kind of display adapter, mouse, and printer to be used. The process takes a few minutes because of the large number of program and program information files that are copied to the hard disk.

One of the utilities copied to the hard disk is a disk cache called SMART-Drive. This program is activated by referencing it in the system CONFIG.SYS file and specifying how much extended memory it is to use; 256 KB is used if no specification is given. With Windows 2.0 it can use either expanded or extended memory, but with Windows/386 it must use extended memory. If configured to use expanded memory, its internal expanded memory manager will conflict with that of Windows/386.

As with all versions of Windows, much of the operation of Windows/386 is controlled by parameters stored in the WIN.INI file. Some of these parameters are determined during the installation process and some are defaults. These parameters can be changed using a text editor; however, they should be changed with care because they are crucial to the operation of Windows/386.

For example, the amount of memory available for use as expanded memory can be set using the EMMSIZE parameter in the WIN.INI file. If this parameter is not set, all available extended memory can be used as expanded memory. This can cause problems if an application requests all available expanded memory, thus leaving no extended memory in which to run other applications. Also, the size of the virtual machine in which Windows 2.0 applications run may be set using the WINDOWMEMSIZE parameter in the WIN.INI file. The default value for this parameter is 640 KB; however, users may wish to set it to a smaller value if memory is limited and there is no need to run many Windows 2.0 applications.

Running Applications

Once installed, Windows/386 is invoked by accessing its directory and entering the command WIN386. This causes an 8086 virtual machine to be created and Windows 2.0 to be loaded into it and executed. Other programs are selected for execution by using the cursor keys or the mouse to select the program, or its program information file (PIF) if it has one, in the usual Windows fashion.

If a Windows application is selected, it is loaded and executed in the Windows 2.0 virtual machine. If a non-Windows application is selected, a new virtual machine is created in accordance with the resource requirements specified in its PIF, and the program is loaded into it and executed. If a PIF is not available for an application, Windows attempts to execute the application in a full-screen window, suspending execution if it is not in the foreground.

PIF files for over fifty popular applications are included with Windows/386, including Lotus 1-2-3, dBASE III, Microsoft Word, and WordPerfect. These files may be modified and new ones created using the Windows PIFEDIT utility. PIFEDIT can be used to indicate a program's name, title, parameters, initial directory, and memory requirements. The PIF file also controls whether a program executes full screen or in a window, whether it executes when it is in the background as well as in the foreground, and whether any other tasks execute when it is in the foreground. This last option is called exclusive mode; a task running in this mode receives all available processor cycles when it is in the foreground, thus assuring maximum performance for the task.

If an exclusive mode application is not in the foreground, Windows/386 allocates the processor to each task using the $1/18$-second system timer tick.

Two-thirds of the processor cycles are allocated to the task in the foreground, and the remaining one-third are allocated equally among those tasks whose PIFs indicate that they are to run in the background as well as in the foreground. Scheduling is done on a preemptive basis: a virtual machine is interrupted once it has executed for its allotted number of timer ticks, and the next virtual machine is allowed to execute in turn. Processor time allocated to the Windows 2.0 virtual machine is allocated among whatever applications it is running, using the Microsoft Windows nonpreemptive scheduling scheme.

The user can view the title bars of applications currently loaded (including full-screen applications) by holding down the Alt key and then repeatedly pressing the Tab key. Releasing the Alt key causes the application whose title bar is visible to become the foreground task. A task can also be made the foreground task by clicking the mouse when the pointer is in the application's title bar. Because the mouse can be used for this as well as cut-and-paste and other Windows functions, an application can only use the mouse when it is running in full-screen mode (and when the DOS device driver for the mouse has been loaded).

Terminate-and-stay-resident pop-up programs, such as Borland's Sidekick, must be initiated in a virtual machine rather than before Windows/386 is started. Once initiated, they may be activated in the usual manner when the machine is in the foreground. If particular pop-up programs are routinely run with a particular application, the user may wish to start the application using a batch file that first loads the pop-up program and then the application.

A task's display can be switched between full screen and window display mode by pressing Alt-Enter, or by using the mouse to select the *Settings . . .* option from its control menu when it is in window display mode. This menu, shown in Figure 13-8, can be used to suspend or resume the processing of a task, or to change the tasking options from their initial values in the PIF file. This menu can also terminate a task. This is a drastic measure, only to be taken when the task cannot be persuaded to exit in a normal manner. In either case, the task's virtual machine is destroyed and control is returned to Windows 2.0.

Applications can share and lock disk files using the MS-DOS SHARE utility. SHARE should be run before loading Windows/386 in order for it to perform correctly. Since DOS applications run in separate virtual machines, protected from each other, Windows/386 detects when more than one application attempts to access the same nonshareable device (such as a communications port). When this occurs, Windows/386 displays a dialog box asking the user to choose which application should gain control of the device. The application not selected to receive control of the device receives a simulated error condition, and must be restarted when the nonshareable device is available for its use, unless it can automatically recover from the device error condition.

After all tasks have terminated, Windows/386 may be terminated by selecting the *Close* option on the control menu of the last active MS-DOS Executive window. An error message is displayed if this option is selected while other vir-

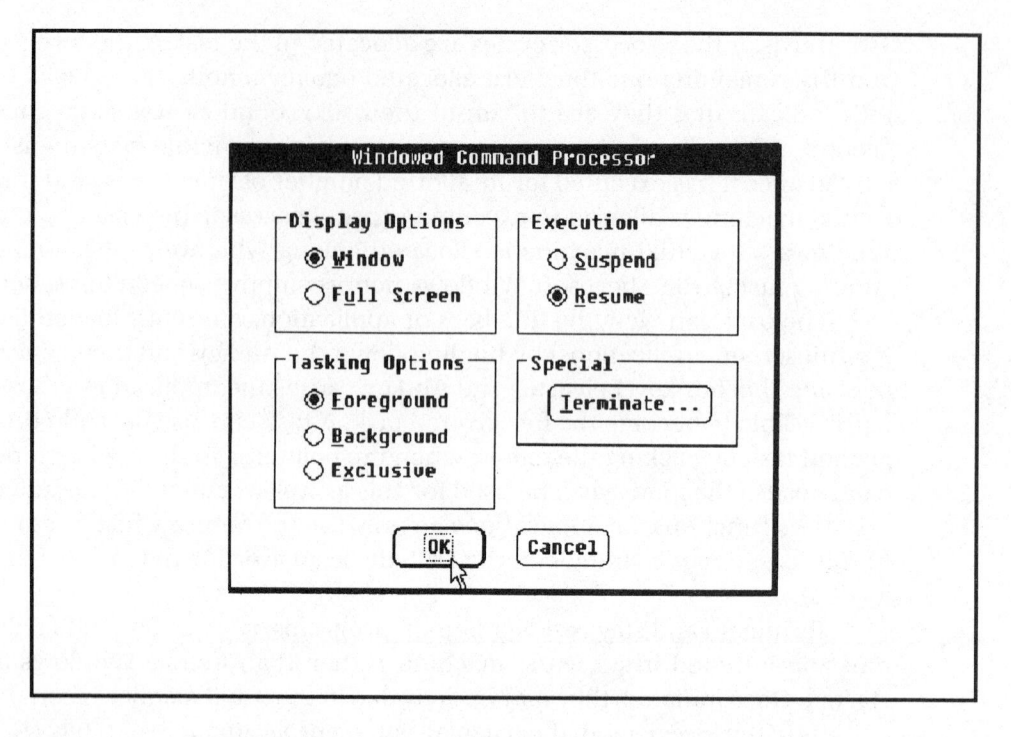

Figure 13-8 Settings Menu. *The operation of a task can be suspended or resumed using this menu.*

tual machines still exist. Control returns to DOS when the MS-DOS Executive window is closed. The SMARTDrive disk cache, if specified in the CONFIG.SYS file, remains available for use.

Summary

Documentation supplied with Windows/386 consists of a Quick Reference Guide and User's Guide for Microsoft Windows 2.0 (which includes documentation on Windows Desktop Applications, Windows Paint, and Windows Write, and a booklet on using Windows/386). The documentation is well illustrated and a marked improvement over previous Windows documentation. The booklet on using Windows/386 explains the operation of the product and provides a good description of the likely causes of error messages.

Windows/386 allows the quick and easy selection and execution of multiple DOS applications using the Windows 2.0 graphic interface. It provides very good isolation of the applications being executed and excellent virtual display support. Applications can be displayed in windows (with the attendant performance degradation) or full screen, with easy switching between the two modes. Some time is required to become proficient in the use of the Windows graphic interface, but once learned, its operation is quick and easy.

Exploiting the Power of the Model 80

The products discussed here can be used with the Model 80 to meet several needs. Concurrent DOS 386 and PC-MOS/386 allow multiple users to share the Model 80 to run DOS applications. DESQview/386 and Windows/386 allow a single user to run multiple DOS applications on the Model 80. Currently and for some time to come, OS/2, with all its capabilities, does neither.

Concurrent DOS 386 and PC-MOS/386 are not 100 percent DOS-compatible, but then again, neither is the OS/2 DOS compatibility environment. Programs that use undocumented features of DOS may not run correctly using either of these operating systems or the OS/2 DOS compatibility environment. Because they work with DOS rather than replacing it, DESQview/386 and Windows/386 provide more DOS compatibility, but they do not provide multi-user capabilities.

The control software products for the Model 80 discussed here represent real alternatives to the current version of OS/2 for users who only require the multi-tasked execution of DOS applications. OS/2 or some other multi-tasking operating system is obviously preferable for the development of systems that require extensive interprocess communications and resource sharing.

The fact that none of these products is currently endorsed by IBM should not detract significantly from their usefulness. However, users who are affected by IBM's efforts to provide consistent applications software interfaces across its product lines through Systems Application Architecture (SAA) may wish to give special consideration to Windows/386. The Windows/386 graphic user interface is very similar to the OS/2 Presentation Manager user interface, which conforms to the SAA Common User Access specifications.

14

Connectivity

The IBM PC was introduced as a computer quite disconnected from the rest of the IBM product line. In addition to featuring a non-IBM CPU and operating system, it used diskettes and file formats that were not compatible with those used by other IBM computers and office systems. Moreover, the communications features provided by the PC's optional asynchronous communications adapter and asynchronous communications program were better suited for communications with non-IBM computers than with IBM host computers.

By the time the Personal System/2 was introduced, this situation had changed dramatically. Adapter boards and support software were available from IBM and numerous third-party vendors to connect the IBM PC to every major IBM computer and office product, and to provide terminal emulation, file transfer, and resource sharing among networked devices. By this time, the connected PC had become so much a part of the work environment that it was clear that the new Personal System/2 family would have to provide all the connectivity features of the PC in order to be successful.

It comes as no surprise, then, that the Personal System/2 family supports all the connectivity features offered by the PC (with the exception of connectivity to certain IBM products that are no longer marketed). This is in sharp contrast to the fact that magnetic media compatibility between the two computer families is limited. The 3.5-inch diskette drives standard on all Personal System/2 models are not compatible with the 5.25-inch diskette drives standard on all PC family machines except the Convertible. IBM provides some devices and facilities for directly exchanging data between PCs and Personal System/2s, but all are less than satisfactory.

Chapter 14

Media and File Interchange

The introduction of 3.5-inch diskette drives as standard equipment on the Personal System/2 machines has produced some difficult buying decisions. Machine performance is not the whole story; media compatibility is also significant. For environments where a single machine is used in isolation, the performance of the machine when running standard applications is paramount. The fact that 3.5-inch diskettes are more expensive than 5.25-inch diskettes is the only additional expense for the machine's continued use.

In a mixed environment where machines are not physically linked, for example through a local area network (LAN), the issues are more complex. It is not unreasonable to have a situation where 5.25-inch, 360 KB and 1.2 MB diskettes as well as 3.5-inch, 720 KB and 1.44 MB diskettes are being used, and data needs to be exchanged among all the machines using them. The general principle used for the purchase of PCs in many corporate environments has been the "trickle down" theory. When new and more powerful machines are purchased, the old machines are given to others in the work group (often support personnel). This system works well for economic reasons, but care must be taken with data exchange. For PCs and ATs, low density diskettes need to be used so that diskettes written on an AT can be read on a PC. If this practice is followed, everyone in the work group can exchange information without complicated procedures.

With the introduction of the Personal System/2 machines, any PC family machine that requires the data written on a Personal System/2 machine's standard diskette drive needs 3.5-inch disk drives added, if the machines are not linked by a LAN or a communications network. The 3.5-inch disk drives available from IBM for installation on the PC family of machines are low density and only format to 720 KB, not 1.44 MB as do the Models 50, 60, and 80. This is similar to the problem of exchanging diskettes between the PC with its standard 360 KB diskette drives and the AT with its standard 1.2 MB diskette drive; but now it occurs twice, between PCs and ATs and between Models 50, 60, and 80 and PCs with 3.5-inch diskette drives. A similar problem occurs if Personal System/2 machines of all calibers are chosen. Models 30 and 25 can only read and write 3.5-inch diskettes that are formatted for 720 KB.

There is no common device that can be added to each machine to remove all media compatibility problems, but the use of either a 5.25-inch external diskette drive on the Personal System/2 machines or a 3.5-inch diskette drive on the PC family of machines alleviates most of the problems. The 5.25-inch external diskette drive offered by IBM for use with the Personal System/2 family can read, write, and format 360 KB diskettes; however, it cannot use 1.2 MB diskettes. It seems likely that someone will offer a 1.2 MB diskette drive for the Personal System/2 to support media interchange with the AT's 1.2 MB diskette drive, even if IBM doesn't.

In order to ease the initial upgrade problems, IBM offers a Data Migration

Facility. This consists of a connector and appropriate data transfer software. The parallel port on the Personal System/2 machines is bidirectional; it can be used for data input as well as output. On the PC family of machines, the parallel port can only be used for output, unless new third-party hardware is purchased. The Data Migration Connector is a black plastic device with two connectors. The 25-pin D-shell connector on one end plugs into the parallel port connector on the Personal System/2 computer; the other end plugs into the Centronics connector on the parallel printer cable of the computer that will be sending data. This is not a straight-through connector that can be fashioned easily. It includes base addressing sensing lines; however, there is no sophisticated circuitry inside. Once the required connections have been made, supplied software can be used to transfer data to the Personal System/2 machine. It is important to note that, although there are many suppliers of printer cables that function adequately with parallel printers, different signals are required to support the transfer of data into a parallel port, so not all cables work. Needless to say, IBM recommends their printer cable, which works satisfactorily.

The IBM Personal System/2 Data Migration Facility comes in a small box that contains the connector adapter, a 5.25-inch diskette with a send program on it, and an 11-page, three-hole-punched manual. The sending system requires DOS 2.0 or higher and the receiving system requires DOS 3.3. The transferred data can be stored on 3.5-inch diskettes or a hard disk. After installing the cable, the receiving computer is set up to receive the data by using the file RECV35.COM, which is on the Reference Diskette supplied with the Personal System/2 machine. The sending machine then uses the send facility that is on the 5.25-inch diskette supplied with the Data Migration Facility. The format of the command used for transferring files is similar to the DOS COPY command. At the DOS prompt, the command COPY35 and parameters are entered. The source file, including any necessary drive or path designation, is specified, followed by the destination drive letter and path. The DOS wildcards * and ? are supported, as in the COPY command, but filenames cannot be changed during the transfer and existing files with the same filename on the destination drive are overwritten. This product supports different versions of DOS at either end, so files such as COMMAND.COM should not be transferred. The DOS REPLACE command can be used on the receiving system to rectify any accidental transfers of files, if backup copies of files are available.

Although the Data Migration Facility provides a relatively inexpensive method of uploading the files from one system to another, it is not quick and the transfer is only one way. Each subdirectory needs to be transferred in turn. A version of XCOPY with its many switch parameters would have been preferable. If the existing PC is to be in use along with the Personal System/2 machine, an internal 3.5-inch drive for the PC and the DOS BACKUP program may be a more cost-effective method in the longer term.

There is no question that 3.5-inch diskettes are a superior product to the 5.25-inch diskettes. They hold more data for their size and are better protected

Chapter 14

in their hard plastic cases. Their drop in price is likely to continue as the demand for them increases. The problem of incompatibility in media across the machines is a short-term handicap. It is bothersome to have to consider what other machines will need in order to read a particular file when it is saved on diskette. On the other hand, hard disks are in common use and data transfer on diskettes is less of an issue than it was a couple of years ago. Small amounts of data can be transferred using the Data Migration Facility, or through communications using the serial port with a modem or a modem eliminator. If large amounts of data must be transferred among machines on a regular basis, many network communications options are available that do not suffer from media incompatibility problems.

Asynchronous Communications

Until the advent of personal computers, asynchronous communications between computers was not common. If computers needed to communicate, this was done using modems and communications protocols that supported the synchronous communication of large blocks of data. The modems and communications hardware involved were expensive to purchase and maintain, and were used mainly for communication with large mainframe computers.

The widespread acceptance of personal computers changed the communications market. Suddenly, there were thousands of computers in homes and in offices that could provide for the computing needs of many people independent of a mini or mainframe computer. There was still a need for communications, however, because personal computer users needed to access data stored on other computers. This was done through the use of asynchronous modems, which allowed dissimilar computers to talk to each other via the existing standard telephone system. The "linking" was already in place, and only the conversion from the information on a phone line to the serial port on a PC was required. Asynchronous modems costing only a few hundred dollars (as opposed to the several thousand dollar cost of synchronous modems) were adequate for the size and speed of data transfers required by most personal computer applications.

Terminal emulation software offered by IBM and others allowed the PC equipped with an asynchronous modem to communicate with mini and mainframe computers, as if it were an asynchronous computer terminal such as an IBM 3101 or DEC VT52. This capability gave the PC user the functionality of a stand-alone computer and an interactive computer terminal for little more than the cost of a terminal alone. Additionally, since most terminal emulation packages supported capturing or downloading data from the host, data obtained from the host could be processed offline on the PC, reducing communications costs and charges for host computer services.

Once asynchronous modems were established as means of communicating between computers, a whole new generation of databases sprang up. Bulletin boards, or electronic message centers, were established in people's basements, allowing this new generation of computer users to exchange the latest tidbits of information. More formal information exchange areas became popular. CompuServe, the Dow Jones News/Retrieval service, and MCI Mail became household names for many. For each general purpose information database that was set up, many more specialized services were offered. Attorneys can now get detailed information on court rulings or patents filed. There are databases on every conceivable subject—travel, astronomy, cars, and even computer programming.

Asynchronous communications with the Personal System/2 machines has not changed much from the PC family of machines. The serial port is still driven by a controller from the same family as the NS 16450 controller on the AT. The NS 16550 or the NS 16550A is used; although this version of the IC offers more features, such as a FIFO mode, they are only implemented on the 20 MHz Model 80 systems.

The communication software that is used in the PC environment can be used with the Personal System/2 machines, usually without modification. Any terminal emulation that was performed with a PC can be done on a Personal System/2 machine with the appropriate communications adapter. For the Micro Channel machines, IBM offers two new adapter boards that can support asynchronous communications—the Dual Async Adapter/A, which has two serial ports, and the Multi-Protocol Adapter/A, which has a single serial port that can be programmed for asynchronous as well as synchronous communication. These boards, as well as the serial port on the Personal System/2 system board, are compatible with the serial ports used on the PC family of computers, and can be used with commonly available asynchronous communications software with no difficulties.

Many personal computers are used initially for stand-alone processing; however, it is usually not long before the user begins to retrieve information from other computers to augment the information contained on the PC. As more and more PCs are in use at the same site or in the same enterprise, the need soon arises to share information stored on personal computers. Initially, this can be accommodated by sharing diskettes or by communicating from one PC to another, but as more people and machines become involved, a means of networking computers for the communication and sharing of information becomes necessary.

Local area networks can be established to connect the computers at a single site, and wide area networks can be established to connect computers at different locations. Establishments that have IBM mini or mainframe computers often use their SNA communications network for the sharing of information among computers, particularly if personal computer users require information from the mini or mainframe host computer. Those without mainframe computers available depend more on local area networks for the communication and sharing of information.

Chapter 14

Local Area Networks

IBM provides a variety of local area network (LAN) products for use with the PC and Personal System/2 computer families. These products support the sharing of resources as well as communication among workstations. Several new LAN products were introduced by IBM concurrently with the April 1987 announcement of OS/2 and the Personal System/2 family. These products include new baseband and broadband PC Network adapters, a Micro Channel version of the Token-Ring Network Adapter, a new version of the PC Local Area Network Program (Version 1.2), and related support and network management software.

A new PC Network adapter, the PC Network Baseband Adapter, was introduced to support small networks. This adapter is available in a /A (Micro Channel) version, as well as a version for use with the PC family and the Personal System/2 Models 25 and 30. The PC Network Adapter II, a version of the original PC Network Adapter designed for use with the higher speed expansion bus used on Models 25 and 30 as well as on PCs, was also introduced.

PC Local Area Network Program

The IBM PC Local Area Network Program Version 1.2 provides support for the Personal System/2 family of computers, as well as higher performance random disk input/output than Version 1.1. It allows connection of PCs and Personal System/2 computers to an IBM PC Network or IBM Token-Ring network.

The IBM PC Network is a network designed to link IBM personal computers together. The original PC Network allows personal computers equipped with PC Network adapters to communicate via cable television (CATV) type coaxial cable, using broadband communications techniques. Up to seventy-two personal computers can be networked together, with each no more than 1000 feet from the PC Network Translator unit, which serves as the head of the broadband network.

PC Network baseband adapters and extenders allow personal computers to be networked together using dual twisted-pair telephone wire (that meets IBM Cabling System type 3 specifications). Up to eight computers can be linked together in a chain. Up to ten chains can be connected to a baseband extender, for a total network size of up to eighty stations. A chain connected to a baseband extender can be up to 400 feet long; stand-alone chains can be up to 200 feet long. Both PC Network implementations use carrier sense, multiple access, collision detect (CSMA/CD) communications protocol.

The IBM Token-Ring network is a higher speed network designed for the interconnection of IBM mainframe and mid-range computers as well as personal computers. The Token-Ring network takes its name from its token-ring access communications protocol. The network can operate at 4 megabits per second, compared to the PC Network's maximum communications rate of 2 megabits per second. Devices networked together can be connected using the several types of media that comprise the IBM Cabling System, including type 3 specification dual twisted-pair telephone wire.

Devices that can be connected directly to the network in addition to personal computers include the IBM mainframe 3720/25 communications controller and 3174 terminal unit controller. Series/1 and System/36 computers can be connected to a Token-Ring network via gateways. Up to 260 devices can be connected together on the same ring and multiple rings can be bridged together.

The LAN program serves as an extension to the personal computer's DOS operating system, allowing the sharing of files and printers and the exchange of messages across the LAN. The LAN program works with DOS 3.3 on the Token-Ring network and both the broadband and baseband versions of the PC Network. It works with DOS 3.2 on the Token-Ring network and the original PC Network, and with DOS 3.1 on the original PC Network.

The LAN program supports printer and file sharing, peer-to-peer communications among PCs and Personal System/2s, multiple servers for file, print, print queue management, and message functions. Message functions include saving, retrieving, and relaying the same message, editing and forwarding received messages, and skipping previously viewed messages. The LAN program is compatible with IBM PC 3270 emulation products, and TopView Version 1.12.

The LAN program can be configured to operate in one of four modes: redirector, server, receiver, or messenger. In redirector mode, the LAN program intercepts all application program calls for disk and printer I/O, and redirects these calls over the network to a personal computer acting as a server for these functions. Once the function has been performed, the LAN program passes the data back to the applications as if the function had been performed on the local computer. A redirector mode station can also send messages to other stations.

The second mode is server mode. Personal computers running the LAN program in server mode respond to network requests for disk I/O and/or printer output. Server requests run in the background, allowing a DOS application to run at the same time, provided sufficient memory and other resources are available.

Receiver and messenger modes are variations on the redirector mode. In receiver mode a station acts as a redirector; in addition, it can receive and log (save) messages as well as send them. A station in messenger mode has the same capabilities as the receiver mode station, and it can receive and relay messages for other stations. The server mode includes the services provided by messenger mode. Stations operating in any of the four modes can use network disks, directories, and printers.

Requirements

The LAN Program works with Personal System/2s and PC family computers with an appropriate network adapter and support software. The adapters mentioned above are supported, including the original PC Network and Token-Ring adapters. Either the IBM Local Area Network Support Program or the PC Network Protocol Driver Program is required for the LAN Program to interface with the network adapter.

The amount of memory required depends on the workstation's network configuration. A redirector configuration requires 50 KB, a receiver requires 68 KB, a messenger requires 160 KB, and a server requires 350 KB if disk caching is used and 228 KB if disk caching is not used. A server workstation must have at least one hard disk. These memory requirements do not include memory required for DOS, LAN adapter device drivers, or applications.

The LAN Program can be run using a series of menus, or commands can be entered directly. A list of available commands is shown in Table 14-1. Commands are provided for starting the LAN Program, starting particular modes of operation, sending and logging messages, and using network devices.

Table 14-1 PC LAN Program Commands. *The LAN Program commands are performed by a program that is loaded from diskette or hard disk each time a command is entered.*

Command	Purpose
NET	Starts PC LAN Program menus
NET CONTINUE	Resumes a particular network function
NET ERROR	Lists network errors that have occurred
NET FILE	Checks status or closes a file
NET FORWARD	Forwards a message to another computer
NET LOG	Controls logging of received messages
NET NAME	Specifies another name for receiving messages
NET PAUSE	Suspends the server, messenger, receiver, disk, or print redirection function
NET PRINT	Prints a file on a network printer
NET SEND	Sends a message
NET SEPARATOR	Specifies if a separator page is to be printed between files
NET SHARE	Controls network sharing of devices and disk directories
NET START	Starts a particular network mode
NET USE	Starts or stops using a device or disk directory on a network computer

The LAN Program incorporates a disk cache feature that can be used on server configurations. This enables Version 1.2 of the LAN Program to provide better random disk I/O performance than Version 1.1. The disk cache can use up to 360 KB of conventional memory or up to 15 MB of extended memory to store previously accessed disk blocks. The default size of the cache is 112 KB (of conventional memory). Subsequent requests for these blocks can then be satisfied at memory access speed rather than disk access speed. This disk cache option is particularly useful for an AT-based server, because IBM does not offer general purpose disk cache software for PC family computers, as it does for the Personal

System/2 Models 50, 60, and 80. Third-party cache software can be used, provided the LAN Program cache feature is disabled.

Packaging and Documentation

The LAN Program is provided in a standard IBM slip-case package. The software is included on three 360 KB, 5.25-inch diskettes, and a single 720 KB, 3.5-inch diskette. Also included are a User's Guide, a Quick Reference card, and an Installation Aid Quick Reference card.

Local Area Network Support Program

The Local Area Network Support Program is a new product that provides an IEEE 802.2 Logical Link Control (LLC) and a Network Basic Input/Output System (NETBIOS) interface for the IBM PC Network as well as the IBM Token-Ring network. The LAN Support Program can be used with the new PC Network adapters (baseband and broadband), the original PC Network Adapter (using IEEE 802.2 emulation), and the original Token-Ring Adapter and Token-Ring Adapter II; both PC bus and Micro Channel (/A) versions of the adapters are supported. It supports the use of one or two network adapters in the same system; only one adapter can be used when the original PC Network Adapter is used.

The LAN Support Program provides an IEEE 802.2 protocol and NETBIOS interface for the new PC Network adapters. It also includes the PC Network Adapter IEEE 802.2 protocol emulation capability. This feature allows the original PC Network Adapter (which used a different protocol implemented in adapter ROM) to communicate with new broadband PC Network adapters that use the LAN Support Program.

The LAN Support Program provides network adapter support interfaces and NETBIOS for the Token-Ring network and the IBM PC Network. It replaces the adapter support interface software (TOKREUI.COM) supplied with the original Token-Ring and Token-Ring II adapters, and the adapter support interface software for the 3270-PC (TOKR3270.COM) supplied with the NETBIOS Program Version 1.1. It also replaces and provides all the functions of Version 1.1 of the Token-Ring network NETBIOS Program (NETBEUI.COM).

The LAN Support Program provides two application programming interfaces. It supports programs written to use the IEEE 802.2 interface, such as IBM Advanced Program to Program Communications/Personal Computer (APPC/PC), and NETBIOS applications such as the IBM Local Area Network Program.

The LAN Support Program is a prerequisite for the use of the Token-Ring Network Adapter/A; it or the PC Network Protocol Driver Program is a prerequisite for the use of the new PC Network adapters (whether PC bus or Micro Channel versions). The PC Network Protocol Program is a separate product which provides NETBIOS support for the PC Network Adapter II and II/A, while providing compatibility with the original PC Network Adapter.

Requirements

The LAN Support Program works with Personal System/2 and PC family computers with an appropriate network adapter. DOS 3.3 is required. Memory requirements depend on the type of adapter and the protocol to be used. When using a Token-Ring adapter, 9 KB is required when using the IEEE 802.2 protocol; NETBIOS and IEEE 802.2 use require 32 KB. When using a PC Network adapter, 33 KB is required when using the IEEE 802.2 protocol; NETBIOS and IEEE 802.2 use require 56 KB. The LAN Support Program Configuration Utility, which can be used to determine the adapters installed and to select the interface to be run, requires 256 KB.

LAN Support Program features are implemented using DOS device drivers. Separate drivers are provided for the common interrupt arbitrator, PC Network, Token-Ring, and NETBIOS support. The configuration utility provided stores the appropriate device drivers on the system and modifies the CONFIG.SYS file to specify that they be loaded each time the system is initialized. The configuration utility determines which drivers should be installed, based on the user's answers to questions about the type of network adapters installed, and whether or not NETBIOS support is required. Users who wish to customize the parameters used with the device drivers or who simply prefer that device drivers be stored in a subdirectory rather than the root directory of the system disk can load the device drivers manually, based on information provided in the User's Guide.

When using the protocol emulator IEEE 802.2 interface on a PC Network adapter, only one network adapter per machine is supported; otherwise up to two supported network adapters can be used in the same computer. Broadband PC Network adapters using the LAN Support Program can communicate only with broadband PC Network adapters that are also using the LAN Support Program.

Packaging and Documentation

The LAN Support Program is provided in a vinyl jacket. The software is provided on a 360 KB, 5.25-inch diskette and a 720 KB, 3.5-inch diskette. Also included is a User's Guide, which provides information on the LAN Support Program, including the PC Network Adapter IEEE 802.2 protocol emulation capability.

PC Network Protocol Driver

The PC Network Protocol Driver is a software product that can be installed on a PC or Personal System/2 to provide a NETBIOS programming interface for the PC Network Adapter II and II/A. It is compatible with programs that use the

NETBIOS interface, such as the IBM PC LAN Program. This driver uses a network protocol that is compatible with the one implemented in adapter ROM on the original PC Network Adapter. Systems using this driver can communicate with systems using the original PC Network Adapter. Systems using the protocol driver can operate on the same network as systems using the LAN Support Program, but the two groups of systems cannot communicate with each other.

The Protocol Driver is installed as a DOS device driver. DOS 3.2 or higher and 64 KB of memory are required. The driver software is provided on a 360 KB, 5.25-inch diskette and a 720 KB, 3.5-inch diskette, along with documentation and a license agreement. An installation utility is included with the device driver software.

LAN Manager

The Local Area Network (LAN) Manager Version 1.0 provides improved network management capabilities for the IBM Token-Ring network and adds network management capability for the broadband IBM PC Network. The LAN Manager operates as either a stand-alone DOS application or as a NetView/PC application. When used with NetView/PC, it can forward operator alerts to Netview running on a System/370 host computer. The LAN Manager can be used for error detection and problem determination for a broadband PC Network, or for single- or multi-ring Token-Ring networks. The LAN Manager provides the same functions as the Token-Ring Network Manager Version 1.1.

The LAN Manager detects and logs problems related to adapters and media; it notifies the operator when permanent and recoverable errors occur. LAN Manager can be configured to monitor the operation of critical stations on a PC Network, and alert the LAN Manager operator with an audible tone and a highlighted indicator if one of them fails to respond.

The network administrator interacts with the LAN Manager through a series of menus. Function keys, including one to request help information, are used to simplify operation. Names can be assigned to network adapters; adapters can be referred to either by address or by name.

Management of a multi-ring Token-Ring network requires the separate Token-Ring Network Bridge Program Version 1.1, to monitor rings remote from the LAN Manager station. Functions for managing a single ring are available for any ring attached to a bridge with which the LAN Manager is communicating. The LAN Manager can monitor the status of any adapter in the network, list active stations on a ring, and test the path between any two adapters.

A dedicated XT, AT, XT 286, or Personal System/2 Model 50, 60, or 80 with 640 KB of memory and a hard disk, running DOS 3.3, is required for use with the LAN Manager. For PC Networks, a PC Network Adapter II or II/A is required; any IBM Token-Ring adapter appropriate for the system is required for Token-Ring use. NetView/PC and NetView for host alerts is required if LAN Manager is used as a NetView/PC application. Stations on an IBM PC Network to be monitored must use the Local Area Network Support Program along with any broadband PC Network adapter, including the original PC Network Adapter. The monitored Token-Ring network may use any application interface or communication software.

The LAN Manager software is provided on six 360 KB, 5.25-inch diskettes and on four 720 KB, 3.5-inch diskettes. A User's Guide and online tutorial are provided along with a license agreement.

LAN Asynchronous Connection Server Program

The Local Area Network Asynchronous Connection Server Program provides asynchronous communications services for LAN-attached PC and Personal System/2 computers, and for terminal and computer devices attached to its system via asynchronous communications adapters. LAN-attached computers can communicate with host computers and other asynchronous devices via the server; asynchronously attached terminals and computers can communicate with host computers and other asynchronous devices and IBM 7171 protocol converters.

The LAN-attached computer or asynchronously attached terminal or computer must operate as the type of terminal or emulated terminal supported by its communications partner. In the case of 3270 data stream communications with the host computer via an IBM 7171, the 7171's communications partner must be an asynchronous terminal (or computer emulating a terminal) supported by the 7171.

The Asynchronous Connection Server can operate on a Token-Ring or broadband IBM PC Network. It supports up to thirty-two ports, at speeds of up to 19,200 bps. It supports auto dial, auto answer, and the Hayes AT modem command set. Multiple servers can operate on the same network along with the IBM Asynchronous Communications Server Program (which only supports two ports).

The Asynchronous Connection Server requires a dedicated AT, XT 286, or Personal System/2 Model 30, with two diskette drives or a hard disk. DOS 3.3 and at least 256 KB of memory is required. It supports the use of up to four IBM Realtime Interface Co-processor Multiport cards, with a total of up to eight ports per card. Up to two serial/parallel adapters can also be used in conjunction with the multiport cards; however, the total number of ports must not exceed the maximum for the computer on which the server is resident. For the AT the maximum is thirty-two, for the XT 286 the maximum is twenty-six, and for the Model 30 the maximum is nine communications ports.

A LAN adapter and appropriate support software is required. If the adapter used in the server is a PC Network Adapter II and it is attached to a PC Network using the original PC Network protocol, the Network Protocol Driver is required for use on the server. Otherwise, the LAN Support Program device drivers are used. One copy per server of the Realtime Control Program DOS Support Version 1.02 is required if the multiport communications adapter is used.

The Asynchronous Connection Server software is provided on a 360 KB, 5.25-inch diskette and a 720 KB, 3.5-inch diskette. Installation and configuration guides are provided, along with a license agreement. The protocol used by the server program is documented in IBM Personal Computer Seminar Proceedings Volume 3, Number 4 (October 15, 1985) (G320-9323), which is available at extra cost.

The Asynchronous Connection Server can operate on the same network as the LAN Manager, Token-Ring Bridge Program Version 1.1, and 3270 Emulation Program Version 3.0. It can also coexist with the older Token-Ring Network Manager, Token-Ring Bridge Program, and 3270 Emulation Program Version 2.0.

5250 Communications

Although many people think only of IBM System/370 computers when they think of IBM host computers, there are thousands of IBM System/36 and System/38 minicomputers in use today. These systems are used in small, medium, and large enterprises, and like practically every computer today, they are being used in conjunction with PCs.

Four different System/36 system units are available: the 5360, 5362, 5363, and 5364 (System/36 PC). The 5362, 5363, and 5364 system units are smaller in size and capacity than the 5360. They can fit beside or under a desk, whereas the 5360 is larger, more the size of a large desk. The 5364 supports up to sixteen locally attached workstations. The 5363 supports up to sixteen local and sixty-four remote workstations. The 5362 supports up to twenty-eight local workstations and sixty-four remote workstations. The 5360 supports up to seventy-two local workstations and sixty-four remote workstations.

The System/38 system unit is approximately the same size and shape as the 5360 system unit. It is available in seven models. All models support up to 256 locally attached workstations; remote workstations can be attached via up to twelve communications lines. With the number of workstations that these computers can support, it is no surprise that a need to connect personal computers to them exists.

System 36/38 Workstation Emulation Program

The System 36/38 Workstation Emulation Program Version 1.0 allows a Personal System/2 Model 50, 60, or 80 to act as a twin-axial, cable-attached workstation with a System/36 or System/38. Attachment of the Personal System/2 to the twin-axial cable is accomplished using the IBM System 36/38 Workstation Emulation Adapter/A. Both products were announced at the same time as the Personal

System/2 family, and are available separately or together as an Installation Convenience Kit.

The Workstation Program allows a Model 50, 60, or 80 to simultaneously emulate one or more 5250 displays and/or one or more 5250 printers. The Workstation Program provides enhanced features beyond those provided by the Enhanced 5250 Emulation Programs available for use on the Personal System/2 Models 25 and 30, and the IBM PC family. The Workstation Program supports any combination of System/3x printer, display, or host graphics sessions, as long as they do not exceed a total of four sessions. A DOS session is also supported; the display and keyboard can be switched among the host sessions and the DOS session using a hot key sequence.

Features

The Workstation Program allows a Model 50, 60, or 80 to emulate a 3196 or 5292-2 workstation, and also perform the functions of a stand-alone Personal System/2 computer. The Personal System/2 may be attached to the System/3x host directly, or remotely via an IBM 5294 Remote Control unit. Due to differences between the Personal System/2 display and that of the emulated terminal, some information may appear differently on the Personal System/2 display than it does on the 3196 and 5292-2 terminals. Magnetic stripe reader and selector pen features are not supported.

Access to all characters and most functions available to the emulated terminals is provided. A mapping of the emulated host terminal keyboard to the Personal System/2 101-key keyboard is provided. The user can modify this mapping to meet particular needs, using an interactive keyboard configuration program.

The Workstation Program allows a printer attached to the Personal System/2 to emulate the functions of a System/3x printer, allowing the System/3x print manager to direct output from host programs to the Personal System/2 printer. Emulated System/3x printers include models 5219, 5224/25, and 5256. Use of the System/36 5224/25 Advanced Printer Function Program with the emulated printers is not supported.

All emulation features are accessible from a single menu. From this menu, the user can start the emulator, start and stop host sessions, run configuration programs, install IBM-supplied updates to the emulator, and stop emulation. An advanced configuration option allows the user to modify the host EBCDIC (Extended Binary Coded Decimal Data Interchange Code) character to Personal System/2 ASCII (American Standard Code for Information Interchange) character translation tables, for both the display and the printer.

The Workstation Program provides graphics features beyond those provided by the Enhanced 5250 Emulation Programs available for use on the Personal System/2 Model 30 and IBM PC family. The 5292-2 terminal emulation feature provides a subset of 5292-2 functions for use with the System/36 and System/38 Business Graphics Utility (BGU) programs. Output from the BGU programs cannot be output on the Personal System/2 printer, but they can be printed on System/3x printers. Output can be plotted using either an IBM Model 6180 or 7372 plotter attached to a serial port on the Personal System/2.

When used with System/38 Release 8.0 Program Change C, the full-screen resolution of the Personal System/2's VGA video system can be used by BGU/38 applications. System/38 GDDM (Graphical Data Display Manager) image orders and Type II characters are supported for all display resolutions, and graphics performance is approximately equivalent to that of the emulated display. BGU/38 or user programs using Presentation Graphics Routines or GDDM/38 can use solid fill pattern on all Personal System/2 displays; BGU/38 provides Host Graphics Workstation support so that overlays such as graphics HELP will not erase the graphics screen contents.

The display of graphics data in a host session requires that a program provided with the Workstation Program be run in the DOS session at the same time. The execution of this program is selected from the Workstation Program menu. The DOS session can be made available for use by other DOS applications at any time by cancelling the graphics program.

The user should not hot key from a DOS application (other than the workstation graphics program) to a host session when the DOS application is using an 8514 display in any mode other than the Color Graphics Adapter (CGA) compatible modes. Such a switch can cause unpredictable results.

The Workstation Program provides several printer support features. It may be initialized to support either PC ASCII code page 437 (the default) or code page 850, at the user's request. Printer function tables are provided to support several IBM printers; these tables can be used to generate tables for non-IBM printers. The size of the printer initialization string has been increased from 32 to 64 characters. A printer function table is provided to support the use of the 3812 printer as a 5219 printer. Use of the bold print and set text orientation features is not supported. A printer function table is provided for the 4202 Proprinter XL.

Requirements

The Workstation Program can be used with the Model 50, 60, or 80 with a System 36/38 Workstation Emulation Adapter/A. One 3.5-inch diskette drive and at least 384 KB of memory is required. A 101-key Enhanced Keyboard and a Personal System/2 display are also required. A System 36/38 Workstation Emulation Attachment Cable is required for attaching the adapter to the System/3x twin-axial cable. It is available separately or as a part of the Installation Convenience Kit.

Printers supported include the IBM 3812-1 Page Printer, 4202-1 Proprinter, 4202 Proprinter XL, 5201-1 Quietwriter, 5201-2 Quietwriter, and the 5216-2 Wheelprinter. The 6180 and 7372 color plotters are also supported.

In order for the Workstation Program to be used with a System/36, the host must be running System/36 System Support Program Release 5.1 (5727-SS1 for a 5360 or 5362 system, or 5727-SS6 for a 5364 system). For use with a System/38, the host must be running System/38 Control Program Facility (5714-SS1) Release 8.0 with Program Change C. DOS 3.3 is required on the Personal System/2.

The Workstation Program is compatible with the following System/3x products and features:

PC Support/36 Release 5.1 (5360/5362)

PC Support/36 Release 5.1 Twinax Attach (5364)

PC Support/38

3278 Emulation via IBM Personal Computer (5360/5362)

3278 Emulation via IBM Personal Computer (5364)

Packaging

The System 36/38 Workstation Emulation Installation Convenience Kit/A Version 1.0 contains the equipment and software required for attachment of the Personal System/2 to a System/3x host. The emulation program is provided on two 720 KB, 3.5-inch diskettes along with a User's Guide. Also included in the kit are a System 36/38 Workstation Emulation Adapter/A and a cable for attaching the adapter to a twin-axial cable. A Reference manual, a Quick Reference and Diagnostics card, and a 3.5-inch diagnostics diskette is provided with the adapter. The Workstation Program, the Emulation Adapter, and the attachment cable can be ordered separately, if desired.

3270 Communications

Whereas many System/3x computers are used in small businesses, larger establishments use System/370 family host computers. These systems with their often huge networks of 3270 terminals are the workhorses of the data processing industry, processing and storing the information that many personal computer users wish to access. A 3270 series terminal, or a PC or Personal System/2 with hardware and software to emulate one, provides access to a complex and flexible communications system that provides interactive access to mainframe resources.

The 3278 and 3279 terminals consist of a large CRT display and terminal logic unit with a detachable keyboard. These terminals have been the standard monochrome and color terminals in the IBM mainframe environment for years. Models 3178, 3179, 3180, 3193, and 3194 are newer tilt-screen displays with PC-type keyboards. The 3178, 3180, 3191, and 3193 terminals are compatible with 3278 terminals, and the 3179 and 3194 terminals are compatible with 3279 terminals.

Connectivity

These terminals, though intelligent devices, do not communicate directly with the System/370 mainframe computer, except on small hosts that have a built-in terminal controller. More commonly, a large number (often hundreds) of terminals are connected to the host via terminal control units, which are connected to communications controllers, which in turn are connected to the host computer, as shown in Figure 14-1.

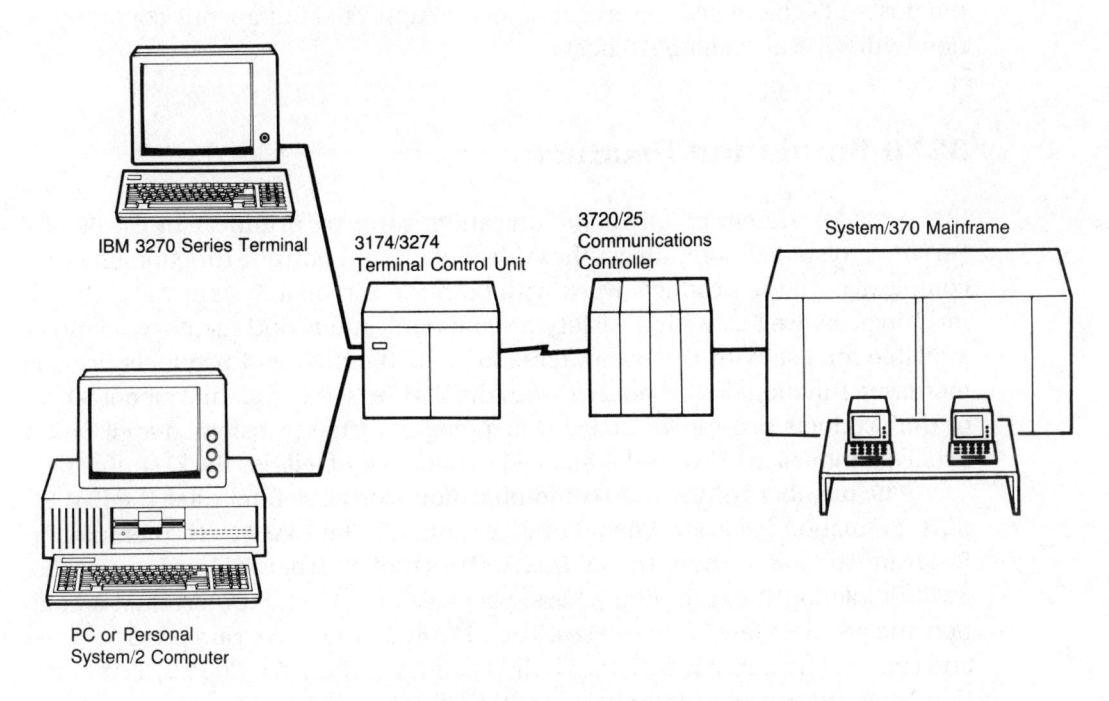

Figure 14-1 3270 Communications Subsystem. *3270 terminals are connected to the System/370 host via terminal control units.*

The 3270 series terminals are connected to a 3174 or 3274 (collectively referred to as 3x74) terminal control unit via coaxial cable, the IBM Cabling System, or some other appropriate wiring system. The 3x74 control units receive message blocks from the host communications controller and transmit them to the appropriate cable-connected terminal. Eight, sixteen, or thirty-two terminals and printers can be connected to the control unit, depending on the particular control unit submodel. Control units can be attached directly to a host system's I/O channel or indirectly via a communications controller that allows multiple control units to be attached to the same I/O channel.

Multiple 3174 and 3274 control units can be attached to 3720/25 communications controllers. Local 3274 control units can be connected to the System/370 I/O channel or to a 3720/25 communications port; remote units are connected to a communications port using modems and communications lines. Remote control units can communicate with the 3720/25 using binary synchronous communications (BSC) protocol or Synchronous Data Link Control (SDLC) protocol. The 3174 supports the same local and remote operation as the 3274, and can also communicate with a 3720/25 via an IBM Token-Ring network. The 3720 can support up to forty-eight communications lines and up to two Token-Ring networks. The 3725 can support up to 256 communications lines and up to eight Token-Ring networks. The 3720/25 are connected to the System/370 host(s) via the host's I/O channel(s). Several submodels support simultaneous communication with several System/370 hosts.

3270 Emulation Products

IBM provides a family of 3270 communications software products for the PC and Personal System/2 computers; these products vary both in functionality and complexity. All the products work with both the Personal System/2 and the PC machines, as well as with a variety of communications and memory adapters available for use with the computers. Some of the high end products provide increased functionality when used with the IBM Personal System/2 Model 80. All of the products provide keyboard remapping, host file transfer, an application program interface (API), and enhanced connectivity facilities (ECF) support.

The members of the 3270 communications software family are the IBM PC 3270 Emulation Program Entry Level Version 1.2, the IBM PC 3270 Emulation Program Version 3.0, and the IBM 3270 Workstation Program Version 1.1. The 3270 Emulation Program Entry Level provides direct connect terminal emulation and requires few PC resources. The 3270 Emulation Program provides local and remote terminal emulation as well as gateway functions. The 3270 Workstation Program provides multi-host, multi-DOS sessions on a PC or Personal System/2—functions previously only available from IBM when using the 3270 PC or 3270 AT with the 3270 PC Control Program.

The software packages support communications with the mainframe via several different types of connections. The Entry Emulation Program allows a PC or Personal System/2 machine to communicate with a 3x74 terminal controller in control unit terminal mode via a co-axial cable. The 3270 Emulation Program and Workstation Program are more versatile, allowing communications via a Token Ring network, an SDLC communications line, or a co-axial cable connection to a 3x74 terminal controller.

3270 Emulation Program Entry Level

The 3270 Emulation Program Entry Level Version 1.2 allows a PC or Personal System/2 to be connected to a 3174 or 3274 terminal control unit, using a 3278/79 Emulation Adapter (installed in a PC family computer or Personal System/2

Model 25 or 30) or a 3270 Connection board (installed in a Personal System/2 Model 50, 60, or 80). This software/adapter combination allows the computer to emulate an IBM 3278 Model 2A display terminal, or a 3279 Model 2A or S2A color display terminal operating in control unit terminal (CUT) mode. The 3174 or 3274 can communicate with the System/370 host via direct channel attachment, a Token-Ring network (3174 only), or via SDLC or binary synchronous communications to a 3720/25 communications controller.

Features

The Entry Emulation Program supports common 3270 terminal features except for the following:

    Alternate cursor

    APL/Text character set

    Attachment to port 0 of a 3174 or 3274 terminal controller

    Cursor blink

    Display of screens larger than 1,920 characters

    Explicit partitions

    Extended data stream and associated keys: programmed symbols, extended highlighting, and 7 colors

    Graphics escape

    Indent key

    Magnetic reader control and accessories

    Mono/dual case switch

    Katakana

    Keyboard click key

    Keyboards other than the 75-key U.S. English ASCII keyboard

    Numeric lock

    RPQs

    Security keylock

    Selector light pen

    Structured field and attribute processing

    Video output

    3270 diagnostic reset dump

    3274 response time monitor for this terminal

Chapter 14

The Entry Emulation Program remains resident in memory once loaded; the user may initiate a DOS task without terminating the Entry Emulation Program or the host session, provided sufficient conventional PC memory is available. Available memory is not a problem when using this program on a Personal System/2, because all Personal System/2 computers have 640 KB of conventional memory, and the Entry Emulation Program itself requires as little as 21 KB of memory. The user switches between the DOS application and the Entry Emulation Program using a "hot key" sequence.

Well-behaved PC programs can operate concurrently with the Entry Emulation Program. Programs that do any of the following cannot operate with the Entry Emulation Program without problems:

- Disabling interrupts, failing to issue an end-of-interrupt (IRET) for a hardware interrupt level, or masking selected interrupt levels for more than 100 milliseconds
- Overlay of DOS or BIOS system memory area
- Programming the 8259 interrupt controller
- Use of cassette interrupt 15H
- Use of interrupt vectors 50H through 57H

With the Entry Emulation Program loaded, up to 535 KB of system memory is available for DOS applications.

An application program interface called the Entry Emulation High Level Language Application Program Interface (EEHLLAPI) is available for software-controlled host communications. The EEHLLAPI provides a subset of the functions provided by the High Level Language Application Program Interface (HLLAPI) Version 3.1, available for use with the 3270 Workstation Program Version 3.1. The EEHLLAPI can be used to develop PC programs that interact with host sessions to automate repetitive tasks, mask complex applications from the user, and consolidate several tasks into one. EEHLLAPI.EXE is a terminate-and-stay-resident utility that is loaded in the PC session once the Entry Emulation Program has been loaded. EEHLLAPI functions can be accessed from IBM BASIC, COBOL, C, and Pascal programs using Language Interface Modules (LIMs) provided with the Entry Emulation Program. Occasional use host graphics support is provided via the Graphical Data Display Manager/PC Link (GDDM/PCLK) Program (described later in this chapter). Enhanced connectivity facilities and national language support are also provided.

IBM announced enhanced connectivity facilities for PC and System/370 computers (ECF) in June 1986. The goal of ECF is to provide improved resource sharing among PCs and System/370 computers and provide a uniform structure for accessing mainframe resources and exchanging data.

The facilities include a set of IBM-provided requester programs for IBM personal computers (PCs and Personal System/2s), and server programs for

Multiple Virtual Storage/Extended Architecture (MVS/XA) and Virtual Machine/ System Product (VM/SP) equipped mainframe hosts. Requesters are PC programs that request data and services from programs on the mainframe host that act as data and services servers. Mainframe server programs accept requests from PC requester programs and perform the file access and processing necessary to provide the requested data or service to the PC.

ECF also includes a consistent interface for communications between requesters and servers. Host systems and 3270 emulation programs perform router functions. Routers provide a Server-Requester Programming Interface (SRPI), which acts as a request interface for requesters or a reply interface for servers.

In addition to file transfer and host database access, the requester/server functions execute host commands and procedures, and provide virtual disk, file, and print services. VM/SP Release 5 or Time Sharing Option Extensions (TSO/E) Release 3 are required on the host System/370, to provide ECF support.

Version 1.2 of the Entry Emulation Program was introduced at the same time as the Personal System/2 family, in April 1987. Prior to the introduction of Version 1.2, ECF facilities were only available to PCs and 3270 PCs running the 3270 Emulation Program Version 3 (see next section) and the 3270 PC Control Program Version 3, respectively.

Users interact with the Entry Emulation Program in U.S. English; however, sixteen different national language keyboards are supported for data entry, display, and file transfer. A parameter may be set in the file transfer program send/receive command to specify the appropriate code page to be used at the host for EBCDIC/ASCII translation. The Entry Emulation Program Version 1.2 supports DOS 3.3 alternate code pages.

Version 1.1 of the Entry Emulation Program (which was introduced before the Personal System/2 family) works with the Personal System/2 Models 25 and 30 without modification, and can also work with the 3270 Connection Micro Channel Adapter used with a program patch on the Personal System/2 Models 50, 60, and 80. This patch file (EE05952) is contained on the diagnostic diskette provided with the 3270 Connection Adapter. The patch is installed by copying the patch file to the Emulation Program system diskette and following patch installation procedures contained in Chapter 7 of the 3270 Emulation Program Entry Level Version 1.1 User's Guide.

Packaging and Documentation

The Entry Emulation is provided in a standard IBM slip-case package. The package includes the software, contained on a single 720 KB, 3.5-inch diskette and on a 360 KB, 5.25-inch diskette, a User's Guide in a three-ring binder, and a Programmer's Guide for the Entry Emulator High Level Language Application Program Interface. Software provided includes the emulation program and file transfer, and ECF programs and utility programs for the keyboard layout redefinition and host session screen colors. A patch utility for the installation of software changes is also provided. Three keyboard templates (one each for the PC XT, AT, and 101-key Enhanced Keyboard) and quick reference cards for the emulation program and EEHLLAPI are included as well.

Installation and Use

The Emulation Program Entry Level Version 1.2 works on a personal computer with at least 128 KB of main memory. Version 3.2 or later of DOS is required; version 3.3 or later is required for use on a Personal System/2. The program itself requires approximately 21 KB of memory; an additional 12 KB is required for the Entry Level High Level Language Application Program Interface, and 21 KB is required for ECF facilities. The remaining system memory is available for concurrent operation of DOS and application programs. DOS 3.3 typically requires 60 KB to 70 KB of memory. A mouse cannot be used concurrently with the Entry Emulation program.

A 3278/79 Emulation Adapter (for the PC or Personal System/2 Models 25 or 30) or a 3270 Connection board (for Personal System/2 Models 50, 60, and 80 computers) and appropriate cabling for connecting the adapter to the 3x74 terminal control unit is also required. Cabling may be either coaxial cable, IBM Cabling System media, or telephone twisted-pair wire. If twisted-pair wire is used, a ROLM 3270 Coax-to-Twisted Pair Adapter (3270-CTPA) or equivalent is required for connection of the adapter to the terminal control unit.

For file transfer to an MVS/TSO or VM/SP system, Release 1 of the IBM 3270 PC File Transfer Program must be installed on the System/370 host mainframe. The CICS/VS 3270 PC File Transfer Program Release 1 must be installed on the System/370 host if transfers are to be made to a CICS/VS (MVS or VSE) system. For file transfer when connected to a 3174 terminal control unit, the file transfer aid must be selected when the 3174 is customized. VM/SP Release 5 or Time Sharing Option Extensions (TSO/E) Release 3 is required on the host System/370 to provide ECF support.

The Entry Emulation Program can be run concurrently with the redirector, receiver, and messenger functions of the IBM PC Local Area Network Program. When used with the LAN Program, the Entry Emulation Program must be loaded after the LAN Program is loaded, and it must be run in RESUME (Alt-R) mode. If a LAN message notification occurs during a host session, any keystroke returns the host screen to the display. The Entry Emulation Program cannot be used on a workstation with a PC Network Baseband Adapter.

The Entry Emulation Program is installed using a batch file named INSTALL, provided on the software diskette. Installation on a diskette system consists of creating a system disk using the DOS FORMAT command, making A the default drive, and then entering the command INSTALL A: B: XX, with the Entry Emulation diskette in drive A and the freshly formatted diskette in drive B. XX in the command is the two-letter identifier for the desired national language keyboard. The INSTALL batch file then copies files, appropriate for the keyboard selected, to the target diskette. This process can be performed on a single or dual diskette drive system, but if performed on a single diskette drive system, the user must insert and remove the source and target diskettes thirty times, once for each file copied.

The User's Guide recommends installing the software on a hard disk by making A the default drive and entering the command INSTALL A: C: XX, where XX is defined as above. This procedure can result in a very messy installation, because the Entry Emulation Program files will be copied to the current default directory on drive C, which more likely than not will be the root directory. A much cleaner installation can be accomplished by creating a directory on the hard disk, making it the default directory, and then executing A: INSTALL A: C: XX from that directory. Using this procedure, all the Entry Emulation files are copied to a single subdirectory, rather than cluttering up the root directory of the hard disk. The directory name should be included in a PATH statement (usually found in the AUTOEXEC.BAT file) to allow the Entry Emulation Program and its utilities to be executed without specifying a directory name on the command line each time one of the programs is executed.

The User's Guide recommends that the user rename the AUTOEXEC.P78 batch file, contained on the software diskette, to AUTOEXEC.BAT so that the Entry Emulation Program is started up each time the system is initialized. Once started, the Entry Emulation Program by default displays an IBM logo screen and requests that the user press any key to continue. Once loaded into memory, it remains resident until the quit function is entered or the system is reinitialized.

The supplied batch file can be used or its contents can be appended to the end of the user's existing AUTOEXEC.BAT file. The file contains the following commands:

```
PC3270
ECHO Press ALT/ESC to switch between host and PC sessions
```

The first command in the batch file loads the Entry Emulation Program. Once the user presses a key to acknowledge the IBM logo screen, the batch file displays a message indicating that the user can switch between the host and PC sessions by pressing Esc while holding down the Alt key. Three other Alt key sequences are provided for use from the host session. They are Alt-S, Alt-R, and Alt-5. Alt-S and Alt-R suspend and resume operations in the PC session, respectively. Users may find it necessary to use the Alt-S key sequence to suspend the active PC application if it writes directly to the video display buffer, rather than using BIOS display writes. Information written to the video display buffer (and thus to the screen), while the host session is displayed, results in a display of overlapped information from both sessions.

If the PC session performs a BIOS write to the screen, the host session screen is stored and the PC session screen is displayed. The user may ignore the information displayed by the PC session, by pressing any key to return the host screen to the display (the keystroke will be sent to the host session). Alternately, the user may press Alt-Esc to switch to the PC session, to perform any action necessary to respond to the information displayed.

The quit function, Alt-5 (using the 5 key located on the numeric keypad), terminates the Entry Emulation Program and returns the storage it was using to DOS. Terminating the Entry Emulation Program also terminates the EEHLLAPI (Entry Emulation High Level Language Application Program Interface) and EESRPI (ECF facilities) programs, if they are in use. The other way to terminate the Entry Emulation Program without turning off the computer is to press Ctrl-Alt-Del while in the PC session. This reinitializes the system, terminating the DOS as well as the host session.

A different batch file (AUTOEXEC.ECF) is provided for use by those who wish to use ECF facilities. This file executes the Entry Emulation Program with the following command:

```
PC3270 E R
```

The E option specifies that the ECF support is to be loaded and the R option specifies that the PC session is to be started in the resume mode, allowing it to execute while the host session is active. The other option that can be specified on the command line is the N option, which specifies that the IBM logo screen not be displayed when the emulator is started. The options may be entered in any order.

Once the emulator has been started, the command EESRPI should be entered to the DOS session, to load the Entry Emulation Server-Requester Programming Interface. To use ECF facilities, the user must start the host router and then the PC requester. After logging onto the host, the appropriate VM/CMS or MVS/TSO command is entered to start the router. Once the host router is active, the user must switch to the PC session to start the requester program. Requesters are provided separately or developed by the user; they do not come with the Entry Emulation Program. The host router can be stopped by pressing PF3 when in the host session.

Individual host or PC session screens can be printed on the PC printer using the PrtSc command. Line by line printing can be initiated by the Ctrl-PrtSc key sequence. A host session screen can be printed on the 3x74 terminal control unit printer by pressing 7 while holding down the Alt key. The Entry Emulation Program does not support a host print session.

File transfer is accomplished using the SEND and RECEIVE programs provided on the Entry Emulation software diskette. Before initiating any file transfer operation, the user should use the host session to direct the mainframe system to send no unsolicited messages (which will interrupt the file transfer unless acknowledged). The user can then switch to the DOS session and enter the SEND or RECEIVE command, with the host and PC filename of the file to be transferred, and any file translation options. File transfer cannot be performed while ECF facilities are in use.

3270 Emulation Program Version 3.0

The PC 3270 Emulation Program Version 3.0 provides more flexible communications with the System/370 host than the Entry Level Program, with a corresponding increase in required PC resources. The Entry Emulation Program allows a PC or Personal System/2 computer connected to a 3174/3274 terminal controller to emulate a 3270 terminal operating in control unit terminal (CUT) mode. The 3270 Emulation Program allows a PC connected to a terminal controller to operate in the more flexible distributed function terminal (DFT) mode, allowing the emulation of a 3278 Model 2A display terminal, or a 3279 Model 2A or S2A color display terminal and/or an IBM 3287 Model 1 or 2 printer. Print output from host applications can be routed to the PC printer or be stored on the PC's disk as an ASCII file. Occasional use host graphics support is provided via Graphical Data Display Manager/PC Link (GDDM/PCLK).

The PC running the Emulation Program can be connected to a 3174 or 3274 terminal control unit via coaxial cable, the IBM Cabling System, or another appropriate wiring system. It can be connected to a System/370 channel-attached 3174 via an IBM Token-Ring network.

Features

The Emulation Program supports common 3270 terminal features, with the same exceptions as the Entry Emulation Program. The user can specify colors to be displayed on a PC color display for host formatted fields. A choice of 16 foreground and 8 background colors is provided. The color of the cursor can be selected from a choice of 8 colors. Colors are selected using a series of menus that allow the user to view the available colors. The following 3287 printer features are not supported:

- Audible alarm
- Control unit signal light
- Index switch
- Mono/dual case switch/light
- Programmed symbols
- Reset switch
- Setup switch
- Test switch/light
- Underscore
- X print error indication

In addition to 3270 terminal emulation, the Emulation Program allows a PC to emulate a 3274 Model 51C control unit with attached terminals and printers. When operating in this mode, the personal computer running the Emulation Program communicates with a 3720/25 communications controller via an SDLC or binary synchronous communications line, or an IBM Token-Ring network.

The Emulation Program emulates common 3274 control unit functions, with the following exceptions:

Configuration C features

Attachment of local or remote R-loop

Decompression of PS load data

Explicit partitions

Extended color

Extended highlighting

Programmed symbols

Structured field and attribute processing

X.21 communications

Entry assist

Network Management Vector Transport (NMVT) alerts

3270 extended data stream (except for file transfer)

The Emulation Program provides enhanced connectivity facilities (ECF) for improved resource sharing and data exchange among IBM personal computers and System/370 host mainframes. The Server-Requester Programming Interface (SRPI) is included with the Emulation Program, to support the development of interconnected host and PC applications. The capability offered through the SRPI provides a subset of IBM's advanced program-to-program communications (APPC) architecture features. This interface provides program-to-program communications over SNA (using LU 2 protocols) and non-SNA connections. It provides PC programmers with a consistent means of requesting and receiving services from mainframe MVS/XA TSO/E and VM/SP host systems.

The interface supports PC programs written in high level languages and/or assembly language. Programs may be written in IBM C/2, Pascal/2, or Macro Assembler/2. Programmers writing such applications need only be concerned with the functions to be performed, and not the communications environment connecting the PC and the host mainframe. Use of the SRPI option in the Emulation Program also requires the use of the alternate task feature (see below), which is mutually exclusive with the operation of IBM TopView 1.1.

Whether communicating with a terminal control unit or a communications controller, the Emulation Program can be used alone or with a network. In stand-alone use, a single 3270 terminal and, optionally, a 3287 printer are emulated. When used with a network, the Emulation Program can act as a gateway to a mainframe host for other PCs on a Token-Ring network or IBM PC Network. The PCs on the network communicate with the host System/370 as 3270 terminals and/or 3287 printers, and the gateway PC performs 3274 terminal control unit functions.

Stand-Alone Use

The Emulation Program, when used in local stand-alone mode, allows a personal computer to be connected to a 3174 or 3274 terminal control unit, using a 3278/79 Emulation Adapter (installed in a PC family computer or Personal System/2 Model 25 or 30) or a 3270 Connection board (installed in a Personal System/2 Model 50, 60, or 80). When used in stand-alone mode, this software/adapter combination allows the computer to emulate an IBM 3278 Model 2A display terminal, or a 3279 Model 2A or S2A color display terminal and/or a 3287 printer. The control unit to which the PC is connected must have the DFT feature installed in it if more than one host session is used. The 3174 or 3274 can communicate with the System/370 host via direct channel attachment, a Token-Ring network (3174 only), or via SDLC or binary synchronous communications to a 3720/25 communications controller.

When used in remote stand-alone mode, the Emulation Program supports the connection of a personal computer to a 3720/25 communications controller using an SDLC Adapter (for the PC or Personal System/2 Models 25 or 30) or a Multi-Protocol Adapter (for Micro Channel equipped Personal System/2 computers) communicating at rates of up to 19,200 bits per second. This configuration emulates a remote 3274 control unit with a 3278/3279 terminal and a 3287 printer attached. When operating in this mode, the personal computer can communicate with the communications controller using SDLC or binary synchronous communications protocols. Because of the strict timing requirements of binary synchronous communications, no other activities should take place on the personal computer during a file transfer.

Once loaded, the Emulation Program remains resident in memory; the user may initiate a DOS task using the alternate task feature without terminating the Emulation Program or the host session, provided sufficient conventional PC memory is available. The user switches between the DOS application and the Emulation Program using a "hot key" sequence. DOS Version 3.1 or later should be used with the Emulation Program. Version 3.2 or later is required if the Token-Ring network is used; Version 3.3 is required for use with a Personal System/2.

While the Entry Emulation Program requires as little as 21 KB of memory, the Emulation Program requires considerably more. When used for connection to a terminal controller in stand-alone mode, the Emulation Program requires 166 KB of memory (in addition to the 60 KB to 70 KB of memory required for DOS). When connected to a 3720/25 communications controller and operating in stand-alone mode, the Emulation Program requires 161 KB; if the PC is con-

Chapter 14

nected to the controller via a Token-Ring network, an additional 7 KB is required for the IBM Token-Ring Adapter Handler Program.

Additional memory is required for each of the following features as follows:

Alternate task (DOS application) support	11 KB
File transfer support	18 KB
General API support	8 KB
SRPI	30 KB
3270 keyboard remap feature	2 KB
3287 printer support	11 KB

If the alternate task feature is used, at least 25 KB of memory must be available.

Network Operation

When used in network mode, the Emulation Program serves as a host mainframe gateway for other PCs on an IBM Token-Ring network or IBM PC Network. The gateway PC can be attached to a 3174 or 3274 control unit, or it can communicate with a 3720/25 communications controller via an SDLC communications line or an IBM Token-Ring network.

When connected to a 3174 or 3274 terminal control unit (via a 3278/79 or 3270 connection adapter), the Emulation Program can act as a local gateway operating in DFT mode, to provide five host sessions for use by itself and the personal computers on the network. Host gateway capability is not supported when connected to a 3174 or 3274 communicating with the mainframe using binary synchronous communications. Each personal computer on the network (including the gateway) can access one host display session and one host printer session, up to the five session maximum for the gateway.

When used as a DFT connected gateway, the Emulation Program requires 200 KB of memory (plus an additional 28 KB if a session is assigned to the gateway). An additional 53 KB of memory is required for the NETBIOS and Adapter Handler Programs, if the IBM Token-Ring network is used.

When communicating with a 3720/25 communications controller via a communications line or an IBM Token-Ring network, the Emulation Program acts as a remote network gateway for up to thirty-two simultaneous host sessions. Each PC on the network (including the gateway) can access one host display session and one host printer session, up to the thirty-two session maximum for the gateway.

The gateway communicates with the 3720/25 communications controller using an SDLC Adapter (for the PC or Personal System/2 Models 25 or 30) or a Multi-Protocol Adapter (for Micro Channel equipped Personal System/2 computers) communicating at rates of up to 19,200 bps.

When acting as a remote gateway, the Emulation Program requires 195 KB of memory, plus 28 KB if a terminal session is assigned to the gateway, and 11 KB if 3287 printer emulation support is used. An additional 53 KB of memory is required for the NETBIOS and Adapter Handler Programs, if the IBM Token-Ring network is used.

Each PC on the network must run the Emulation Program in order to communicate with the System/370 host via the gateway. The Emulation Program can be run concurrently with the redirector, receiver, messenger, and server functions of the IBM PC Local Area Network Program. When running on a network station, the Emulation Program requires 156 KB, plus 11 KB if a 3287 printer is emulated, and 53 KB (as noted above) if the IBM Token-Ring network is used.

The Emulation Program can act as a gateway for a PC Network even if it is connected to the 3720/25 communications controller via a Token-Ring network. In this configuration, the gateway must contain both a Token-Ring adapter and a PC Network adapter.

The Emulation Program uses PC interrupt level 2 for the 3278/79 Emulation Adapter, interrupt levels 3 and 4 for the SDLC Adapter, and interrupt level 2 or 3 for the PC Network Adapter II or Token-Ring Network II Adapter on PCs and Personal System/2 Models 25 or 30. On the Personal System/2 Models 50, 60, and 80 it uses the corresponding interrupt levels with the 3270 Connection, Multi-Protocol Adapter/A, PC Network Adapter II/A, and Token-Ring Network Adapter/A. Use of these interrupts by other system devices (such as serial ports) can cause interrupt conflicts and unpredictable results.

Configurations that use the SDLC or Multi-Protocol adapters can include only one network adapter, and its interrupt level must be set to 2 (the factory default). Configurations that use a 3278/79 Adapter or 3270 Connection can include only one network adapter, and its interrupt level must be set to 3. If a network station running the Emulation Program contains more than one network adapter, the program will use the functioning network card contained in the lowest numbered expansion slot.

The Emulation Program is provided on a 3.5-inch diskette and two dual-sided, 5.25-inch diskettes. Keyboard reference cards are provided as well as a User's Guide, a System Planner's and User's Reference, and a Quick Reference card.

Installation and Use

The Emulation Program is installed using the SETUP program provided. This program copies the appropriate files onto a diskette or user-specified directory on a hard disk; the directory must be created prior to running SETUP.

Once installed, the program is initiated using the PSC batch file. This file executes the Emulation Program, which results in the display of an IBM logo screen. Pressing any key causes a menu to be displayed, from which the user can initiate communications or create a communications profile describing the method of communications desired. (See Figure 14-2.) Profiles are created by making menu selections. Items to be selected include configuration for stand-alone, network, gateway, or gateway with network station.

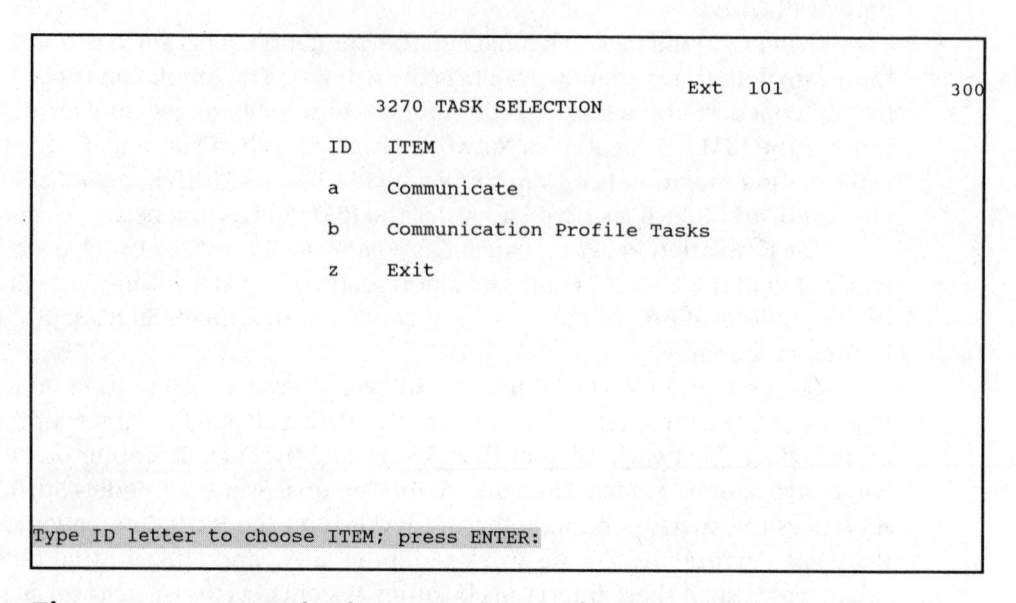

```
                                                  Ext  101          300
                    3270 TASK SELECTION
               ID   ITEM
               a    Communicate
               b    Communication Profile Tasks
               z    Exit




Type ID letter to choose ITEM; press ENTER:
```

Figure 14-2   3270 Task Selection Menu. *From this menu, users can initiate 3270 communications or create or revise the 3270 communications profile.*

Once the configuration has been selected, parameters appropriate for the configuration can be changed using additional menus. These include communications method, modem and line profiles, and the number of gateway sessions, if gateway mode is used, and whether a DOS session is to be provided. All menus show allowable values for items, and the default values. Figure 14-3 shows the Communication Profile Tasks menu.

When the user exits the Communications Profile menu, any changes made are saved, and the user is instructed to restart the Emulation Program. Once this is done and communication is initiated from the main menu, operation is similar to that of the Entry Level Emulator Program.

GDDM-PCLK

GDDM-PCLK Version 1.1 is a personal computer program that allows a PC or Personal System/2 computer to act as a System/370 Graphical Data Display Manager workstation. Use of GDDM-PCLK requires an IBM 3270 terminal emulation program on the PC and the GDDM PC Link feature (GDDM PCLKF) on the host

```
Chg Profile
                                                        Ext   101                300
                        COMMUNICATION PROFILE TASKS

                                   YOUR         POSSIBLE
     ID    ITEM                    CHOICE       CHOICES

     a     Configuration           1            1 = Standalone Station
                                                2 = Network Station
                                                3 = Gateway
                                                4 = Gateway with
                                                    Network Station
     b     Communication Attachment 2           1 = SDLC        2 = DFT
                                                3 = IBM Token-Ring Network
     c     Alternate Tasks         1            1 = Yes         2 = No
     d     3270 Keyboard           101          1 - 999
     e     3270 Keyboard Remap     2            1 = Yes         2 = No
     f     Create or Revise Communication Setup
     g     Create or Revise Modem and Line Description
     h     Create or Revise Gateway Setup

     z     Return to Task Selection

Type ID letter to choose ITEM; press ENTER:
```

Figure 14-3 Communication Profile Tasks Menu. *This menu is used to select the configuration to be used for 3270 communications.*

System/370. GDDM-PCLK can be used with host applications using GDDM Version 2.2 running in an MVS (MVS or MVX/XA TSO ACF/VTAM), VM (VM/SP and VM/XA SF CMS), or VSE environment.

When connected to a GDDM host application, GDDM-PCLK supports the display, print, and plot of graphics information. The program provides no standalone graphics editing or data creation facilities; printing and plotting of graphics at the PC can be immediate or deferred.

User interaction with the host GDDM application takes place through the terminal emulator host session; actual display of graphics results is performed in a DOS session. GDDM PCLK converts the presentation space graphics vectors received from the host to graphics information that can be displayed, printed, or plotted on the PC. When GDDM PCLK is used with the IBM PC 3270 Emulation Program Entry Level or the IBM PC 3270 Emulation Program, it is not possible to run other DOS applications. When used with the 3270 Workstation Program, it's use is limited to a single DOS session at one time. Each such session requires one DOS session and one host session.

GDDM user control functions can be used when running a GDDM application. When using the 3270 Emulation Program Entry Level or 3270 Emulation Program, however, the user must hot key to the DOS session to see the graphics results. Use of the hot key is not necessary when using the 3270 Workstation Program Version 1.1. GDDM-PCLK supports a graphics cursor and can use any mouse normally used for PC applications. Graphics from the cursor keys or a mouse can be sent through the 3270 terminal emulator to the host GDDM application. The graphics cursor support does not include string and stroke input.

GDDM PCLK is provided on one 5.25-inch and one 3.5-inch diskette. A GDDM-PCLK guide is provided with the software. Upon installation on the personal computer, during its first use with a GDDM host application, GDDM PCLK will download from the System/370 host any updates applicable to the PC on which it is installed. It will likewise transfer any required updates when GDDM PCLKF libraries on the System/370 host are updated.

GDDM PCLK runs on a Personal System/2, AT, or PC (including the 3270 PC) with at least 512 KB of memory, using DOS 2.1 or later and the 3270 Emulation Program Entry Level Version 1.1 or 1.2, 3270 Emulation Program Version 3, or 3270 Workstation Program Version 1.1. The host system to be accessed must be running GDDM Version 2.2 PCLKF.

Display adapters and video modes supported on the Personal System/2 Models 50, 60, and 80 include VGA (640 by 480 pixels, 16 colors) and 8514/A (640 by 480 pixels, 16 colors and 1024 by 768 pixels, 16 colors). Display adapters and modes supported include the CGA (640 by 200 pixels, 2 colors), 64 KB EGA (640 by 200 pixels, 16 colors), 128 KB EGA (640 by 350 pixels, 16 colors), and Personal System/2 Model 30 MCGA (640 by 480 pixels, 2 colors). Use of the Personal System/2 Display Adapter (640 by 480 pixels, 16 colors) is supported when used on a Personal System/2 Model 30 or PC family computer.

Printers supported include the IBM 3852 Color Jetprinter, 4201, 4201-2, 4207, and 4208 Proprinters, 5152 Graphics Printer, 5182 Color Printer, and 5201 and 5202 Quietwriters. Plotters supported include the IBM 6180, 6184, and 6186 color plotters. These plotters are supported only when attached to the system via an RS232 serial port. A plotter cannot be used on an SDLC-attached system, because of hardware interrupt conflicts.

Response times for the display of alphanumeric data are comparable to those observed when running the terminal emulation without GDDM-PCLK. GDDM-PCLK is recommended for occasional use graphics applications, because graphics response times will be slower than for 3270 graphics displays, particularly for graphics with filled areas. GDDM-PCLK is mainly useful for the output of previously generated graphics.

3270 Workstation Program

Whereas the previously described emulation programs allow one DOS session and at most two host sessions per personal computer, the 3270 Workstation Program supports multiple DOS and multiple host sessions on a single PC or Personal System/2 computer. This program can be used on an XT, AT, and all Personal System/2 models, as well as on a 3270 PC and 3270 AT. Previously, this multi-DOS session, multi-host session capability was only available from IBM using the 3270 PC Control Program, which was available for use only on 3270 PCs and 3270 ATs. The Workstation Program replaces the 3270 PC Control Program, which is no longer available.

Connectivity

The Workstation Program provides up to four host sessions, six DOS sessions, and two notepad sessions, all of which can be active at the same time. (See Figure 14-4.) The Workstation Program uses up to 2.25 MB of system memory, using the IBM Personal System/2 80286 Expanded Memory Adapter/A, the 2 MB Expanded Memory Adapter, or the 3270 PC Expanded Memory Adapter. Lotus Intel Microsoft Expanded Memory Specification (LIM EMS) device drivers are provided that support one EMS and virtual disk session, or up to six virtual disk sessions. LIM EMS 3.2 and a subset of LIM EMS 4.0 functions 1 through 15 are provided.

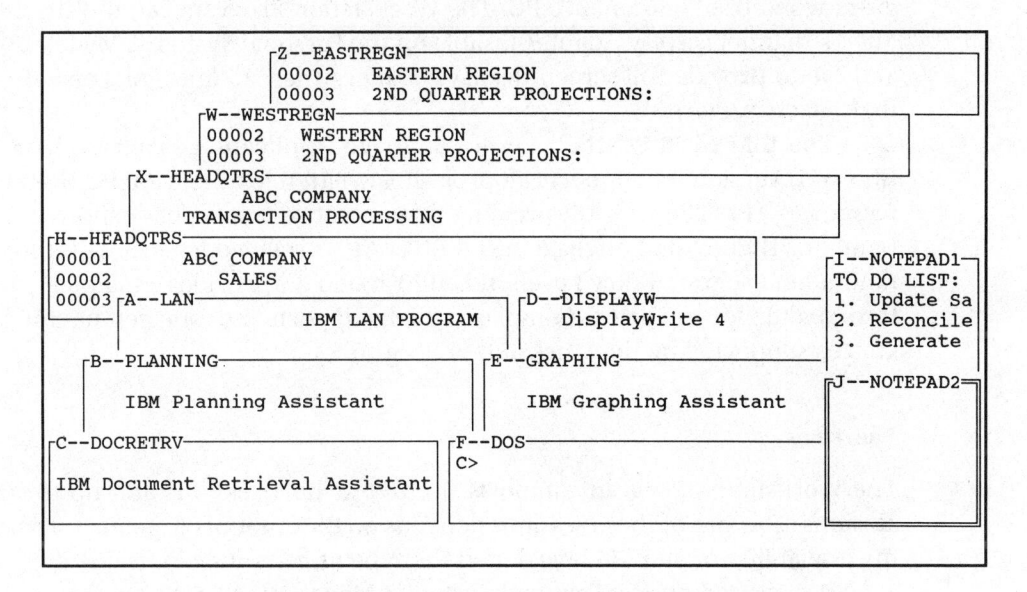

```
┌─────────────────────────────────────────────────────────────────────────┐
│               ┌Z--EASTREGN──────────────────────┐                       │
│               │00002    EASTERN REGION          │                       │
│               │00003    2ND QUARTER PROJECTIONS:│                       │
│           ┌W--WESTREGN──────────────────────┐                           │
│           │00002    WESTERN REGION          │                           │
│           │00003    2ND QUARTER PROJECTIONS:│                           │
│       ┌X--HEADQTRS───────────────────┐                                  │
│       │      ABC COMPANY             │                                  │
│       │      TRANSACTION PROCESSING  │                                  │
│ ┌H--HEADQTRS───────────────┐                                            │
│ │00001      ABC COMPANY    │                                            │
│ │00002      SALES          │                          ┌I--NOTEPAD1──┐   │
│ │00003 ┌A--LAN────────────────────┐ ┌D--DISPLAYW─────┐│TO DO LIST:  │   │
│ │      │      IBM LAN PROGRAM     │ │  DisplayWrite 4││1. Update Sa │   │
│ │      │                          │ │                ││2. Reconcile │   │
│ │      │                          │ │                ││3. Generate  │   │
│    ┌B--PLANNING───────────────┐  ┌E--GRAPHING─────────┐                 │
│    │   IBM Planning Assistant │  │ IBM Graphing Assistant│ ┌J--NOTEPAD2┐│
│ ┌C--DOCRETRV──────────────────┐ ┌F--DOS──┐                              │
│ │ IBM Document Retrieval Assistant│ │C>   │                             │
│ │                              │ │        │                             │
│                                                                         │
└─────────────────────────────────────────────────────────────────────────┘
```

Figure 14-4   3270 Workstation Display. *Up to four host sessions, six DOS sessions, and two notepad sessions can be displayed.*

When used on the Personal System/2 Model 80, the Workstation Program uses the Model 80's 80386 paging hardware and the system's extended memory to emulate up to 2 MB of expanded memory. When used on the Model 80, it uses the 80386's I/O control capabilities to provide a logical display buffer for DOS applications. This allows the concurrent execution of ill-behaved DOS applications that write directly to the screen buffer.

The Workstation Program supports the connection of a PC or Personal System/2 to a 3x74 terminal control unit in CUT or DFT mode, but CUT mode allows only one host session. Connection to the IBM 9370 workstation subsystem controller in CUT or DFT mode is also supported. The Workstation Program can be run concurrently with the redirector, receiver, messenger, and server functions of the IBM PC Local Area Network Program being used with either the IBM Token-Ring network or IBM PC Network. Support is provided for communications with a host

System/370 via an IBM Token-Ring network attached 3725 communications controller or 3174 terminal control unit. The LAN Support Program is required for Token-Ring communications with a System/370 host.

Occasional use graphics is provided without requiring special host graphics hardware on the personal computer. This support is provided using GDDM-PCLK on the personal computer, to convert host vector graphics to PC raster-scan graphics. GDDM-PCLKF is available at extra cost and requires the availability of the GDDM-PCLKF software on the System/370 host computer. See *GDDM-PCLK*, this chapter, for additional information. The graphics capabilities provided by the 3270 Control Program are also provided with the Workstation Program when used with a 3270 PC. The Workstation Program can use the 8514/A Micro Channel Display Adapter (usable on the Personal System/2 Models 50, 60, and 80) to provide full-screen displays of more than 25 lines, or lines of more than 80 characters.

The IBM 3270 PC High Level Language Application Program Interface (HLLAPI) Version 3.1 supports all programs written for the 3270 PC High Level Language API. The HLLAPI provides a superset of the functions in the Entry Emulator High Level Language API (EEHLLAPI) provided for use with the 3270 Emulation Program Entry Level. HLLAPI Version 3.1 provides a Language Interface Module for the IBM C Compiler, which supports the fourteen national languages supported by the Workstation Program.

Features

The Workstation Program supports up to four host sessions and up to six PC sessions. The size of these sessions depends on the amount of memory available, the availability of an EMS board, and the type of computer in use.

When used on the Personal System/2 Model 30, 3270 PC, and supported PCs, up to 2.25 MB of memory is available when using the 2 MB Expanded Memory Adapter or 3270 PC Expanded Memory Adapter. When using these adapters, one DOS application can be up to 592 KB in size and the other five can be up to 384 KB in size. Use of expanded memory requires that system board memory above 256 KB be removed or disabled. The Workstation Program stores DOS applications in expanded memory to support their time-shared use of the CPU. Because each application is always resident in (expanded) memory, control of the CPU can be rapidly switched among them.

Using the expanded memory drivers provided with the Workstation Program, expanded memory can be made available to DOS applications. The amount of memory used by applications decreases the amount of expanded memory available to contain active DOS sessions. Thus applications requiring large amounts of expanded memory can limit the number and size of DOS sessions.

When used with the Personal System/2 Models 50 and 60, a maximum of 2 MB of memory is available when using the Personal System/2 80286 Expanded

Memory Adapter/A. (This board provides 2 MB of memory, but when it is used, the 1 MB of memory on the system board is disabled.) Up to six concurrent DOS sessions are possible. Each can be up to 512 KB minus 4 KB times the number of DOS sessions in size, limited by available memory.

When the Workstation Program is used on the Model 80 with more than 2 MB of memory, using XMA emulation, all six DOS sessions may be up to 512 KB minus 4 KB times the number of DOS sessions in size. On Models 50 and 60, all six DOS sessions cannot be this large, because for six sessions each session can be up to 488 KB, and six times 488 KB is greater than 2 MB.

The maximum size for DOS sessions is reduced by the amount of memory required for network and other device drivers. If the LIM EMS driver is used, the size of all DOS sessions is reduced by 64 KB. The effect of this is not severe for DOS applications that use expanded memory, because while they lose access to 64 KB of conventional memory, they gain access to a larger amount of expanded memory.

Packaging and Documentation

The Workstation Program Version 1.1 is provided in an IBM slip-case package, which includes the software and documentation. The software is contained on four 720 KB, 3.5-inch diskettes and five 360 KB, 5.25-inch diskettes. Three manuals are provided: a User's Guide, a Problem Determination Guide and Reference manual, and a Setup and Customization Guide. Three versions of a quick reference card and three keyboard templates (one each for the 3270 PC, 101-key Enhanced, and XT/AT keyboard) are provided. The three versions of the quick reference card correspond to the three different types of keyboards, as well.

Installation and Use

The Workstation Program is a complex product, and its installation requires considerable planning and effort. The Setup and Customization Guide, which leads the user through the process, is nearly an inch thick. It does, however, cover all the topics required—from display window layout to configuration of terminal unit controllers to work with the Workstation Program. Program information files (PIFs) are even provided to support the execution of popular DOS applications in windows.

Once installed, the Workstation Program is a useful and flexible product. The Workstation Program provides quality host communications capabilities combined with the convenient simultaneous execution of well-behaved DOS applications. This, combined with a windowing system that allows the easy transfer of information between sessions, makes it useful for anyone who needs to execute DOS applications along with sophisticated 3270 communications support.

Many of the features of the Workstation Program are being incorporated

Chapter 14

into the OS/2 Extended Edition Communications Manager. The machines that have the speed necessary to provide full benefit from the Workstation Program (the AT and Personal System/2 Models 50, 60, and 80) can also run OS/2. As more OS/2 applications become available, most users who require the type of functions offered by the Workstation Program will probably want to use the new generation of software products available only under OS/2, and will thus choose to use OS/2 instead of the Workstation Program. Nonetheless, the Workstation Program may find significant use among the millions of 8088/8086-based computers (including the Models 25 and 30) that cannot run OS/2.

What Next?

IBM offers an elaborate array of products for the communication and sharing of information among all types of computers—mainframe, mini, and personal. Its inclusion of a serial port on each Personal System/2 and the incorporation of numerous database and communications facilities into the extended edition of OS/2 indicate how important IBM believes connectivity is, or should be.

IBM will continue to offer communications products that attempt to provide maximum connectivity between its diverse line of computer and office products. Other manufacturers, unencumbered by this obligation, will continue to develop products that work well for specific areas of the communications market. The acceptance of IBM's extended edition of OS/2 and its communications features will determine the extent to which other vendors have to make their products compatible with IBM's communications products in order to be successful. IBM, for its part, will continue to develop and market products that allow users even better connectivity with other computers, many of which will also be developed and marketed by IBM.

Appendixes

A

Performance Testing

In order to provide a balanced view of the performance of computers it is necessary to test them with a variety of different applications, running each test on each machine under similar circumstances. To this end, Personal System/2 machines were compared with an IBM XT and AT 339 (8 MHz), the AST 10 MHz Premium/286 machine, and a Compaq Deskpro 286 and Deskpro 386/20. Wherever possible, the machines were tested with and without a math coprocessor. If a vendor sold the machine with a cache program as a standard feature, that machine was tested with and without the cache installed. A variety of good quality disk caching programs, which provide a comparable improvement in performance in a machine that is not supplied with a cache program, are available for a nominal cost.

The various machines that were used for evaluation were tested using sample word processing, spreadsheet, database manager, and graphics applications. A program was also assembled using each machine, to assess its performance in a program development environment. Word processing performance was tested using WordPerfect Version 4.2 from WordPerfect Corporation, Orem, Utah. The spreadsheet used was Lotus 1-2-3 Version 2.01 from Lotus Development Corporation, Cambridge, MA. The database manager used was dBase III PLUS Version 1.1 from Ashton-Tate, Torrance, CA. PC-KEY-DRAW from OEDWARE, Columbia, MD was used to test graphics performance, and the assembler used was Microsoft Macro Assembler Version 5.0, from Microsoft, Redmond, WA.

The word processor test consisted of loading a large file, (310,996 bytes) from the various input drives available. On the XT this was the hard disk and the 5.25-inch diskette drive. On the Model 25 this was the 3.5-inch diskette drive. On the Model 30 the file was loaded from the hard disk and the 3.5-inch diskette drive. On the other Personal System/2 machines the file was loaded from a RAM disk, the hard disk, and the 3.5-inch diskette drive, from disks that had been formatted at each of the available densities—720 KB and 1.44 MB. The other machines, which have a 5.25-inch diskette drive that can read 360 KB or 1.2 MB diskettes, were tested by loading from the different diskettes, the hard disk, and a RAM disk. The time required to load a file from the hard disk was not affected by the presence of a cache, as the data loaded had not been accessed previously in the test.

Once loaded, times were taken to assess the speed required to move around in a document and to perform a RAM-intensive search and replace. The time taken to jump from the top of the document to the bottom and from the bottom of the document to the top was measured. The search and replace test required the word processor to change all occurrences of the letter j with the letter k. None of these tests were dependent on the presence of the math coprocessor.

The most popular spreadsheet for the PC environment is Lotus 1-2-3. This program was used to assess the performance of RAM-intensive operations that make use of the math coprocessor, as well as typical spreadsheet functions, because the program can make use of a math processor if it is present in a machine. Three different spreadsheets were used. The first consisted of 10,001 cells, arranged in 100 columns by 100 rows. Each cell, except the first, contained the formula A plus 1, where A was the previous cell number. Cells at the top of a column referred to the contents of the cell at the base of the previous column. The contents of the first cell were changed from 0 to 1 and the time for all of the cells to be recalculated was measured.

The second spreadsheet was the same size and arrangement, but the formula was changed so that the recalculation involved multiplication and division, as well as the use of a constant π. The odd and even numbered columns contained the formulas $A \times \pi$ and $A \div \pi$, respectively, where A was the value of the previous cell. Again, the cell that was at the top of the column referred to the cell at the base of the previous column. The contents of the first cell were changed from 1 to 2, and the time for a recalculation was measured.

The third spreadsheet contained a more complicated formula. In order to contain the spreadsheet completely in conventional RAM, as the first two simple spreadsheets were, it had to be reduced in size. There were 50 columns and 100 rows of information. Each cell contained the formula: the square root of the natural logarithm of $(A^2 + 1)$, where A was the contents of the previous cell. Again, the cell at the top of a column referred to the cell at the base of the previous column, and the first cell was changed from 1 to 2 for timing purposes. Obviously, the recalculation of this more complicated spreadsheet took longer than the simple addition or multiplication spreadsheets.

The time to load the first spreadsheet was also measured. This time was dependent on the location of the data, so it was timed while being loaded from a RAM disk, from each of the removable disk drive densities, and from the hard disk with and without the cache installed.

The database manager used has also gained immense popularity as the database manager standard in the industry. There are many more sophisticated database managers, but for most office applications dBase III PLUS is more than equal to the task. The test performed involved sorting a reasonably sized database of 1287 records, each with five different fields. The database contained information on a user disk library; the fields were the disk title, disk number, its type, and two description fields. The amount of time required to sort the database by title was measured, as was the time to perform a sort on disk ID. Although Ashton-Tate recommends a specific setting for the buffers and files in the CONFIG.SYS file, the machine's defaults were used for test purposes. A sort procedure involves disk activity, so the tests were timed using a RAM disk, each of the diskette densities, and the hard disk with the cache on and off.

A graphics program was chosen to evaluate the relative performances of the video systems on the various machines. PC-KEY-DRAW is a CGA-based graphics program that is capable of many drawing functions that exercise a system's video capabilities. It comes with a sample macro program, which was replayed for the timing tests. In order to run the macro the disk is accessed, so the test was timed using a RAM disk, each of the diskette densities, and the hard disk with the cache on and off. The macro includes demonstrations of drawing lines, circles, arcs, and polygons as well as text in various bit-mapped fonts. Images are scaled both up and down, and rotated, and area fills and color palette changes are shown.

A sample program assembly was also performed for the benchmarks. The VDISK.LST program that is supplied with DOS 3.2 was used. The listing was converted to assembly language source code by removing the listing information that had been added when it was assembled. The time to load MASM and assemble the VDISK.ASM was timed, using the time mark program from the Norton Utilities Version 4.0. A batch file—which started time mark, loaded MASM and compiled VDISK without generating list or cross reference files, and stopped time mark—was executed from the DOS prompt. On completion, time mark indicated the number of seconds that it had been active. The average of several tests was used for a more accurate measurement.

Each Personal System/2 machine is compared with the most comparable machines in its respective chapter. For example, in Chapter 2 the Model 30 is compared with the Model 25, the Model 50, the XT, and the AT. An analysis of the different results obtained is given in the relevant chapter for each machine, and the full results for all of the machines are given in the following tables. Table A-1 shows the results for the non-disk-dependent tests, including and excluding the math coprocessor where appropriate. Table A-2 shows the RAM disk results, Table A-3 the hard disk results with cache on and off as appropriate, Table A-4 the high density diskette results, and Table A-5 the low density diskette results.

The Personal System/2 machines varied in performance across a wide range. The Model 25's performance was better in general than the PC, and the 20 MHz Model 80 outperformed the other tested machines, except for the Compaq Deskpro 386/20. However, the market is changing rapidly and there will always be faster and better machines available. The raw performance of a machine is far less critical than the cost for the required performance, and this can only be assessed when the situation in which the machine is to be used is known. There is room in the marketplace for quality machines that have different performance-to-cost ratios.

Table A-1 Non-Disk Dependent Results. *The Personal System/2 machines encompass a wide range of performances. Equivalent third-party machines that are PC or AT compatible are available as alternatives.*

| Test Machine | Equipment | WordPerfect ||| Lotus 1-2-3 |||
		Move Top to Bottom	Move Bottom to Top	Search and Replace	Recalc—Add	Recalc—Multiply	Recalc—Mix
XT	No math	109	82	129	18	30	200
	Math	109	82	129	18	18	19
Model 25	No math	48	37	58	8	14	91
	Math	48	37	58	8	8	9
Model 30	No math	49	38	58	8	14	91
	Math	49	38	58	8	8	9
AT 339	No math	28	21	32	5	8	45
	Math	28	21	32	5	5	7
Compaq 286	No math	28	21	34	5	7	45
	Math	28	21	34	5	5	7
Model 50	No math	22	16	27	4	6	36
	Math	22	16	27	4	4	5
Model 60	No math	22	16	27	4	6	36
	Math	22	16	27	4	4	5
AST 10 MHz	No math	16	12	20	3	4	28
	Math	16	12	20	3	3	4
16 MHz Model 80	No math	12	9	14	2	3	20
	Math	12	9	14	2	2	3
20 MHz Model 80	No math	11	8	13	2	3	16
	Math	11	8	13	2	2	2
Compaq 386/20	No math	9	6	10	2	2	13
	Math	9	6	10	2	1	2

All results are in seconds.

Performance Testing

Table A-2 RAM Disk Results. *IBM's VDISK was used to construct a RAM disk in extended memory for test purposes. This uses the protected mode of the 80286 microprocessor.*

Test Machine	WordPerfect Load File	Lotus 1-2-3 Load File	dBASE III PLUS Sort on Title	dBASE III PLUS Sort on Disk ID	PC-KEY-DRAW Run Sample Macro	MASM Assemble File
AT 339	11	22	9	10	520	8
Compaq 286	9	22	10	11	473	8
Model 50	9	17	7	8	406	6
Model 60	9	17	8	8	406	6
AST 10 MHz	10	16	7	8	384	5
16 MHz Model 80	5	9	4	5	277	3
20 MHz Model 80	5	8	4	4	249	3
Compaq 386/20	4	6	3	3	189	2

All results are in seconds.

Table A-3 Hard Disk Results. *The performance of the machines was dependent on the performance of the hard disk as well as the system architecture.*

Test Machine	Equipment	WordPerfect Load File	Lotus 1-2-3 Load File	dBASE III PLUS Sort on Title	dBASE III PLUS Sort on Disk ID	PC-KEY-DRAW Run Sample Macro	MASM Assemble File
XT	No cache	39	90	88	92	1441	40
Model 30	No cache	25	44	32	34	655	16
AT 339	No cache	12	29	25	26	526	11
Compaq 286	No cache	11	28	21	22	477	9
Model 50	No cache	10	27	29	29	409	7
	Cache	10	20	20	21	405	7
Model 60	No cache	10	26	24	21	406	7
	Cache	10	19	22	21	406	7
AST 10 MHz	No cache	8	17	25	24	388	9
16 MHz Model 80	No cache	7	13	26	26	283	4
	Cache	7	11	22	21	277	4
20 MHz Model 80	No cache	7	13	19	18	249	3
	Cache	7	10	12	12	249	4
Compaq 386/20	No cache	5	7	12	12	189	2
	Cache	4	6	10	11	185	3

All results are in seconds.

Table A-4 High Density Diskette Results. *The performance of machines that can use both low density and high density diskettes varied, depending on the media used.*

Test Machine	WordPerfect Load File	Lotus 1-2-3 Load File	dBASE III PLUS Sort on Title	dBASE III PLUS Sort on Disk ID	PC-KEY-DRAW Run Sample Macro	MASM Assemble File
AT 339	29	112	204	223	583	19
Compaq 286	33	112	171	169	543	21
Model 50	29	128	204	207	469	20
Model 60	29	128	204	204	468	20
AST 10 MHz	20	108	199	223	465	16
16 MHz Model 80	28	125	196	194	339	18
20 MHz Model 80	28	125	194	193	309	16
Compaq 386/20	44	107	171	172	334	20

All results are in seconds.

Table A-5 Low Density Diskette Results. *All of the tested machines were able to read the lower density diskettes, even though some do not support the higher density.*

Test Machine	WordPerfect Load File	Lotus 1-2-3 Load File	dBASE III PLUS Sort on Title	dBASE III PLUS Sort on Disk ID	PC-KEY-DRAW Run Sample Macro	MASM Assemble File
XT	73	159	232	239	1480	48
Model 25	50	140	219	224	697	36
Model 30	54	140	219	224	704	36
AT 339	43	115	193	184	572	30
Compaq 286	43	115	181	182	546	27
Model 50	45	134	209	211	480	22
Model 60	45	134	209	211	465	23
AST 10 MHz	40	112	176	180	468	26
16 MHz Model 80	39	125	209	202	337	20
20 MHz Model 80	45	125	195	197	317	25
Compaq 386/20	45	111	184	185	337	28

All results are in seconds.

Vendor Addresses

Ashton-Tate
20101 Hamilton Avenue
Torrance, CA 90502-1319

Lotus Development Corporation
55 Cambridge Parkway
Cambridge, MA 02142

Microsoft Corporation
16011 NE 36th Way
Redmond, WA 98073-9717

OEDWARE
PO Box 595
Columbia, MD 21045

Peter Norton Computing, Inc.
2210 Wilshire Boulevard, Suite 186
Santa Monica, CA 90493

WordPerfect Corporation
288 West Center Street
Orem, UT 84057

B

Benchmark Machines

The machines that were chosen for comparison with the IBM Personal System/2 machines included the best compatible machines available. Compaq Computer Corporation has become the leader in compatible manufacturers, and two of their machines were selected for evaluation. The 8 MHz Compaq Deskpro 286 and the Compaq Deskpro 386/20 running at 20 MHz were used.

AST Research, Inc. has become well known as the manufacturer of top quality add-in boards for the PC and AT. Their AT-compatible machine, the AST Premium/286, was chosen because it runs at 10 MHz with no-wait-state access to memory. This has been done by other manufacturers but not with the same level of compatibility.

The IBM XT and IBM AT 339 (8 MHz) were also used for comparison. The features of these IBM machines are described in Chapter 1.

Compaq Deskpro 286

This 80286-based AT compatible has features and performance that are very similar to the IBM AT 339. It has a hard disk and a single 5.25-inch diskette drive that reads and writes both 1.2 MB diskettes and 360 KB diskettes. The system has 640 KB of RAM, and the microprocessor runs at 8 MHz. The RAM is accessed in this machine with one wait state, giving a cycle time of approximately 375 nanoseconds. The math coprocessor runs at 5.3 MHz, two-thirds the speed of the CPU.

AST Premium/286

The AST AT compatible provides better performance than the AT 339. It has a hard disk and a single 5.25-inch diskette drive that reads and writes both 1.2 MB diskettes and 360 KB diskettes. The system RAM can be directly accessed by the

10 MHz microprocessor at zero wait states, giving a cycle time of approximately 200 nanoseconds. The math coprocessor runs at 8 MHz.

Compaq Deskpro 386/20

The successor to the Compaq Deskpro 386 is the Deskpro 386/20. This 80386-based machine runs at 20 MHz, and its 80387 coprocessor also runs at 20 MHz. Due to various architectural improvements, this machine outperforms the 20 MHz Model 80 in many instances. Some of these features include a cache memory controller and 32 KB of associated static RAM, a hard disk with an interleave of 1 to 1 or 2 to 1 depending on the disk option, and 32-bit memory architecture. The tested machine had a hard disk and a 5.25-inch diskette drive that can read and write 1.2 MB and 360 KB diskettes.

Summary

All of these machines have expansion slots that are AT compatible. Features such as a serial port and a parallel port, which have been added as standard on the Personal System/2 machines, are also standard on these machines. Their power supplies are rugged enough to accommodate filled expansion slots, unlike the original PC.

These machines are real alternatives to the Personal System/2 machines. They do not have the IBM label, but they come from manufacturers that have solid reputations. They are cost effective in many situations and will be a part of the future of personal computers. They are, however, using an expansion slot architecture that has reached its limit for expansion. The future will encompass both types of machines and the most favored remains to be seen.

Vendor Addresses

AST Research, Inc.
2121 Alton Avenue
Irvine, California 92714-4992

Compaq Computer Corporation
20555 FM 149
Houston, Texas 77070

C

80286/80386 Instruction Sets

80286 Instruction Set

Data Transfer Instructions

MOV	Move byte or word
PUSH	Push word onto stack
POP	Pop word off stack
PUSHA	Push all registers onto stack
POPA	Pop all registers from stack
XCHG	Exchange byte or word
XLAT	Translate byte
IN	Input byte or word
OUT	Output byte or word
LEA	Load effective address
LDS	Load pointer using DS
LES	Load pointer using ES
LAHF	Load AH register from flags

SAHF	Store AH register in flags
PUSHF	Push flags onto stack
POPF	Pop flags off stack

Arithmetic Instructions

ADD	Add byte or word
ADC	Add byte or word with carry
INC	Increment byte or word by 1
AAA	ASCII adjust for addition
DAA	Decimal adjust for addition
SUB	Subtract byte or word
SBB	Subtract byte or word with borrow
DEC	Decrement byte or word by 1
NEG	Negate byte or word
CMP	Compare byte or word
AAS	ASCII adjust for subtraction
DAS	Decimal adjust for subtraction
MUL	Multiply byte or word unsigned
IMUL	Integer multiply byte or word
AAM	ASCII adjust for multiply
DIV	Divide byte or word unsigned
IDIV	Integer divide byte or word
AAD	ASCII adjust for division
CBW	Convert byte to word
CWD	Convert word to doubleword

String Instructions

MOVS	Move byte or word string
INS	Input bytes or word string

Instruction Sets

OUTS	Output bytes or word string
CMPS	Compare byte or word string
SCAS	Scan byte or word string
LODS	Load byte or word string
STOS	Store byte or word string
REP	Repeat
REPE/REPZ	Repeat while equal/zero
REPNE/REPNZ	Repeat while not equal/not zero

Shift/Rotate Logical Instructions

NOT	"NOT" byte or word
AND	"AND" byte or word
OR	"Inclusive OR" byte or word
XOR	"Exclusive OR" byte or word
TEST	"Test" byte or word
SHL/SAL	Shift logical/arithmetic left byte or word
SHR	Shift logical right byte or word
SAR	Shift arithmetic right byte or word
ROL	Rotate left byte or word
ROR	Rotate right byte or word
RCL	Rotate through carry left byte or word
RCR	Rotate through carry right byte or word

Program Transfer Instructions

JA/JNBE	Jump if above/not below nor equal
JAE/JNB	Jump if above or equal/not below
JB/JNAE	Jump if below/not above nor equal
JBE/JNA	Jump if below or equal/not above
JC	Jump if carry
JE/JZ	Jump if equal/zero

JG/JNLE	Jump if greater/not less nor equal
JGE/JNL	Jump if greater or equal/not less
JL/JNGE	Jump if less/not greater nor equal
JLE/JNG	Jump if less or equal/not greater
JNC	Jump if not carry
JNE/JNZ	Jump if not equal/not zero
JNO	Jump if not overflow
JNP/JPO	Jump if not parity/parity odd
JNS	Jump if not sign
JO	Jump if overflow
JP/JPE	Jump if parity/parity even
JS	Jump if sign
CALL	Call procedure
RET	Return from procedure
JMP	Jump
LOOP	Loop
LOOPE/LOOPZ	Loop if equal/zero
LOOPNE/LOOPNZ	Loop if not equal/not zero
JCXZ	Jump if register CX = 0
INT	Interrupt
INTO	Interrupt if overflow
IRET	Interrupt return

Processor Control Instructions

STC	Set carry flag
CLC	Clear carry flag
CMC	Complement carry flag
STD	Set direction flag
CLD	Clear direction flag
STI	Set interrupt enable flag

CLI	Clear interrupt enable flag
HLT	Halt until interrupt or reset
WAIT	Wait for BUSY not active
ESC	Escape to extension processor
LOCK	Lock bus during next instruction
NOP	No operation
LMSW	Load machine status word
SMSW	Store machine status word

High Level Instructions

ENTER	Format stack for procedure entry
LEAVE	Restore stack for procedure exit
BOUND	Detect values outside prescribed range

80386 Instruction Set

Data Transfer Instructions

MOV	Move operand
PUSH	Push operand onto stack
POP	Pop operand off stack
PUSHA	Push all registers onto stack
POPA	Pop all registers from stack
XCHG	Exchange operand, register
XLAT	Translate
MOVZX	Move byte or word, dword, with zero extension
MOVSX	Move byte or word, dword, sign extended
CBW	Convert byte to word, or word to dword

CWD	Convert word to dword
CWDE	Convert word to dword extended
CDQ	Convert dword to qword
IN	Input operand from I/O space
OUT	Output operand to I/O space
LEA	Load effective address
LDS	Load pointer into D segment register
LES	Load pointer into E segment register
LFS	Load pointer into F segment register
LGS	Load pointer into G segment register
LSS	Load pointer into S (stack) segment register
LAHF	Load A register from flags
SAHF	Store A register in flags
PUSHF	Push flags onto stack
POPF	Pop flags off stack
PUSHFD	Push EFlags onto stack
POPFD	Pop EFlags off stack
CLC	Clear carry flag
CLD	Clear direction flag
CMC	Complement carry flag
STC	Set carry flag
STD	Set direction flag

Arithmetic Instructions

ADD	Add operands
ADC	Add with carry
INC	Increment operand by 1
AAA	ASCII adjust for addition
DAA	Decimal adjust for addition

SUB	Subtract operands
SBB	Subtract with borrow
DEC	Decrement operand by 1
NEG	Negate operand
CMP	Compare operands
AAS	ASCII adjust for subtraction
DAS	Decimal adjust for subtraction
MUL	Multiply double/single precision
IMUL	Integer multiply
AAM	ASCII adjust after multiply
DIV	Divide unsigned
IDIV	Integer divide
AAD	ASCII adjust before division

String Instructions

MOVS	Move byte or word, dword string
INS	Input string from I/O space
OUTS	Output string to I/O space
CMPS	Compare byte or word, dword string
SCAS	Scan byte or word, dword string
LODS	Load byte or word, dword string
STOS	Store byte or word, dword string
REP	Repeat
REPE/REPZ	Repeat while equal/zero
REPNE/REPNZ	Repeat while not equal/not zero

Shift/Rotate Logical Instructions

NOT	"NOT" operands
AND	"AND" operands
OR	"Inclusive OR" operands

XOR	"Exclusive OR" byte or word
TEST	"Test" byte or word
SHL/SHR	Shift logical left or right
SAL/SAR	Shift arithmetic left or right
SHLD/SHRD	Double shift left or right
ROL/ROR	Rotate left/right
RCL/RCR	Rotate through carry left/right

Bit Manipulation Instructions

BT	Bit test
BTS	Bit test and set
BTR	Bit test and reset
BTC	Bit test and complement
BSF	Bit scan forward
BSR	Bit scan reverse
IBTS	Insert bit string
XBTS	Exact bit string

Program Transfer Instructions

SETCC	Set byte equal to condition code
JA/JNBE	Jump if above/not below nor equal
JAE/JNB	Jump if above or equal/not below
JB/JNAE	Jump if below/not above nor equal
JBE/JNA	Jump if below or equal/not above
JC	Jump if carry
JE/JZ	Jump if equal/zero
JG/JNLE	Jump if greater/not less nor equal
JGE/JNL	Jump if greater or equal/not less
JL/JNGE	Jump if less/not greater nor equal

JLE/JNG	Jump if less or equal/not greater
JNC	Jump if not carry
JNE/JNZ	Jump if not equal/not zero
JNO	Jump if not overflow
JNP/JPO	Jump if not parity/parity odd
JNS	Jump if not sign
JO	Jump if overflow
JP/JPE	Jump if parity/parity even
JS	Jump if sign
CALL	Call procedure/task
RET	Return from procedure
JMP	Jump
LOOP	Loop
LOOPE/LOOPZ	Loop if equal/zero
LOOPNE/LOOPNZ	Loop if not equal/not zero
JCXZ	Jump if register CX = 0
INT	Interrupt
INTO	Interrupt if overflow
IRET	Interrupt return/task
CLI	Clear interrupt enable
SLI	Set interrupt enable

Processor Control Instructions

HLT	Halt until interrupt or reset
WAIT	Wait for BUSY# negated
ESC	Escape
LOCK	Lock bus
NOP	No operation

Protection Model

SGDT	Store global descriptor table
SIDT	Store interrupt descriptor table
STR	Store task register
SLDT	Store local descriptor table
LGDT	Load global descriptor table
LIDT	Load interrupt descriptor table
LTR	Load task register
LLDT	Load local descriptor table
ARPL	Adjust requested privilege level
LAR	Load access rights
LSL	Load segment limit
VERR/VERW	Verify segment for reading or writing
LMSW	Load machine status word (lower 16 bits of CR0)
SMSW	Store machine status word

High Level Instructions

ENTER	Set up parameter block for entering procedure
LEAVE	Leave procedure
BOUND	Check array bounds

Index

1dir + DOS interface (Bourbaki), 334
5.25-inch External Diskette Drive Adapter, 299
36/38 Workstation Emulation Adapter/A, 28
300/1200 Internal Modem/A, 28, 295
386/DOS Extender (Phar Lap Software), 416
3174 terminal unit controller, LAN support for, 431
3270 Emulation Program Entry Level, 442–443
3270 PC File Transfer Program, 446
3270 terminal communications, 440–460
    AIX emulation of, 391–392
3270 Workstation Program, 298, 456–460
3270/25 communications controller, LAN support for, 431
3278/79 Emulation Adapter, 446, 451, 453
3363 Optical Disk Drive, 300–301
4201 Proprinter II, 302, 303, 344
4207 Proprinter X24, 304
4216 Personal Pageprinter, 302
5173 PC Network Baseband Extender, 292
5250 terminal communications, 437–440
6157 Streaming Tape Drive, 301, 303
8042 keyboard controller (Intel), 90–91, 174, 263
8086 microprocessor (Intel), 2
    compatibility of, 8–9, 11
    for Model 25, 19, 66
    for Model 30, 33–35, 51–52
8087 math coprocessor (Intel), 290
    for Model 25, 66
    for Model 30, 18, 35, 52, 53, 61
    for XT, 7
8088 microprocessor (Intel), 2
    for Model 30, 18
    for XT, 7
8259A interrupt controller (Intel), 88–89, 169–170
8503 Monochrome Display, 23, 131, 277–279
8512 Color Display, 23, 131, 279–280
8513 Color Display, 23–24, 131, 280–282
8514 Color Display, 23–24, 282–283
8514/A Display Adapter, 22–23, 131
80186 microprocessor (Intel), 296
80286 Expanded Memory Adapter/A, 298
80286 Memory Expansion Option, 27, 297
80286 microprocessor (Intel)
    for AT, 8–9
    instructions for, 473–477
    Micro Channel support for, 14
    for Model 50, 20, 153, 160, 163–165
    for Model 60, 20, 207, 217
    for Personal System/2 family, 24, 80–81
80287 math coprocessor (Intel), 9–10, 290–291
    installation of, 191–192, 225
    for Model 50, 165–166
    for Model 60, 207
    for Personal System/2 family, 80

80386 Memory Expansion Option, 27, 297
80386 microprocessor (Intel), 11–12
    instructions for, 477–482
    Micro Channel support for, 14
    for Model 80, 21, 80, 82, 255–261, 265
80387 math coprocessor (Intel), 11, 24, 290–291
    for Model 80, 80, 255, 257

A

ABIOS
    for Micro Channel, 120, 123–124
    for Model 50, 20, 167
    with OS/2, 346
ABIOS Transfer Convention, 123
ABIOSCall service, 346
ABIOSCommonEntry service, 346
ACDI (asynchronous communications device interface), 357
Action bars with windows, 354
Active page, interrupts for, 137
Adapter boards
    installation of, 191, 225
    with Micro Channel, 111
    for PC, 3–5
    POST for, 124–126
    problems with, 226
Add a program option (DESQview/386), 415
Add a Program Title option, 374–376
Add-in boards
    for Micro Channel, 27–28
    for Model 30, 26–27
    third-party, 313–315
Address bus for Micro Channel, 96, 98–100, 105
Addresses of vendors, 469, 472
Addressing with 80386, 83
ADDTASK command (PC-MOS/386), 410
Adobe Systems, PostScript by, 302
Advanced diagnostics, 190, 224, 229–235
    for Model 30, 54–55
Advanced function and text modes for 8514/A adapter, 287–288
Advanced Program-to-Program Communications, 356, 450
Advanced Transducer Devices
    LAN by, 315
    modem by, 310
AIX C, 390
AIX Personal System/2 operating system, 29
    connectivity with, 388–391
    features of, 385–388
    installation of, 394
    market position of, 394–395
    terminal support by, 388

483

Aldus, PageMaker by, 302
All-points-addressable graphics, 278
    with 8514/A adapter, 289-290
    with Presentation Manager, 351
    with VGA, 147
Alphanumeric mode with 8514/A adapter, 287-288
ANSI command (OS/2), 380-381
APPEND command, 320
Application Development Toolkit for AIX, 391
Application programming interface
    DESQview, 416
    OS/2, 347
    Windows, 334
Applications, OS/2, installation and running of, 374-382
Arbitration
    levels of, 116
    Micro Channel bus for, 97
    multi-device, 117-119
    priority signals for, 103
Area fill with 8514/A adapter, 289-290
Arithmetic instructions
    80286, 474
    80386, 478-479
Ashton-Tate, dBase III by, 463-468
Aspect ratio
    of 8513 monitor, 281
    of 8514 monitor, 282
    with MCGA, 22
AST Research products
    AST-3270/CoaxIIA, 310
    AST Premium/286, in performance tests, 463-468, 471-472
    AST Premium/286 memory board, 308
    AST-VGA board, 313-314
    OS/2 version, 335
Asynchronous communications
    BIOS interrupts for, 121
    boards for, 294-296
    connectivity for, 428-429
    device interface for, 357
    See also Serial ports
ATTRIB command, 320
Attributes
    with MCGA, 43-44
    with VGA, 148-149
Audio subsystem 89-90, 104
    for Model 50, 173
Auto restart power supply, 246
AUTOEXEC.BAT file, 326, 332-333
    with Concurrent DOS 386, 400
    with Emulation Program Entry Level, 447
    with OS/2, 365, 369, 380
AUTOINST command (DESQview/386), 415
Automatic Configuration Utility, 112-115
Auxiliary video extension signals, 106-107

B

Background, OS/2 sessions in, 340
Background color, MCGA, 44
BACKUP command, 320, 323
Backup configuration option, 112, 185, 186
Backup the Reference Diskette option, 182

Barrel shifter with 80386, 259
Baseband LAN boards, 292, 430
BASIC, IBM, 325, 331
BASIC Compiler/2, 359-360
Basic input output system. *See* BIOS
BASIC ROM for PC, 3
Battery, replacement of, 51, 192, 225
Battery-backed RAM, 167-168, 178, 217, 253
BCD arithmetic with coprocessors, 290-291
Bi-directional ports, 315
BIOS (basic input output system)
    for DOS 3.30, 323-324
    for MCGA, 41-42
    for Micro Channel, 119-124
    for Model 25, 66
    for Model 50, 20, 167
    for Model 80, 258
    for OS/2, 346
    for PC, 2-3
    for Personal System/2 family, 80, 84
    for VGA, 135-146, 285
BIOS Interface Technical Reference manuals, 120, 194-195, 236
BIT-BLT transfer with 8514/A adapter, 289
Bit manipulation instructions, 480
Bit map characters, 288
Bit map/cursor editor, 364
Bourbaki, 1dir + by, 334
Bourne shell for AIX, 386, 393
Broadband LAN boards, 293-294
Bus signals for Micro Channel, 96-107

C

C
    for AIX, 390
    for OS/2, 359-361
    for Personal System/2, 16
C/2, 359-361
C shell for AIX, 386, 393
CAD/CAM applications, monitors for, 283
CALL command, 320
CALL interface with Presentation Manager, 351
CALL-RETURN interface with OS/2, 347
Cardfile utility (Concurrent DOS 386), 404
CBIOS
    for Micro Channel, 120-123
    for Model 50, 20, 167
    with OS/2, 346
Central Arbitration Control Point, 117-119
CGA. *See* Color Graphics Adapter
Change Active Partition option, 331
Change configuration option, 183-185, 224, 370
Change Configuration utility, 112
Change power-on password option, 187
Change Program Information option, 376
Character generators
    interrupt for, 141-143
    for MCGA, 41, 43
    for VGA, 135-136
Character I/O, interrupts for, 138, 144
Character modes for 8514/A adapter, 287-288
Character size with MCGA, 22
CHCP command, 321

Index

CHIPSlink IC, 310
CHKDSK command, 333, 369
CLASS command (PC-MOS/386), 412
Clones
    for AT, 11
    and Micro Channel, 96
    for PC, 1, 5-7
    for Personal System/2, 276, 315-316
    and VGA, 286
Close option (Windows/386), 421
CMD.EXE file, 348, 376-377
CMOS memory for AT, 10
Cobol for Personal System/2, 16
COBOL/2, 359, 361-362
Code page switching, 333, 344
Color
    on 8513 Color Display, 281
    with MCGA, 44, 22
    for PC, 4
    with VGA, 129
Color Graphics Adapter
    MCGA compatibility with, 42
    for PC, 3-4
    and VGA, 22
Color palette
    interrupts for, 138-140, 143
    with MCGA, 41
Color registers, MCGA, BIOS calls for, 41-42
Columbia Data Products computers, 6
COMMAND.COM file, 318-319
    compared to CMD.EXE, 348
Commands
    DOS, 319-322, 401-402
    PC LAN Program, 432
    for PC-MOS/386, 408-409
Communication products, 294-296
    third party, 309-311
Communications Manager, 29, 355-358
Compaq Computer Corp. products, 7
    Deskpro 286, 10-11, 463-468, 471
    Deskpro 386, 11-12, 14, 463-468, 472
    Deskpro desktop computer, 8
    OS/2 version, 335
    portables, 5-6, 10-11
Compatibility
    of 80286, 163
    of BIOS, 120-123
    of diskette drives, 26, 177, 213, 298-300, 426-428
    and Micro Channel, 95-96
    of OS/2 and DOS, 347-349
    of PC compatibles, 5-7
    of PC software, 13
    of Personal System/2, 16-17
    of VGA, 136
    See also Performance and compatibility
Complementary metal oxide silicon memory, 10
Concurrent DOS 386 (Digital Research, Inc.), 399-406
CONFIG.SYS file, 324, 326, 332-333
    and disk cache program, 198-199
    with OS/2, 365, 369-371, 373
    with PC-MOS/386, 408
    TRACEBUF in, 382

CONFIG.SYS file—cont
    with Windows/386, 419
Configuration, setting of, 83
    on Model 25, 74
    on Model 50 and 60, 182-190, 192-194, 225
    on Model 80, 246, 265-266
    with OS/2, 370-373
    RAM for, 180
Connectivity, 425
    with 36/38 Workstation Emulation Adapter/A, 28
    for 3270 communications, 440-460
    for 5250 communications, 437-440
    with AIX, 391-394
    for asynchronous communications, 428-429
    with Communications Manager, 355-357
    for LANs, 430-437
    for media and file interchange, 426-428
Continuously Variable Slope Delta, 26, 61
Control signals for Micro Channel, 97
Control software for Model 80, 397-398
    Concurrent DOS 386, 399-406
    DESQview/386, 413-417
    PC-MOS/386, 406-413
    Windows/386, 417-422
Copy an option diskette option, 187-189, 191, 225
COPY command with Data Migration Facility, 427
Copy protected diskettes and Model 30, 55
Country information
    codes for, 325, 332-333
    COUNTRY command for, 321, 326, 333
    DOS for, 322
    with Emulation Program Entry Level, 445
    with OS/2, 344
Create DOS Partition option, 327, 329
Create Extended DOS Partition option, 329-330
Create Logical DOS Drive(s) option, 331
Create Primary DOS Partition option, 328-329, 331
CREATEDD utility, 383
CREF utility, 362
CRT controller, VGA, 134
    registers for, 142, 149-150
Ctrl-A for advanced diagnostics, 190, 224
Cursor, interrupts for, 137, 142-143
CVSD (Continuously Variable Slope Delta), 26, 61
Cycle times for Personal System/2 family, 83
Cylinders on hard drives, 329-330

D

DASDDRVR.SYS file, 324
Data bus for Micro Channel, 96, 100-101, 105
Data latch register, 40
Data migration, 55
    Data Migration Facility for, 26, 177, 300, 426-428
    parallel port controller for, 92
Data transfer instructions
    80286, 473-474
    80386, 477-478
Database Manager for OS/2, 357-358
Database managers in performance tests, 463-465, 467-468
DATE command, 321, 326, 333

Date and time, setting of, 83, 185, 188
DBase III PLUS (Ashton-Tate) in performance tests, 463–468
DEBUG utility (PC-MOS/386), 412
Dell Computer Corp., Personal System/2 clones by, 315
Design features of Personal System/2, 16–18
Deskpro 286 (Compaq), 10
    in performance tests, 463–468, 471
Deskpro 386 (Compaq), 11–12, 14
    in performance tests, 463–468, 472
Deskpro desktop computer (Compaq), 8
DESQview/386 (Quarterdeck Office Systems), 413–417
DETACH command (OS/2), 381
DevHlp interface, 346
Device drivers
    DOS for, 324
    with OS/2, 345–347
Device Test menu, 230–231
Diagnostic manuals, 195, 236
Dialog boxes, 354
Dialog editor, 364
Digital Research, Inc. products
    Concurrent DOS 386, 399–406
    GEM operating environment, 334
Digital-to-analog converters
    with MCGA, 45, 51
    with VGA, 135
Direct memory access controller. *See* DMA Controller
Disk cache program
    for Model 50, 195–200
    for Model 80, 266–267
    with PC LAN Program, 432
    in performance tests, 463
Disk Operating System, Version 3.30
    enhanced user interfaces for, 333–334
    features of, 319–322
    history of, 317–319
    installation and use of, 324–333
    packaging and documentation for, 322–323
    updates for, 323–324
Disk operating systems
    with Concurrent DOS 386, 401–402
    DESQview/386 services for, 415
    with OS/2, 347–349, 365, 368, 372, 380
Diskette drives
    BIOS interrupts for, 121
    compatibility of, 26, 177, 213, 298–300, 426–428
    controllers for, in Personal System/2 family, 91–92, 254, 257
    DOS for, 320
    error codes for, 227–228
    external, 298–300
    for Model 25, 66–68, 73–74
    for Model 30, 48–50
    for Model 50, 155, 158, 176–177
    for Model 60, 20–21, 209, 212–213, 220–221
    for Model 80, 21–22, 244, 248, 251–252, 264
    for PC, 4–5
    for Personal System/2, 16–17, 24–25
    problems with, 226

Diskette drives—cont
    removal of, in Model 30, 48–50
Diskette gate array for Model 30, 38
Display adapters
    Display Adapter 8514/A, 283, 286–290
    GDDM-PCLK support for, 456
    for Model 30, 283–286
    Personal System/2 Display Adapter, 22, 283–284
DISPLAY.SYS file, 322, 333
DMA controller, 85–86
    for AT, 9
    for Model 30, 37–38
    for Model 50, 169
Documentation
    for BASIC Compiler/2, 360
    for Concurrent DOS 386, 405
    for DESQview/386, 417
    for DOS, 322–323
    for Emulation Program, 445–446
    for Model 25, 74
    for Model 30, 53–54
    for Model 50, 194–195
    for Model 60, 236
    for Model 80, 267
    for OS/2, 363–364
    for PC-MOS/386, 412
DOS. *See* Disk operating systems
DOS Command Prompt option, 376, 380
DOS Merge command (AIX), 387
Dots, interrupts for, 139
DR EDIX Editor (Concurrent DOS 386), 404
Dual Async Adapter/A, 28, 294, 429
    error codes for, 228
Dump facility for OS/2, 382–383
DV.BAT file (DESQview/386), 415
Dynamic window operations, 354

E

ECF. *See* Enhanced Connectivity Facilities
Education market, Model 25 for, 63
Electronic mail with AIX, 393–394
EMI (electromagnetic interference), 45, 97–98, 107–109
Emulation. *See* Connectivity
ENDLOCAL command (OS/2), 381
Enhanced Color Display, 10
Enhanced Connectivity Facilities, 356, 358
    with Emulation Program, 450
    with Emulation Program Entry Level, 444–445, 448
Enhanced Keyboard, 10, 17, 19–21
Enhanced user interfaces for DOS 3.30, 333–334
Enhancements and PC compatibility, 13
Equipment, video, interrupt for, 145
Errors
    for configuration, 193–194, 224–225
    OS/2 recording of, 382–383
    POST codes for, 226–229
    Reference Diskette messages for, 125
ESDI (enhanced small device interface), 26, 214, 220, 264
    error codes for, 228–229

Index

EVGA board (Everex Systems, Inc.), 313–315
EXCEPT command (PC-MOS/386), 408
EXE2BIN file, 318, 323
Executive System, Xtree by, 334
EXEMOD utility for compilers, 360–362
Expanded Memory Adapter
    for 3270 Workstation Program, 27
    Micro Channel version, 38
Expansion boards for Micro Channel, 94–96
Expansion memory for DESQview/386, 414
Expansion slots
    for Model 25, 19
    for Model 30, 18, 51, 55
    for Model 50, 20
    for Model 80, 21, 250–251, 256
    for PC, 5
    for XT, 7–8
Extended Graphics Adapter, 10, 22, 344
Extended OS/2, 335–336, 355–359
Extended partitions with FDISK, 367
Extension signals on Micro Channel, 105–107
External diskette drives, 298–300
External hard drives, 309
EXTPROC command (OS/2), 381

F

FASTOPEN command, 320
FDISK.COM file, 327, 366–367
Features
    of AIX, 385–388
    of DESQview/386, 416–417
    of DOS 3.30, 319–322
    of Emulation Program, 449–451
    of Emulation Program Entry Level, 443–445
    of Model 50, 153–155
    of Model 60, 207–209
    of Model 80, 244–248
    of PC-MOS/386, 407, 411–412
    of Windows/386, 417–419
    of Workstation Program, 438–439, 458–459
File Manager utility (Concurrent DOS 386), 402, 403
Filing System with Presentation Manager, 354–355
Filling areas with 8514/A adapter, 289–290
FIXBIOS file, 324
Floating-point arithmetic, coprocessors for, 290–291
Flush cache function, 196
Fonts, 141–143
    with 8514/A adapter, 288
    editor for, 364
    VGA, 133, 135–136
Foreground, OS/2 sessions in, 339–340
    with DOS programs, 380
Foreground color, MCGA, 44
FORMAT command, 319, 321, 326, 332, 369
Format fixed disk option, 232, 234
Format Track function, 197
Format Unit function, 197
Formatting of partitions, 327–333, 368–369
FORTRAN for AIX, 388–389
FORTRAN/2, 359, 362
Functional description
    of Model 25, 66–67
    of Model 30, 33–40

Functional description—cont
    of Model 50, 162–178
    of Model 60, 218–223
    of Model 80, 257–265
Functionality and status interrupt, 145
FXFER file transfer utility (AIX), 392–393

G

Gate arrays
    for Model 25, 71–72
    for Model 30, 35–38
GDDM-PCLK program, 454–456, 458
GEM operating environment (Digital Research, Inc.), 334
General purpose registers, VGA, 152
Get cache statistics function, 196
Global descriptor table
    with 80386, 260
    with OS/2, 337–338
Graphics
    controller for, VGA, 134–135
    modes for, 43–44
    monitor for, 281
    in performance tests, 463, 465, 467–468
    registers for, 151
    systems for, 22–23
    user interface for, with Presentation Manager, 350
    with VGA, 134–135, 142, 147
    with Workstation Program, 438–439
GRAPHICS command, 321
Guide to Operations manual, 53, 74

H

Hard disks
    for AT, 10
    controller for, in Model 30, 38–39
    DOS for, 318
    error codes for, 228
    external, 309
    formatting of, 232–235
    installation of DOS on, 327–333
    installation of Emulation Program Entry Level on, 447
    installation of OS/2 on, 364–373
    for Model 25, 71
    for Model 30, 55
    for Model 50, 155, 159, 176–177
    for Model 60, 20, 209, 214, 220–222
    for Model 80, 21–22, 244–246, 252–253, 264–265
    and Move the computer option, 190
    for PC, 7
    performance tests using, 467
    for Personal System/2, 16, 25–26
    for XT, 8
Hardware Maintenance Reference manuals, 194–195, 236
Hardware Maintenance Service manuals, 194, 236
Help functions
    with Concurrent DOS 386, 402

487

Help functions—cont
    with OS/2, 372, 374, 377–378
    with PC-MOS/386, 410
Hercules monographics boards, 278
Hierarchical file directories, 318
High density diskette mode, 92, 176
    performance tests using, 468
High Function Terminal driver in AIX, 386
High level instructions
    80286, 477
    80386, 482
High resolution graphics for AT, 10
High Speed Adapter/A, 28
History
    of DOS, 317–319
    of Personal System/2, 1–15
Host connection boards, 294

I

IBM AT, 8–11
    in performance tests, 463–468
IBM Cabling System, 431
IBM PC, 1–7
    software for, 13–14
IBM PC Convertible, 14–15, 319
IBM XT, 7–8
    in performance tests, 463–468
IBMBIO.COM file, 318, 348
IBMCACHE program. *See* Disk cache program
IBMDOS.COM file, 318, 348
IDEAcomm 3278/MC (IDEAssociates), 310, 311
IDEAcomm 5251/MC (IDEAssociates), 310
IEEE floating point standard, coprocessors for, 291
Index Sequential Access Method, 360, 361
INmail/INed/INnet/FTP program (AIX), 393–394
INSTALL batch file with Emulation Program Entry
    Level, 446–447
INSTALL command, 323
Installable device drivers, 318
Installation
    of AIX, 394
    of applications, with OS/2, 374–376
    of Concurrent DOS 386, 400–401
    of DESQview/386, 414–415
    of DOS 3.30, 324–333
    of Emulation Program, 453–454
    of Emulation Program Entry Level, 446–447
    of Model 30, 53–54
    of Model 50, 179–194
    of Model 60, 223–226
    of Model 80, 265–266
    of OS/2, 364–373
    of PC-MOS/386, 408
    of Windows/386, 419–420
    of Workstation Program, 459–460
Installation Diskette, 366, 369
Instructions, microprocessor
    80286, 473–477
    80386, 477–482
INT 10H BIOS functions, 121
    with MCGA, 41–42
    for VGA, 136–146
INT 12H BIOS functions, 122

INT 13H BIOS functions, 121
INT 14H BIOS functions, 121
INT 15H BIOS functions, 121, 123
INT 16H BIOS functions, 122–123
INT 21H BIOS functions, 322, 346
INTERACTIVE Systems Corp., AIX by, 385
Interface cables, problems with, 226
Interprocess communications with OS/2, 342–343
Interrupts
    for 80287, 87
    BIOS, 120–123
    level-sensitive, 116–117
    MCGA, 41–42
    on Micro Channel, 103–104
    for Model 30, 38, 39
    for Model 50, 169–170
    with OS/2, 344
    system, 88–89
    VGA, 136–146
Introducing OS/2 program, 373–374
I/O
    for OS/2, 343
    support gate array for, 36–38
    system board ports for, 171–172
IOCtl interface, 345

J

JOIN command, 319

K

KEYB command, 321, 326
Keyboards
    for AT, 10
    BIOS interrupts for, 122–123
    controller for, in Model 50, 174
    enhanced, 10, 17, 19–21
    error codes for, 227
    errors on, and POST, 180–181, 224
    foreign, 321, 325–326, 332, 392, 445
    layouts of, with OS/2, 344
    for Model 25, 19, 67
    for Model 30, 37
    for Model 50, 20, 153–154, 178
    for Model 60, 21, 208–209, 222–223
    for Model 80, 244, 262–263
    passwords for, 91, 154, 174, 179, 187, 223, 266
    for Personal System/2 family, 90–91
    repeat speed of, 85, 187
Keylock function, 154, 156, 209, 266

L

LAN Asynchronous Connection Server Program, 296
LAN boards, 291–294
    for Personal System/2, 16
Lancard A-II (Tiara Systems, Inc.), 310
Languages
    for AIX, 388–391
    foreign, with 8514/A adapter, 288
    for OS/2, 359

Index

Languages—cont
    See also Keyboards, foreign
Laser printer controller card, 301–302
Last chance to stop option, 234
Learn about the computer option for Model 50, 181–182
Level-sensitive interrupts, 116–117
Light pen, interrupts for, 137
LIM memory expansion, 27
Line control register, 40
Line drawing with 8514/A adapter, 289
Line status register, 39
Linear address mapping, 44
Linear Predictive Coding, 26, 61
Local Area Network Asynchronous Connection Server Program, 436–437
Local Area Network Manager, 435–436
Local Area Network Support Program, 433–434
Local area networks, 358, 430–437
    Communications Manager for, 356
Local descriptor tables
    with 80386, 260
    with OS/2, 337–338, 340
Log or display errors option, 230
Logical DOS drives, 328, 331, 367–369
Lotus 1-2-3 (Lotus Corp.), 13
    in performance tests, 463–468
Low density diskette mode, 92, 176
    performance tests using, 468
LPC (Linear Predictive Coding), 26, 61

M

Macro Assembler (Microsoft) for Personal System/2, 16
    in performance tests, 463, 467–468
Macro Assembler/2, 359, 362
MainLink II (Quadram Corp.), 310, 311
MAKE facility, 360–362
Matched Memory, 106, 256, 267
Math coprocessors
    compiler support for, 360, 362
    error codes for, 228
    in performance tests, 463–464, 466
    with Personal System/2 family, 24, 86–88
    problems with, 226
    See also 8087 math coprocessor; 80287 math coprocessor; 80387 math coprocessor
Matrox, high resolution card by, 323
Maynard Electronics, tape drive board by, 309
MCGA (Multi-Color Graphics Array), 22
    compared to VGA, 285
    for Model 25, 19, 66
    for Model 30, 17, 40–45
    for Personal System/2, 17
MDA (Monochrome Display and Printer Adapter), 3, 22, 277
Media and file interchange, connectivity for, 426–428
Memory
    80386 management of, 259, 261–262
    for AIX, 386
    error codes for, 227
    for LANs, 432, 434

Memory—cont
    OS/2 management of, 337–338
    OS/2 requirements for, 365
    for Personal System/2 family, 83–84
    privilege levels of, 340-341
    VGA, 133
    for video, 143, 146–148
    for Workstation Program, 439
    See also RAM; Read-only memory
Memory expansion boards, 297–298
    for 8514/A, 287
    for Micro Channel computers, 27
    for Model 50, 160
    for Model 60, 217
    for Model 80, 258
    for PC, 5
    for Personal System/2, 16
    third-party, 306–309
Memory manager with Windows/386, 419
Menus with Concurrent DOS 386, 404–405
Message switching with Presentation Manager, 352
Micro Channel
    add-in boards for, 27–28
    architecture of, 14, 17, 94–119
    for Model 50, 168–169
Micro Channel 3270 Connection Adapter, 294
MicroRAM memory board (Tecmar), 306–307
Microsoft products
    Macro Assembler, 16, 463, 467–468
    MS DOS, 317
    Windows/386, 417–422
MIDI (Musical Instruments Digital Interface), 27
MODE command, 321
Model 25, 14–15, 19, 63–65
    disk drives for, 25
    functional description of, 66–67
    options for, 78
    performance of, 74–78
    physical description of, 67–74
    setup and configuration of, 74
Model 30, 14–15, 18–19, 31–32
    add-in boards for, 26–27
    advanced diagnostics for, 54–55
    compatibility and performance of, 55–61
    disk drives for, 25
    display adapters for, 283–286
    functional description of, 33–40
    hard disks for, 25–26
    installation and setup of, 53–54
    MCGA for, 41–45
    options for, 61
    physical description of, 45–52
    size of, 17
Model 50, 14–15, 20
    disk cache for, 195–200
    documentation for, 194–195
    features of, 153–155
    functional description of, 162–178
    hard disks for, 26
    installation and use of, 179–194
    performance and compatibility of, 200–205
    physical description of, 155–162
    security in, 178–179
    size of, 17

489

Model 50—cont
    warranty and service for, 194
Model 60, 14-15, 20-21
    documentation for, 236
    features of, 207-209
    functional description of, 218-223
    hard disks for, 26
    installation and use of, 223-226
    performance and compatibility of, 237-241
    physical description of, 210-218
    problem diagnosis for, 226-235
    security of, 223
    size of, 17
    warranty and service for, 235-236
Model 80, 14-15, 21-22, 243
    documentation for, 267
    features of, 244-248
    functional description of, 257-265
    hard disks for, 26
    installation and configuration of, 265-266
    performance and compatibility of, 267-274
    physical description of, 248-257
    security in, 266-267
    size of, 17
    See also Control software for Model 80
Modems
    300/1200 Internal Modem/A, 28, 295
    2400-baud (Advanced Transducer Devices), 310
    for asynchronous communications, 428-429
    for Micro Channel, 28
    registers for, 39, 40
    serial port controller for, 94
Modes, video, interrupts for, 136-137, 139
Monitors, 23-24, 276-283
    for Model 25, 67
    for Model 80, 244
    third-party, 311-312
    for VGA, 130-131
Monochrome Display and Printer Adapter, 3, 22, 277
MOS command (PC-MOS/386), 411
Mouse, 296-297
    error codes for, 228
    for Personal System/2, 24
    with Presentation Manager, 352
Move the computer option for Model 50, 190
MS DOS, 317
Multi-Color Graphics Array. *See* MCGA
Multi-device arbitration, 117-119
Multi-Protocol Adapter/A, 27-28, 296, 429
Multi-tasking
    with 80286, 9, 165
    with 80386, 11-12
    with Concurrent DOS 386, 399-406
    with DESQview/386, 413-417
    on Model 50, 20
    operating environments for, 334
    with OS/2, 338-343
    with PC-MOS/386, 406-413
    saving video for, 146
    and SPOOL and START commands, 381-382
    with Workstation Program, 456-457
Multiple application sessions, 378-379

Multiplex interrupts with OS/2, 344
Multi-protocol adapter codes, 228
Multi-protocol boards, 294, 296
Music for Model 30, 27

N

NEC Multisync monitor, 280, 312
NETBIOS (Network Basic Input/Output System)
    LANs for, 434-435
    PC-MOS/386 emulation of, 412
Network operation of Emulation Program, 452-453
Network server password mode, 179, 187, 223, 266
NLSFUNC command, 321, 322, 333
Non-disk performance tests, 466
Nroff utility (AIX), 387
NS 16650 serial port controller, 93-94, 175, 429
Numbering system for Personal System/2, 17

O

OEDWARE, PC-KEY-DRAW by, 463, 465, 467-468
ONLY command (PC-MOS/386), 408
Open Window menu (DESQview/386), 414, 415
Operating environments, 334
    for Model 50, 161-162
    for Model 60, 217
Operating System/2, 16, 28-29, 335-336
    Extended Edition, 355-359
    installation of, 364-373
    languages for, 359-363
    memory required by, 298
    for Model 80, 244
    packaging and documentation for, 363-364
    system architecture of, 337-355
    use of, 373-383
Operating System Transfer Convention, 123-124
Operating systems
    for Model 80, 244
    for Personal System/2, 28-29
    See also Operating System/2
Optical disk drives, 21, 25, 300-301
Optimization with AIX, 390-391
Orchid Technology products
    memory board, 308
    Orchid Designer VGA board, 313-314
    RAMQuest 50/60, 306
OS/2. *See* Operating System/2
OS/2 Command Prompt option, 376
Overscan register, interrupt for, 139

P

Packaging
    of DOS 3.30, 322-323
    of Emulation Program Entry Level, 445-446
    of OS/2, 363-364
Page mapping, 82
PageMaker (Aldus), 302
Pageprinter, 27, 29
Paging with 80386, 259-262
Parallel ports
    controller for, 92-93, 175

Index

Parallel ports—cont
    error codes for, 227
    for Model 30, 40, 55
Partitioning of hard drives, 327–333, 366–368
Pascal
    for AIX, 389–390
    for Personal System/2, 16
Pascal Compiler/2, 359, 363
Password protection
    for databases, 357–358
    for keyboards, 91, 154, 174, 179, 187, 223, 266
    power-on, for POST, 125–126
PATCH command, 381
PC 3270 Emulation Program, 449–454
PC DOS, 28, 317–318
PC-KEY-DRAW in performance tests, 463, 465, 467–468
PC LAN software, 16
PC Local Area Network 1.2, 362, 363
PC Local Area Network Program, 430–433
PC-MOS/386 (Software Link, Inc.), 406–413
PC Music Feature for Model 30, 27
PC Network, 291–294
    COBOL/2 support for, 362
    DOS for, 319
PC Network Adapter II, 27, 293
PC Network Baseband Adapter/A, 292
PC Network Protocol Driver, 293, 433–435
Performance and compatibility
    of Model 25, 74–78
    of Model 30, 55–61
    of Model 50, 200–205
    of Model 60, 237–241
    of Model 80, 267–274
    testing procedures for, 463–468
Personal System/2 computers, 79
    BIOS for, 119–124
    compatibility of, 14–18
    design features of, 16–18
    Micro Channel for, 94–119
    POST for, 124–126
    system board for, 80–94
    See also specific models
Phar Lap Software, 386/DOS extender by, 416
Physical description
    of Micro Channel, 107–111
    of Model 25, 67–74
    of Model 30, 45–52
    of Model 50, 155–162
    of Model 60, 210–218
    of Model 80, 248–257
    of video system, 132–136
PIF files
    for Windows/386, 420
    with Workstation Program, 459
Pipelining with 80386, 259
Pipes
    with PC-MOS/386, 412
    with processes, 342
Planar control register, 37
Pointing device port, 296–297
Portables
    Compaq, 10–11
    PC, 5–6

Ports, 17
    bi-directional, 315
    for Model 25, 19
    for Model 30, 18, 37
    for Model 50, 20
    for Model 80, 21
    for PC, 7
    See also Parallel ports; Serial ports
POS (Programmable Option Select)
    and Micro Channel, 112–116
    and POST, 125
    setup procedure for, 84, 92
POST (power on self test)
    and 80286 mode setting, 173
    for Micro Channel, 124–126
    for Model 25, 66
    for Model 50, 167, 179–180
    for Personal System/2 family, 84
    and POS, 113, 125
    for problem diagnosis, 226–229
PostScript (Adobe Systems), 302
Power-on password mode, 125–126, 178–179, 187, 223, 266
Power supplies
    for 8503 monitor, 278
    for 8512 monitor, 280
    for 8513 monitor, 281
    and Micro Channel, 108–110
    for Model 50, 177–178
    for Model 60, 212, 222
    for Model 80, 22, 246, 248, 251, 262
    for PC, 7
    for Personal System/2, 17
    problems with, 226
    for XT, 7
Prepare Drive C for DOS option, 232
Presentation Manager, 339, 349–355, 363–364
Priam, external hard drives by, 309
PRINT command (OS/2), 28
PRINT command (PC-MOS/386), 411
Printer control register, 40
Printer Manager (Concurrent DOS 386), 404
Printer status register, 40
PRINTER.SYS file, 322, 333
Printers, 29, 301–305
    GDDM-PCLK support for, 456
    for Personal System/2, 16
    and Workstation Program, 438–439
Priority levels, execution, 341–342
Privilege levels, memory, 340–341
Problem diagnosis on Model 60, 226–235
Procedures, window, 354
Processes, multi-tasking, 340–343
Processor control instructions
    80286, 476–477
    80386, 481
Professional Graphics Controller, 22–23, 131, 277
Program Selector menu for OS/2, 340, 349, 373–375, 377–379
Program transfer instructions
    80286, 475–476
    80386, 480–481
Programmer Guide for OS/2, 363–364
Programmer Toolkit (OS/2), 351, 363

491

Proprinter printers, 29
Protected virtual-address mode, 163
    for 80286, 9, 80-81, 163-164
    for 80287, 11, 86
    for 80386, 258-260
    switching from, 173
Protection model instructions (80386), 481-482
Publishing SolutionPac Option/A, 302

Q

Quadram products
    MainLink II, 310
    Quadmeg PS/Q memory board, 306, 307
    QuadVGA board, 313-314
Quarterdeck Office Systems, DESQview products by 334, 413-417
Queues with processes, 342
Quick Reference manuals, 179, 194, 236, 267
Quietwriter III, 29, 304-305, 344

R

RAM (random-access memory)
    for AT, 9-10
    for Model 25, 19, 66, 71
    for Model 30, 33, 35, 52
    for Model 50, 20, 153, 160, 166-168, 180
    for Model 60, 208, 216-217
    for Model 80, 245-246, 258, 261-262
    for PC, 2-3, 7
    for Personal System/2 family, 83-84
    for VGA, 133-134
    for XT, 7
RAM disk drivers, 318
    performance tests using, 467
RamQuest 50/60 memory board (Orchid Technology), 306
Read-only memory
    for AT, 9
    for BIOS, 2-3, 120
    for Model 25, 66, 71
    for Model 30, 51-52
    for Model 50, 160, 166-167
    for Model 60, 217
    for Model 80, 257-258
    for PC, 2
    for Personal System/2 family, 83-84
    and POST, 124-125
Read Sectors disk cache function, 196
Real-address mode, 163
    of 80286, 8-9, 80-81
    of 80287, 86
    of 80386, 11, 82, 258-259
    switching to, 173
Real-time clock
    for Model 30, 18, 37, 52
    for Model 50, 20, 166-168
    for Model 60, 21, 216-217
    for Model 80, 21
    for Personal System/2 family, 83-84
Realtime Interface Coprocessor Multiport Adapter, 295

Rear panel connectors
    for Model 25, 68-69
    for Model 30, 46-48
    for Model 50, 156-157
    for Model 60, 210-212
    for Model 80, 249-251
RECEIVE program (Emulation Program Entry Level), 448
RECV35.COM file, 427
Redirection, 318
Reference Diskette
    disk cache program on, 195-200
    error messages on, 125
    for Model 50, 179-182, 187-193
    for Model 60, 223-224
    for Model 80, 265-266
    for password setting, 178-179
    and POS, 112, 113
Regenerative buffer, 141-143
Registers
    for 80386, 259
    color, interrupt for, 140-141
    for DMA controller, 85-86
    with MCGA, 41
    with VGA, 148-152
Reliability and system board size, 50-51
Remapping of ROM, 258
Remove disk cache option, 199-200
REPLACE command, 319, 323
    with Data Migration Facility, 427
Resolution of 850x monitors, 278-282
RESTORE command, 320-321
Restore configuration option, 112, 185, 266
ROM. *See* Read-only memory
Rotate instructions
    80286, 475
    80386, 479-480
RS232 serial ports, 294
    for Model 30, 39
    for Model 50, 175
Run automatic configuration option, 185, 186, 224
Run tests continuously option, 231

S

SAA (Systems Application Architecture), 349-350
$SALUT instruction, 362
Scan codes, keyboard, for Model 80, 263
Scheduling, multi-tasking, 341-342
Scheduling options, with DESQview/386, 415
Scissoring with 8514/A adapter, 290
SCP 86-DOS, 317
Scratch register, 40
Screen groups, 339-340
Scrolling, interrupts for, 137
Seattle Computer Products, 317
Sector buffers for disk cache program, 195, 197
Security
    with 8042 controller, 91
    for keyboards, 91, 154, 179, 187, 223, 266
    with PC-MOS/386, 412
Seequa computers, 6
Segment descriptor tables, 260, 337-338, 340
SELECT command, 325-326, 332

Index

Semaphores with processes, 342, 346
SEND program (Emulation Program Entry Level), 448
Sequencer, VGA, 134
    registers for, 150–151
Serial ports, 28
    controller for, 93–94, 175
    error codes for, 228
    for Model 30, 39–40, 55
    and OS/2 installation, 369–370
Server-Request Programming Interface, 356–357, 450
Session management, 339–340
Set configuration option, 182–183, 265
Set features option, 113, 185, 187
Set keyboard speed option, 187
Set passwords option, 187, 188
SETLOCAL command, 381
SETUP program (Emulation Program), 453
SHARE utility (Windows/386), 421
SHELL program (WordPerfect), 334
Shift instructions
    80286, 475
    80386, 479–480
Signals with processes, 343
SMARTDrive (Windows/386), 419
Software Link, Inc., PC-MOS/386 by, 406–413
Sony MultiScan monitors, 280, 312
SORT/MERGE module for COBOL/2 compiler, 361
Source code control system for AIX, 391
Space Saving keyboard, 19
Speakers, 89
    for Model 30, 52
    for Model 50, 157–158
    for Model 60, 215
    for Model 80, 253
Specify your own partitions option, 366–367
Speech adapter board, 26, 61
Speed. *See* Performance and compatibility
SPOOL command (OS/2), 381–382
SPOOL command (PC-MOS/386), 411
Spool support screen, 371
Spreadsheets, 8
    in performance tests, 463–468
SRPI (Server-Request Programming Interface), 356–357, 450
ST-506 disk controller, 214, 220
STACKS command, 319, 322
Standard OS/2, 335–336
Start a Program menu, 373–376
START command (OS/2), 381
Starter Diskette
    for Model 25, 74
    for Model 30, 53–54
STARTUP.CMD file, 373, 376
Startup menu for Concurrent DOS 386, 402–404
Status, video, interrupt for, 145
STB systems, 306
Stop error log option, 230
String instructions
    80286, 474–475
    80386, 479
Structured Query Language for Database Manager, 357–359

Subdirectory Tree Display option (Concurrent DOS 386), 402
SUBST command, 319
Switch to a Running Program menu, 376, 378–379
SYS command, 322
SYSDISK.COM menu, 366
System 36/38 Workstation Emulation Adapter/A, 294
System 36/38 Workstation Emulation Program, 437–440
System architecture
    of OS/2, 337–355
    for PC, 2–3
System boards
    error codes for, 227
    for Personal System/2 family, 80–94
    problems with, 226
    See also specific computer models
System Board Memory Expansion Kit, 297
System Board Memory Expansion Option, 298
System checkout option, 230
System control ports, 171–174
System interrupts, 88–89
System menus with windows, 353–354
System passwords, setting of, 185
System services BIOS interrupts, 121
System support gate array for Model 30, 36
System timers
    for Model 30, 37
    for Model 50, 170–171
    with Personal System/2 family, 89
System unit
    for Model 25, 69–74
    for Model 30, 48–52
    for Model 50, 157–162
    for Model 60, 212–218
    for Model 80, 251–253
Systems Application Architecture, 349–350

T

Tandy computers, 6–7
    Personal System/2 clones by, 315
Tape backup, 309
Task Manager with Presentation Manager, 355
TCP/IP (AIX), 393
Technical Quick Reference Card for DOS 3.30, 323
Technical Reference guides, 194, 236, 267, 323, 364
Tecmar, MicroRAM by, 306–307
Teletype mode, interrupt for, 139, 144
Terminal emulation. *See* Connectivity
Terminal support by AIX, 388
Test error message, 233
Test selection option, 230–231
Test start message, 233
Test the computer option, 179, 229, 266
Text Formatting System with AIX, 387
Text modes
    with 8514/A adapter, 287–288
    with CGA, 4
    with MCGA, 43
    with VGA, 133, 135, 146–147
Third-party vendors for PCs, 1, 305–315, 469, 472
Threads, multi-tasking, 340–343

493

Tiara Systems, Lancard A-II, 310
TIME command, 321, 326, 333
Time slices, 341–342
Timers
    for Model 30, 37
    for Model 50, 170–171
    with Personal System/2 family, 89
Title bars with windows, 353
Token-Ring Adapter/A, 292–293
Token-Ring network, 293–294
    AIX support for, 393
    DOS for, 319
    LAN support for, 430–431, 433–435
Token-Ring Network Adapter/A, 28
Token-Ring Network Bridge Program, 435
Token-Ring PC Adapter for Model 30, 27
Token-Ring Starter Kit/A, 293–294
Tone generator, 89
TopView operating environment, 334
TRACE command (OS/2), 382
TRACEBUF statement, 382
TRACEFMT command (OS/2), 382
Transfer control bus for Micro Channel, 96
Transfer cycles on Micro Channel, 100
Translation lookaside buffer, 261
Troff utility (AIX), 387
Tutorial
    for Model 30, 53–54
    for Model 50, 181
    for Model 80, 265

U

UNIX operating system. *See* AIX Personal System/2
    operating system
Update menu option, OS/2, 374–376
Updates for DOS 3.30, 323–324
Usability Services, AIX, 386
User interfaces
    for AIX, 386
    for DOS 3.30, 333–334
User Shell, Presentation Manager, 351–352
User's Guide, 322, 363
User's Reference manuals, 322, 363

V

VDISK.SYS file, 318
Vendors, 305–315, 469, 472
VGA, 22, 344
    for 8514/A adapter, 287
    for Model 30, 26, 283–284
    for Model 50, 20, 154–155, 160, 174–175
    for Model 60, 21, 209
    for Model 80, 21
    for Personal System/2, 17
    physical description of, 132–136
    video display modes of, 129, 136–146
Video
    BIOS interrupts for, 120

Video—cont
    buffers for, in PC, 3
    error codes for, 228
    extension signals for, 106–107, 246
    interrupts for, 121, 136–146
    and multi-tasking, 146
    switching of, 144
    third-party boards for, 312–313
    See also MCGA; VGA
Video Graphics Array. *See* VGA
Video services with OS/2, 344
Video system architecture, 127–131
    hardware registers for, 148–152
    memory arrangement in, 146–148
    for Model 30, 40–45
    physical description of, 132–136
    video modes for, 136–146
View configuration option, 183, 370
View disk cache setting options, 199, 200
View error log option, 230
VisiCalc, 13
VS FORTRAN, 388–389
VS PASCAL, 389–390

W

Wang computers, 6–7
Warranty and service, 194, 235–236
Western Digital, hard disk controller by, 10
WIN386 command (Windows/386), 420
Windows, 344
    with AIX, 386–387
    with Concurrent DOS 386, 404–405
    with Presentation Manager, 350, 352–354
Windows/386 (Microsoft), 417–422
WMENU command (Concurrent DOS 386), 404–405
Word processors, 8
    in performance tests, 463, 466–468
WordPerfect Corp. products
    SHELL program, 334
    WordPerfect, 13, 463–468
WordStar, 13
Workstation Host Interface Program (AIX), 391–393
Write-once-read-many-times disk drives, 21, 25,
    300–301
Write Sectors function with disk cache program,
    196
WSETUP.BAT file (Concurrent DOS 386), 405

X

X-Windows with AIX, 386–387
XCOPY command, 319
Xtree (Executive System), 334

Z

Zuckerlink LAN (Advanced Transducer Devices), 315

The Best Book of: dBASE II® /III®
Ken Knecht

For readers who already know how a data base performs, this time-saving guide explores the tricks that make dBASE II and dBASE III respond to their needs. In an effort to get the most out of dBASE II and dBASE III and apply these systems to specific business needs, it describes how to detect and correct errors, sort files, create new and useful programs, and manipulate data. The conversational style makes this book easy-to-read and an enjoyable, rewarding way to master dBASE II/III.

Topics covered include:

- Getting Started
- Creating a Simple Database
- More About dBASE II Data
- Dealing with the Entire Database
- Creating Reports
- dBASE II Command Files
- Advanced Programming
- An Annotated Program
- Quickcode
- Putting Quickcode to Work
- dGRAPH
- dBASE II—In Conclusion
- Introducing dBASE III—System and Program Specifications
- More dBASE III—Report, Variables, Label Commands, New Functions, The File Conversion Program

256 Pages, 7½ x 9¾, Softbound
ISBN: 0-672-22349-X
No. 22349, $21.95

The Best Book of: Framework™
Alan Simpson

Here are practical examples and applications to unleash Framework's frames, word processor, spreadsheet, and other features. This reference illustrates how to access national information systems; interface with WordStar®, dBASE II® /III®, and Lotus® 1-2-3®; use Framework macros; and program with FRED™, Framework's programming language. It is clear, concise, and perfect for any first-time user of Framework.

Topics covered include:

- Getting Started—Installing Framework
- Getting Organized with Frames
- Exploring Spreadsheets
- Exploring Graphics
- Exploring Database Management
- Exploring Outlines
- Advanced Word-Processing Techniques
- Spreadsheet Functions and Formulas
- Sample Spreadsheets
- Advanced Graphics Techniques
- Advanced Database Management Techniques
- Form Letters and Mailing Labels
- Printing Framework Reports
- Framework Communications
- Communicating with the World
- Interfacing with Other Software Systems
- Macros
- Programming with Fred

362 Pages, 7½ x 9¾, Softbound
ISBN: 0-672-22421-6
No. 22421, $21.95

Best Book of: Symphony®
Alan Simpson

This book introduces Symphony. It illustrates how to create, edit, and format documents with word processing; figure taxes and amortize loans with the spreadsheet; display graphs and produce slides with graphics; and create a mailing list with the data base manager. It also shows how to transfer spreadsheets and programs to other computers and interface with other programs like Lotus® 1-2-3® and dBASE II® /III.®

Topics covered include:

- Getting Started
- Exploring Word Processing
- Exploring the Spreadsheet
- Exploring Graphics
- Exploring Database Management
- Exploring Windows
- Advanced Word Processing Techniques
- Spreadsheet Functions and Formulas
- Sample Spreadsheets
- Advanced Spreadsheet Techniques
- Advanced Graphics
- Advanced Database Techniques
- A Mailing System with Form Letters and Mailing Labels
- Printing Symphony Reports
- Symphony Communications
- Communicating with the World
- Transferring Symphony Files
- Macros
- Programming Symphony

396 Pages, 7½ x 9¾, Softbound
ISBN: 0-672-22420-8
No. 22420, $21.95

Personal Publishing with PC PageMaker®
Terry M. Ulick

Here is everything you need to know about PC PageMaker to design publications. It shows you how to select and use type, work in multicolumn and multipage layouts, create graphs, and merge text with graphic elements.

This handbook introduces all the equipment needed to assemble a personal publishing system to create professional-looking typeset journals, advertisements, brochures, and other publications.

Hands-on instruction at the terminal, numerous visual examples, and a detailed explanation of typesetting terms provide the information necessary to help the beginning to intermediate PC user produce attractive copy.

Topics covered include:

- Assembling a Personal Publishing System
- Selecting the Right Hardware and Software
- Pages on the IBM®
- Electronic Page Assembly
- Working with Type
- PostScript™ and LaserJet Plus™ Typestyles
- Formatting Type
- Working with PageMaker
- Building Master Pages
- Placing Elements on a Page
- Adding Graphic Elements
- Linking PageMaker Files
- Printing Page Files
- High-Volume Printing
- Multicolored Pages
- Grids and Sample Pages

304 Pages, 7½ x 9¾, Softbound
ISBN: 0-672-22593-X
No. 22593, $19.95

Visit your local book retailer, use the order form provided, or call 800-428-SAMS.

The Best Book of: Lotus® 1-2-3®, Second Edition
Alan Simpson

This is *the* book for beginning 1-2-3 users. Written as a tutorial, this book steps readers through the various functions of the program and shows them how to use Version 2.0 in today's business environment.

Divided into four sections—worksheets, graphics, database management, and macro—each chapter within a section is designed to allow newcomers to be productive right away. With each new chapter, the users' skills are honed for faster and more spontaneous use of the software.

Topics covered include:

- Creating and Worksheet
- Functions and Formulas
- Formatting the Worksheet
- Copying Ranges
- Editing, Displaying, and Managing Worksheet Files
- Practical Examples
- Creating Graphs and Printing Graphs
- Database Management and Sorting
- Tables and Statistics
- Macros
- Custom Menus

350 Pages, 7½ x 9¾, Softbound
ISBN: 0-672-22563-8
No. 22563, $21.95

Lotus® 1-2-3® Financial Models
Tymes, Dowden, and Prael

With this book, Lotus 1-2-3 users will learn to create models for calculating and solving personal and business finance problems using Version 2.0! This book shows how to avoid frustration and delay in setting up individual formulas and allows users to perform any spreadsheet calculation in minutes.

This revision of the popular *1-2-3® from A to Z* includes numerous models or templates for the Lotus user, each preceded by a brief explanation of how the model works and how it can be altered. More advanced business models and an increased emphasis on macros as well as the inclusion of Version 2.0 make this book ideally suited to the business user.

Whether calculating a complex statistical analysis or a home heating analysis, readers can use this book as the one source for the guidance and tools needed to do spreadsheet calculations in the most efficient manner possible.

Topics covered include:

- Personal Models
- Business Models
- Advanced Business Models

Elna Tymes is an experienced computer book author in California with a number of successful books to her credit including *Mastering Appleworks* and *1-2-3 from A to Z.*

300 Pages, 7½ x 9¾, Softbound
ISBN: 0-672-48410-2
No. 48410, $19.95

The Best Book of: WordPerfect®
Vincent Alfieri

From the author of the best-selling *Mastering WordStar®*, this hands-on tutorial provides step-by-step explanations of all WordPerfect features from simple to advanced.

Perfect for beginners, it offers clear and extensive coverage of merge-printing; explores powerful writing aids such as automatic table of contents creation, index generation, and user-defined macros; and contains a thorough discussion of printer setups and the new laser printers.

Over 75 practical examples with screen illustrations are included.

Topics covered include:

- Essential Formatting
- Printers and Printing
- Laser Printers
- Spelling and Hyphenation Tips
- Searching and Replacing Text
- Special Printing Effects
- Using the Math Features
- Working with Two Documents
- Different Formats in a Document
- Pagination and Page Formatting
- Text and Column Blocks
- Enhancing the Finished Product
- Fill-In Documents
- Macros
- File Management
- Sorting by Paragraphs, Lines, Words, and Numbers
- Merge Printing: Form Letters, Form Mailings, Fill-In Forms, Advanced Uses, and Beyond Merge Printing

464 Pages, 7½ x 9¾, Softbound
ISBN: 0-672-46581-7
No. 46581, $21.95

The Best Book of: WordStar® Features Release 4
Vincent Alfieri

The Best Book of: WordStar includes all of the more than 125 enhancements and revisions found in Release 4.

Readers who want to get the most from WordStar, the most popular word-processing program ever published, are guided step-by-step through a series of real-life problems—from opening a file to sending personalized letters and from simple correspondence to complex tables.

This text shows how to import and export WordStar files to and from 1-2-3 and dBASE III PLUS.

Topics covered include:

- Print Effects and Printing
- Formatting Essentials
- Moving, Copying, and Deleting Blocks
- Finding and Replacing
- Keyboard Macros and Other Shortcuts
- Fun with Multiple Formats
- Editing Features
- Working with DOS and File Management
- Headers, Footers, and Page Numbering
- Controlling the Page
- Boilerplates, Format Files, and Document Assembly
- Math Maneuvers
- A Form Letter Course
- Conditional Merge Printing
- Automated Document Assembly
- Bibliophilic Occupations
- Working with Laser Printers

576 Pages, 7½ x 9¾, Softbound
ISBN: 0-672-48404-8
No. 48404, $19.95

Visit your local book retailer, use the order form provided, or call 800-428-SAMS.

Hard Disk Management Techniques for the IBM®
Joseph-David Carrabis

This is a resource book of in-depth techniques on how to set up and manage a hard disk environment directed to the everyday "power user," not necessarily the DOS expert or programmer.

Each fundamental technique, based on the author's consulting experience with Fortune 500 companies, is emphasized to help the reader become a "power user." This tutorial highlights installation of utilities, hardware, software, and software applications for the experienced business professional working with a hard disk drive.

Topics covered include:

- Introduction to Hard Disks
- Hard Disks and DOS
- Backup and What You Need to Know
- Service and Maintenance
- Setting Up a Hard Disk
- Organizing a Hard Disk
- Hard Disk Managers
- Utilities to Manage Hard Disks, Find Files, UNERASE Files, Recover Damaged Files, Speed Up Disk Access, and Restore and Backup Disks
- Maintenance Utilities
- Security Utilities

472 Pages, 7½ x 9¾, Softbound
ISBN: 0-672-22580-8
No. 22580, $22.95

IBM® PC AT User's Reference Manual
Gilbert Held

Includes everything you need to know about operating your IBM PC AT—how to set the system up, write programs that fully use the AT's power, organize fixed-disk directories, and use IBM's multitasking TopView.

Includes a BASIC tutorial for beginners and includes several fixed disk organizer programs—all clearly described, explained, and illustrated.

Topics covered include:

- Hardware Overview
- System Setup
- Storage Media and Keyboard Operation
- The Disk Operating System
- Fixed Disk Organization
- BASIC Overview
- Basic BASIC
- BASIC Commands
- Advanced BASIC
- Data File Operation
- Text and Graphics Display Control
- Batch and Shell Processing
- Introduction to TopView
- Appendices: ASCII Code Representation, Extended Character Codes, BASIC Error Messages, Programming Tips and Techniques

453 Pages, 7 x 9¼, Softbound
ISBN: 0-8104-6394-6
No. 46394, $29.95

IBM® PC & PC XT User's Reference Manual, Second Edition
Gilbert Held

Expanded to include the more powerful PC XT, this second edition contains the most up-to-date information available on the IBM PC. From setup through applying and modifying the system, this book continues to provide users with clear, step-by-step explanations of IBM PC hardware and software—complete with numerous illustrations and examples.

Highlights of the second edition include instructions for using DOS 3.1 and upgrading a PC to an XT; information on the customized hardware configuration of the PC and XT; explanations on how to load programs on a fixed disk and how to organize directories; and material on available software, including compilers.

Topics covered include:

- Hardware Overview
- System Setup
- Storage Media and Keyboard Operation
- The Disk Operating System
- Fixed Disk Organization
- BASIC Overview
- BASIC Commands
- Data File Operations
- Text and Graphics Display Control
- Batch Processing and Fixed Disk Operations
- Audio and Data Communications
- Introduction to TopView

496 Pages, 7 x 9¼, Softbound
ISBN: 0-672-46427-6
No. 46427, $26.95

The Waite Group's Desktop Publishing Bible
James Stockford, Editor
The Waite Group

Publish high-quality documents right on your desktop with this "bible" that tells you what you need to know—everything from print production, typography, and high-end typesetters, to copyright information, equipment, and software.

In this collection of essays, experts from virtually every field of desktop publishing share their tips, tricks, and techniques while explaining both traditional publishing concepts and the new desktop publishing hardware and software.

Topics covered include:

- Publishing Basics: Traditional Print Production, Conventional Typography, Case Studies in Selecting a Publishing System, and a Comparison of Costs for Desktop and Conventional Systems
- Systems: The Macintosh, PC, MS-DOS, An Overview of Microsoft Windows, Graphics Cards and Standards, Monitors, Dot and Laser Printers, UNIX, and High-End Work Stations
- Software: Graphics Software, Page Layout Software, Type encoding Programs, PostScript, and JustText
- Applications: Newsletters, Magazines, Forms, Comics and Cartooning, and Music

480 Pages, 7½ x 9¾, Softbound
ISBN: 0-672-22524-7
No. 22524, $24.95

Visit your local book retailer, use the order form provided, or call 800-428-SAMS.

The Best Book of: Microsoft® Works for the PC
Ruth K. Witkin

This step-by-step guide uses a combination of in-depth explanations and hands-on tutorials to show the business professional or home user how to apply the software to enhance both business and personal productivity.

Clearly written and easy to understand, this book explains how to use such varied applications as the word processor with mail merge, the spreadsheet with charting, the database with reporting, communications, and integration. For each application the author provides a detailed overview of the hows and whys followed by practical examples that guide the reader easily from idea to finished product. Quick-reference charts, summaries, and end-of-chapter questions and answers enhance the learning process, and each example is illustrated at important developmental stages.

Topics covered include:

- Spreadsheet Essentials
- Exploring the Spreadsheet Menus
- About Formulas and Functions
- Charting Your Spreadsheet
- Exploring the Chart Menus
- Database Essentials
- Filling a Database
- Exploring the Database Menus
- Word Processor Essentials
- Exploring the Word Processor Menus
- Integration Essentials
- Communications Essentials
- Exploring the Communications Menus

350 Pages, 7½ x 9¾, Softbound
ISBN: 0-672-22626-X
No. 22626, $21.95

The Best Book of: WordPerfect® Version 5.0
Vincent Alfieri

From formatting a simple letter to setting up time-saving macros, this book unravels all of the new features and updates of Version 5.0. With detailed explanations, clear examples, and screen illustrations, it teaches first-time users how to use WordPerfect and helps advanced users learn shortcuts and tips for using it more efficiently.

Covering every new feature and re-examining the old ones, this step-by-step tutorial allows readers to take a self-paced approach to word processing.

Topics covered include:

- Fonts for Special Effects
- Printing and Formatting
- Document and File Management
- Speller and Thesaurus
- Working with Forms
- Headers, Footers, and Page Numbering
- Working with Multiple Text Columns
- Hard Space and Hyphenation
- Footnotes, Endnotes, and Automatic Referencing
- Table, List, and Index Generation
- Merge Features
- For Power Users—Macros, Sorting, Graphics
- The Laser Revolution

700 Pages, 7½ x 9¾, Softbound
ISBN: 0-672-48423-4
No. 48423, $21.95

Macro Programming for 1-2-3®
Daniel N. Shaffer

This book reveals the uses of 1-2-3 Release 2's powerful built-in programming language. With Release 2's expanded macro language, 1-2-3 users can better automate and control routine or tedious worksheet functions, saving time, effort, and keystrokes.

The author shows how to create more accurate, more powerful worksheets that are also easier to use. Concentrating on the practical application of macros, this book gives the reader step-by-step hands-on experience through examples and exercises.

Topics covered include:

- Using Range Names
- Using the Automatic Typing Features of 1-2-3
- Using the Programming Features of 1-2-3
- Easy Data Entry, Results Display, Report Printing, Graphing, Data Storage, and Retrieval
- Making Your Worksheets More Powerful
- Appendices: Summary of Macro Commands, Using the New String Functions, Differences Between Release 1A and 2.0

304 Pages, 7½ x 9¾, Softbound
ISBN: 0-672-46573-6
No. 46573, $19.95

dBASE III PLUS™ Programmer's Library
Joseph-David Carrabis

Written for intermediate to advanced programmers, this book shows how to quickly build a library of dBASE III PLUS code that can be reused many times with only minor modifications. The author reveals universal patterns for editing, adding, deleting, finding, and transferring records in databases, so that dBASE III PLUS tools can be adapted to a variety of applications.

The book has two major parts: The first provides kernels of code that can be used in applications as varied as dental records keeping, fundraising, small business management, and newsletter subscription systems. The second part shows the code necessary to handle such systems, and how kernels and modular programming help you set them up.

Topics covered include:

- What Is a Library?
- The Basic Kernels of Code
- Database Designs
- Advanced Kernels
- Inventory Systems
- Client/Personnel Record Keeping
- Subscription and General Accounting Systems
- Appendices: Clipper Versions of dBASE III Plus Listings, and dBASE III Plus Commands, Functions, and Abbreviations

536 Pages, 7½ x 9¾, Softbound
ISBN: 0-672-22579-4
No. 22579, $21.95

Visit your local book retailer, use the order form provided, or call 800-428-SAMS.